Modern Acoustics and Signal Processing

More information about this series at http://www.springer.com/series/3754

The ASA Press

The ASA Press imprint represents a collaboration between the Acoustical Society of America and Springer dedicated to encouraging the publication of important new books in acoustics. Published titles are intended to reflect the full range of research in acoustics. ASA Press books can include all types of books published by Springer and may appear in any appropriate Springer book series.

ASA PRESS

Murray Campbell • Joël Gilbert • Arnold Myers

The Science of Brass Instruments

ASA PRESS

Modern Acoustics and Signal Processing

Springer

Murray Campbell
School of Physics and Astronomy
University of Edinburgh
Edinburgh, UK

Joël Gilbert
Laboratoire d'Acoustique de l'Université du Mans - CNRS
Le Mans, France

Arnold Myers
Royal Conservatoire of Scotland
Glasgow, UK

ISSN 2364-4915 ISSN 2364-4923 (electronic)
Modern Acoustics and Signal Processing
ISBN 978-3-030-55684-6 ISBN 978-3-030-55686-0 (eBook)
https://doi.org/10.1007/978-3-030-55686-0

This Springer imprint is published by the registered company Springer Nature Switzerland AG
The registered company address is: Gewerbestrasse 11, 6330 Cham, Switzerland

Preface

When a musician is presented with an unfamiliar instrument, the natural reaction is to investigate its properties and musical possibilities. If the instrument in question is a member of the brass family, the player can attempt to sound it by buzzing the lips against the mouthpiece, searching out the most easily playable notes and investigating its timbral and expressive possibilities. Experimenting systematically in this way is quite close to the approach that would be adopted by a scientist asked to investigate the properties of the instrument. Initial experiments could be carried out by subjecting the instrument itself to laboratory tests of various types, but to arrive at a complete understanding of its purpose and possibilities, it is necessary to study the way in which the player interacts with the instrument.

Our understanding of how brass instruments work has developed greatly over the last few decades. Valuable coverage of brass instruments is available in several textbooks on musical instrument acoustics, including *Fundamentals of Musical Acoustics* (Benade 1976), *The Musician's Guide to Acoustics* (Campbell and Greated 1987), *The Physics of Musical Instruments* (Fletcher and Rossing 1998) and *Acoustics of Musical Instruments* (Chaigne and Kergomard 2016), but to our knowledge this is the first monograph dealing exclusively with instruments sounded by lip vibration. It includes the reviews of scientific research into many aspects of the behaviour of brass instruments undertaken in a number of centres worldwide and reported on a lively conference circuit and in peer-reviewed journal articles. Areas in which important advances have been made include the action of the player's lips, the effects of nonlinear sound propagation in instrument tubing, the radiation of sound from brasswind bells, instrument taxonomy and computer modelling as an aid to manufacturing and a tool for musical synthesis.

This book is intended to be intelligible and useful to the large number of brass instrumentalists who take a scientific interest in their instruments and how they are played. It is also intended to serve as an academic and research text, and more advanced treatments of several topics are included in sections labelled 'Going Further'.

The book also includes a detailed discussion of historically important types of brass instrument, both for their intrinsic interest and because many have been revived for 'period instrument' performance; the commonly used modern instruments are presented alongside successful models of previous generations. This material is complementary to the published works on brass instrument history such as *Brass Instruments: Their History and Development* (Baines 1976), articles in *The Grove Dictionary of Musical Instruments* (Libin 2014) and *The Cambridge Encyclopedia of Brass instruments* (Herbert et al. 2019) and the many articles in the *Historic Brass Society Journal*. A comprehensive bibliography provides references for readers wishing to pursue particular topics in more depth.

The term 'Labrosone' was introduced by the organologist Anthony Baines (1976, p. 40) and is used for all instruments in which sound is generated by the application of the player's lips to the instrument and blowing in such a way that the lips vibrate and admit an intermittent or fluctuating stream of air into the instrument. The terms 'lip reed' and 'brass instrument' are commonly used, although as discussed in Chap. 3 the action of the player's lip differs somewhat from that of the reeds in woodwind and free-reed instruments, and although brass is the most commonly used material, many labrosones are made of other materials. All practically used labrosones are instruments of tubular construction, and the behaviour of the air column is intrinsic to their operation. One can postulate other instruments being excited in the same way, such as a cavity supporting a Helmholtz resonance vibration mode, but non-tubular air/lip-excited instruments do not seem to have been developed.

Within the broad field of labrosones, some are what may be called 'European' brass instruments made for musical purposes. These include instruments made in the European tradition on other continents, but not purely signalling instruments or instruments of folk traditions that have not been to some extent integrated into the mainstream of art music. The properties of European brass instruments made for musical purposes are that the instrument is designed to play a range of notes, typically over one to four octaves; the instrument is designed to allow notes to be played, which are intended to be 'in tune' with the prevailing conventional framework of pitch standards and temperament; and that the instrument is normally assigned a nominal pitch (Myers 1994), i.e. is *in* a key such as $B^\flat$.

By analogy with woodwind instruments, a term that is not normally used in describing whistles and folk flutes and reed pipes, the instruments satisfying these criteria may be termed 'brasswinds'. Brasswinds include all the brass instruments used in orchestras and bands playing from written music, with the addition of instruments only occasionally so used, but built with this possibility, e.g. bugles in $B^\flat$, and alphorns in $G^\flat$. The instruments of the Russian horn band, each designed to play a single note, can be considered as a group to meet the criteria.

This book aims to explain the science underlying the behaviour of brasswinds. There will be inevitably some of the many hundreds of kinds of brasswinds past and present with some degree of established identity which we will not mention; however, the types most commonly used in music making since the period of the Renaissance will be treated. Although many of the experimental and theoretical studies that are described were carried out for the specific case of the trombone, the basic scientific principles that have emerged from these studies are applicable with minor modifications to trumpets, horns, tubas and other common brasswinds. We will also discuss some instruments such as the didgeridoo (which has a sophisticated playing technique but is not intended to be assigned a nominal pitch) and ancient instruments such as the bucina, carnyx, lituus and lur. While we do not describe individually all the many kinds of labrosone used worldwide, many of the general principles we explain will be applicable to the broad field of labrosones.

It goes without saying that the material presented in this monograph is the fruit of decades of work by very many colleagues who are part of the vibrant and friendly international community of musical acousticians. The text evolved during a 20-year-long collaboration between the acoustics laboratories at Edinburgh and Le Mans and has been enriched by the input of generations of postgraduate students. The authors are enthusiastic brass players as well as professional scientists, and many musical colleagues have also contributed greatly to our research in this field and in the preparation of the book.

The text is organised in three parts. Part I consists of two chapters, which review two different but complementary perspectives: the musician's view of brass instrument behaviour, and the scientific approach to the topic. These chapters provide a non-mathematical introduction to the detailed acoustical model of brass playing, which is unfolded in Part II. Mathematical methods, including the use of calculus, are used when necessary in Part II, but each of the four chapters in this part begins with a non-mathematical summary of the main ideas and results. Part III consists mainly of a detailed presentation of the taxonomy of brasswinds past and present, illustrated by many photographs from the Edinburgh University Collection of Historic Musical Instruments, the Royal Conservatoire of Scotland and other sources.

Edinburgh, UK
Le Mans, France
Edinburgh, UK
June 2020

Murray Campbell
Joël Gilbert
Arnold Myers

Acknowledgements

Throughout our work of planning and writing this book, we have enjoyed the support of our respective institutions: the University of Edinburgh, the University of Le Mans and CNRS and the Royal Conservatoire of Scotland.

An incalculable benefit has been the stimulating insights, always generously shared by our professional colleagues, who have provided encouragement and allowed us to draw on the fruits of their hard work. Some have also been so kind as to read over and comment on the drafts of this book.

We are grateful to all the colleagues and institutions who have allowed use of their photographs and artworks to illustrate our text, in particular the Edinburgh University Collection of Historic Musical Instruments (denoted in captions by 'EU') and the Royal Conservatoire of Scotland (denoted in captions by 'RCS').

Finally, we acknowledge the assistance given by the many musicians and institutions who have allowed study of their brass instruments, enabling us to populate the scatter diagrams in Chap. 7. These include: Alistair Braden, Karl Burri, John Chick, Jean Clamens, Niles Eldredge, William Giles, Maximilian Goldgruber, Bruno Kampmann, Graeme Lee, Jeremy Montagu, Guy Oldham, Martin Schmid, Gerhard Stradner, Frank Tomes, Jean-Claude Verdié, John Webb; Stearns Collection, University of Michigan, Ann Arbor; Trompetenschloss, Bad Säckingen; Historisches Museum, Basel; Musikinstrumenten-Museum SIMPK, Berlin; Birmingham Conservatoire; Museum of Fine Arts, Boston; Royal Pavilion, Brighton; Musical Instrument Museum, Brussels; Schloss Burgdorf; Musikhistorisk Museum, Copenhagen; Edinburgh University Collection of Historic Musical Instruments; Accademia, Florence; Viadrina Museum, Frankfurt (Oder); Burrell Collection, Glasgow; Royal Conservatoire of Scotland, Glasgow; Steierisches Landesmuseum, Graz; Gemeentemuseum, The Hague; Händelhaus, Halle; Leblanc Corporation, Kenosha, Wisconsin; Musica Kremsmünster; Musikinstrumenten-Museum, University of Leipzig; Music Museum, Lisbon; Boosey and Hawkes, London; Horniman Museum, London; Royal College of Music, London; Victoria and Albert Museum, London; Musikinstrumenten-Museum, Markneukirchen; Civic Museum, Modena; Bayerisches Nationalmuseum, Munich; Deutsches Museum, Munich; Stadtmuseum, Munich; Metropolitan Museum of Art, New York; Newtongrange Silver

Band; Conservatoire Collection, Nice; Germanisches Nationalmuseum, Nuremberg; Bate Collection, University of Oxford; Musée de l'Armée, Paris; Musée de la Musique, Paris; Musical Instrument Museum, Phoenix, Arizona; National Museum of Czech Music, Prague; National Museum of Musical Instruments, Rome; State Museum of Theatre and Music, St. Petersburg; Swedish Museum of Performing Arts, Stockholm; Hirsbrunner, Sumiswald; Ringve Museum, Trondheim; Musikwissenschaftliches Institut, Tübingen; Royal Military School of Music, Kneller Hall, Twickenham; National Music Museum, University of South Dakota, Vermillion; Accademia Filarmonica, Verona; Kunsthistorisches Museum, Vienna; Technisches Museum, Vienna; National Museum of American History (Smithsonian), Washington; Museum Bellerive, Zürich; National Museum of Switzerland, Zürich.

Contents

Part I
The Musician's Experience and the Scientific Perspective

When a brass player is asked to evaluate an unfamiliar instrument, the first step is usually to explore the potential of the instrument by playing it. Arpeggios, scales, crescendos and wide interval leaps are typical tests in which the musician judges whether the instrument can meet the demands of professional performance. An experienced player can usually recognise a poor instrument within a few seconds, although a comparison of the qualities of two well-made instruments might require several much longer playing sessions. The judgements of the performer are subjective: although most players would probably agree in recognising a poor instrument, it is not uncommon to find disagreements among musicians about the relative merits of high-quality instruments.

A scientist approaching the task of evaluating a brass instrument will probably start by trying to identify objectively measurable properties of the instrument. These could include straightforward physical parameters such as length and weight or more subtle features like the frequencies of the internal resonances of the instrument's air column. The scientific perspective must also include an understanding of the way in which the player interacts with the instrument in performance. To achieve this, mathematical models have been devised which represent the behaviour of the player's vocal tract and lips.

This first part of the book consists of two short chapters which introduce and review these two different but complementary perspectives. Chapter 1 outlines the musician's view of brass instrument behaviour, while Chap. 2 discusses the scientific approach to the topic. The aim is to set the scene for the detailed acoustical model of brass playing which is unfolded in Part II.

Chapter 1
The Musician's Experience of Brass Instruments

This chapter presents an introduction to the study of brass instruments from the point of view of the players whose lips waken silent horns to passionate musical life. Section 1.1 takes a brief look at a few aspects of the rich and varied cultural history of brasswind instruments. Section 1.2 reviews the different aspects of a player's interaction with a modern brass instrument, including the pitches of playable notes, the range of dynamics which can be readily achieved and the responsiveness of the instrument to subtle changes in embouchure or blowing pressure. Section 1.3 discusses the ways in which musicians verbalise their perceptions of instrumental behaviour and some of the unsuspected complications that arise when performers or listeners subjectively assess the musical qualities of instruments.

1.1 Creating Music from Lip Vibration: Labrosones Through the Ages

For most people, the description 'brass instrument' conjures up an image of a musical wind instrument with a shining brass bell. The brass band illustrated in Fig. 1.1 consists of instruments, including cornets, trombones, saxhorns and tubas (Herbert 1997), which are made from brass or a similar metal. The photograph in Fig. 1.2 shows the swing band lineup of the famous Glenn Miller Orchestra, including four trumpets and four trombones which are also made from brass. All these instruments are classified both musically and acoustically as members of the brass family. However although the three saxophones in the front row of the swing band are made from brass, they have single reed mouthpieces and are therefore considered to be members of the woodwind class. The distinguishing character of the brass instrument family is not the material of construction but the fact that the instrument is made to sound by buzzing the lips in the mouth aperture of the instrument. The word 'labrosone', which has been adopted by organologists

M. Campbell et al., *The Science of Brass Instruments*, Modern Acoustics and Signal Processing, https://doi.org/10.1007/978-3-030-55686-0_1

Fig. 1.1 A brass band

to denote the family of lip-excited instruments, will be used in this book when the colloquial description 'brass instrument' seems inappropriate or potentially misleading.

1.1.1 Labrosones from Found Objects

At some stage in the unrecorded millennia of prehistory, the dramatic effect of the lip buzz on a suitable resonator was discovered. The first ever brass player may have been experimenting on a Pacific shore, buzzing into the rudimentary mouthpiece formed on a conch shell by breaking off its end. The use of these naturally occurring hollow shells as labrosones (Petiot and Tavard 2008) was certainly well-established at least 3000 years ago, as evidenced by the discovery in 2001 of a cache of 20 polished and decorated *Strombus galeatus* marine shell trumpets with carefully shaped mouthpieces in a pre-Inca site in Peru. Figure 1.3a shows one of these original shells being played by Professor Perry Cook, a member of the team which researched their acoustical and musical properties (Cook et al. 2010; Kolar et al. 2012; Kolar 2014).

The use of animal horns as labrosones is recorded in some of the earliest Biblical texts. In the Book of Leviticus, describing events at a time around 1300 BCE but probably written several centuries later, Moses receives a command from God that once in every 50 years, a jubilee year should be proclaimed:

Fig. 1.2 The Glenn Miller Orchestra (c.1941). Source: the Official Ray Anthony Band website

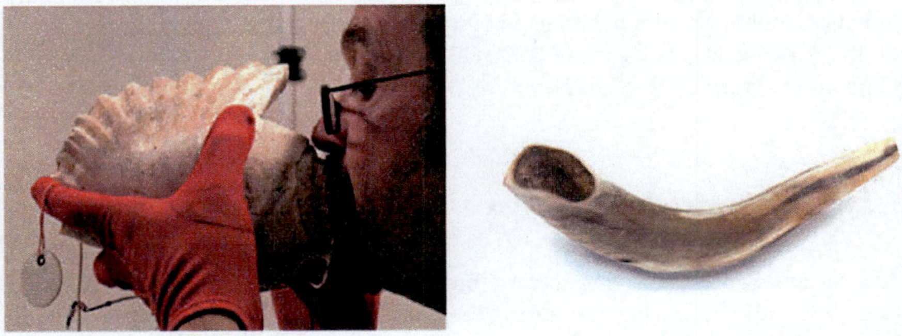

(a) (b)

Fig. 1.3 (**a**) Perry Cook playing a conch shell found at Chavín de Huántar, Peru. Photograph: José Luis Cruzado Coronel. (**b**) Shofar (Israel, 1997), EU (3757)

In the seventh month on the tenth day of the month, on the Day of Atonement, you are to send the ram's horn throughout your land to sound a blast. Hallow the fiftieth year and proclaim liberation in the land for all its inhabitants. (Leviticus 25: 9–10, Revised English Bible)

Fig. 1.4 A didgeridoo played by Ben Lange. Photograph: Kate Callas, www.phys.unsw.edu.au/jw/didjeridu.html

The ram's horn, also known as the shofar (Fig. 1.3b), continues to play an important role in Jewish religious ceremonies.

The Australian didgeridoo, like the conch and the ram's horn, is based on a found object with dimensions suitable for a labrosone resonator. In the case of the didgeridoo, the resonator is made by termites hollowing out the dead heartwood of a Eucalyptus tree. In aboriginal mythology the didgeridoo has been played for many millennia, although rock paintings in Northern Australia do not contain depictions of the instrument until around 1000 CE. Figure 1.4 illustrates a contemporary didgeridoo played by Ben Lange.

1.1.2 Early Metal Labrosones

The earliest known surviving metal labrosones are the two short trumpets found in Egypt in 1922 during the excavation of the tomb of Tutankhamun. These instruments, one of silver and one of gilded bronze, date from the fourteenth century BCE. Figure 1.5a is a photograph of the silver trumpet with its accompanying solid wooden core, taken by Harry Burton at the time of the excavation.

What sounds might a modern brass player coax from this iconic instrument, which lay mute and entombed for more than 3000 years? Current museum conservation protocols forbid playing experiments on such unique and fragile instruments, but standards were less strict in the early decades of the twentieth century. In 1939 Bandsman James Tappern of the British regiment of 11th Hussars was invited to play the Tutankhamun trumpet in a live BBC broadcast from the Cairo Museum

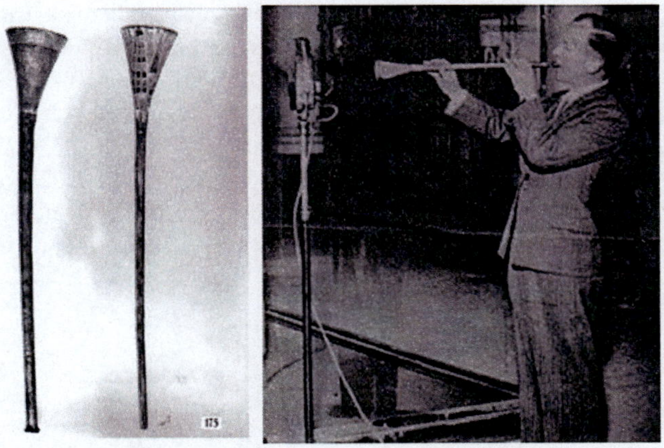

(a) (b)

Fig. 1.5 (**a**) Photograph of the Tutankhamun silver trumpet (left) with its protective wooden core (right). Copyright: Griffith Institute, University of Oxford. (**b**) James Tappern playing the Tutankhamun trumpet in a 1939 BBC broadcast

Fig. 1.6 Part of the fanfare played by James Tappern on the Tutankhamun trumpet (actual pitch)

(Fig. 1.5b). Tappern inserted a modern trumpet mouthpiece into the instrument and was able to play a three-note fanfare (Fig. 1.6).

It is dangerous to conclude that these pitches were accessible to the original Egyptian performers, however, since the addition of the mouthpiece radically changes the instrument. The eminent English trumpeter Philip Jones was permitted to sound the silver trumpet in the Cairo Museum as late as 1975, presumably using the original wide-bored entrance aperture. In a letter now held in the Department of Egyptian Antiquities at the British Museum, he reported his reaction:

> Its sound was not exactly melodious, as the bore in relation to its length is quite out of proportion, but it was probably the most thrilling experience I shall have as a trumpeter.

An insight into the use of trumpets in this period is provided by a passage in the Biblical Book of Numbers:

> The Lord said to Moses: 'Make two trumpets of beaten silver and use them for summoning the community and for breaking camp. When both are sounded, the whole community is to muster before you at the entrance to the Tent of Meeting. If a single trumpet is sounded, the chiefs who are heads of the Israelite clans will muster. When a fanfare is sounded, those encamped on the east side are to move off. When a second fanfare is sounded, those encamped on the south are to move off. A fanfare is the signal to move off. When you convene the assembly, a trumpet call must be sounded, not a fanfare.' (Numbers 10: 1–7, Revised English Bible).

While the exact meanings of the Hebrew words translated here as 'trumpet call' and 'fanfare' are uncertain, it is clear from this passage that the trumpeters were required to sound a number of signals which were audibly different. If only one pitch was sounded on the trumpets, as some scholars have suggested, variations in rhythm could have been used to differentiate the signals.

In 1797 three pairs of bronze lurs dating from around 700 BCE were discovered at Brudevaelte in Denmark; one of these pairs is illustrated in Fig. 1.7a. Subsequently many other instruments of this type have been found, mainly in Scandinavia.

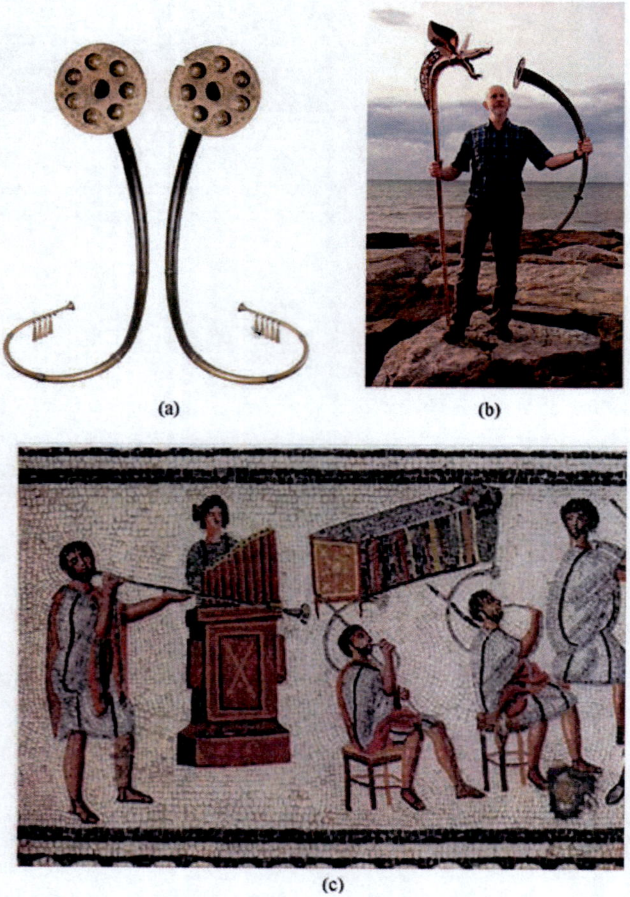

Fig. 1.7 (**a**) A pair of bronze lurs found on Brudevaelte Moor in 1787 (National Museum of Denmark 8115 and 8114, photograph: Roberto Fortuna and Kira Ursem, courtesy of National Museum of Denmark). (**b**) John Kenny holding copies of the Tintignac carnyx (left) and the Loughnashade horn (right) (photograph: Francesco Marano). (**c**) Roman tuba (left) and two cornua (right) being played with a hydraulis (organ), from a Roman floor mosaic, c. second century CE, Zliten, Libya. Photograph: Nacéra Benseddik, licensed under Creative Commons 1.0

By the beginning of the Common Era, different designs of brass instrument had emerged in Celtic regions of northern Europe; Fig. 1.7b illustrates modern copies of the Tintignac carnyx and the Loughnashade horn. Long conical trumpets are found in Asian regions and are illustrated in Mayan paintings. Labrosones in common use in the Roman Empire included the long straight tuba and the curved cornu, depicted on the mosaic in Fig. 1.7c. Some of these instruments are discussed in more detail in Sect. 9.1.

1.1.3 Labrosones in Renaissance and Baroque Music

By the end of the fifteenth century, the trumpets used for courtly ceremonials and military signalling had developed the folded shape shown in Fig. 1.8a. The bends on these instruments have little effect on their musical behaviour (see Sect. 4.7.7). The acoustically significant length of a brass instrument is its sounding length, defined as the axial length measured from the entrance plane, following the mid-bore line round bends in the tubing, and terminating in the plane of the bell end.

Until the invention of the valve in the early nineteenth century, a trumpet with a fixed sounding length was capable of playing only a set of approximately harmonic natural notes (see Sect. 1.2.2). Natural trumpet ensembles nevertheless developed a sophisticated repertoire using instruments of the same sounding length. A modern natural trumpet ensemble specialising in this repertoire is illustrated in Fig. 1.8a, while Fig. 1.8b shows an example of music written for such a group. Some members of the ensemble traditionally specialised in playing the lower natural notes (marked *Basso* and *Vulgano* on the score), while others mastered the art of *clarino* playing which opened up the possibility of playing diatonic melodies in the high register. Virtuoso sonatas and obbligato solo parts were also written for expert clarino players.

Fixed length natural trumpets were incapable of playing chromatically or even diatonically in lower registers. To meet this need, a trumpet with a sliding mouthpipe was developed, probably in the early fifteenth century. Although slide trumpet players were not part of the regular trumpet ensemble, they found an important role in the 'alta capella', a three-piece dance band consisting of two shawms and a slide trumpet.

By the mid-fifteenth century, the slide trumpet had evolved into an instrument with a U-shaped forward-facing slide. From its inception this instrument was known in Italian as the 'trombone' . Fifteenth-century French sources use the term 'sacqueboute', which is possibly a combination of the words 'sacquer' (to pull) and 'bouter' (to push) (Carter 2012, 2019). The contemporary English term 'sackbut', evidently a variant of the French word, is now used as a generic description of pre-eighteenth-century trombones. The sackbut in the style of the sixteenth century shown in the centre of Fig. 1.9 has very little difference in its basic design and functioning from the trombone played by Glenn Miller.

(a)

(b)

Fig. 1.8 (**a**) The Swiss natural trumpet ensemble Trummet, photograph: Daniela Niedhammer. (**b**) The score of the opening toccata in Monteverdi's Opera *Orfeo*

Fig. 1.9 The three authors playing (left) a serpent made by Keith Rodgers, (centre) a sackbut made by Frank Tomes and (right) a cornett made by Keith Rodgers

The cornett was the natural partner to the sackbut in Renaissance wind ensembles, since it was also capable of chromatic playing with a range of well over two octaves. The cornett illustrated on the right of Fig. 1.9, a modern copy based on a seventeenth-century original, is an approximately conical wooden tube with six finger holes and one thumb hole. The cornett was the supreme virtuoso wind instrument in the early seventeenth century; after a period of neglect in the nineteenth and early twentieth centuries, it has once more regained its place as an indispensible element in performances of the music of composers such as Giovanni Gabrieli, Claudio Monteverdi and Heinrich Schütz.

The instrument illustrated on the left of Fig. 1.9 has played a more modest musical role than the cornett and sackbut, but nevertheless deserves an honourable mention in our brief survey of the varied world of labrosones. The serpent emerged in northern France at the end of the sixteenth century, and its original function was to accompany the singing of plainchant in churches. It is a wide-bored conical tube more than 2 m in length and in its simplest form has six finger holes. Its serpentine shape makes it possible for the player to cover the six holes with three fingers of each hand, but the irregular spacing of the holes and their small diameter relative to the diameter of the main bore introduce acoustical complications which make the instrument hard to play in tune. Nevertheless, it continued to be in use in churches and military bands for more than two centuries and even appeared in the scores of Berlioz and Wagner (see Sect. 7.8).

1.1.4 The Nineteenth-Century Labrosone Revolution

In 1814 the Prussian musician Heinrich Stölzel applied for patent protection for a design of piston valve which allowed the player of a brass instrument to change the sounding length of the instrument almost instantaneously by inserting or removing a section of tubing. Before this time orchestral trumpet and french horn players usually carried sets of crooks of different lengths to perform this function, but these crooks took at least several seconds to change. By depressing a piston which increased the sounding length by around 6%, the player could flatten any of the natural notes of the instrument by a semitone. A set of three valves, adding different tube lengths which could be used in combination, allowed a trumpet or horn to play a chromatic scale from an augmented fourth below the second natural note upwards.

Within a few decades, several improved valve designs had been patented, including the Périnet piston valve used in most modern trumpets, cornets and flugelhorns and the Riedl rotary valve now standard on french horns. New types of labrosone, such as the saxhorns and the bass tuba, exploited the ability of valved instruments to play chromatically in their lower registers. The simplifications in playing technique afforded by valved brass instruments encouraged the development of amateur brass bands, and the popularity of the brass band movement in turn provided a mass market for brass instrument manufacturers.

A modern brass quintet, such as the group shown in Fig. 1.10, usually includes two trumpets, a french horn, a trombone and a tuba. The roots of all of these instruments can be traced back to the early labrosones described in Sects. 1.1.1 and 1.1.2, but the current form of each instrument has been influenced and moulded by musical and technological developments over the years. The addition of valves in the nineteenth century has perhaps been the most dramatic change: even the slide trombone now frequently has a thumb-operated rotary valve which lowers the basic pitch of the instrument by a fourth. In the twentieth century, the average bore

Fig. 1.10 Brass quintet

diameter of most orchestral brass instruments increased significantly, with musical consequences which will be discussed in Sects. 4.6 and 6.1.

The labrosone journey has certainly not reached its terminus. After exploring and describing the musical and scientific behaviour of today's brass instruments, our book concludes in Sect. 9.2 with some thoughts about the possible next steps in that journey.

1.2 The Musician's Interpretation of the Brass Playing Experience

When a skilled musician plays a brass instrument, the production of a good sound can appear to an observer to be almost effortless. A complete beginner attempting to play a brass instrument by placing it to the lips and blowing will typically expend great muscular effort and lung power to produce even an uncontrolled and unmusical note. In fact, sound production in brass instruments requires skill rather than effort. The physical aspects of playing need to be easy so that the musician can concentrate on the artistic aspects of playing – phrasing, ensemble and expression – with minimal attention to producing notes. The training and extensive practice which are necessary to make the transition from complete beginner to experienced musician allow a player to perform without concentrating on technique. In this section, we analyse aspects of playing which musicians recognised to be important in learning and teaching but are not always conscious of during performance.

1.2.1 Musical Pitch Notation

In describing or analysing musical performance, it is frequently necessary to identify the pitch of a played note. This can be done graphically using musical staff notation, but it is convenient to have a symbolic method of identifying the pitch class and octave of a particular note. Several different systems for labelling pitches can be found in the literature on musical instruments (Campbell et al. 2004). In this book frequent use is made of International Pitch Notation (IPN), whose relationship to staff notation is illustrated in Fig. 1.11. An 88-note piano keyboard has a pitch range from A0 to C8 in IPN; 'middle C' is C4. The pitch standard to which orchestral instruments tune is specified by the frequency of the note A4. The current international standard pitch is defined as A4 = 440 Hz, although in historically informed performances, other pitch standards including A4 = 415 Hz and A4 = 465 Hz are commonly adopted (Haynes 2002).

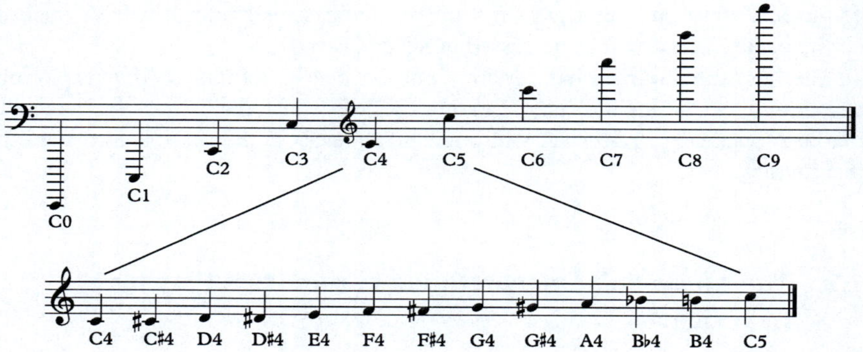

Fig. 1.11 International pitch notation

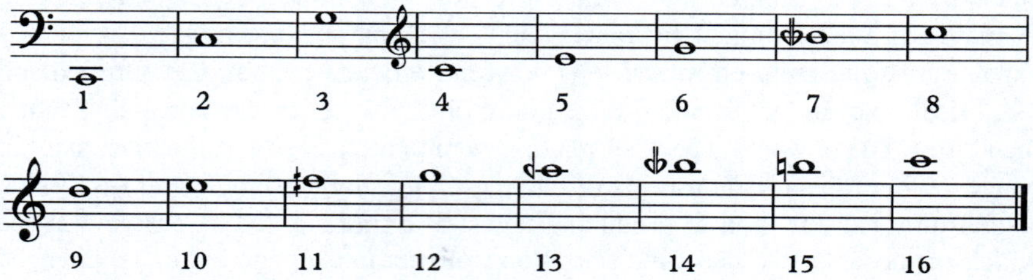

Fig. 1.12 The pitches of the first 16 members of a harmonic series with fundamental C2

1.2.2 Natural Notes and Harmonics: The Musician's View

When a trumpet is sounded without operating any valves, the player finds that a number of notes at different pitches can be easily obtained by varying the lip tension and blowing pressure. These notes are often referred to by musicians as 'harmonics'. Underlying this description is the idea of a harmonic series of frequencies, each member of which is an integer multiple of the first member (sometimes called the 'fundamental'). For example, the first few members of a harmonic series with a fundamental of 110 Hz are 110, 220, 330, 440 Hz, etc.

Figure 1.12 shows the pitches of the first 16 members of a harmonic series with fundamental 65.4 Hz (pitch C2). Although the difference in frequency between successive members of the series remains constant at 65.4 Hz, the pitch intervals become smaller as the series is ascended. Many of the notes shown in Fig. 1.12 are members of a diatonic C major scale, but the 7th, 11th, 13th and 14th are not close to any of the pitches on a piano keyboard. The 'three-quarter flat' sign on the 7th harmonic is used here to indicate that the note is significantly flatter than B$^{\flat}$4; in fact it is 231 cents below C5. Similarly the 'quarter sharp' sign on the 11th harmonic is used as a reminder that this note is slightly sharper than F4 (in fact, 151 cents below G5), while the 'quarter flat' on the 13th harmonic indicates that it is slightly flatter than A4 (in fact, 139 cents above G5).

Musical experience confirms that the easily sounded notes on a well-made brass instrument usually correspond closely in pitch to members of a harmonic series, whose fundamental pitch is basically determined by the tube length of the instrument. However the pitches of the playable notes are strongly influenced by the resonance frequencies of the instrument's air column, and we will see in Sect. 4.3 that in realistic brass instruments, these resonance frequencies are never exact members of a harmonic series. For that reason we will avoid using the term 'harmonic' to describe one of the easily playable notes on a brass instrument, employing instead the more neutral term 'natural note'.

The natural notes are not absolutely fixed, but can be raised or lowered in pitch to some extent by changing the breath pressure and/or lip muscle tension. The degree to which the natural notes can be moved varies from one instrument to another and from one dynamic to another: the pitch of loud notes on a trombone can be varied very little, while moderately quiet notes on the serpent can be moved through several semitones. Players often think in terms of a 'pitch centre', meaning the pitch of a natural note at which the response of the instrument is easiest and sound is produced most efficiently. Musicians use the term 'slotting' to indicate the experience in which feedback from a brass instrument pulls the player's lip vibrations up or down in frequency until the vibration falls into a 'slot' at the pitch centre of the note (see Sect. 2.2.2).

1.2.3 Nominal Pitches of Brass Instruments

The pitches of the natural notes available on a brass instrument are commonly indicated by attaching a pitch letter to the instrument name. In the *Symphony No. 5 in C minor* by Beethoven, for example, the score includes parts for two trumpets described as 'trombe in C' (Fig. 1.13a). The trumpets for which Beethoven was writing in the early years of the nineteenth century were valveless instruments capable of playing only the natural notes of the fixed length tube. A comparison of the pitches in Fig. 1.13a with the pitches shown in Fig. 1.12 confirms that Beethoven's trumpet in C was an instrument with a total tube length of around 2.4 m, with natural notes close to a harmonic series with fundamental pitch C2.

The score of Bartók's *Concerto for Orchestra* also includes parts for 'trumpets in C'. The twentieth-century trumpets scored for by Bartók were very different from the instruments familiar to Beethoven: they were valved trumpets capable of fully chromatic playing, as the excerpt reproduced in Fig. 1.13b shows. The tube length of the twentieth-century trumpet in C was around 1.2 m, only half that of the early nineteenth-century natural trumpet, and its natural notes corresponded to a harmonic series with fundamental pitch C3.

To distinguish between different instruments with the same pitch letter but different lengths, it is possible to extend the nomenclature to include a reference to the octave of the nominal fundamental. Using this system, the Beethoven trumpet is described as a 'trumpet with nominal fundamental C2', while the Bartók trumpet

Fig. 1.13 Parts for trumpets in C in orchestral scores by Beethoven and Bartók. (**a**) Ludwig van Beethoven: *Symphony No. 5 in C minor* (1808), 4th Movement. (**b**) Béla Bartók *Concerto for Orchestra* (1944), 1st Movement. Reproduced by permission of Boosey and Hawkes Music Publishers Ltd

is a 'trumpet with nominal fundamental C3'. An alternative approach is to specify the approximate length of the instrument: the Beethoven trumpet could be called a 'trumpet in 2.4 m C' and the Bartók trumpet a 'trumpet in 1.2 m C'. The system based on length has been widely adopted in the organological literature, but using imperial instead of metric units, the Beethoven trumpet is 'in 8-ft C', while the Bartók trumpet is 'in 4-ft C'. The system using imperial units of nominal length is adopted when appropriate in this book and used systematically in Chap. 7. The relationship between different methods of identifying the nominal pitch of instruments is shown in Table 7.1.

1.2.4 Compass

The lower and higher limits of musically useful natural notes depend on the instrument, the choice of mouthpiece and the player's skill and stamina. As we will see in Chap. 7, some instruments (especially those used to supply a bass in an ensemble) are designed to exploit the lower members of the series of natural notes, whereas other instruments (especially 'natural' instruments without slides, keys or valves used melodically) are designed to exploit the higher members of the series of natural notes.

We saw in Fig. 1.12 that the series of natural notes does not form a complete musical scale, although the octave from the 8th to the 16th notes includes eight notes that come close to forming a diatonic scale. Some instruments have a repertoire restricted to a subset of this series: bugle calls, for example, use the second to the sixth members of the series with a compass of one and a half octaves. Baroque natural (valveless) horn and trumpet parts employ a much wider compass extending to the 24th natural note, more than three octaves. Particularly on the horn, subtle adjustment of the position of the player's hand in the bell is used to improve the intonation of 'out-of-tune' members of the series, such as the 7th, 11th and 14th. Hand technique, in which the bell is partially closed by the hand, makes it possible for the player to sound additional notes on the natural horn, required in some passages (see Sect. 4.5.6).

Various mechanisms have been devised to fill the gaps in the series of natural notes. They shorten or lengthen the sounding length of the tube: this has the effect of shifting all the natural notes up or down. The simplest device is to drill toneholes in the side of the tube in such a way that they can be covered by the player's fingertips or covered by keys, as on many woodwind instruments. We will see in Sect. 4.4 that the seven finger-holes of a cornett or eleven keys on an ophicleide allow a complete chromatic compass from the lowest natural note upwards. Other instruments such as the keyed bugle mostly have fewer keys, but have a regular playing range only from the second natural note upwards.

All trombones and most valve instruments (modern band and orchestral brass-wind included) employ mechanisms which lengthen the air column. A trombone with a slide which can be extended incrementally by distances which lower the natural notes by six successive semitones has a complete chromatic compass from the second natural note (slide fully extended) upwards. The commonest system of valves has three valves, conventionally described as first, second and third, which lower the natural notes by two, one and three semitones, respectively. One reason for the placement of the semitone valve second is that the loops of additional tubing can often be most conveniently arranged with the shortest in the middle of the cluster. These three valves used singly or in combination lower the natural notes by up to six semitones and also give a complete chromatic compass from the second natural note (three valves operated) upwards.

Some instruments with only one or two valves are used in repertoire for which an incomplete compass is sufficient; other instruments (such as tubas) have more than three valves and a chromatic compass from the lowest natural note upwards. The mechanism of a valve can be equally arranged to shorten an air column: although most valves are 'descending', some instruments employ 'ascending' valves. The fourth valve of a modern french horn can either lower the basic pitch from 9-ft B$^\flat$ to 12-ft F or the opposite, depending on the model and the setting chosen by the player.

1.2.5 *Intonation Control*

Playing in tune is the concern of every musician, but what 'in tune' means is not always clear-cut. Equal temperament is widely used in tuning modern keyboard instruments and so is generally necessary when brass and keyboard instruments play together. Equal temperament is also adopted in much ensemble playing, but the pitch level of an ensemble can drift in the course of performance, and musicians have to remain in tune. Even when equal temperament is adopted, a purer tuning of some chords is often desirable. In historically informed performance of music earlier than the mid-nineteenth century, a non-equal temperament is usual, but a musician may still find it aesthetically desirable to 'bend' the pitch of certain notes.

The pitches of an instrument's natural notes are unlikely to coincide exactly with the pitches judged to be 'in tune', so accommodation is necessary, and can be achieved in several ways. Firstly, the natural notes can be raised or lowered in pitch to some extent by changing breath pressure and/or lip muscle tension. As we will see in Sect. 6.3, the player's windway, from the lungs to the lips, is part of the complete instrument plus player system and can in some cases be used to influence intonation. The degree to which a player can bend notes depends on the characteristics of the instrument and on the playing dynamic. In some situations, it can be difficult to remain in tune through a crescendo. It is unrealistic to expect a brasswind maker to supply an instrument that is always naturally in tune: the best that can be expected is for an instrument to be able to be played in tune with little difficulty by an experienced musician in regular perfomance situations.

To a large extent, placing the pitches of the natural notes to allow good intonation is a matter of the instrument maker's skill. However, there are some specific devices which can help significantly. Firstly, most brass instruments are equipped with a tuning-slide. Because musicians' playing characteristics vary, instrument makers build brass instruments so that an average player adopting the intended pitch standard (such as A4 = 440 Hz) will find an optimum tube length with the tuning-slide extended by a centimetre or two. In addition to the main tuning-slide, valve instruments have valve tuning-slides on the valve loops. In the case of finger-hole and keyed instruments such as the cornett and the ophicleide, the natural notes are relatively easy to bend, and playing in tune requires the player to have a clear conception of what is required, to be in command of the fingering possibilities and to exercise firm control. On both tonehole and valved brass instruments, alternative fingerings are often available giving different tunings for the same note; these can be used to adapt intonation to suit different musical contexts. In the case of slide instruments such as the trombone, pitch is easily variable. Although for a beginner the trombone is thought of as having seven slide positions, for an advanced player, the slide placement is adjusted for each of the natural notes. With the trombone slide fully closed, tuning of the natural notes by slide movement can only be downwards, something to be allowed for in adjusting the tuning-slide before playing. On some models of trombone, there are spring buffers ('touch springs') at the top of the slide so the slide can be moved upwards to a small extent from its rest position to cope with flat natural notes.

The biggest problems in intonation arise with valved instruments. A natural (valveless) instrument can, with skill in designing the bore profile, be made to allow good intonation for much of the compass; extending this to the bore profile when each of three or more valves operated is more difficult. The major problems set in when valves are used in combination. If a valve adds the correct amount of tubing to lower the instrument by n semitones, it will not add enough tubing to lower the instrument by n semitones when another valve is in use at the same time. The calculations are more complex than the simple proportions often presented, since valves increase the amount of cylindrical tubing in the windway and affect the bore profile as well as the air column length. Typically, a three-semitone third valve used in combination with a five-semitone fourth valve will give a lowering of nearer seven rather than eight semitones. Various ways of overcoming this problem have been devised (Myers 1997b), some of amazing mechanical complexity (Fig. 1.14). With small instruments, it is often enough to tune the third valve to lower the pitch by slightly more than three semitones and then to avoid using it on its own – the player can then 'lip' any wayward notes up or down sufficiently for reasonably good intonation. Fitting the third valve tuning-slide (sometimes the first as well) with a finger-ring or sprung lever ('trigger') so that it can be moved by the player (at least in slow-moving passages) is common with trumpets and cornets. Some tubas are designed so that a tuning-slide can be manipulated in performance, and some models of large instruments such as euphoniums and tubas have large numbers of valves (five or six) allowing the player some flexibility of fingering. A tuba might, for instance, have two valves nominally giving a semitone but with one adding more tubing than the other, allowing the player to choose an air column length, giving good intonation from the 32 possible fingerings (five valves) or 64 fingerings (six valves). A 'compensating' valve system (see Sect. 8.4) is used on some french horns and many euphoniums and tubas; the 'full double' principle makes for a heavy

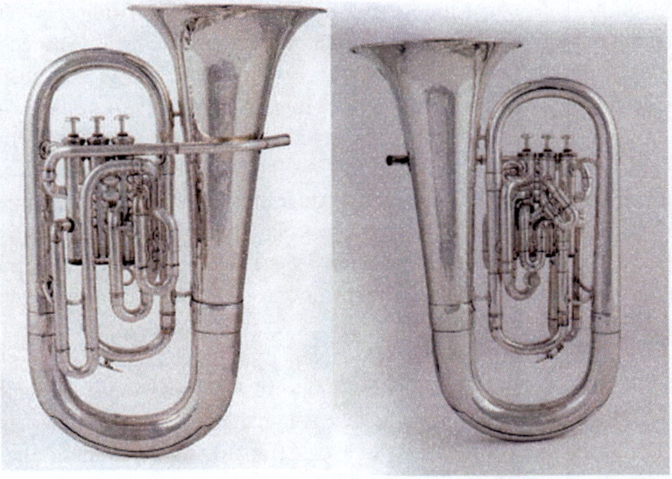

(a) (b)

Fig. 1.14 Example of a complex valve system designed to improve intonation when valves are used in combination: Euphonium, three-valve 'Synchrotonic' system, with tuning-slides for each valve and additionally for each pair of valves operated together. (**a**) Front view. (**b**) Back view. (J. Higham Ltd, Manchester, probably c1923), EU (6381) instrument, since three valves have two full-length valve loops; as a result it has rarely been used on tubas, but it is the preferred valve system for modern french horns.

1.2.6 Dynamic Range

In simple terms, blowing harder produces a louder sound (see Sect. 2.1). Even for advanced players, increasing air pressure in the vocal tract and mouth is the principal means to higher dynamics, but varying the dynamic level while remaining in tune is a skill that requires practice. An important aspect of technique is developing the ability to initiate and sustain very quiet notes; the upper limit to playing dynamics, however, is largely dictated by the instrument design. We will see in Sect. 6.1 that transformation of sound energy into heat and in some cases shock wave formation ultimately limit the acoustic output of an instrument. The common experience of musicians is that wide-bore instruments can be played at a higher dynamic level than narrow-bore instruments, and this is also explained in Sect. 6.1.

1.2.7 Timbre

It is a truism that playing with a 'good tone' is an attribute of a fine musician and that a high-quality instrument is important for this. Timbre is certainly an aspect of brass playing in which instrument design and performance technique are inseparably involved. It is inescapable for beginner and advanced musician alike that a note on a brass instrument played at a high dynamic level sounds brighter than the same note on the same instrument played quietly (see Sect. 2.2.3); the reasons for this are discussed in Chaps. 3 and 6. This is closely connected with the overall differences between types of instrument: flugelhorns, cornets and trumpets (all in $4\frac{1}{2}$-ft $B^\flat$) can be easily distinguished when played at moderate or high dynamics but are difficult or impossible to tell apart in very quiet playing. The same can be said of euphoniums, french horns and trombones an octave lower. The varieties of brasswind are discussed in Chap. 7.

For a given instrument played at a given dynamic, a musician has a limited scope for influencing the timbre. There are some factors, however, which can be controlled to produce a sound quality that is considered optimal. The first of these is choice of mouthpiece: in nearly all cases, choosing an instrument and choosing a mouthpiece are separate exercises. When a required note can be obtained by more than one fingering of a finger-hole or keyed instrument, musicians generally prefer the timbre obtained by a fingering giving the longest sounding length; when alternatives of valve or slide operation give the required note, the shortest sounding length is generally preferred. However, opening tone-holes on a finger-hole or keyed instrument is not generally a matter of opening the holes one-by-one from the bell;

the best intonation and timbre often require a complex fingering pattern known as 'cross-fingering' or 'fork fingering'.

With valve and slide instruments, most notes can be played with more than one fingering or slide position: the shortest sounding length gives the greatest security in hitting the desired note in a series of natural notes since the natural notes higher in the series are closer together. However, a player may use a longer alternative valve fingering or slide position on occasion to achieve a different quality of sound, to achieve a smooth transition from one note to the next or to reduce valve or slide movement.

Timbre is also dependent on placement of the mouthpiece on the lips, on lip muscle tension and on the configuration of the vocal tract. Brasswind pedagogy places great importance on breathing techniques in producing a 'good tone'. A large part of the perceived quality of sound is in the way notes are initiated: the 'starting transients' in brass playing, which in many cases depend on the use of the tongue, separate beginners from experienced players.

We can also mention vibrato here: although strictly speaking independent of timbre, vibrato is heard as an aspect of tone quality. On brass instruments, vibrato is more of a matter of pitch variation than dynamic variation. It is achieved by inducing an oscillation in air pressure in the mouth with a frequency of a few Hz, variation in lip muscle tension (in some cases involving moving the whole instrument backwards and forwards) or both; on instruments with a slide such as the trombone, it is commonly achieved by slide movement.

A concept often discussed by musicians is 'projection', meaning the ability of a musician to be heard throughout an auditorium even when partially masked by other sounds. There is no way this can be achieved which is independent of dynamics and timbre. The use of vibrato increases the salience of a solo voice over an orchestral accompaniment (Deutsch 2013). In brass playing, vibrato is not always appropriate, and at a given dynamic level, producing a brighter timbre can be the only means to increase projection.

1.2.8 Blowing Pressure and Air Flow

The normal breathing process involves the regular repetition of inhalation and exhalation. In the inhalation phase, the lung volume increases, and air flows in through the nose (and the mouth if it is open); in the exhalation phase, the lung volume reduces and air flows out again. To sound a note on a brass instrument, the player expands the lung cavity by breathing in, usually through both mouth and nose; in exhalation, the soft palate is normally raised to close the aperture connecting the back of the throat to the nasal cavity, so that all the air flows through the player's lips into the mouthpiece of the instrument.

A complex array of muscles controls the expansion and contraction of the lungs (Watson 2019). These muscles do not operate directly on the lungs, but on the surrounding chest walls to which the lung surfaces are attached. The principal muscle active in inspiration is the diaphragm, which separates the chest cavity from the abdominal cavity. When the diaphragm contracts, it pulls the floor of the chest cavity downwards; the resulting increase in lung volume is enhanced by intercostal muscles which pull the ribs upwards and outwards. When these muscles are relaxed, the elasticity of the lungs causes them to contract, raising the internal lung pressure, and resulting in an outflow of air. The increase in lung pressure can be enhanced by the action of abdominal muscles which push the chest floor upwards and a second group of intercostal muscles which pull the rib cage downwards. To maintain a smooth flow of air, the effects of these expiratory muscles can be partially counteracted by the first group of inspiratory intercostal muscles. The use of respiratory muscles in brass performance is discussed further in Sect. 6.3.4.

Achieving good control of blowing pressure and air flow is essential for any wind instrumentalist, and there has been some disagreement over the best way to teach this aspect of brass playing technique (Dudgeon et al. 1997). Some instructors focus on the regulation of the mouth pressure by the respiratory muscles, while others encourage students to think more about the creation of a smooth air flow into the instrument. The flow of air into the mouthpiece during the sounding of a note is not strictly speaking smooth, since the lip aperture through which the air passes is opening and closing regularly at the frequency of the note. The modulated flow which leaves the lips can be thought of as a steady mean flow plus an acoustic flow fluctuating around this mean value at the playing frequency. Although it may be pedagogically helpful to encourage students to focus on a column of air moving through the instrument and out of the bell, the work of the air flow in generating sound is complete once it has passed through the lips and entered the mouthpiece. Even if the flow is then diverted so that it does not pass into the instrument (Sect. 2.1.5), acoustic pressure waves transmit the sound from the mouthpiece down the air column to the bell and beyond.

1.2.9 Resistance and Playing Effort

The resistance to the steady flow of air through a tube is defined as the pressure difference between its ends divided by the rate of air flow through it. This steady flow resistance increases as its internal diameter is reduced, so a trumpet mouthpiece with a throat diameter of 3.7 mm has a much higher steady flow resistance than a tuba mouthpiece with a throat diameter of 8.4 mm. The difference in steady flow resistance can be qualitatively experienced by taking a deep breath and allowing the breath to exhale naturally through open lips into each mouthpiece. The initial lung pressure and the volume of air taken in during inspiration will be approximately the same in each case; expiry through the tuba mouthpiece will be complete in around a second, but will last several seconds through the trumpet mouthpiece. When the mouthpieces are inserted into their respective instruments, the expiry times are increased by the additional resistance of the instrument tubing, but the same qualitative difference is evident.

The situation is substantially different, however, when the lips of the player form an embouchure and sound a note. High-speed filming of the motion of brass players' lips has shown that over most of the playing range of orchestral brass instruments, the average open area of the aperture between lips is much less than the open area of the mouthpiece throat (see Sect. 3.1.3). The resistance presented to the mean flow of air is therefore determined mainly by the player's embouchure rather than the mouthpiece throat. The relatively large lip opening area used for a low-pitch note on a particular instrument corresponds to a relatively low mean flow resistance, while for a high-pitch note on the same instrument, the lip opening aperture will be much smaller and the mean flow resistance larger. The practical consequence is that the supply of air in the lungs is used up more quickly for low notes than for high notes. To play sustained notes with a low mean flow resistance, embouchure players sometimes use a circular breathing technique, playing continuously without interrupting the sound by breathing in through the nose while maintaining mouth pressure using cheek muscles. This technique is an essential feature of performance on the Australian didgeridoo (Sect. 6.3).

In discussing the playing properties of an instrument, musicians sometimes employ the word 'resistance' to describe the amount of effort required to sound a note. The characteristic which is being evoked by this use of the word is not the mean flow resistance, which determines when the air supply is exhausted, but a different property of the instrument related to the ease with which the player can sound the note. There does not seem to be a strong consensus among players about the precise nature of this property, which we could call the 'sounding resistance', but it seems likely that it is related to the minimum blowing pressure which is needed to start the lips vibrating. The hypothesis that the musician's experience of sounding resistance can be identified with the scientific concept of threshold mouth pressure is discussed further in Sects. 2.1.3 and 5.4.3.

1.2.10 Responsiveness and Rapid Articulation

The recording of Joe Oliver's *West End Blues* by Louis Armstrong and His Hot Five in 1928 is recognised as a milestone in the history of jazz. The electrifying solo trumpet cadenza at the start of the recording, transcribed in Fig. 1.15, demonstrates

Fig. 1.15 Trumpet cadenza played by Louis Armstrong in *West End Blues* (1928)

Armstrong's mastery of articulation in both relatively slow and extremely fast playing. Instruments capable of meeting the demands of this type of performance are often described as responsive, reflecting the sensation of the player that the instrument is an understanding and co-operative partner in the musical task.

The response of an instrument can be defined more technically as the speed with which the instrument reacts to the initiation of a note by the player (Herbert et al. 2019). Any brass instrument sound begins at a very low level when the player's lips start to vibrate and builds up to its steady level over a number of vibration cycles. The period of initial growth is called the starting transient (see Sect. 2.1.1). On a highly responsive instrument, the starting transient is short, and the threshold pressure is low, allowing the player to sound the note promptly and with little effort.

Special techniques, such as double and triple tonguing, have been evolved for rapidly interrupting and initiating air flow. Finger movement and trombone slide movement generally impose greater limitations, and factors such as the mass and distance of travel of piston valves and trombone slides can be significant in determining the maximum possible articulation speed on larger brass instruments.

1.2.11 Wrap, Directivity and Ergonomics

'Wrap' means of the particular way the tubing is arranged: which parts are coiled or straight (Fig. 1.16), the angle of the bell axis to the mouthpiece axis, etc. Many instruments, such as the tenor trombone, have a conventional wrap used in the vast majority of instruments; other instruments such as tubas display considerable variety. Historically, the diversity has been even greater, with some instruments such as valve trombones having the bell directed either forwards, upwards or sideways (see Fig. 1.17). Since high-frequency components are radiated more strongly along the axis of the bell of the instrument (see Sect. 4.6.6), trumpets in a band marching towards a listener sound much brighter than they do if the band is marching away. On the whole, however, the bell direction does not seem to be a consideration of great importance to musicians. This may be because much performance takes place in fairly lively rooms, in which reflections even out inequalities in dynamic levels resulting from directional sound radiation. However some composers (notably Mahler) occasionally ask orchestral brass players to perform with 'bells up' (i.e. pointed directly out towards the audience), presumably to enhance the brightness of the direct sound.

Fig. 1.16 Two horns in the hands of two of the present authors. The instrument on the right is a true french horn made for acoustical demonstration purposes without any turns by the London maker John Köhler in 1856 for the University of Edinburgh. The instrument on the left is a horn of close to equal tube length but in conventional french horn wrap. Not only does the straight horn lack a tuning-slide and require a second player for any hand-in-the-bell playing, its sheer impracticality justifies the time and skill needed to bend tubing and design instruments ergonomically. Photograph: Antonia Reeve

1.3 Subjective and Objective Evaluation of Brass Instrument Quality

The relationship between performer and musical instrument is a very intimate and personal one. A beginning brass player may start with a student instrument chosen by a teacher, but as experience is gained, the player is likely to spend considerable time and effort in trying to find an instrument which promises a more fulfilling musical partnership. This quest frequently involves discussions with other players about the relative merits of different models. A specific vocabulary has evolved to describe particular performing characteristics which are generally agreed to be important; some of these characteristics were already identified in Sect. 1.2. There is however evidence, presented in Sect. 1.3.2, that not all of the terms commonly used by players to describe instrumental qualities are in fact useful. Expert performers often disagree about the merits of particular features of brass instrument design and construction; one possible explanation for this lack of unanimity is advanced in Sect. 1.3.3.

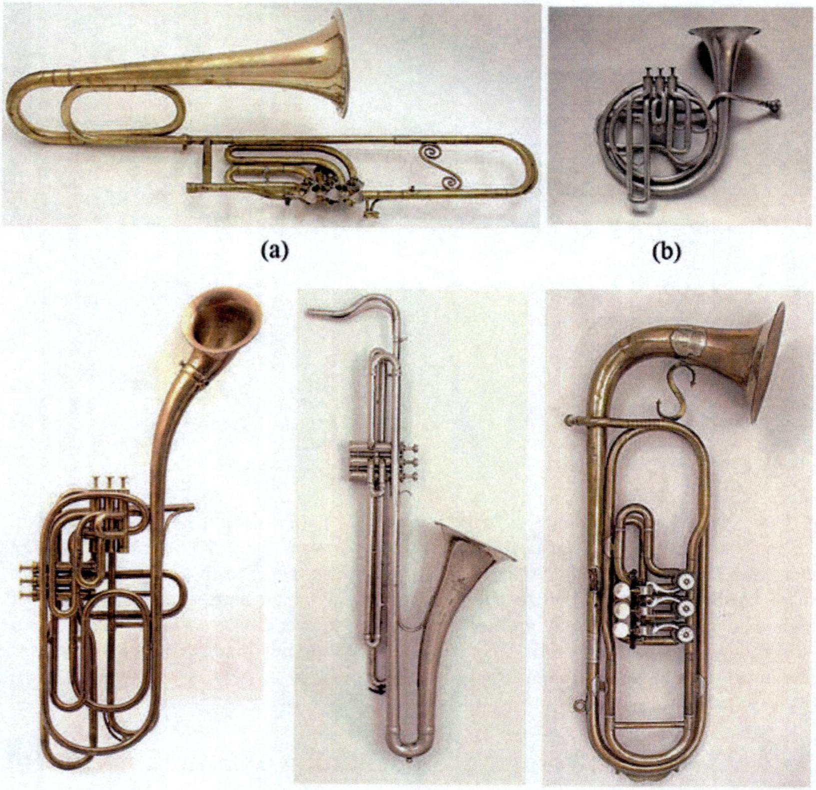

(a) (b)

(c) (d) (e)

Fig. 1.17 Tenor valve trombones with bells directed forwards, upwards and sideways. (**a**) Baritone trombone (Seltmann, Philadelphia, USA, late nineteenth century), EU (3832). (**b**) Ventil trombone (Distin & Co, London, 1873), EU (4631). (**c**) Trombone with *pavillon tournant* – bell can be rotated to face different directions (Adolphe Sax, Paris, 1864), EU (5864). (**d**) Tenor Normaphon (probably Richard Oskar Heber, Markneukirchen, c 1928). EU (4253). (**e**) *Armeeposaune* (Karl Schamal, Prague, c 1880), EU (3472)

1.3.1 Sound Quality and Playability

The primary purpose of any musical instrument is to generate sound, and it is on the properties of the radiated sound that a listener will judge the quality of the instrument. For the player, too, the quality of the sound output is paramount. 'The point is to try to sound great when you play', said the renowned brass instructor Arnold Jacobs (Kelly 1998); if the sound reaching the ears of the performer is judged to be great, the instrument will have passed its most important test.

The performer, however, has an additional set of criteria against which the success or failure of an instrument must be judged. An expert player can make a good sound on almost any instrument, but this may involve all sorts of compensations and adjustments which are concealed from the audience but add greatly to the effort of the performance. The term ‘playability’ is now widely used by performers and manufacturers to encompass the features of an instrument which make it easy to play. The word seems to have been coined by Stevenson (1881) in an early essay entitled ‘Child’s Play’ and was adopted as a convenient term to describe the playing properties of violins by Woodhouse (1993). In his discussion of violin playability, Woodhouse makes an important point which is also valid for brass instruments: an instrument considered to be easy to play will not necessarily rate highly on the sound quality scale. An obvious example can be found in the choice of a mouthpiece for high register playing: a smaller cup can make high notes less tiring, but this can be at the expense of fullness of sound.

1.3.2 Descriptive Terms Used by Musicians to Describe Brass Instrument Behaviour

In 2002, Matthias Bertsch and colleagues in Vienna instituted an extensive research project on player evaluation of trumpets (Bertsch and Waldherr 2005; Bertsch et al. 2005). Over 250 playing tests were performed by 55 trumpeters, mostly professionals or advanced students from 3 different countries. A player performing the test was given one of a number of trumpets in a darkened room, to prevent visual identification; the test consisted of 40 questions assessing the sound quality and playability of the instrument. Each player carried out the test several times, usually being presented with a different instrument in each test. Some of the tests were however repeats on the same instrument, allowing the reliability of the judgements to be evaluated.

Bertsch et al. found that although individual players were able to rate consistently the overall quality of the tested trumpets on a 5-point scale from ‘professional instrument’ to ‘beginner instrument’, they did not all agree on which instrument was their own personal preference. This finding reinforces the personal and subjective nature of the players’ judgements: it seems clear that not all players were using the same criteria in evaluating overall quality.

Some of the test questions probed specific features of instrumental performance. One important aspect of playability is the ease with which a particular note can be lipped up or down from its median pitch; notes which have little pitch flexibility are described as ‘strongly centred’ or ‘slotted’. Judgements of this property were consistent, showing that this is a well-understood and robust criterion of quality.

Players were also consistent in answering questions about how quickly a note could be started or repeated (see Sect. 1.2.10). General questions on the ‘response’ of the instrument were however answered very inconsistently, implying that a number of different factors, varying from test to test, may have been influencing the players’ interpretation of this word. A similar lack of reliability was found in judgements of ‘resistance’; it may be that the player sensation described by this term is too dependent on the technique of the player to be useful in characterising the instrument.

In evaluating the quality of an instrument, players also consider the nature of the radiated sound. Judgements of 'sound power' were found by Bertsch et al. to be consistent, but estimates of timbral quality were less reliable. Timbre is a multidimensional quality, and many different semantic scales have been used to describe different tone colours. Sounds can be rated on a scale from 'dirty' to 'clean', from 'narrow' to 'wide' or from 'solid' to 'hollow' (Bismarck 1974). One of the most commonly used timbral scales spans the range from 'dark' to 'bright'; this scale is related to the balance between low- and high-frequency components in the sound spectrum (see Sect. 2.2.3). A surprising finding in the tests of Bertsch et al. was that trumpet players were very inconsistent in judging the 'brightness' of a particular instrument. Ratings on a scale from 'colourless' to 'brilliant' were more reliable.

What conclusion can be drawn from this brief survey of some of the terms used by brass players in comparing instruments? The response of a player to a particular instrument is bound to depend very heavily on the player's background, training, experience and musical taste. Short-term factors, such as whether a different instrument has been played just before the test, can have a biasing effect on judgements of particular qualities of playability or timbre. Some of the descriptive words used by musicians do not have well-defined and universally accepted meanings, and such terms should be treated with caution when discussing and evaluating brass instruments.

1.3.3 Biases in Quality Evaluation of Musical Instruments

In the tests described in the previous section, the experimenters were careful to ensure that the players were unable to have clear sight of the instruments being played. The reason for this was that clues such as a maker's trademark or a characteristic shape or colour have been shown to exert a significant influence on the player's assessment of an instrument. A classic example of this phenomenon was reported more than 30 years ago by Smith (1986), then working for the London brass manufacturer Boosey & Hawkes. As part of a study of the perceptual significance of wall vibrations in brass instruments (see Sect. 6.6), ten professional trombonists were asked to play and evaluate a number of tenor trombones, identical except for the bell sections. All the bells were made on the same mandrel to ensure that they had the same internal dimensions, and the mass distributions of the instruments were equalised. Visual cues were eliminated by blindfolding the players (Fig. 1.18).

Most of the bells were made from yellow brass, which is an alloy of 70% copper and 30% zinc. A pure copper bell was also included in the test set. Brass instrument manufacturers typically claim that a red brass bell, with 90% copper in the alloy, gives a darker, warmer sound than a yellow brass bell, and this opinion is shared by many players. In the blind test, the trombone with a pure copper bell was not identified as having a noticeably different timbre from those with yellow brass bells;

Fig. 1.18 Blindfold test of a trombone (Smith 1986)

however Smith reported that 'when subsequently played in non-blind tests it gained magical properties'.

Similar evidence of the importance of visual cues in influencing players' judgements of musical instruments has emerged in recent comparison tests on old Italian and newly manufactured violins (Fritz et al. 2012, 2014). Many violinists are convinced that instruments made by Italian master craftsmen such as Antonio Stradivari in the seventeenth and eighteenth centuries have particular qualities of playability and timbre which cannot be reproduced by the best modern makers. Carefully conducted blind tests by Claudia Fritz and colleagues have shown that even very experienced professional performers are incapable of distinguishing between old and new instruments without visual or other non-auditory cues.

Are trombonists misled, then, when they judge that an instrument with a red brass bell has a warmer sound than a yellow brass or silver bell? The answer to this question is more subtle than might at first appear. In 1976 Harry McGurk and John MacDonald published an article in the journal *Nature* in which they showed that the perception of speech sounds could be changed by visual cues (McGurk and MacDonald 1976). They dubbed the sound track of a speaker repeating the syllable 'ba' on to a film of the same speaker mouthing the syllable 'ga'. Subjects who listened to the audio track while watching the film reported hearing a third syllable, 'da'. An example of this test, performed by Patricia Kuhl, is available online (Kuhl 2020). It is remarkable that, even when the illusion is explained, it is usually impossible for the listener to hear the 'correct' sound 'ba' while watching the incompatible visual cue.

This fusing of auditory and visual speech cues, described as the McGurk Effect, is an example of a broader phenomenon known as cross-modal interference (Campbell 2013, 2014b). The perception of speech is described as bimodal because two different modalities (hearing and seeing) are involved. In playing a musical instrument, there is a third modality, described as haptic feedback, conveyed by the player's sense of touch. In a brass instrument, haptic feedback comes from the vibrations sensed by the player's lips and fingers, but also from the muscles associated with movements of pistons or slides. The perception of the player is likely to be influenced by a complex interaction of the sensations delivered through these three modalities. In the case of the copper-belled trombone, the mental processing of the purely auditory information may be modified by the visual association between the colour of copper and the idea of warmth. In that case, the player may indeed hear a 'warmer' sound, even if there is no objectively measurable difference in the sound spectrum.

Not everyone hears the McGurk effect. The process by which inputs from different modalities are combined by the brain, described as multimodal or multisensory integration, is still a subject of vigorous research in neuroscience, but it is known that it occurs to differing extents in different people. This may partly explain why there are such vehement disagreements among brass players about issues like the influence of bell material on timbre, since the influence of the visual cues on the auditory experience will not be the same for every player.

Chapter 2
The Scientist's Perspective on Brass Instrument Behaviour

The scientist's approach to the study of a musical instrument is inevitably different from that of a musician interested primarily in its performing properties, although the two points of view are complementary. The first step in the classic scientific method is to observe, measure and record the behaviour of the object under investigation. A simple theory (or 'model') is then developed which is hoped to explain the main features of the observed phenomena. The theory is expressed in a set of mathematical equations, whose solutions are used to make predictions of the behaviour of the studied object which can be compared with the experimental measurements. In the case of a brass instrument, for example, it may be possible to predict the sound output from the instrument under specified playing conditions. If the predictions match the measurements well, the theory can be considered adequate. Usually, however, there are divergences between theoretical predictions and experimental measurements, which show that the theory needs refinement. Some of the simplifications are then revisited with the aim of developing a more realistic theory which better accords with reality.

A major difficulty in applying the physicist's method to the study of a brass instrument is the presence of the musician in the system. As a consequence, the most elementary models of brass instrument performance make use of simplifications and approximations which may seem very crude from the musical viewpoint, but which nevertheless allow interesting and useful conclusions to be drawn. A first overview is given in the present chapter; more details are presented in Part II.

2.1 Scientific Measurements of Brass Instrument Behaviour

The scientific study of any aspect of the behaviour of the world usually begins with observations and measurements of the phenomena involved. When the subject of study is the behaviour of a brasswind musical instrument, an obvious starting

M. Campbell et al., *The Science of Brass Instruments*, Modern Acoustics and Signal Processing, https://doi.org/10.1007/978-3-030-55686-0_2

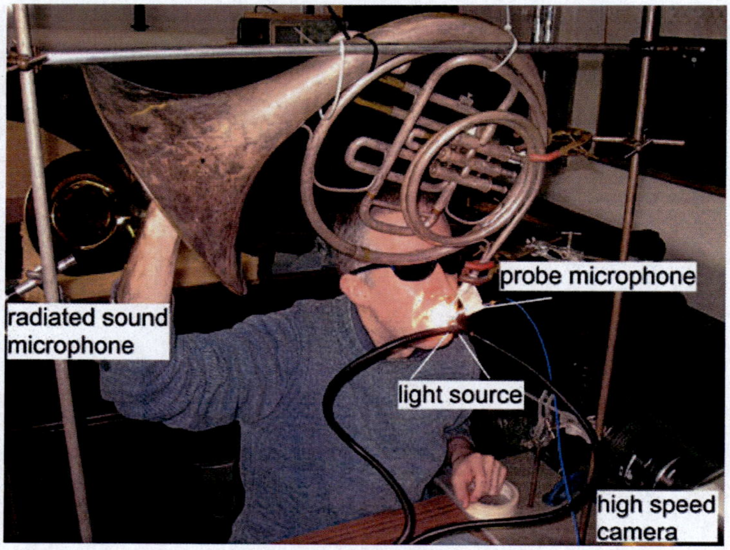

Fig. 2.1 Scientific measurements of a horn player (Logie 2012)

point is the recording and analysis of the sound generated by the instrument. The recording microphone may be placed at various points in the room in which the instrument is played; it is also possible to record the sound at different points inside the instrument and even inside the mouth of the player. Sensors other than microphones can also provide useful information: force sensors can measure the pressure exerted by the lips on the mouthpiece, laser distance sensors can track the movements of valves or slides, accelerometers can sense the vibrations of the bell and high-speed video cameras can record the motion of the lips in a transparent mouthpiece (Fig. 2.1). Under certain circumstances, it is even possible, using optical techniques, to obtain pictures of the sound wavefronts radiated from a brass instrument bell (see Sect. 4.6.5).

Measurement techniques of the types mentioned in the previous paragraph have been used in many experimental studies of the behaviour of brass instruments. Three different methods of creating sound in the instrument have been frequently employed. In the first method, the instrument is sounded in the normal way described in Chap. 1: the lips of a human player are pressed against the mouthpiece of the instrument, an excess pressure is generated in the mouth by the player's lungs and the resulting air flow between the lips sets them vibrating. In the second method, the human player is replaced by an artificial mouth: the lips are typically water-filled rubber tubes, and the mouth is a metal box in which an excess pressure is created by an air pump. In the third method, there are neither real nor artificial lips; an electronically generated signal is fed to a loudspeaker connected directly to the input of the instrument.

Each type of experiment has its own advantages and disadvantages, which are discussed fully in Chaps. 3 and 4. It is important to note at the outset, however, that scientists and musicians agree that the behaviour of a musical instrument can only be fully understood when the nature of the human player's involvement with the instrument is taken into account. The ultimate goal for scientists working on the musical acoustics of a brass instrument is to explain fully how the instrument functions in a musical performance, which normally involves four linked and interacting elements: the player, the instrument, the performance space and the listener. Progress towards this goal will require much further research, including contributions from physics, physiology, neuroscience, psychoacoustics and room acoustics. In the case of brass instruments, the interaction between the player and the instrument is crucial even at a very basic level, since the sound generating mechanism is provided by the lips of the player.

To illustrate the nature of the scientific approach to brass instruments, we review briefly some of the topics which will be studied in more detail in following chapters, outlining the methods of measurement used to gain scientific understanding of the phenomena involved.

2.1.1 Sound Radiated from a Brass Instrument

First we look at the sound radiated from a trombone during a musical performance. The signal illustrated in Fig. 2.2b was recorded by a microphone 50 cm in front of the bell of a tenor trombone. Just under 1 second after the recording started, the player sounded the note F4, shown in staff notation in Fig. 2.2a. The note was maintained at a constant piano level for around 1 second; the player then made a crescendo to forte, followed by a diminuendo. Since a calibrated instrumentation microphone was used in the recording, the signal shows the changes in the pressure p_{rad} of the air due to the sound wave radiated by the trombone.

During the sounding of the note, the pressure was rising and falling several hundred times every second. On the time scale in Fig. 2.2b, it is impossible to distinguish these very rapid variations in the sound signal, which are all contained within the solid blue area. To see the detail of the pressure changes corresponding to the sound of the trombone, it is necessary to look at a much smaller time segment of the signal. The red box in Fig. 2.2b marks a time interval of 230 ms starting at $t = 0.92$ s. The zoom into this time interval shown in Fig. 2.2c reveals how the sound begins. The tiny wiggles in the horizontal blue line around $t = 0.93$ s are the first signs of a developing note. At this stage, the signal is fluctuating between +5 mPa and −5 mPa, so the peak-to-peak pressure amplitude is only 10 mPa. This amplitude grows, at first slowly and then more rapidly; by $t = 1.03$ s it has reached a value of 0.46 Pa which remains more or less constant for around 1 s. The initial part of the sound, in which the amplitude is continuously increasing, is called the 'starting transient'; in musical terms this corresponds to the 'attack' of the note. The

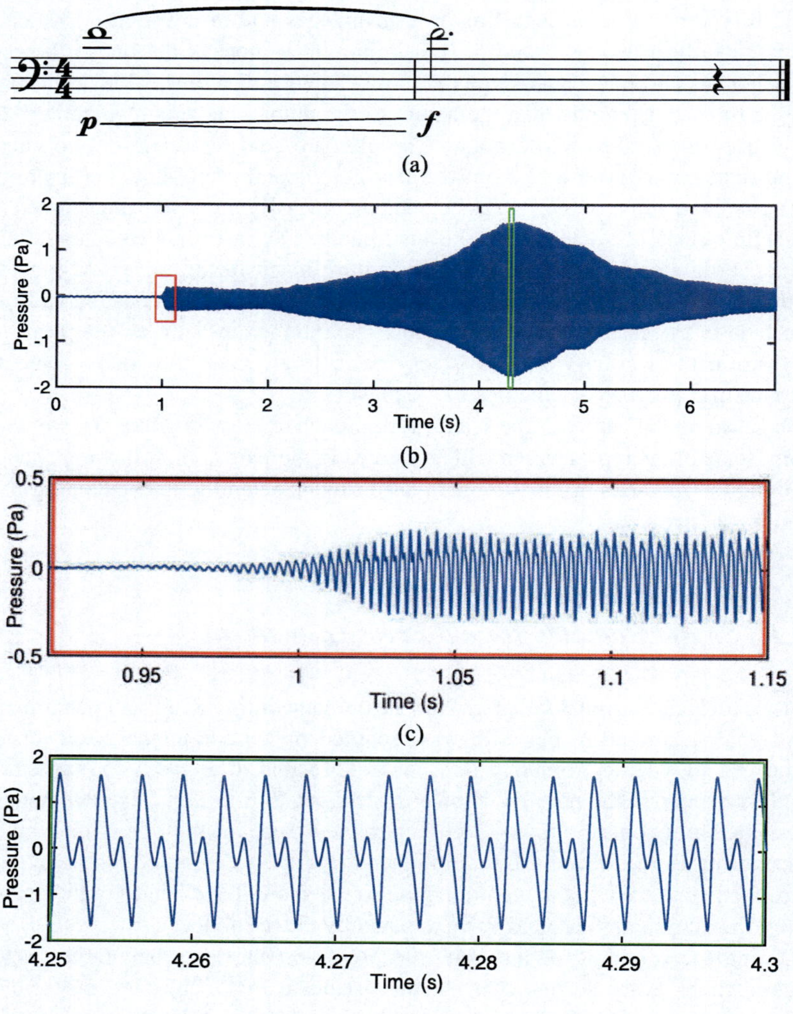

Fig. 2.2 (**a**) The note F4 played on a tenor trombone: staff notation. (**b**) The radiated acoustic pressure recorded on axis 50 cm outside the bell. (**c**) Zoom to 230 ms of the radiated pressure signal starting at $t = 0.92$ s. (**d**) Zoom to 50 ms of the radiated pressure signal starting at $t = 4.25$ s

following part, in which the note is sustained at a constant amplitude, is described as the 'steady state'.

The nature of the starting transient has an important influence on how the sound is perceived by a listener (Grey and Moorer 1977). In the example shown in Fig. 2.2c, the period of rapid growth lasts around 60 ms, which is a fairly typical value for many orchestral brass instruments (Luce and Clark 1967). An instrument judged by the player to be highly responsive is likely to be one on which it is easy to play notes with short starting transients (see Sect. 1.2.10). This type of clean attack is favoured when the frequencies of the air column resonances supporting the note are harmonically aligned but can be disrupted by unwanted reflections due to sharp bends, sudden changes in cross-section or partial blockages in the instrument bore (Benade 1976).

The peak-to-peak amplitude of the pressure signal shown in Fig. 2.2b starts to grow again after $t = 2$ s, when the crescendo begins. A zoom into a 50- ms-long section of the signal starting at $t = 4.25$ s, marked by the green box in Fig. 2.2b, is shown in Fig. 2.2d. At this point the sound had reached the maximum level of the crescendo, and the peak-to-peak amplitude of the pressure signal has risen to 3.2 Pa. In the time interval displayed, the pressure has a waveform characterised by a large positive peak followed by a small negative peak, a small positive peak and a large negative peak; this pattern repeats almost exactly every 2.86 ms. The repetition time is called the period T of the signal. The number of times that the waveform repeats in 1 second is called the frequency f. For the note played in this example

$$f = 1/T = 1/0.00286 = 350\,\text{Hz}. \tag{2.1}$$

As a first approximation, the repetition frequency of the signal can be associated with the perception of pitch: a doubling of frequency makes the pitch rise by an octave. The relationship between the amplitude of the signal and the musical dynamic level is clear from Fig. 2.2b: the increase of loudness in the crescendo is correlated with the increase in the vertical width of the signal, which corresponds to the difference between maximum and minimum values of the pressure. The form of the pattern of the repeating section of wave in Fig. 2.2d is associated with the timbre of the sound. In reality, however, the relationship between the perceptual quantities (pitch, loudness, timbre) and the objective quantities (frequency, amplitude, waveform) is more complex: for example, the pitch of a note of fixed frequency can in some circumstances change significantly if either the amplitude or the waveform is modified (Fastl and Zwicker 2007; Roederer 2009; Hartmann 2013).

Before the start of the note in Fig. 2.2c, the value of the pressure is shown as 0. It is important to realise that what this graph illustrates is the acoustic pressure, which is the change in the steady pressure of the atmosphere due to the arrival of the sound wave. In mathematical terms

$$p_{\text{tot}} = p_{\text{atmos}} + p_{\text{ac}} \tag{2.2}$$

where p_{tot} is the total pressure at some instant, p_{atmos} is the steady atmospheric pressure that would exist in the absence of a sound wave at that time and p_{ac} is the pressure change due to the presence of the sound wave. Since $p_{atmos} \simeq 100{,}000\,\text{Pa}$ and the peak-peak amplitude of p_{ac} is less than 4 Pa even in the forte part of the note illustrated in Fig. 2.2, the variations due to acoustic pressure would be too small to be visible on a plot of total pressure. Frequently in discussion of acoustic phenomena, acoustic pressure is simply described as pressure; it is usually clear from the context whether the meaning is intended to be total pressure or acoustic pressure.

The quantity plotted as a function of time in Fig. 2.2b–d is the instantaneous value of the acoustic pressure $p_{ac}(t)$. In describing the strength of the signal, it is common to use the root mean square (rms) pressure, defined between two times t_1 and t_2 as

$$p_{rms} = \sqrt{\int_{t_1}^{t_2} \frac{p_{ac}(t)^2}{(t_2 - t_1)}\,dt}. \tag{2.3}$$

The strength of an acoustic signal is often described using the sound pressure level (SPL) scale. The SPL value L_p in decibels (dB) is related to the rms acoustic pressure by the equation

$$L_p = 20\log_{10}\left(\frac{p_{rms}}{p_0}\right), \tag{2.4}$$

with the reference pressure $p_0 = 0.00002\,\text{Pa}$. An increase of p_{ac} by a factor of 10 corresponds to an increase of L_p by 20 dB. The relationship between SPL levels in decibels and the perceived loudness of the sound depends on many factors, including the frequency content of the sound and the acuity of the listener's hearing, but the difference between pianissimo and fortissimo on a brass instrument can reach around 40 dB.

2.1.2 *Sound Measured Inside a Trombone Mouthpiece*

Having examined the properties of the sound waves radiated from a brass instrument, the natural next step for the scientist is to investigate the nature of the pressure changes taking place inside the instrument during the performance. Figure 2.3a illustrates the signal recorded by a miniature microphone monitoring the acoustic pressure inside the mouthpiece of the trombone on which the note shown in Fig. 2.2a was played. At first sight this signal may look similar to the radiated pressure signal in Fig. 2.2b, but there is in fact a dramatic difference: in this case the pressure scale is in kPa, and the pressure amplitude in the mouthpiece is several orders of magnitude

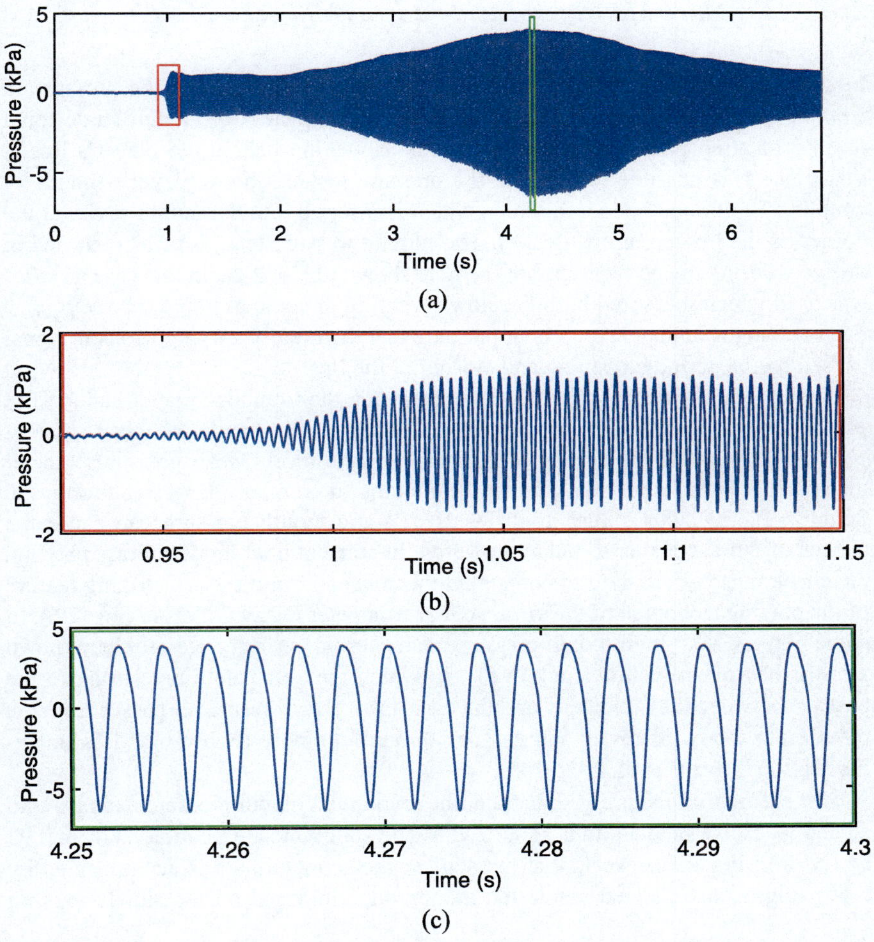

Fig. 2.3 (**a**) The acoustic pressure inside the mouthpiece of the trombone sounding the note F4 illustrated in Fig. 2.2a. (**b**) Zoom to 230 ms of the mouthpiece pressure signal starting at $t = 0.92$ s. (**c**) Zoom to 50 ms of the mouthpiece pressure signal starting at $t = 4.25$ s

greater that the pressure in the radiated sound. The peak-to-peak pressure amplitude at $t = 1.1$ s inside the mouthpiece is 2.8 kPa, compared to 0.46 Pa 50 cm outside the trombone bell; at $t = 4.3$ s the mouthpiece peak-to-peak pressure amplitude is 10.6 kPa, while the radiated pressure amplitude is 3.3 Pa. It is a remarkable feature of all wind instruments that the radiated pressure amplitude is a tiny fraction of the internal pressure amplitude.

2.1.3 Pressure Measured Inside a Brass Player's Mouth

Figure 2.3c records that during the sounding of a trombone note the pressure in front of the player's lips was rising and falling with a peak-to-peak amplitude equal to 10% of atmospheric pressure. What was happening behind the player's lips at that time? It is possible to measure the pressure inside a brass player's mouth by coupling a microphone to a thin tube which is inserted into the mouth cavity at the corner of the lip aperture (Fig. 2.4). The mouth pressure (p_{mouth}) was recorded in this way during the performance of the note shown in Fig. 2.2a. In this case a sensor was used which measured both the slowly varying or static pressure (above p_{atmos}) due to changes in the player's lung pressure and the rapidly varying acoustic signal caused by the periodic opening and closing of the lips.

Figure 2.5a shows that to make the note sound, the trombone player had to raise p_{mouth} above a threshold value of 1.7 kPa. During the crescendo the mean value of p_{mouth} was increased, reaching 3.7 kPa at the forte level. Mean pressures greater than 20 kPa have been measured in the mouths of trumpet players (Schwab and Schultze-Florey 2004). Such high levels of static mouth pressure can cause the cheeks of a brass player to bulge outwards. In conventional modern brass playing, the cheek muscles are used to counteract this tendency, but it was a striking feature of the playing technique of the virtuoso jazz trumpeter Dizzy Gillespie (Fig. 2.6). In some players, very high mouth pressures can cause a performance problem known as velopharyngeal insufficiency (VPI), in which the soft palate fails to close the passage between the oral cavity and the nasal cavity. As a result, air passes from the mouth into the nasal cavity and escapes through the nose (Schwab and Schultze-Florey 2004; Evans et al. 2010).

Just before $t = 1$ s in Fig. 2.5, a the line tracing the mouth pressure expands into a band whose vertical width is about 1 kPa. The zoom into a time interval from 0.92 to 1.15 s in Fig. 2.5b reveals that the start of the sounded note is accompanied by a growing acoustic signal inside the mouth, superimposed on the slowly varying

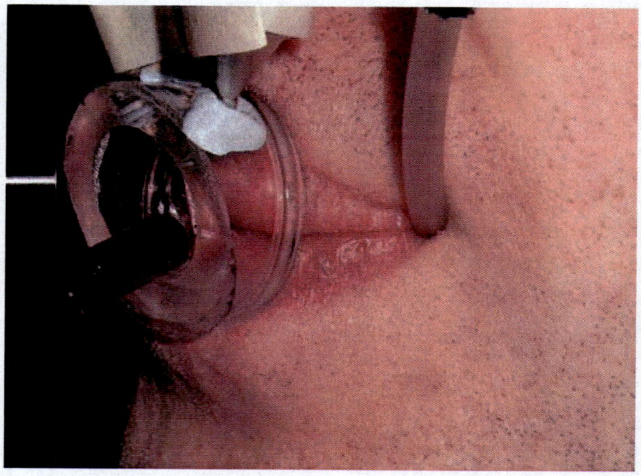

Fig. 2.4 Pressure measurement probe tube in the mouth of a horn player (Stevenson 2009)

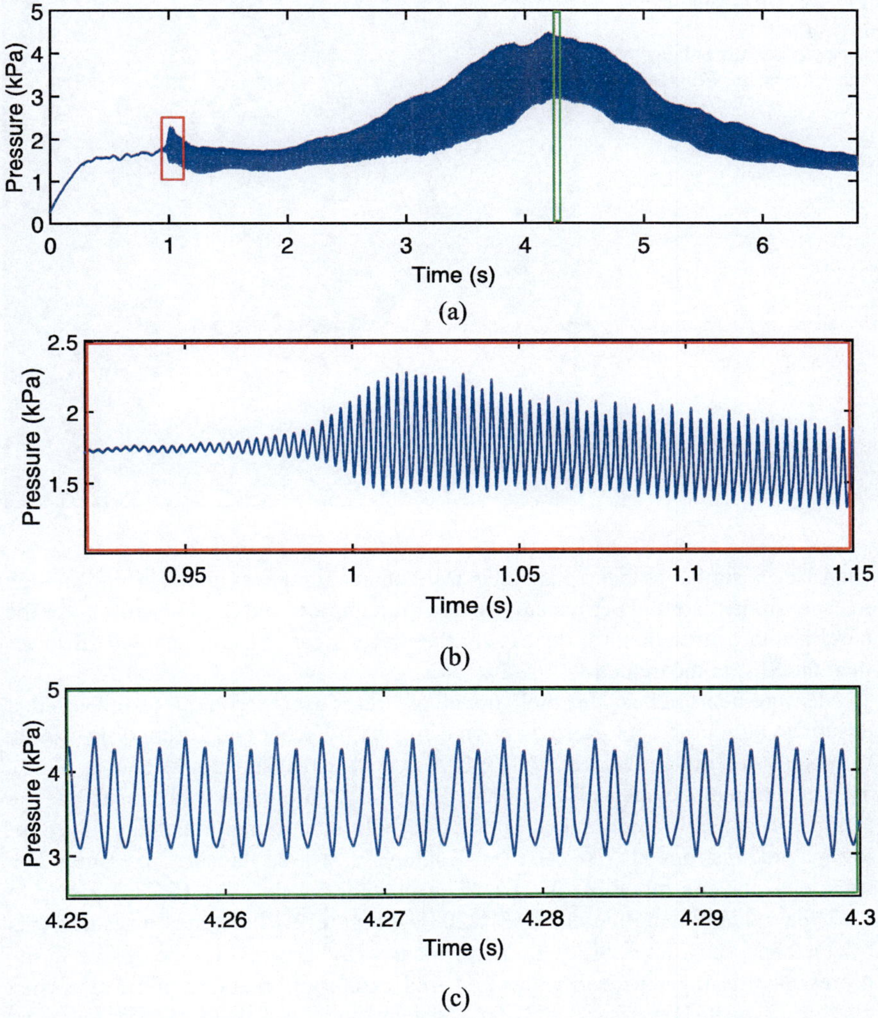

Fig. 2.5 (**a**) Pressure measured in the mouth of a trombonist playing the note F4 illustrated in Fig. 2.2a. (**b**) Zoom to 230 ms of the mouth pressure signal starting at $t = 0.92$ s. (**c**) Zoom to 50 ms of the mouthpiece pressure signal starting at $t = 4.25$ s

mouth pressure. The frequency of this signal is 350 Hz, which is the same as the frequency of the acoustic pressure variations in the mouthpiece and the radiated sound. The amplitude and waveform of the acoustic pressure signal in the mouth can be modified by changes in the upper windway of the player (see Sect. 6.3). The zoom to a time interval from 4.1 to 4.25 s shows that, for the note studied here, the acoustic pressure measured in the trombonist's mouth when playing forte had a peak-to-peak amplitude of 1.4 kPa, corresponding to an SPL above 140 dB. This

Fig. 2.6 Dizzy Gillespie playing with characteristically bulging cheeks. Source: Willis Music

sound level, similar to that measured in front of an aircraft jet engine, is high enough to cause immediate and permanent damage to an unprotected ear. Fortunately for the aural health of trombonists, the SPL at the player's ears is more than 40 dB lower than the SPL in the mouth.

Measurements of near-threshold mouth pressures using techniques similar to that described here have been carried out on trumpets (Fletcher and Tarnopolsky 1999), trombones (Fréour and Scavone 2013; Gilbert et al. 2018) and tuba (Bouhuys 1965). This experimental evidence shows that players of high-pitched instruments like the trumpet in general require much higher threshold mouth pressures than players of low-pitched instruments like the tuba. Bouhuys measured threshold pressures in a bass tuba player's mouth of 330 Pa when playing the note A1 (55 Hz), rising to 1.73 kPa when playing the note A3 (220 Hz). Figure 2.7 illustrates measurements by Fletcher and Tarnopolsky of performances by several trumpet players; the professional trumpeter identified as GC used a threshold pressure of 1.8 kPa when playing G3 (196 Hz), rising to 7 kPa when playing C6 (1047 Hz). The tendency for threshold mouth pressure to rise with playing pitch is reproduced by computer simulations of trombone performance described in Sect. 5.4.3.

It was suggested in Sect. 1.2.9 that the musician's judgement of ease of playing of a note is inversely related to the sounding resistance represented by the threshold blowing pressure. Support for this hypothesis is offered by measurements on the performing properties of saxophone reeds by Petiot et al. (2017): a strong correlation was found between the soft-hard scale on which the sounding resistance of different reeds was judged by saxophonists and the threshold pressures measured in the mouths of the performers. Accepting the hypothesis, it can be said that the sounding resistance of the bass tuba is overall much lower than that of the trumpet, but the resistance of each increases with the frequency and pitch of the played note. It should be noted, however, that the measured threshold pressure for playing a

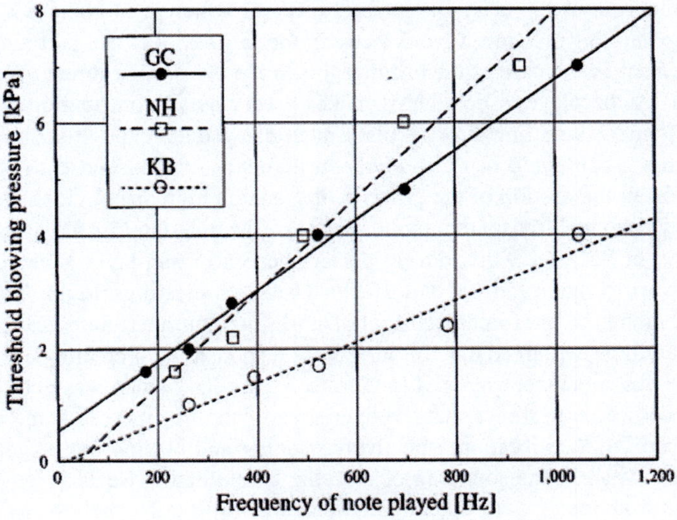

Fig. 2.7 Threshold mouth pressures used to sound various pitches by a professional orchestral trumpeter (GC) and two experienced amateurs (NH and KB) (Fletcher and Tarnopolsky 1999)

note with frequency around 200 Hz was almost identical on the two instruments, suggesting that the sounding resistance, which is related to the perceived ease of playing, is primarily determined by the fundamental frequency of the note rather than by properties of the instrument on which it is played.

When a player is sounding a steady note, the acoustic signals measured in the player's mouth, at different points along the instrument tube, at the player's ears and at the back of a concert hall have very different amplitudes and waveforms. The one thing which remains constant as the sound wave travels from the mouthpiece down the instrument tube and radiates into the surrounding environment is the repetition frequency of the signal, which is equal to the lip vibration frequency.

2.1.4 Lip Vibration and Air Flow: The Valve Effect Sound Source

What is the nature of the sound source in a brass instrument? Is the sound caused directly by the vibration of the lips of the player? Everyone is familiar with the idea that a 'buzzing' sound can be made by blowing air between the almost closed lips in the absence of any instrument. A young child imitating the sound of a tractor is in fact taking the first step to playing a trombone, while the imitation of a bee could be an indication of a budding trumpeter. It is important, however, to recognise that there is a fundamental difference between the sound made by twanging an elastic band held between the fingers and the sound made by buzzing the lips. In the former case, the motion of the elastic strip forces the air which is in contact with it into motion, generating a sound wave directly in the same way as the vibrating cone of a loudspeaker. In the case of the buzzing lips, there is a similar direct displacement of the air by the player's lips, but it makes a very small contribution to the total sound. To understand how a brass instrument can radiate enough sound power to carry across a battlefield or thrill a concert audience of thousands, it is helpful to study in detail the motion of the player's lips using a high-speed video camera and a specially designed transparent mouthpiece, as illustrated in Fig. 2.1.

We saw in Sect. 2.1.3 that during performance a brass player generates mouth pressures up to and even beyond 10 kPa (10% of an atmosphere). This excess pressure behind the lips forces a current of air through the aperture between them. In Chap. 3 it will be explained that this air flow is responsible for creating a situation in which the lips spontaneously start to vibrate. Figure 3.4, which was obtained using a high-speed camera, shows that the open area of the lips increases and decreases regularly during the vibration, effectively opening and closing the player's mouth once every cycle. In the sounding of the note F4 studied in Sects. 2.1.1–2.1.3, the lips of the trombonist were opening and closing with the frequency of the played note (350 Hz). In consequence the flow of air from the mouth into the mouthpiece was periodically interrupted, and the modulated air flow injected a high level of sound power into the instrument.

Vibrating lips operate like a valve which is periodically opened and closed to control the air flow; sound generators of this type are described as valve effect sources. The vibrating cane or plastic reeds in woodwind instruments such as the clarinet, saxophone, oboe and bassoon also operate as valve effect sound sources. Industrial sirens based on the valve effect are the most powerful acoustic sources available and are widely used in alarm sounders such as the horns on trains (Fig. 2.8).

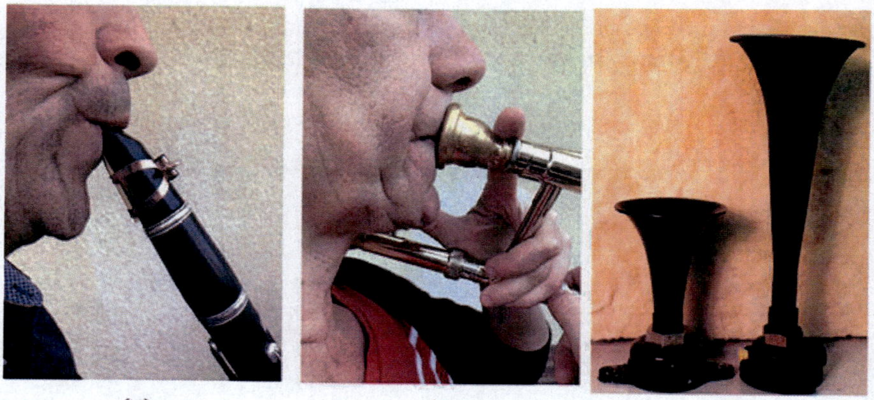

Fig. 2.8 Three valve effect sound sources: (**a**) clarinet mouthpiece with single reed; (**b**) trombone mouthpiece with player's lips; (**c**) train siren horns

2.1.5 *Is Air Flow Through the Instrument Tube Important?*

The flow of air between the lips is essential to the sound production in a brass instrument, since it results in the lip vibration which generates sound power in the mouthpiece. What about the air flow in the rest of the instrument? Air flow velocity is more difficult to measure than pressure, but Elliott and Bowsher (1982) recorded the velocity of the air flow in a specially modified trombone mouthpiece with a hot wire probe inserted in the throat (the narrowest section). Elliott and Bowsher's measurement of the air flow velocity during the playing of a low note ($B^{\flat}2$) is shown in Fig. 2.9.

As expected, Fig. 2.9 shows that the flow through the throat has a periodic vibration, dropping to zero once every cycle as the lips close. However there is clearly a mean positive flow into the main bore. The mean flow can also be observed emerging from the bell using the optical technique of schlieren photography explained in Chap. 6: it is responsible for the cloud-like turbulent flow in Fig. 6.5.

To the surprise of many brass players, it turns out that the mean flow through the instrument has practically no effect on its acoustic and musical performance. This has been elegantly demonstrated by Richard Smith using the ingeniously modified trombone mouthpiece illustrated in Fig. 2.10. A conventional mouthpiece was sawn across near the base of the cup and reassembled with a thin elastic membrane (a

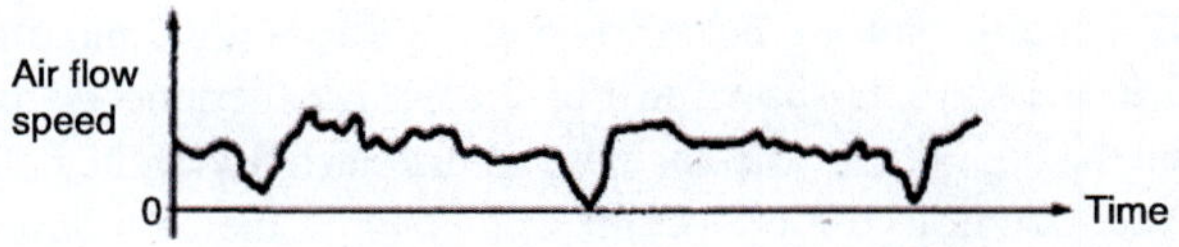

Fig. 2.9 Air flow velocity in the throat of a trombone mouthpiece (Elliott and Bowsher 1982)

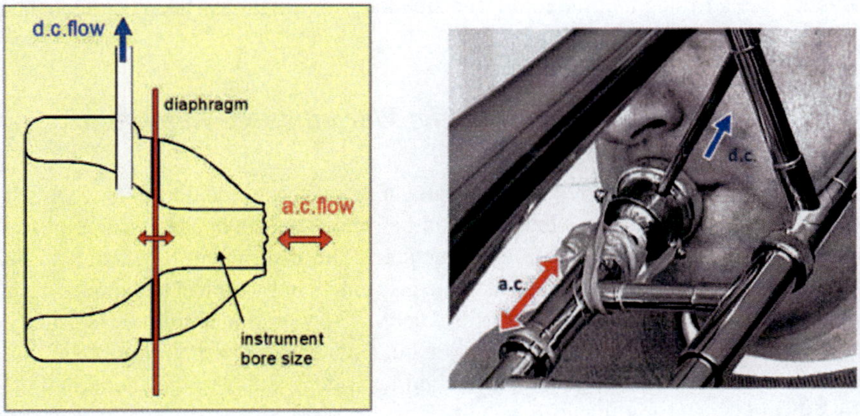

Fig. 2.10 A modified mouthpiece designed to demonstrate that a trombone can be played without any mean air flow through the instrument. Courtesy of Richard Smith

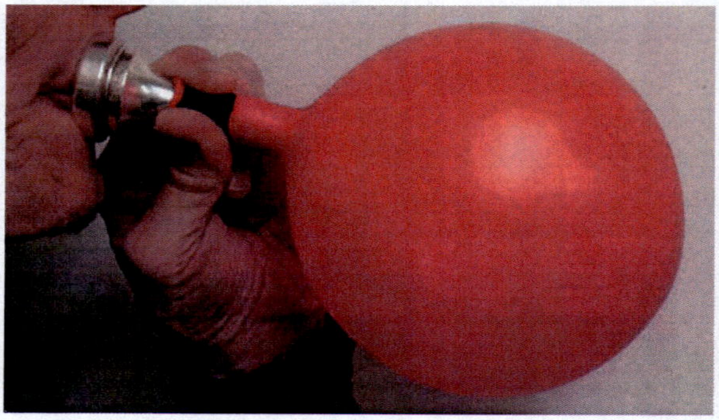

Fig. 2.11 Experiment illustrating independence of mean flow and acoustic wave propagation

modified condom) stretched across the gap, preventing any mean flow of air into the mouthpiece throat. If that were the only modification the instrument would be impossible to sound, since the player would be unable to create the air flow through the lips which is necessary to initiate their vibration. However the mouthpiece is also fitted with a long and narrow side tube, allowing the mean flow to escape upstream of the sealing membrane. The trombone can be played normally, with only a slight reduction in the ease of sound production. This demonstration shows that the transmission of acoustic waves in the instrument is independent of the mean flow through it.

The independence of acoustic wave propagation and mean air flow can also be very simply demonstrated by buzzing notes on a trombone mouthpiece with a balloon sealed to the stem as shown in Fig. 2.11. An experienced trombonist can easily overcome the additional resistance due to the elasticity of the balloon and can play tunes on the mouthpiece for several seconds while the balloon gradually fills with air due to the mean flow. Although all the air which has left the player's mouth during this time remains trapped inside the balloon, the sound waves will have radiated freely throughout the room in which the experiment has been conducted.

2.1.6 Is Sound Radiation from the Vibrating Bell Important?

Musicians and scientists agree that the walls of a brass instrument vibrate when the instrument is played. These vibrations can be sensed by the hands and lips of the player, which are in contact with the structure of the instrument; they can also be measured by accelerometers attached to the instrument or by optical techniques such as interference holography and laser vibrometry. The vibration amplitude is usually largest in the flaring bell section of the instrument, and it is tempting to view this substantial area of vibrating metal as a significant contributor to the sound output. Does this idea have any scientific validity?

The air in contact with a loudspeaker cone undergoes alternating compressions and expansions when the cone vibrates, and this disturbance of the air results in the radiation of a sound wave. This is an example of a vibroacoustic sound source. Almost all members of the stringed and percussion instrument families can be described scientifically as vibroacoustic musical sound sources. On a violin, for example, a string is set into vibration by bowing or plucking; the forces exerted by the string on the bridge generate vibrations of the extended surfaces of the body of the violin, and radiation from these surfaces provides almost all the sound output of the instrument. In a similar way, the soundboard of a piano is the vibroacoustic source of the sound radiation.

Wind instruments, including the brass instrument family, are described as aerophones, a technical term which reminds us that the transmission of wave energy in these instruments is primarily through sound waves in air rather than through structural vibrations. Since powerful sound waves already exist in the air column within a brass instrument tube, there is no need for a vibrating surface to act as a sound source: an aperture in the tube which allows some of the sound energy to escape will act as a powerful source of radiated sound waves. In most brass instruments, the bell is the only opening, and this aperture is the source of almost all the radiated sound.

It is the established scientific view that, while the vibrating bell does constitute a vibroacoustic sound source in normal brass instruments, it is so weak in comparison with the direct transmission of sound waves through the bell aperture that it is unlikely to make a detectable contribution to the sound of the instrument. Many musicians still find it difficult to accept that this can be true. To attempt to convince them, an experiment suggested by Peter Hoekje was undertaken at the Laboratoire d'Acoustique de l'Université du Mans. The experimental setup is illustrated in Fig. 2.12. A small accelerometer was mounted on the bell of a trumpet to record its vibration while the instrument was being played. The sound radiated from the bell was also recorded with a microphone. The player was then replaced by a

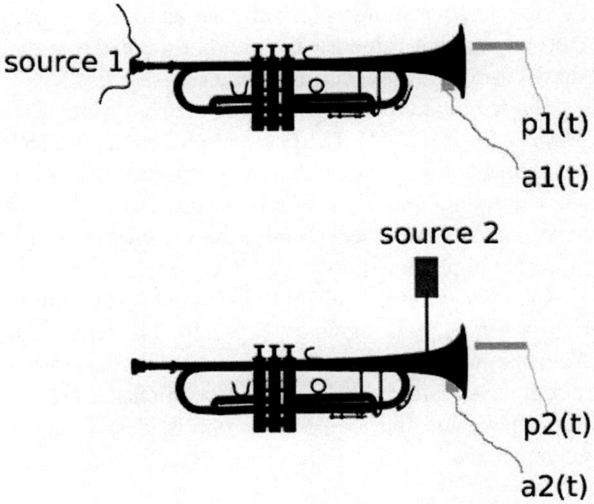

Fig. 2.12 Experiment to investigate the contribution of wall vibration to the total sound power radiated by a played trumpet. Source 1: a human player. Source 2: a mechanical shaker

mechanical shaker, which was attached to the instrument body and driven by a computer generated signal which created the same amplitude of bell vibration as that measured during the playing test. Again the radiated sound was measured. Since in the second test there were no sound waves in the instrument air column, the only sound source was the vibroacoustic one. It was found that the sound pressure level was around 40 dB lower than the level when the player was performing.

Although direct sound radiation from the bell is probably insignificant, this does not mean that wall vibrations have no effect on the radiated sound. Recent research has demonstrated that under some circumstances coupling between acoustic and structural resonances can significantly change the internal sound field, thus indirectly modifying the radiated sound. It must also be borne in mind that the player's perception could be different from that of a listener if wall vibrations transmitted mechanically from the mouthpiece to the lips provide additional feedback. A full discussion of this topic is given in Sect. 6.6.

2.1.7 Warming Up a Brass Instrument

We conclude this section by considering an aspect of brass instrument behaviour familiar to all players: the dependence of the playing properties of the instrument on temperature. When a player picks up an instrument which has been for some time on a stand in a rehearsal room, the internal air column will be at the temperature of the room, which we will assume to be 20 °C. The musician usually starts a playing session by 'warming up' the instrument, which typically involves playing arpeggios over an increasingly wide pitch range. Air enters the mouthpiece from the musician's mouth close to the body temperature of 37 °C. After a few minutes of steady warm-up, a stable temperature profile is created in the instrument, decreasing from near 37 °C at the mouthpiece to near 20 °C at the bell. It is difficult to measure this temperature profile precisely, but an idea of its nature can be obtained using a thermal imaging camera which allows the external temperature of the metal to be shown using infrared thermography (Fig. 2.13).

The temperature scale, shown at the right side of the image in Fig. 2.13, ranges between 26 and 19 °C. The photograph was taken a short time after warming up the trombone, but it is evident that the instrument has already cooled down significantly since the mouthpiece receiver is only at 26 °C. A second trombone which had not been played can be seen behind the warmed-up instrument; as expected, it has a uniform temperature close to 20 °C.

The temperature profile of the warmed-up instrument, derived from the information in Fig. 2.13, is shown in Fig. 2.14. The temperature shows a global decrease from the mouthpiece to the bell, apart from a large peak around 1.5 m from the mouthpiece. This is the position at which the left hand of the player holds the instrument, and the temperature rise is clearly due to additional heating by the player's hand.

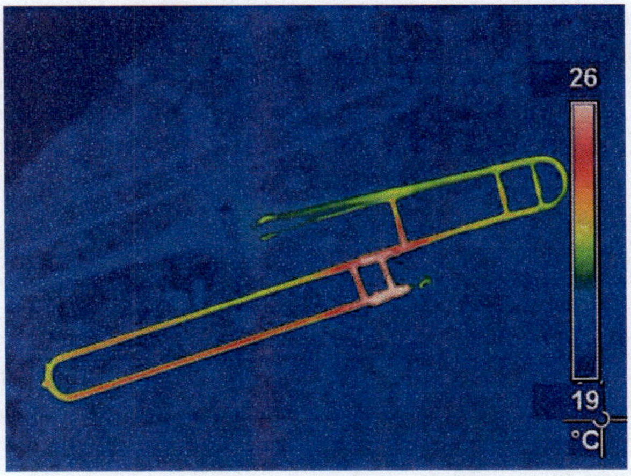

Fig. 2.13 Infrared thermography image of a trombone after warm-up. Adapted from Gilbert et al. (2006)

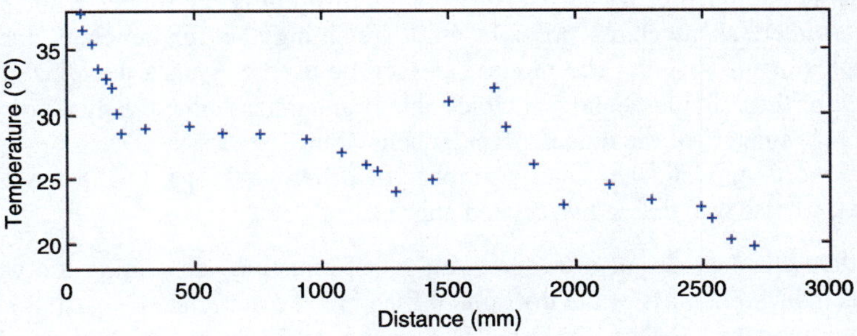

Fig. 2.14 Temperature profile along a warmed-up trombone, estimated by infrared thermography. Adapted from Gilbert et al. (2006)

Although Fig. 2.13 shows that a brass instrument is literally warmed up by the preparatory exercises of the musician, these exercises also have an important physiological purpose. A runner preparing for a race engages in a sequence of activities designed to increase the flow of blood to the leg muscles. In the brass player's case, the most important muscles are those controlling the embouchure. Since the blood flow is restricted if the lips are pressed against a cold surface, a metal mouthpiece is commonly warmed up by the player before starting the playing exercises.

2.2 An Approach to Modelling Brass Instruments

The progress of all branches of physics relies on an interplay between experimental observation of phenomena and the development of theories which explain the phenomena. The theoretical approach uses models which are based on fundamental laws and principles, and the predictions of these models must then be compared with reality as revealed by experiments. A good theoretical model will not only explain previous observations but will predict behaviours and outcomes that have not yet been observed and can be experimentally validated. A model may also be useful to optimise the behaviour of a system, as long as the target for optimisation can be defined. In the case of a brass instrument, for example, a good physical model of the instrument could help a maker to optimise the intonation of the instrument, provided that the ideal intonation was clearly defined.

2.2.1 The Scientific Case for Simplified Models

To explain the essential features of the functioning of brass instruments, acousticians have developed models based on a number of simplifying hypotheses. This follows the classic approach of the physicist, which is to simplify the model of the system being studied as much as possible while retaining the fundamental operating principles of the system. The model can then be used to gain a deep insight into the role of these fundamental principles and to understand how the most important control parameters of the model affect its behaviour.

The word 'model' can mean a number of different things, but in the present context, we use it to define two related approaches.

1. A theoretical model of a trumpet is a set of equations describing the ways in which different parts of the instrument function and are related.
2. A numerical model of a trumpet is a computer program which is capable of solving the equations numerically and predicting how the instrument will behave under given circumstances.

Numerical models typically rely on close-grained discretisation of the object to be studied in both time and space and therefore require the rapid solution of a very large number of equations. The development of efficient methods for solving such equations together with the continuing growth in the memory capacity and calculating power of computers has meant that it is now possible to test proposed modifications of a brass instrument by making the appropriate changes to a numerical model rather than to an instrument in the real world. It has to be borne in mind, however, that even with unlimited computational power, the validity of a numerical calculation depends on the quality of the model equations.

The discussion in Sect. 2.1 of various aspects of brass instrument behaviour suggests that a simplified model can be constructed by dividing the system into three sub-systems (see Fig. 2.15):

1. the musician, represented by the lungs and trachea, the mouth cavity and the lips;
2. the instrument including the mouthpiece;
3. the external environment into which the sound radiates.

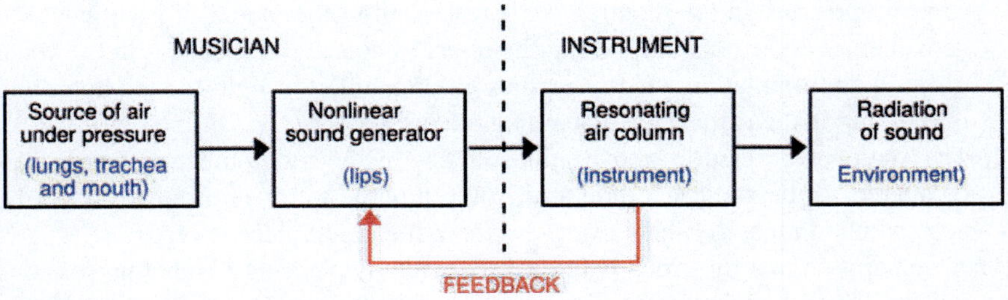

Fig. 2.15 A simplified model of a brass instrument and player

We note in passing that in this type of model, the mouthpiece is considered to be an integral part of the instrument.

Although at first sight the division of the player-instrument system into two subsystems consisting of player and instrument may seem straightforward, it is important to recognise that these two subsystems are strongly coupled together by feedback from the instrument to the player: a pressure change in the mouthpiece modifies the rate at which air flows through the lips, which in turn changes the pressure in the mouthpiece (see Sect. 2.2.2). It is interesting to consider the related case of the clarinet, another valve-effect wind instrument in which the role of the valve is played by the single cane reed of the instrument. The reed is attached to the clarinet mouthpiece, although it is damped by the lower lip of the player, so in this case the first subsystem includes part of the instrument as well as the player. Since the two systems are strongly coupled the division between them is in fact rather arbitrary.

The feedback loop linking the musician and instrument subgroups is indicated by an arrow in Fig. 2.15. There is no similar arrow marking a feedback from the external environment to the player or the instrument, and from the physicist's point of view, no such feedback need be considered. Every musician knows well that the acoustics of the room in which a performance is given certainly affects the manner of performance, and in that sense there is indeed a feedback from the environment to the player. It is not however of the same order of importance as the internal feedback between musician and instrument, and does not play a role in the physics of the system.

It is in reality impossible to express all the complexities of the functioning of a musical instrument in a system of equations, and modelling is always based on approximations that make it possible to simplify drastically the physical system under study. Nevertheless, the approximations must be made in a way that conserves the essential characteristics of the instrument. In modelling wind instruments in general, and brass instruments in particular, the following approximations make good starting points for models.

1. We saw in Sect. 2.1.3 that the acoustic pressure in the mouth is small compared to the static overpressure, and it is also usually much smaller than the acoustic pressure in the mouthpiece. In a first approximation, we can assume that the

acoustic pressure in the mouth is negligible, and in the model, it is set equal to zero. This is equivalent to ignoring the effect of acoustic resonances in the vocal tract, an approximation which becomes questionable for high-pitched notes and is certainly inadequate when discussing the didgeridoo.
2. The static overpressure (above atmospheric) in the mouthpiece is very small compared to the static overpressure in the mouth, and it is usual to make the approximation that the static overpressure in the mouthpiece is zero.
3. A first approximation which is frequently made by physicists is that no sound is radiated from the instrument. To a musician this must appear to be a shockingly crude simplification, since it neglects the musical sound production which is the instrument's sole purpose! Nevertheless this approximation provides a good basis for modelling the playable notes on a brass instrument.
4. In Sect. 2.1.6 it was pointed out that the sound pressure levels in the air column of a brass instrument can be high enough to generate wall vibrations which can be felt by the player. In normal performing conditions, the influence of these wall vibrations on the sound heard by the listener in a concert hall is very slight, and in simplified treatments, the approximation is made that the walls are completely rigid.

2.2.2 *Coupled Systems and Feedback Loops*

In the block diagram of the brass instrument model shown in Fig. 2.15, the progress of the sound signal from its generation at the player's lips through the instrument and into the external environment is shown symbolically by a series of arrows pointing from left to right on the lines linking the subsystems. We noted in Sect. 2.2.1 that Fig. 2.15 included a second line linking the musician and instrument subsystems, with an arrow pointing from right to left. This indicates symbolically that the two subsystems are coupled in a feedback loop: the first subsystem in such a loop behaves differently depending on whether or not the second subsystem is present. Any brass player will be familiar with the musical experience described by the rather technical language of the previous sentence: the ability to buzz the lips at a given pitch depends strongly on whether or not the lips are attached to the instrument. A continuous glissando can be performed when the lips are vibrating against the rim of an isolated mouthpiece, but when the mouthpiece is connected to the rest of the instrument, only certain lip vibration frequencies can be easily sustained. The pitches corresponding to these frequencies, which seem to be preferred by the instrument, are called its natural notes (Sect. 1.2.2).

Figure 2.16 shows the first six natural notes of a B$^\flat$ trumpet with no valves operated. The figure shows the sounding pitches; in a performer's score, the notes are usually transposed a tone higher. It is an instructive exercise for a trumpet player to sound the pitch G4 by buzzing the lips on a mouthpiece detached from the instrument and to attempt to continue playing this note while inserting the mouthpiece into the trumpet. It is very difficult to sustain the lip vibration at the pitch

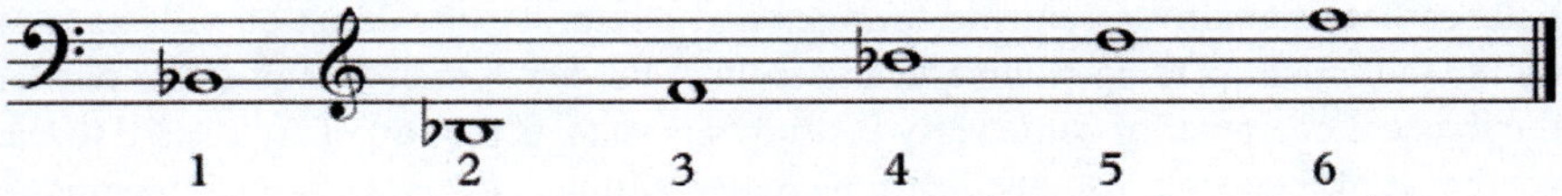

Fig. 2.16 The first six natural notes of the B$^\flat$ trumpet (sounding pitches)

G4 once the coupling to the instrument is made, because this note is intermediate between the third and fourth natural notes of the trumpet (F4 and B$^\flat$4). The feedback from the instrument exerts a strong guidance on the lips towards a vibration at either F4 or B$^\flat$4. If the lips are buzzed at either of these pitches while the mouthpiece is inserted, the sound is stabilised and reinforced by the coupling to the instrument.

The feedback arrow in Fig. 2.15 indicates the return of an acoustic signal from the instrument resonator to modify the pressure difference across the lips. The way in which this pressure difference controls the rate of air flow into the mouthpiece is discussed in detail in Chap. 3. The strong preference of the lips to vibrate at one of the natural note frequencies is due to the fact that the air flow is not simply proportional to the pressure difference. This lack of simple proportionality is expressed mathematically by describing the equation relating pressure and flow (Eq. 3.27) as nonlinear, and the lips are shown in Fig. 2.15 as a nonlinear sound generator.

The crucial importance of this nonlinearity can be understood by considering a second experiment in which a trumpeter again buzzes the sounding pitch G4 into a mouthpiece. This time the mouthpiece is not inserted into a trumpet but instead is held near a microphone connected to an amplifier and loudspeaker. If this apparatus is of sufficiently high quality it will behave like a linear amplification system, reproducing the sound more loudly but exerting no control over the player's lips. A continuous glissando will be just as easy with the amplification as without it. The existence of preferred natural notes is proof that in normal performance a brass instrument does not behave like a linear amplifier.

When the feedback from the instrument provides firm guidance towards a well-defined pitch, the note is sometimes described by the musician as 'well centred'. The strength and pitch accuracy of the feedback depend on the instrument and on the playing range. In the higher registers of most brass instruments, the strength of feedback diminishes substantially; in the highest pitch range, accessible only to virtuoso performers, there is no significant feedback, and it is possible to play continuous glissandi. In the lower registers of modern brass instruments, notes are typically well centred, and a poorly centred note may indicate a lack of feedback due to a leak in the instrument.

Powerful feedback from the instrument can force the player's lips towards vibration at a particular pitch. This strong 'slotting' (Sect. 1.2.2) is clearly only a good thing if the enforced pitch is that desired by the musician. A player may wish to 'bend' a note slightly for expressive reasons or to achieve perfect intonation in an ensemble; it is important that the feedback is not so strong that the player finds it impossible to make the necessary changes in lip vibration frequency. The ideal balance between strong centring of notes and flexibility of intonation will depend on the individual playing technique and style of the performer. A significant part of the musical character of some early instruments such as the serpent is related to the weakness of feedback from the instrument, demanding a different playing technique to control notes which are not well centred (see Sect. 1.2.2).

2.2.3 *Natural Notes and Harmonics: The Scientific View*

The natural notes which can be easily sounded on a brass instrument are commonly described as 'harmonics' by musicians (Sect. 1.2.2). From the scientist's viewpoint, the natural notes themselves are not strictly speaking harmonics, but each natural note contains within itself an almost exact harmonic set of frequencies. To explain this apparently paradoxical statement, we need to understand how the musical concepts of pitch and timbre are related to the scientific description of a musical sound.

A periodic variation in any quantity is one that repeats itself regularly. The time interval between two successive repetitions is the period τ, and the repetition frequency is $f_R = 1/\tau$. If the variation in pressure due to a sound wave is at least approximately periodic, it is perceived as having a definite pitch (see, e.g., Campbell and Greated 1987). An oboe player sounds a standard A to which other members of an orchestra can tune by generating a periodic pressure fluctuation with a repetition frequency $f_R = 440$ Hz.

Although the waveform of a pitched sound is periodic, it is not normally a simple sine wave. Figure 2.17 illustrates the pressure signal recorded when the third natural note F4 was played on a B$^\flat$ trumpet. The period is the time interval $\tau = 2.87$ ms between successive sharp downward spikes, corresponding to the frequency $f_R =$ 349 Hz of this note. Fourier's theorem tells us that any signal can be analysed into a set of component sine waves with different frequencies, amplitudes and phases (see Sect. 4.1.4). If the signal is periodic, the frequencies of these Fourier components are

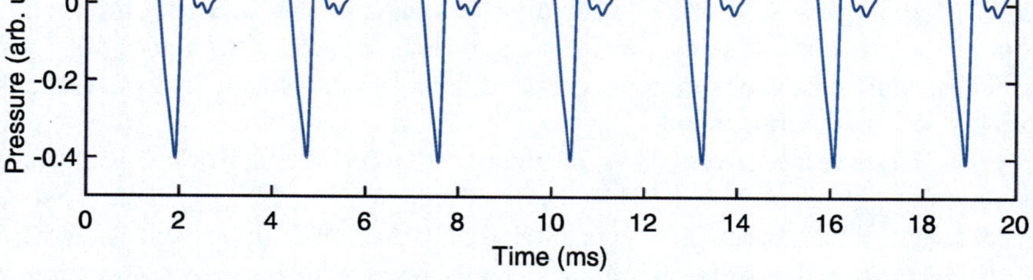

Fig. 2.17 The pressure waveform of a note with pitch F4 played on a B$^\flat$ trumpet. Courtesy J.-F. Petiot

members of an exact harmonic series, whose fundamental is the repetition frequency f_R. In other words, the signal shown in Fig. 2.17 is equivalent to a simultaneous set of sine wave signals with frequencies f_R, $2f_R$, $3f_R$, ...

The mathematical process required to derive the amplitude and phase of the components of a known time signal is called Fourier analysis (Hartmann 2004). A graph showing the component amplitudes as a function of frequency is described as the Fourier spectrum (or simply the frequency spectrum) of the signal. Figure 2.18a shows the frequency spectrum of the trumpet sound whose waveform is shown in Fig. 2.17. In the frequency range illustrated, only the first four harmonics appear, but the series continues well above 10 kHz.

A tuning meter will report that the trumpet note has a pitch F4 and a frequency 349 Hz. While it is true that this is the frequency of the first harmonic in the spectrum, the second, third and fourth harmonics have amplitudes only slightly smaller than the first. Why do we not also hear the pitches of these components? The reason is that the perception of pitch by the human hearing system does not rely on either the first harmonic or the strongest harmonic but involves a complex and not yet fully understood set of signal processing stages which uses the complete frequency spectrum. If a harmonically related series of frequencies is detected, the pitch is associated with the fundamental of that series, even if the first harmonic is weak or indeed absent (Campbell and Greated 1987).

We turn now to discussing the relationship between the harmonic frequency components of the note and the resonances of the instrument on which it is played. The acoustic pressure variation in the mouthpiece of a brass instrument during playing was discussed in Sect. 2.1.2, and the corresponding fluctuation of the air flow into the mouthpiece was illustrated in Fig. 2.9. The ratio of acoustic pressure to acoustic volume flow in the mouthpiece is called the input impedance of the instrument (see Sect. 4.1.6). An input impedance curve, in which the input impedance $Z(f)$ is plotted as a function of frequency, is a useful way of illustrating the acoustical response of an instrument. Each peak in the input impedance curve corresponds to one of the acoustic resonances of the air column inside the tube.

The input impedance curve for a B$^\flat$ trumpet is shown by the blue line in Fig. 2.18b. Frequencies corresponding to exact harmonics of the nominal fundamental pitch B$^\flat$2 are marked by vertical red lines. The acoustic resonances are evidently quite close to the nominal harmonic series, although the correspondence is not exact. The glaring exception is the first peak, which is far below the value $f_R = 116.5$ Hz of the first harmonic of B$^\flat$2.

The green vertical lines in Fig. 2.18b mark the frequencies of the harmonics of F4 which were observed in Fig. 2.18a. To sound this note, the player chooses a lip setting which favours vibration close to the third acoustic resonance; the acoustic feedback from this resonance strengthens and stabilises the vibration. Because of the approximately harmonic nature of the acoustic resonances, the sixth resonance is in the right place to support the second harmonic of F4; similarly, the ninth and twelfth resonances lend support to the third and fourth harmonics, respectively. The collaboration of several acoustic resonances in reinforcing the sounding of a note was recognised by Bouasse (1929) and Benade (1968), and we describe the

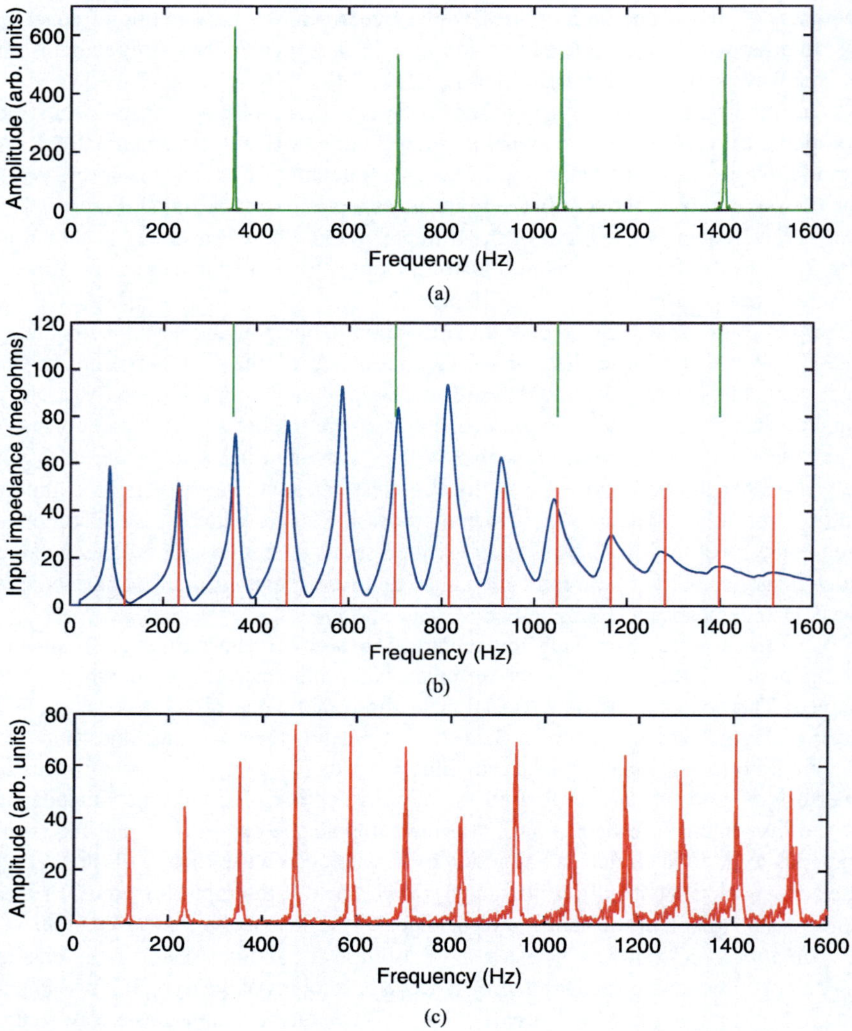

Fig. 2.18 (**a**) Frequency spectrum of trumpet note F4. (**b**) Input impedance curve for a B^b trumpet. Green lines: harmonics of F4. Red lines: harmonics of B^b2. (**c**) Frequency spectrum of trumpet note B^b2 (pedal note) (Color figure online)

corresponding requirement of an approximately harmonic set of resonances as the 'Bouasse-Benade prescription' (see Sect. 5.4.5).

Figure 2.18c shows the frequency spectrum for the note B^b2 played on the trumpet. This 'pedal note' is difficult to sound in tune, because there is no acoustic resonance to support the first harmonic of the mouthpiece pressure. There is however the possibility of support for the higher harmonics, each of which lies not

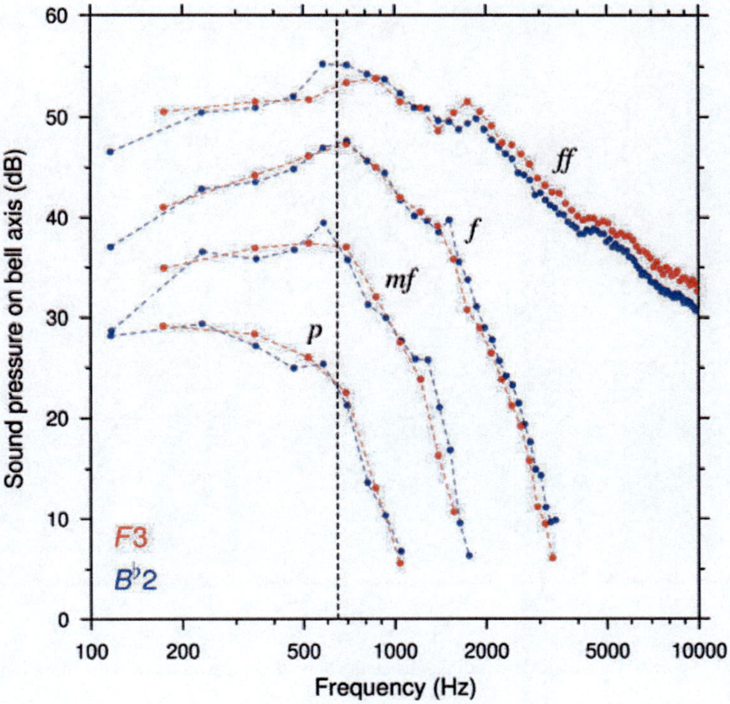

Fig. 2.19 Spectra of notes played on a trombone at different dynamic levels. R. W. Pyle (unpublished)

far from an acoustic resonance. The sound production mechanism in pedal notes is discussed further in Sect. 5.4.4.

To illustrate the relative strengths of the harmonics in different parts of a frequency spectrum, it is convenient to extract the peak amplitude values and plot them as as a function of frequency. Figure 2.19 shows such a plot for notes of two different pitches played at several dynamic levels on a trombone. The dashed line joining the peak amplitudes of each note is described as the 'spectral envelope'. A comparison of the changing shapes of the spectral envelopes with increasing dynamic level reveals several important and characteristic features of the sound of the trombone. At the piano level, only the first few harmonics have significant amplitudes, with a precipitate falloff in amplitudes for harmonics above 650 Hz. For dynamic levels of mezzoforte and above, the amplitudes rise to a peak around 650 Hz and then fall away. The relative growth in the strength of the high-frequency harmonics with increasing dynamic level is described as 'spectral enrichment'. The corresponding increase in brightness of timbre as the instrument is played more loudly (see Sect. 1.3.2) is found in every wind instrument. The spectacular enrichment of the spectrum above 5000 Hz at the fortissimo level is however specific

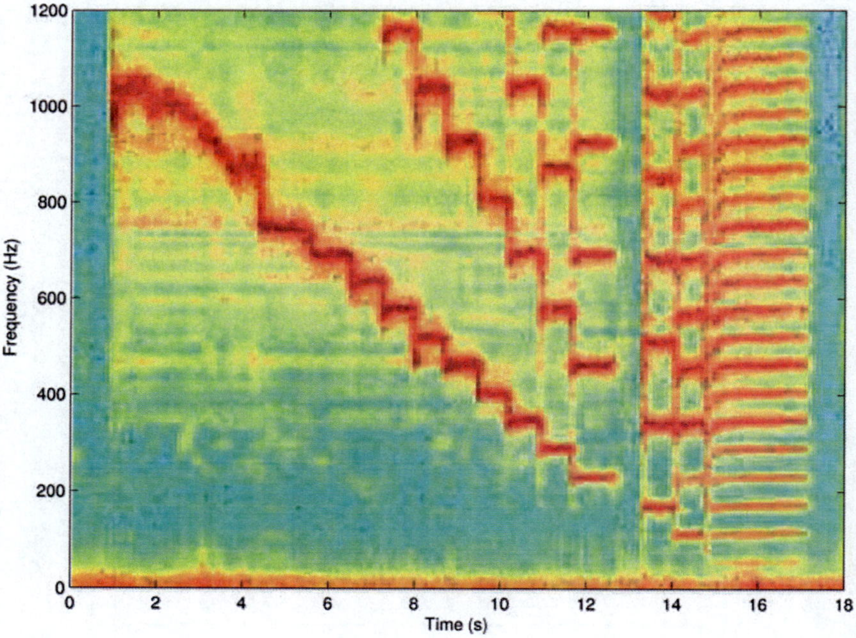

Fig. 2.20 Spectrogram of a descending glissando played on a trombone with the slide in first position

to the brass instruments and is linked to the 'brassy' timbre which characterises trumpets, trombones and french horns (see Sect. 6.1).

The spectrogram provides an alternative means of displaying the spectral content of sounds. In this type of display, time is plotted on one axis and frequency on the other axis; amplitude (or sometimes acoustic power) is shown by a grey or colour scale. Figure 2.20 illustrates a performance in which a virtuoso trombonist played a descending glissando over more than five octaves, from C6 to B♭1.

The horizontal axis in the spectrogram represents the passage of time during the glissando. The total time interval of 18 s is divided into time windows short enough that the signal can be assumed to have a constant repetition frequency during each window. In each window, the amplitudes of the components in the frequency range from 0 to 1200 Hz are calculated and represented along the vertical frequency axis in Fig. 2.20 using a colour scale.

A red line descending from the upper left of the spectrogram shows how the fundamental frequency f_R of the signal drops during the glissando. Starting at just over 1000 Hz, f_R falls continuously until around 850 Hz, since in this frequency region, there are no significant resonances in the instrument and there is therefore no feedback to the lips. Below 850 Hz (a pitch between G♯5 and A5), f_R falls in a series of steps: the almost horizontal part of each step corresponds to the frequency of an instrument resonance which attempts to centre the played pitch on the corresponding natural note.

Towards the right-hand side of the spectrogram in Fig. 2.20, other descending stepped traces appear. These represent the higher-frequency components which Fourier analysis tells us are also present in the sound. At $t = 12$ s, for example, the natural note B$^\flat$3 is played, and the fundamental $f_R = 233$ Hz is shown as the lowest horizontal red line. Above this line there are several more horizontal red lines, at frequencies which are integer multiples of 233 Hz. Since these frequencies are Fourier components of a periodic signal, they are exact members of a harmonic series whose first member is f_R.

The lowest note in the glissando, which begins around $t = 15$ s, is the pedal note B$^\flat$1. It is striking that the fundamental $f_R = 58$ Hz appears much weaker than any of the 19 higher harmonics shown in the spectrogram. Reasons for this characteristic of the pedal note are given in Sect. 5.4.4.

2.2.4 *Self-Sustained Oscillations*

Making use of Approximation No. 1 in Sect. 2.2.1, we can model the musician playing a brass instrument as a system that transforms a constant excess pressure in the player's mouth into an oscillating pressure in the radiated sound. A system that transforms a constant quantity into an oscillating one in this way is described as a self-sustained oscillator or auto-oscillator. In brass instruments the vibrating lips of the player are the agents which transform the constant mouth pressure into the oscillating downstream pressure through the valve effect discussed fully in Chap. 3.

Self-sustained oscillators generate different types of oscillation regime. The most important type from the musical viewpoint is the periodic regime, an example of which was described in Sect. 2.1.1. For a given fingering or slide position on a brass instrument, each natural note corresponds to a different periodic regime of oscillation. Although the regimes of oscillation are the result of a complex interaction between the vibrating lips of the musician and the air column in the musical instrument, the oscillation frequencies which define the pitches of the natural notes are largely determined by the geometrical property of the instrument tube known as the bore profile. This is a curve showing how the internal radius (or diameter) of the tube depends on the axial length from the mouthpiece to the plane of the bell. For purposes of illustration, the tube is usually assumed to be unwrapped, with the axis shown as a straight line. Figure 2.21 illustrates typical bore profiles for a trombone and a euphonium.

Other types of regimes of oscillation exist, including quasi-periodic oscillations. These are associated with musical sounds such as multiphonics and flutter tonguing, as well as some of the less musical sounds which are involuntarily generated by beginners.

We saw in Sect. 2.2.3 that it is possible for an experienced brass player to sound notes with repetition frequencies well above those of the significant input

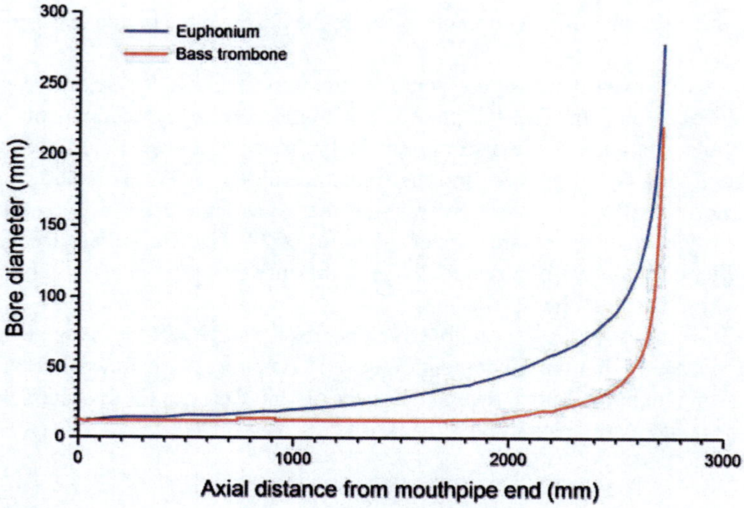

Fig. 2.21 Bore profiles of a euphonium and a bass trombone

impedance peaks of the instrument. For such very high notes, the system 'player-brass instrument' can no longer be treated as a single self-sustained oscillator. Although the vibrating lips are in self-sustained oscillation (see Sect. 3.4), there is no feedback from the air column, which is driven in forced oscillation. The continuous glissando sweep illustrated in the high frequency range of the trombone spectrogram in Fig. 2.20 resembles the normal behaviour of the singing voice, in the production of which the vibrating vocal folds excite the vocal tract in forced oscillation without experiencing significant acoustic feedback.

2.2.5 *The Wind Instrument Paradox*

The acoustic pressure amplitude in the sound wave radiated by a brass instrument is remarkably small in comparison with the acoustic pressure amplitude in the interior of the instrument. As a specific example, the measurements described in Sect. 2.1.1 and Sect. 2.1.2 show that the SPL of the radiated sound measured 50 cm outside the bell of a trombone is around 70 dB lower than the SPL measured in the mouthpiece cup. Since the purpose of the instrument is to radiate musical sound, it would at first sight appear desirable to redesign the instrument so that most of the sound energy is radiated. We saw however in Sect. 2.2.2 that the creation of a stable periodic regime of oscillation in a valve effect source like the lips requires a strong feedback from the resonating air column of the instrument. The strength of this feedback relies on the fact that most of the sound energy arriving at the bell is reflected back into the instrument. We thus have the paradox, which applies also to other types of wind instrument, that the condition for strong sound generation (efficient reflection back towards the lips of sound arriving at the bell) is opposed to the condition for strong sound radiation (efficient transmission into the environment of sound arriving at the bell).

The resolution of this paradox comes through understanding the way in which the reflection and transmission of a sound wave at the bell depend on the frequency of the wave. This topic is discussed in detail in Sect. 4.3. A characteristic feature of the acoustical behaviour of a flaring bell is that it is very efficient at reflecting low-frequency waves back into the instrument but very efficient at radiating high-frequency waves into the environment. The frequency above which reflections become negligible is known as the cutoff frequency of the bell. For a trombone, the cutoff frequency is typically around 800 Hz. Consider the note $B^\flat 3$ with a repetition frequency of 233 Hz, which is discussed in Sect. 2.2.3. The fundamental component of the sound wave is almost completely reflected when it arrives at the bell, giving a strong feedback to strengthen the lip vibration. The internal sound wave also contains frequency components which are harmonics of 233 Hz, and the higher harmonics are radiated with very high efficiency. To summarise, before the sound wave can radiate (most efficiently by its high harmonics), it must be generated by the lip vibration (assisted mainly by the low harmonics).

Part II
Acoustical Modelling of Brasswinds

Part II consists of four chapters which review in detail the physics underlying a performance on a brass instrument. The buzzing lips of the player, which are the source of the sound, are discussed in Chap. 3. Propagation of acoustic waves within the bore of the instrument and into the surrounding space is the principal topic of Chap. 4. The theoretical development in these two chapters is based on several approximations: propagation of sound waves inside the bore of the instrument is described using linear acoustics, resonances of the player's windway are not taken into account and the walls of the instrument are assumed to be perfectly rigid. Accepting these approximations allows us to derive an elementary model of a brass instrument under playing conditions, which is presented in Chap. 5. In this chapter, periodic solutions of the model are discussed, making use in part of a nonlinear dynamical system representation. In Chap. 6, the approximations are challenged, leading to some important refinements of the model. The theory of nonlinear propagation is presented in order to explain the brassy character of brass instruments under loud playing, the role of resonances of the player's windway in performance is discussed, and the influence of wall vibrations on the radiated sound and the perception of the performer are addressed.

Chapter 3
Buzzing Lips: Sound Generation in Brass Instruments

This chapter is concerned mainly with the vibrating lip sound source, which is the defining feature of every instrument in the brass family. Section 3.1 identifies some of the major muscles which control the embouchure of a brass player and presents the results of experimental studies which reveal the nature of the lip motion when a note is sounded. Measurements on human players have been supplemented by studies using artificial lips to generate the sound in a brass instrument, and some insights derived from these experiments are outlined. Section 3.2 describes the simplest possible mathematical model of the lip valve, often called the one degree of freedom model. The equation describing the behaviour of this model includes a term representing the net force exerted on the lips by the surrounding air; the subtle fluid dynamics of the air flow through the lips turns out to be of crucial importance in understanding why the lip vibration generates a fluctuating pressure in the mouthpiece. The mechanical behaviour predicted by the simple lip model is compared with experimental measurements on artificial and human lips in Sect. 3.3. In Sect. 3.4 some of the processes through which air flow generated by the player's lungs can initiate and sustain the vibration of the lips are explored. An equation relating the volume flow of air through the lips to the pressure difference across them is derived in Sect. 3.5; this equation is an important constituent of the elementary model of brass playing constructed in Chap. 5.

3.1 The Nature of Lip Vibration

3.1.1 The Brass Player's Embouchure

To get a well-controlled buzz from the lips, it is necessary to use the muscles which form the embouchure to obtain the correct lip configuration. Figure 3.1 illustrates some of the facial muscles which are important in achieving a good embouchure

M. Campbell et al., *The Science of Brass Instruments*, Modern Acoustics and Signal Processing, https://doi.org/10.1007/978-3-030-55686-0_3

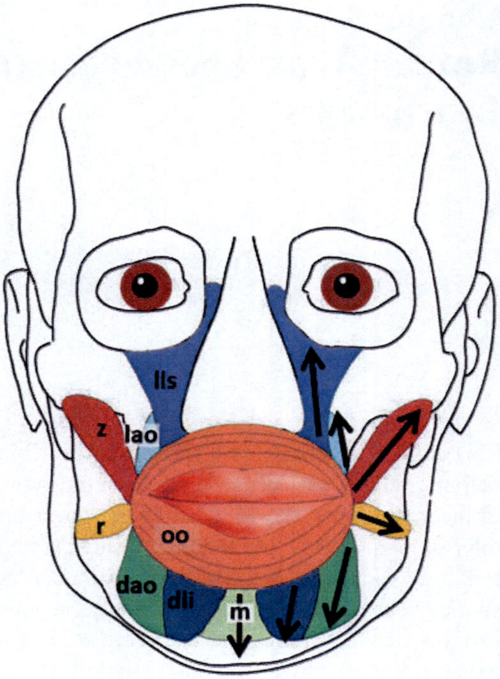

Fig. 3.1 The principal muscles controlling the embouchure: orbicularis oris (oo), depressor anguli oris (dao), depressor labii inferioris (dli), levator anguli oris (lao), levator labii superioris (lls), mentalis (m), risorius (r), zygomaticus (z). Courtesy of Alan Watson

for brass playing. The *orbicularis oris* muscle runs round the entire circumference of the mouth. It merges into the muscular wall of the cheek to form the *buccinator* (not marked in Fig. 3.1). Contracting this muscle puckers the lips, as in pouting or whistling. When the *zygomaticus* muscles are contracted, the corners of the lips are pulled laterally, as in smiling. The *levator* and *depressor* muscles also shown in Fig. 3.1 control the vertical positioning of the upper and lower lips.

In his classic book on the art of brass playing, Farkas (1962) refers to the long-running debate among brass players about the ideal embouchure: should it be ‘smiling’ or ‘puckered’? The fact that the *buccinator* muscle is named after the player of the ancient Roman trumpet called the bucina reflects a common view that the brass player’s embouchure is essentially a smiling lip configuration, and it is certainly true that a degree of lip tension is necessary in order to achieve a stable buzz. The use of the *orbicularis oris* is also important, however, in controlling the mass of tissue active in the lip vibration. Farkas’s recommendation is that all the muscles should be employed in a balanced way to create a ‘puckered smile’ embouchure (Fig. 3.2).

The final stage in preparing the embouchure for performance on a brass instrument comes when the lips are pressed against the rim of the mouthpiece. The degree of pressure which should be applied between the lips and the mouthpiece is another controversial topic in brass pedagogy. Some teachers advocate the ‘no pressure method’, but Farkas points out that at least a minimum pressure is necessary to

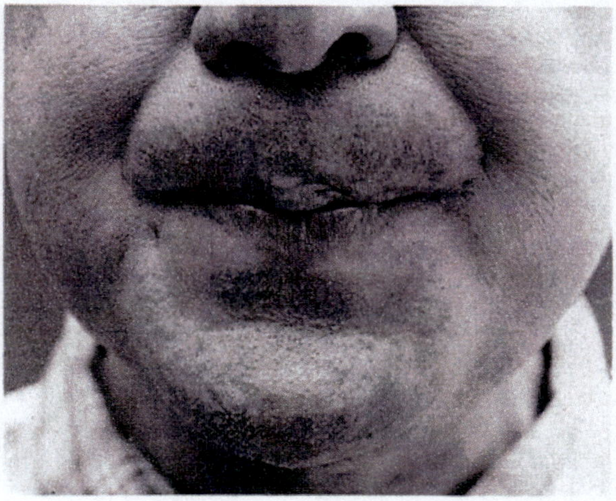

Fig. 3.2 The puckered smile embouchure recommended by Farkas (1962)

seal the lips hermetically to the mouthpiece and to give the player a sense of security and contact with the instrument. In playing an ascending scale, it is necessary for the musician to adapt the embouchure to increase the natural lip resonance frequency. Part of this adaptation normally involves some increase in the static lip pressure on the mouthpiece. The variation of the force exerted on the mouthpiece by three trumpeters playing a variety of exercises has been studied by Petiot (2003). For all the players, the force was found to increase with rising pitch, but the maximum force applied by a moderate-quality amateur was between two and three times greater than the largest force used by a highly experienced professional.

3.1.2 Experimental Studies of Vibrating Lips

Having chosen a suitable embouchure, the brass player sounds a note by using the muscles controlling the lungs to create a sufficiently high blowing pressure. The mouth overpressure is defined as the difference between the mean air pressure in the mouth and the atmospheric pressure outside the lips. For a given well configured embouchure, there is a critical value of the mouth overpressure, typically between 1 and 2 kPa, which corresponds to the threshold for lip oscillation. If the overpressure is below the threshold value, the lips will not vibrate; the only sound will be the gentle hiss corresponding to the wide frequency band turbulent noise generated as the steady mean flow emerges through the lip opening. When the overpressure is raised above the threshold value, the lips spontaneously start to vibrate, and a musical note with a definite pitch is generated.

The first photographic study of the motion of a brass player's lips during performance was reported in a classic paper by Martin (1942). Using a specially constructed transparent mouthpiece and stroboscopic illumination, Martin obtained

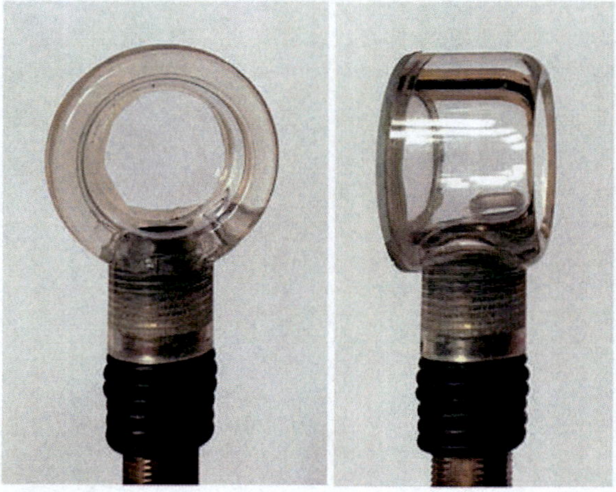

Fig. 3.3 Transparent mouthpiece used for high-speed video recording of in plane lip motion during trombone playing

sequences of images of a cornet player's lips at different stages in a cycle of vibration. Similar techniques have subsequently been used by various researchers to study the motion of lips in the mouthpieces of horns (Yoshikawa 1995; Yoshikawa and Muto 2003) and trombones (Copley and Strong 1996; Richards 2003; Bromage 2007; Boutin et al. 2015b).

A transparent trombone mouthpiece specially designed to allow the open area between the lips to be clearly visualised is shown in Fig. 3.3. In this mouthpiece, based on a design by Ayers (1998, 2001), the stem connecting the cup to the instrument is at right angles to the blowing direction, and a flat window at the base of the cup allows the lips to be viewed from the front. The cup volume and rim diameter are similar to those of a Denis Wick 6BS trombone mouthpiece.

The use of a high-speed camera to record the motion of a brass player's lips during playing was discussed briefly in Sect. 2.1.4. Figure 3.4 illustrates the nature of the lip motion recorded by Bromage et al. (2010) when a trombone player sounded the pedal note $B^\flat 1$, with a frequency of 58.3 Hz, using the mouthpiece shown in Fig. 3.3. The sequence of 14 images in Fig. 3.4, reading from top to bottom in the left-hand column and then from top to bottom in the right-hand column, shows the gradual closing and then re-opening of the aperture between the lips during one cycle of the vibration. The time interval between successive images was 1.32 ms, so that the complete cycle was recorded in 17.2 ms.

3.1.3 *Time Dependence of the Lip Opening Area*

In describing the motion of the lips, it is helpful to make use of the rectilinear co-ordinate system shown in Fig. 3.5. The z axis is the direction of flow of air through the lip opening; the x axis lies along the mid-line between the lips when

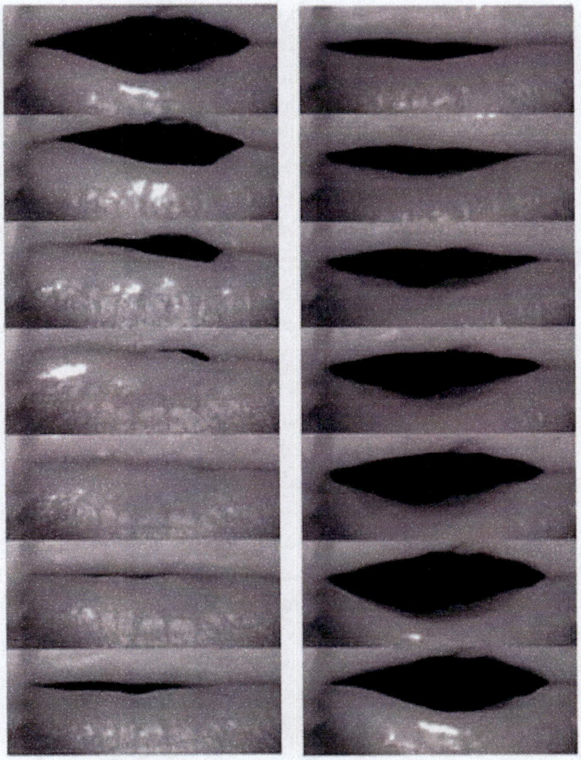

Fig. 3.4 A cycle of motion of a brass player's lips (frontal view, obtained with a high-speed camera and transparent mouthpiece). The time sequence of the images is from top to bottom in the left-hand column and then from top to bottom in the right-hand column. From Bromage et al. (2010)

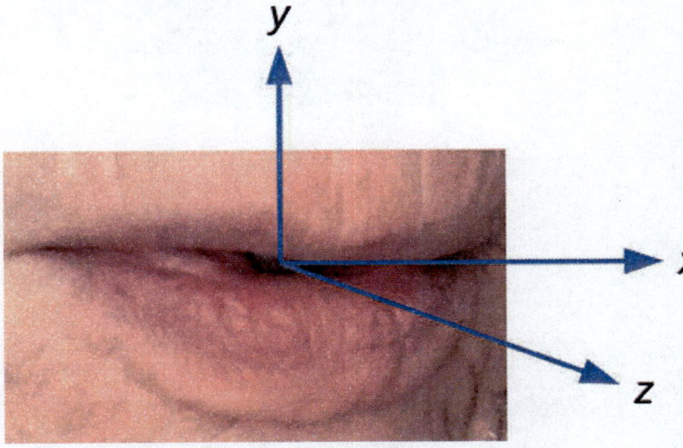

Fig. 3.5 Cartesian axes used to describe lip motion: x (lateral), y (transverse) and z (axial)

slightly open, and the y axis is perpendicular to both x and z axes. The absolute orientation of this set of axes is determined by the angle at which the player holds the instrument, but to simplify the discussion, the three mutually perpendicular axes will be described as 'lateral' (x), 'transverse' (y) and 'axial' (z).

Figure 3.4 illustrates the lip motion for a low-pitch note played loudly on a trombone. The width of each image is 25 mm, which is approximately the internal diameter of the mouthpiece rim. The gap between the lips has a maximum value of around 5 mm and a minimum value close to zero (in the fifth image). The shape of the lip opening changes significantly during the cycle, and at some stages (images 3 and 4), there is a strong right-left asymmetry.

Video recordings of many different brass players have been made with apparatus similar to that used in obtaining Fig. 3.4. Since each brass player has a unique set of lips, it is not surprising that images from performances of the same note by different players differ in detail. In particular, the degree of lateral asymmetry can vary substantially from one player to another, even although the radiated sounds are almost indistinguishable. The explanation for the fact that the same sounds can be generated by different shapes of lip opening lies in the nature of the valve effect source, discussed in detail in Sect. 3.5.1. The acoustic pressure generated in the mouthpiece cup depends on the volume flow of air through the lip aperture, and this volume flow is controlled mainly by the effective area between the two lips rather than the shape of the opening.

The open area (projected on to the xy plane) can be estimated from images of the type shown in Fig. 3.4 by defining a threshold of illumination below which a pixel is assumed to be in the lip opening. All pixels below the threshold are then set to black and all pixels above to white. The result is an image like that in the upper part of Fig. 3.6. The pixels in this image are then summed to give the total open area (Bromage et al. 2010).

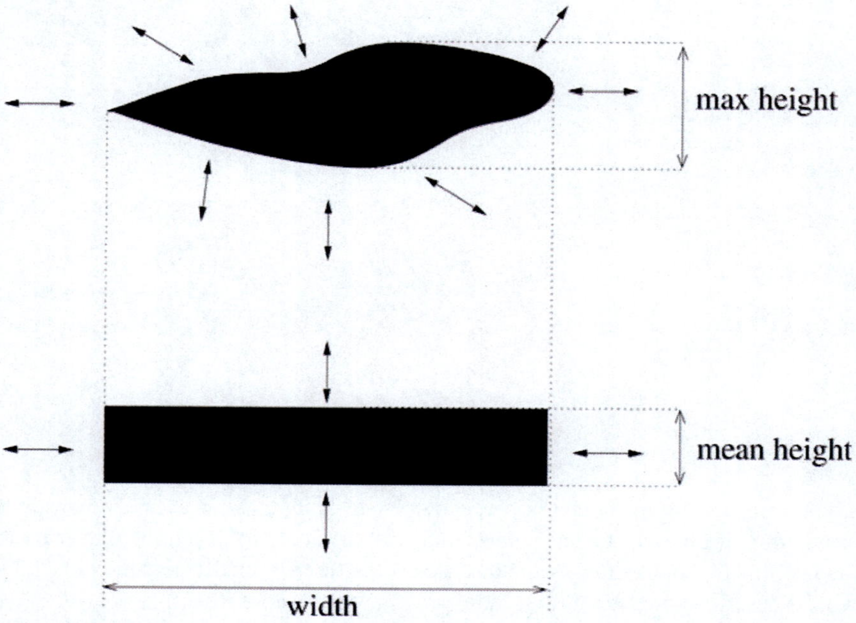

Fig. 3.6 Thresholded image of the lip open area (upper diagram); equivalent rectangle (lower diagram). From Bromage et al. (2010)

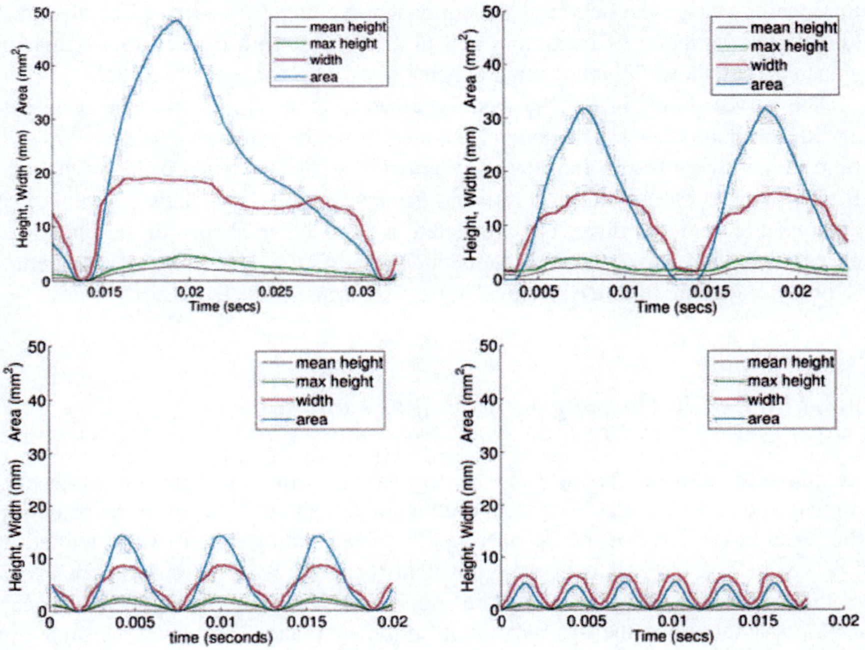

Fig. 3.7 Opening height, width and area of the lips of a trombonist playing B$^\flat$1 (top left), B$^\flat$2 (top right), F3 (bottom left) and F4 (bottom right). From Bromage et al. (2010)

The variation of the lip opening area with time for a trombonist playing four different pitches at a constant mezzoforte dynamic is illustrated in Fig. 3.7 (cyan lines). For the note F4, the highest pitch shown, six complete cycles of vibration are recorded. The variation of open area with time is almost sinusoidal, as observed in the early experiments on the cornet by Martin (1942). As the pitch decreases, the area variation with time becomes less sinusoidal, and the maximum open area increases. Only one cycle is shown for the lowest note, B$^\flat$1. The maximum lip opening area of 49 mm^2, which occurs less than a third of the way through the cycle, is almost 10 times the maximum area for the F4 note.

The width of the lip opening, measured along the x axis, is shown by the magenta lines in Fig. 3.7. For the highest pitch, the variation of width with time is also approximately sinusoidal. As the pitch of the played note falls, the maximum width increases, but the limitation imposed by the fact that the lips are pressed against the rim of the mouthpiece becomes obvious for the lowest pitches. When the note B$^\flat$1 is being played, the lip opening width increases abruptly from 0 to around 20 mm at the beginning of the cycle; it remains close to this width for most of the cycle before decreasing rapidly back to zero.

The maximum height of a lip opening image is defined as the distance between two lines parallel to the x axis which touch the upper and lower limits of the image, as shown in Fig. 3.6. The equivalent rectangle has the same width and area as the lip opening image; the height of the equivalent rectangle is described as the mean height. The variation of maximum and mean heights with time is approximately sinusoidal, as shown by the green and black lines, respectively, in Fig. 3.7.

The curves shown in Fig. 3.7 were obtained in experiments on a single experienced trombone player. The general features of the dependence of lip opening on time seen in these results have been observed in studies of many different players. It should be borne in mind, however, that every player develops an individual embouchure, and this diversity is reflected in the detailed shapes of the open area curves. A skilled player can also deliberately modify the nature of the lip vibration to obtain a specific timbre or technical effect (Norman et al. 2010).

3.1.4 The Lip Opening Area-Height Function

In computer simulations of brass instrument playing, the lips are frequently represented as an oscillating system with a single degree of freedom, representing the mean height $h(t)$ of the lip opening. In the co-ordinate framework defined in Fig 3.5, $h(t)$ is the height of the equivalent rectangle representing the open area of the lips and is measured along the y direction in the xy plane. This approach does not assume that the lips move only in the xy plane. The 'swinging door' lip model described in Sect. 3.2.3 implies a substantial component of lip motion in the z direction, and the experiments described in Sect. 3.1.5 confirm that the tip of a trombone player's lip can move almost 10 mm into the mouthpiece when playing a fortissimo pedal note. It is assumed however that the volume flow of air into the mouthpiece is primarily determined by the area $S(t)$ of the lip opening projected on to the xy plane, which is perpendicular to the flow direction.

Is the area $S(t)$ proportional to the mean height $h(t)$? Such a linear relationship will only hold if the width w of the equivalent rectangle is constant throughout the cycle of vibration. For single-reed woodwind instruments like the clarinet and saxophone, this is a reasonable approximation, since the reed opens and closes uniformly across its width. In this case the open area is in fact a rectangle whose constant width is equal to the width of the slotted opening in the mouthpiece under the reed tip. The nature of the area variation in this case is illustrated by the graphical construction in Fig. 3.8a. The lower rectangle is fixed, while the upper dotted rectangle oscillates vertically. The overlapping shaded region represents the opening, whose change in area is directly proportional to the vertical displacement.

Measurements such as those shown in Fig. 3.7 demonstrate that the assumption of constant width is not in general valid for the lips of a brass player. The open area is given by the product of two time-varying quantities:

$$S(t) = w(t)h(t).$$

The time-varying width of the lip opening can be incorporated into a single degree of freedom model if an additional assumption is made about the relationship between

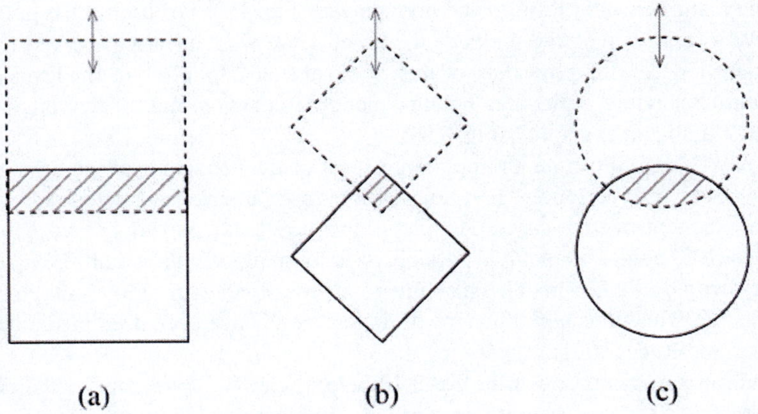

Fig. 3.8 Three illustrations of opening areas: (**a**) a 'moving rectangular' area (the overlap of two rectangles), (**b**) a 'moving diamond' area (the overlap of two diamonds), (**c**) a 'moving circle' area (the overlap of two circles). From Bromage et al. (2010)

$h(t)$ and $w(t)$. For example, Fig. 3.7 shows that for the pitch F4, $w(t) \simeq 0.16\,h(t)$; in this case the area-height function takes the form

$$S(t) = 0.16\,h(t)^2.$$

The graphical construction in Fig. 3.8b, showing the open area as the overlap of two diamonds, illustrates such a quadratic relationship.

A more general approach is to postulate a power law relationship between $S(t)$ and $h(t)$:

$$S(t) = S_0 \left(\frac{h(t)}{h_0}\right)^q, \tag{3.1}$$

where S_0 and h_0 are reference values of the lip opening area and mean height, respectively, and q is an exponent to be empirically determined. The two examples previously discussed are special cases of this relationship: the 'moving rectangular' area illustrated in Fig. 3.8a corresponds to $q = 1$, while the 'moving diamond' area illustrated in Fig. 3.8b corresponds to $q = 2$. The 'moving circle' area shown in Fig. 3.8c corresponds to the intermediate value $q = 1.5$.

In a classic study of sound generation in brass instruments, Elliott and Bowsher (1982) assumed that the lip opening could be described by a linear area-height function, equivalent to the choice $q = 1$. Later work on trombone sound synthesis showed that choosing $q = 2$ led to realistic brass instrument sounds (Msallam et al. 2002). Extensive investigations carried out by Bromage et al. (2010) using a high-speed digital camera and a transparent trombone mouthpiece showed that exponents could be found from just below 1 to significantly greater than 2, depending on the pitch and dynamic level of the note played (Fig. 3.9). For high-pitch notes the relatively small height and width of the lip aperture were usually observed to vary in proportion, resulting in values of the exponent $q \simeq 2$ for all dynamic levels. This quadratic behaviour of the area-height function was also found for low-pitch notes played quietly: an example is Fig. 3.9a.

Figure 3.9b, c illustrate an important aspect of the behaviour of the lip aperture for low notes played loudly. It is not possible to fit an area-height function with a single exponent in such cases. The plots for a pedal $B^{\flat}1$ played *mf* and *ff* have a characteristic dogleg shape: it is possible to fit these plots with two different slopes, the first with $q \simeq 2$ for small height values and the second with $q \simeq 1$ for high height values. Area functions with this type of dogleg slope have been used in simulations (Vergez and Rodet 2001a).

The measurements shown in Fig. 3.10 help to clarify the reason for the dogleg appearance of the area-height curve in low-pitch playing. Figure 3.10a shows the variation of height, width and area of the lip aperture during one cycle of the note $B^{\flat}1$ played *mf* on a trombone. The curves are similar to those shown in Fig. 3.7, although the measurements were made on two different trombonists. The dashed vertical lines divide the period of the cycle into three time intervals. During the first interval (red markers), the width increases; during the second (green markers), it remains approximately constant; and during the third (blue markers), it decreases. To simplify the discussion, these intervals are described as 'opening', 'mid cycle' and 'closing'.

The corresponding intervals are identified by the same marker colours in Fig. 3.10b, which is the corresponding area-height plot. The opening interval is fitted by Eq. 3.1 with an exponent $q = 2.2$; the mid-cycle interval requires $q = 1$, while the closing interval $q = 2.7$ gives the best fit. The linear behaviour in the mid-cycle interval is explained by the constancy of the width of the lip aperture, which is probably due to the constraining effect of the mouthpiece rim.

3.1.5 *Two-Dimensional Motion of the Brass Player's Lips*

The front view of the lips, shown in Fig. 3.4, gives valuable information about the modulation of the air flow from the player's mouth into the mouthpiece. It is however important to recognise that the lips can also have a significant component of motion along the direction of the air flow.

Figure 3.11 shows a transparent mouthpiece designed to view motion of the vibrating lips in the *yz* plane. A Kelly bass trombone mouthpiece was modified by planing one side of the cup and attaching a flat optical window which gave good visibility over around half the cup diameter. A PCB microphone was also connected by a short tube to the backbore of the mouthpiece to monitor the downstream pressure signal.

The first image in the sequence in Fig. 3.12, reading from left to right, shows the phase of the vibration in which the lips momentarily close completely. The

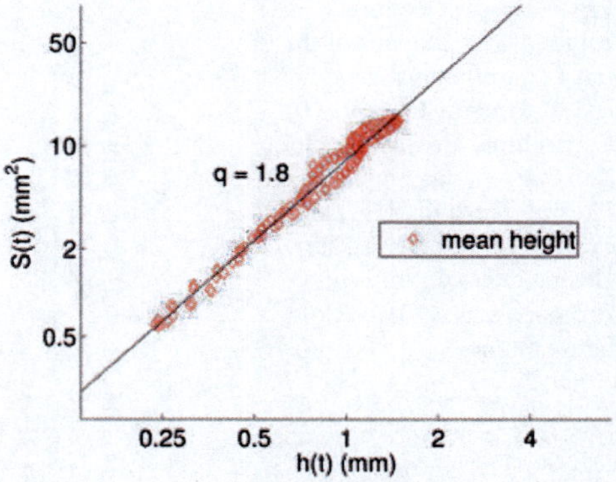

(a) *p*

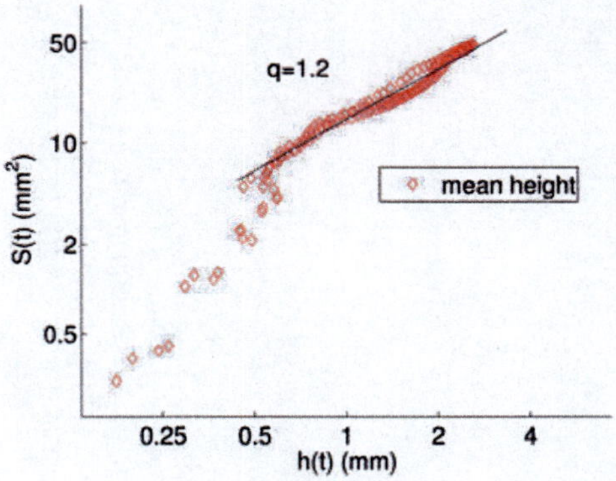

(b) *mf*

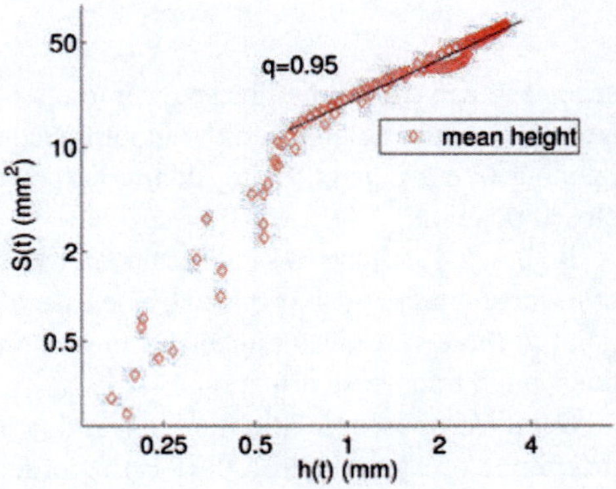

(c) *ff*

Fig. 3.9 Logarithmic area-height plots for the note $B^{\flat}1$ played by a trombonist at dynamic levels (**a**) piano (*p*), (**b**) mezzoforte (*mf*), (**c**) fortissimo (*ff*). From Bromage et al. (2010)

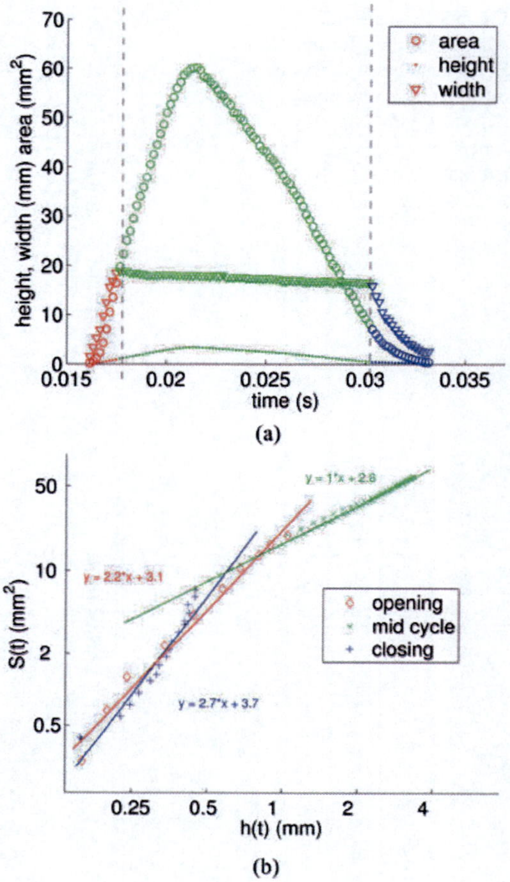

Fig. 3.10 (**a**) Opening height, width and area of the lips of a trombonist playing $B^{\flat}1$ at dynamic level *mf*. (**b**) Logarithmic area-height plot. Red markers, opening interval; green markers, mid cycle interval; blue markers, closing interval. From Bromage et al. (2010) (Color figure online)

second image shows the lips starting to open transversely (vertically in the image), but also moving axially into the mouthpiece cup. Subsequent images show the lips opening like a pair of swing doors, before returning in the seventh image to the closed position.

Figure 3.13 illustrates the scale of the motion of the lips in both axial and transverse directions. For this extreme case of a low note played very loudly, a fixed point at the centre of the upper lip moves almost as far in the axial direction as it does in the transverse direction.

Using stroboscopic illumination and a mouthpiece with a fibre-optic probe inserted, Copley and Strong (1996) recorded the motion of a trombonist's upper lip in both transverse and axial directions for five different played pitches. The trajectories of a central point on the lips throughout a vibration cycle at each pitch are shown in Fig. 3.14. The notes were all played at a comparable forte dynamic level. It is evident that the amplitude of lip motion decreases as the pitch rises. Axial motion is most noticeable at the lowest pitches, while at higher pitches, the motion is

Fig. 3.11 Transparent mouthpiece used for high-speed video recording of out-of-plane lip motion during trombone playing. From Stevenson (2009)

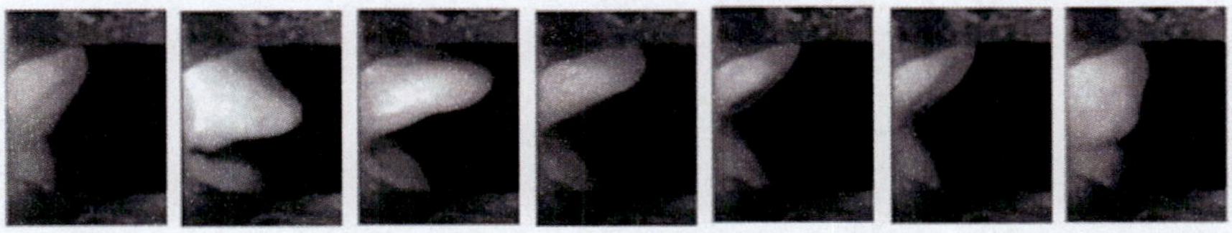

Fig. 3.12 A cycle of motion of a brass player's lips (side view, obtained with a high-speed camera and transparent mouthpiece). From Stevenson (2009)

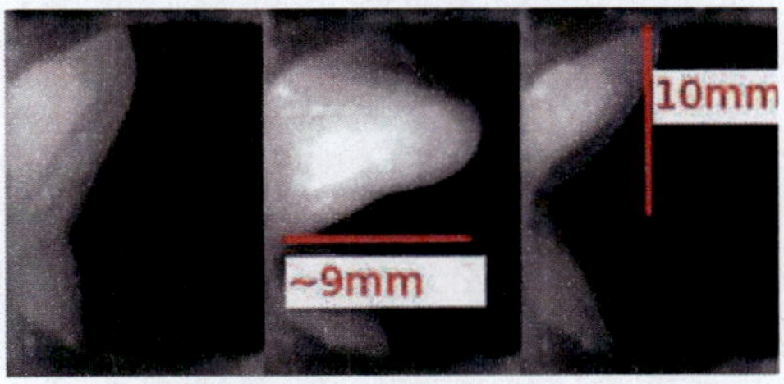

Fig. 3.13 Images of three phases in the vibration cycle of the lips of a trombonist playing the note $B^\flat 1$ fortissimo, side view. Left: closure. Centre: partial opening. Right: maximum opening. From Stevenson (2009)

largely in the transverse direction. Similar trajectories illustrating the anticlockwise cyclical motion of the lips of trombonists have been recorded by Boutin et al. (2015b).

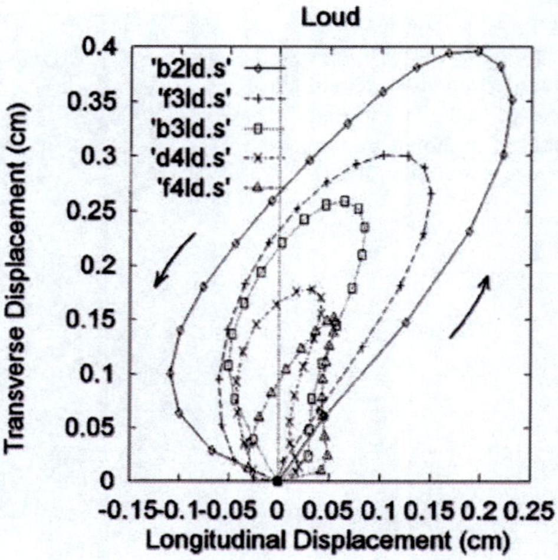

Fig. 3.14 Trajectories of a central point on the upper lip of a trombonist playing five different notes: $B^{\flat}2$ (open diamond), F3 (plus symbol), $B^{\flat}3$ (open square), D4 (times symbol), F4 (open triangle). Reproduced from Copley and Strong (1996) with the permission of the Acoustical Society of America

3.1.6 Experiments with Artificial Lips

On 24 February 1874, John E. Fowler was granted a patent by the United States Patent Office (Fowler 1874) for an invention which he described as:

> …an Improvement in Mouthpieces for Musical Instruments.

Fowler's invention was a modified cornet mouthpiece, provided with:

> …an artificial lip, which is made of rubber or other suitable material, which is arranged and connected with other devices in such a manner that, while being used, it may have different degrees of tension or hardness applied to it, and thus affect the tone or sound of the instrument.

Various other inventors have patented brass instrument mouthpieces in which the lips of the human player are replaced by strips of elastic material, but these innovations have not attracted the interest or approval of serious brass players. Artificial lips have however served a useful function in pedagogical demonstrations in which brass instruments are sounded using an air pump or compressed air source; the famous horn player Philip Farkas recounts that as a young bugler in the 1920s, he was intrigued by an exhibit of this type in the Chicago Museum of Science and Industry (Farkas 1962).

By the mid-1950s, the manufacturer C.G. Conn Ltd. was using artificial lip excitation in acoustical measurements on brass instruments (Kent 1956). In the last 20 years, artificial lips have been widely employed to study different aspects of brass instrument behaviour, including the nature of the excitation mechanism, the influence of bore profile changes on the pitch and timbre of the radiated sound and the interaction between acoustic and structural resonances. It has been found that realistic brass instrument sounds can be obtained using artificial lips made of thin latex rubber tubes filled with water (Gilbert and Petiot 1997; Vergez and Rodet 1997;

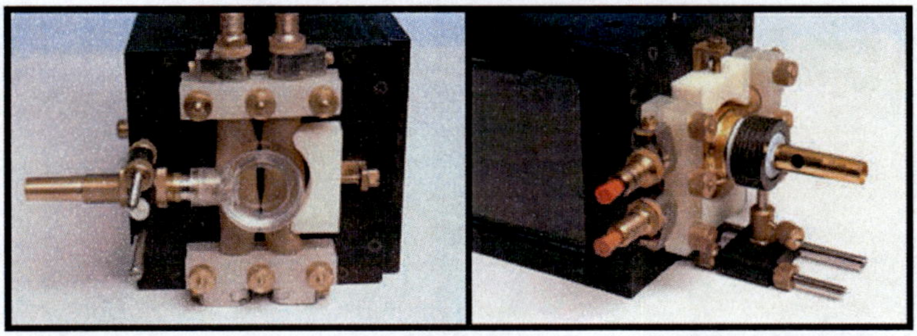

Fig. 3.15 An artificial mouth for sounding brass instruments. Left view: fitted with a transparent mouthpiece for high-speed filming. Right view: fitted with a conventional trombone mouthpiece with an optional shunt hole in the backbore (Richards 2003)

Gilbert et al. 1998; Cullen 2000; Ehara et al. 2001; Bromage et al. 2003; Petiot et al. 2003; Vilain et al. 2003). An artificial mouth using latex lips is illustrated in Fig. 3.15.

It is evident that an artificial embouchure using cylindrical latex tubing is incapable of reproducing fully the complex behaviour of the human embouchure described in Sect. 3.1.1. The artificial lips are therefore only of passing interest to the musician. From the scientific perspective, however, it is very valuable to be able to study the physics of artificial lips, since despite their simplicity they are capable of generating musically acceptable sounds on brass instruments. Many aspects of the complicated relationship between the player and the instrument have been clarified by experiments using artificial mouths.

Figure 3.16 illustrates the nature of the lip motion recorded using a high-speed camera when the artificial mouth sounded the note F3 on a tenor trombone using the transparent mouthpiece shown in Fig. 3.3. The sequence of nine images in Fig. 3.16 shows just over one cycle of opening and closing of the aperture between the lips. The variation of open area is similar to that seen when a human player plays the same pitch.

Figure 3.17 shows measurements of the trajectory in the yz plane of a fixed point near the centre of the upper artificial lip when the note F3 was sounded on a tenor trombone (Richards 2003). These measurements were carried out using a high-speed camera which could be rotated to different viewing angles, taking advantage of the long-term stability of the artificial lip excitation. The note was played at a musical forte level, corresponding roughly to the 'loud' dynamic level generated by the human player whose lip trajectories were measured by Copley and Strong (Sect. 3.1.5). The curve for the note F3 in Fig. 3.14 has some similarities with the artificial lip trajectory in Fig. 3.17. In both cases the motion near the tip of the upper lip, starting from the point at which $y = 0$, can be thought of as outward and upward, then backward and downward, with a final forward motion to the point of minimum transverse displacement. In each case the maximum transverse (y) displacement is significantly larger than the maximum axial (z) displacement. However the shapes of the trajectories are noticeably different, and the displacements of the human lip are around 50% greater than the corresponding displacements of the artificial lip.

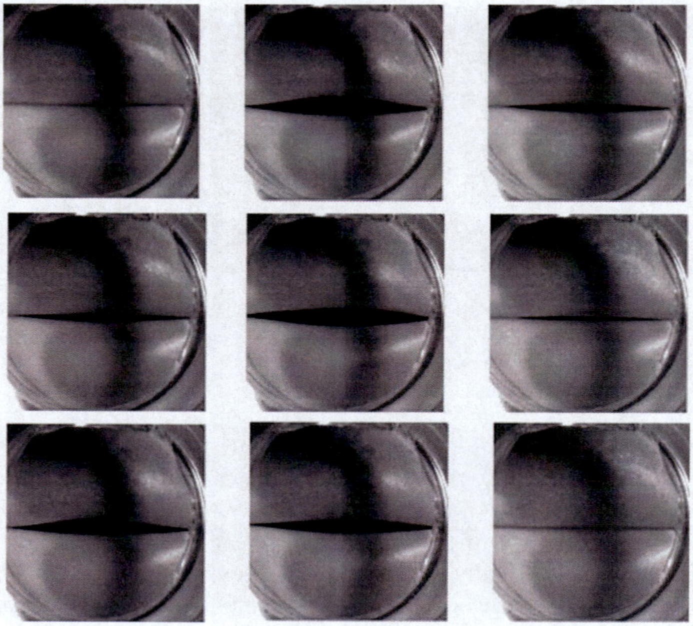

Fig. 3.16 A cycle of motion of an artificial lip vibration (frontal view, obtained with a high-speed camera and transparent mouthpiece). The time sequence of the images is from top to bottom in the left-hand column, the central column and then the right-hand column. From Richards (2003)

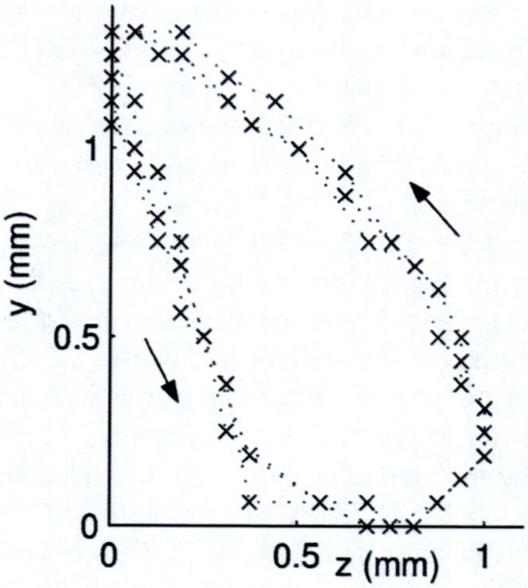

Fig. 3.17 Trajectory of a central point on the latex rubber lip of an artificial mouth playing the note F3 forte on a trombone (Richards 2003)

3.2 An Equation of Motion for the Lips

The lips of a brass player are flexible, continuous structures subjected to aerodynamic forces which depend on the pressure field around them. The vibrating area of each lip is limited by the rim of the mouthpiece against which it is pressed. Many of the material properties of the lips are difficult to measure experimentally, and the internal forces exerted by the muscles of the player's embouchure are in general unknown. A complete mechanical model of this vibrating system with many degrees of freedom would be extremely complex and would include a large number of parameters whose values could not be accurately determined. A more fruitful approach to understanding the essential physics of the vibrating lip is to begin with a highly simplified model, as outlined in Sect. 2.2.

In Sect. 3.2.1 a model is developed in which the complex multidimensional lip motion revealed by the experimental studies described in Sect. 3.1 is represented by the time dependence of a single spatial variable. A 'sliding door' model, in which the lip vibration is assumed to take place only in the vertical plane, is described in Sect. 3.2.2. A 'swinging door' model, which allows for the observed two-dimensional motion of the lips (Sect. 3.1.5), is discussed in Sect. 3.2.3.

3.2.1 A One-Mass Model of the Lips

The simple lip model with only one degree of freedom (1DOF) is frequently described as a one-mass model, since in it the distributed mass of the moving section of the lip is concentrated into a single mass m (Fig. 3.18). To begin with, we assume that the single degree of freedom corresponds to motion along the

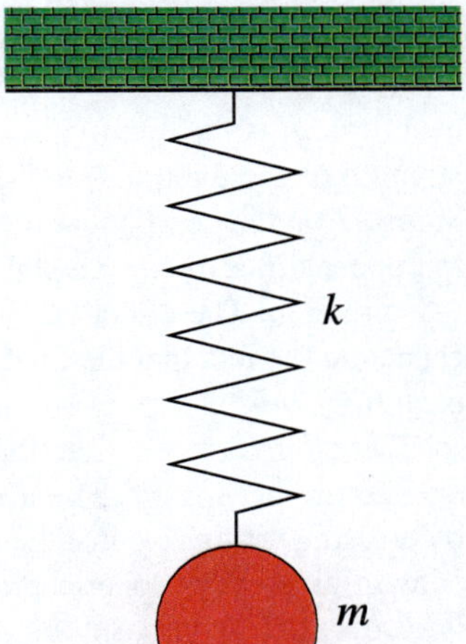

Fig. 3.18 The one-mass model of a single degree of freedom oscillator with mass m and stiffness k

y axis. The forces generated by the elasticity of the player's embouchure are represented by a single spring with stiffness k, which reacts to any displacement of the mass from its equilibrium position y_{eq} by applying a proportional restoring force $F_R = -k(y - y_{eq})$.

If the mass in Fig. 3.18 is pulled away from its equilibrium position and then released, it will oscillate at its natural resonance frequency, which is

$$f_r = \frac{\omega_r}{2\pi} = \frac{1}{2\pi}\sqrt{\frac{k}{m}}. \tag{3.2}$$

The amplitude of the oscillations will gradually diminish as the energy of motion is reduced by viscothermal losses and internal friction in the spring. This process is described as damping. In a similar way, the tensed lip of a brass player's embouchure would vibrate at its natural resonance frequency if plucked. The player is capable of changing this resonance frequency over several octaves by adjustments of the embouchure (see Sect.3.1.1). The oscillations of the plucked lip would die out more quickly than those of a mass on a spring because the internal friction in the lip is much greater than in a metal spring.

In the operation of the valve effect source, the most important feature of the lip vibration is not the motion of one lip, but rather the modulation of the open area S between the two lips. In Sect. 3.1.3 the equivalent rectangle was defined: at any time t, this rectangle has width $w(t)$ and height $h(t)$, such that $S(t) = w(t)h(t)$. Making the additional simplifying assumptions that the width is constant and that the motion of the lower lip is a mirror image of the motion of the upper lip, it is possible to model the simultaneous motion of the two lips as one oscillator with a single degree of freedom (along the y axis) (Cullen et al. 2000). The equation of motion for this 1DOF oscillator can be written as

$$\frac{d^2h}{dt^2} + \frac{\omega_l}{q_l}\frac{dh}{dt} + \omega_l^2(h - h_{eq}) = \frac{F}{m}, \tag{3.3}$$

where ω_l is the mechanical lip resonance frequency (assumed to be the same for each lip), q_l is the quality factor of the lip resonance (inversely proportional to the strength of the damping) and h_{eq} is the separation between the lips when they are at rest. F is the component of the total external force acting in the $+y$ direction on the upper lip; a force of equal magnitude is assumed to act on the lower lip in the $-y$ direction. The effective mass m is equal to half the effective mass of each lip, reflecting the fact that the lip displacement h in Eq. 3.3 is twice the displacement of each lip.

The external force F acting on the brass player's lips arises from the pressure exerted on the lips by the surrounding air. When the player forms an embouchure by pressing the lips against the mouthpiece rim, mouth and mouthpiece both initially contain air at atmospheric pressure. Since the internal pressure of the lip tissue is also equal to atmospheric pressure, there is no net pressure across any part of the lip surface exposed to the air. To initiate the note, the player increases

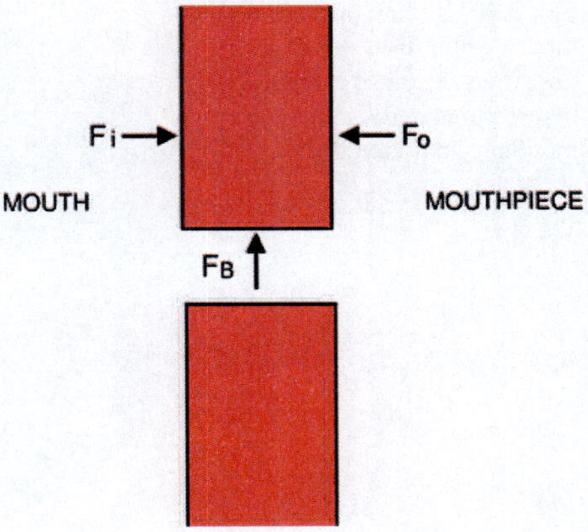

Fig. 3.19 Forces exerted on the upper lip by the surrounding air

the mouth pressure above atmospheric. The resulting imbalance between mouth and mouthpiece pressures results in unbalanced external forces acting on each lip surface, as illustrated in Fig. 3.19.

3.2.2 *The Sliding Door Lip Model*

The relationship between the pressure field acting on the lips and the y component of the external force depends on assumptions made about the mechanical behaviour of the lips and the nature of the flow through the lip aperture. It was noted in Sect. 3.1.5 that the lip vibration of brass players appears to be predominantly transverse (along the y axis) for high-pitch notes but predominantly axial (along the z axis) for low-pitch notes. We start with the simple picture of the lips as a pair of sliding doors moving only in the y direction. This model is clearly not able to reproduce the behaviour of human lips when playing low pitches but is more successful in describing high register playing (Adachi and Sato 1995).

Figures 3.19 and 3.20 illustrate the forces exerted by the air pressure on the surfaces of the upper lip. The mouth pressure p_{m} acts on the inner surface of the lip, which lies in the xy (vertical) plane. Since this surface is perpendicular to the z axis, the force F_{i} acts in the $+z$ direction. Similarly, the mouthpiece pressure p acts on the outer lip surface, causing a force F_{o} in the $-z$ direction. If the inner and outer lip surfaces have the same area S_{io}, there will be a net force on the lip in the $+z$ direction equal to

$$F_{\mathrm{io}} = F_{\mathrm{i}} - F_{\mathrm{o}} = (p_{\mathrm{m}} - p)S_{\mathrm{io}} = \Delta p S_{\mathrm{io}}. \tag{3.4}$$

Since F_{io} has no component acting in the y direction, it plays no role in the dynamics of the sliding door mode. There is, however, a third force, labelled F_{B} in Fig. 3.19, which acts in the $+y$ direction. This is due to the pressure within the

Fig. 3.20 Sliding door lip model. Solid lines show directions of forces exerted by pressures in the mouth and mouthpiece; dotted lines show directions of lip motion

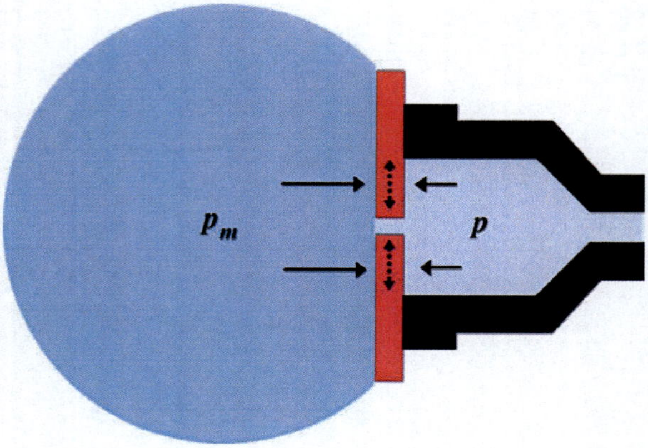

lip channel acting on the lower surface of the upper lip. In the simplified case of a rectangular lip channel of length l, width w and height h, the pressure p_{lc} inside the lip channel exerts a force on the surface $S = lw$ of magnitude

$$F_{\mathrm{B}} = lwp_{\mathrm{lc}}. \tag{3.5}$$

In the absence of air flow through the lip channel, the mouth, lip channel and mouthpiece are all at atmospheric pressure, and there is no net force on the lip. To sound a note, the player raises the mouth pressure by an amount p_{m} (see Sect. 2.1.3), generating a volume flow of air from the mouth through the lips into the mouthpiece. There is an important difference between the flow patterns entering and leaving the lip channel. The air from the mouth converges smoothly as it enters the lip channel; the speed of the air flow increases substantially, and there is a large drop in pressure. At the exit the flow separates and forms a turbulent jet; the kinetic energy of the jet is dissipated in the mouthpiece, but the pressure change from lip channel to mouthpiece is very small. We can therefore make the assumption that $p_{\mathrm{lc}} \simeq p$.

The air flow from the mouth through the lip aperture and into the mouthpiece is discussed in detail in Sect. 3.5. The pressure drop at the entrance to the lip channel is described by the Bernoulli equation (3.19),, and the transverse force F_{B} on the surface of the lip channel is often called the Bernoulli force.

In Fig. 3.20 , the forces F_{i} and F_{o} are shown by the long and short solid arrows, respectively; F_{B} is not shown but acts in the y direction. When a note is sounded, there are large fluctuations in the mouthpiece pressure $p(t)$ (see Sect. 2.1.2); in the simplified model discussed in this chapter, the smaller pressure fluctuations which occur in the player's mouth are neglected (see Sect. 6.3). As a result of the fluctuations in F_{B}, the lips are driven into oscillation along the y axis, indicated by dashed arrows in Fig. 3.20. The mean pressure in the mouthpiece is also raised above atmospheric because of the resistance of the instrument tube to steady flow. This change is small and can to a first approximation be neglected, implying that for this sliding door model, the equilibrium lip opening height h_{eq} is independent of the mouth pressure.

The expression for F_B given by Eq. 3.5 can be substituted in Eq. 3.3 to give a version of the equation of motion of the lip containing only the two dependent variables $h(t)$ and $p(t)$:

$$\frac{d^2h(t)}{dt^2} + \frac{\omega_l}{Q_l}\frac{dh(t)}{dt} + \omega_l^2(h(t) - h_{eq}) = \frac{lwp(t)}{m} = \frac{p(t)}{\mu}, \tag{3.6}$$

where $\mu = m/lw$ is an effective mass per unit area of the vibrating lips.

The simplified model described above does not account for collisions between the lips. If the amplitude of vibration reaches the value h_{eq}, the lips come into contact during the inward part of the vibration cycle. A strong deceleration force can be added to Eq. 3.6 to incorporate this behaviour (Adachi and Sato 1995).

3.2.3 The Swinging Door Lip Model

The experimental results described in Sect. 3.1.5 demonstrate clearly that the simple sliding door model is inadequate to describe the motion of a brass player's lips during performance. The trajectories of a trombonist's lip in Fig. 3.14 confirm that motion along the z axis becomes increasingly important for lower-pitched notes: for the pedal $B^\flat 1$ whose lip motion is illustrated in Fig. 3.13, the vertical and axial lip displacements are comparable in magnitude. In such cases the lip motion resembles an outward swinging door more than a sliding door, and a simple 'swinging door model' has been developed to describe this motion (Adachi and Sato 1995).

The swinging door lip is illustrated schematically in Fig. 3.21. The lip is viewed as a flap hinged at its junction with the mouthpiece rim. The force $F_{io} = F_i - F_o$ generates a torque on the lip; under the action of which, it rotates about the hinge

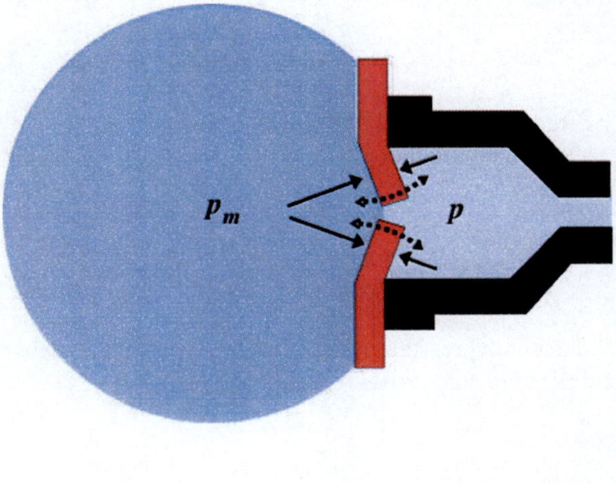

Fig. 3.21 Swinging door lip model. Solid lines show directions of forces exerted by pressures in mouth and mouthpiece; dotted lines show directions of lip motion

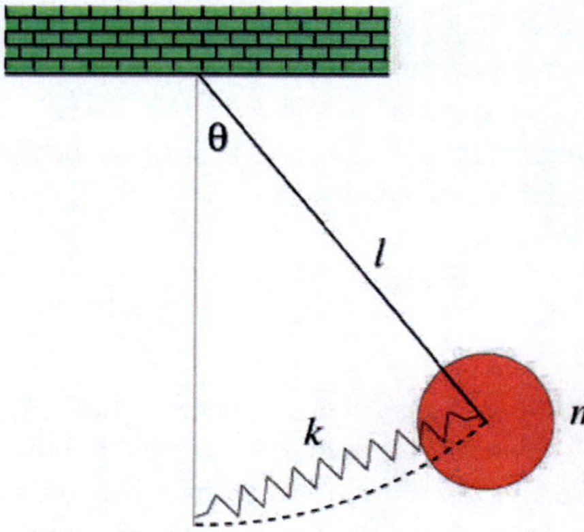

Fig. 3.22 One degree of freedom model of a swinging door lip

as indicated by the dashed lines in Fig. 3.21. We assume for simplicity that when $F_{\text{io}} = 0$, the lips lie in the vertical plane with a separation h_0. The swinging motion of the lips takes place in two dimensions, and a model with two or more dimensions is necessary to describe all the effects of this type of lip valve. It is however possible to incorporate the swinging door picture into the framework of the elementary 1DOF model by making the further simplifying assumption that the lips rotate without deformation (Adachi and Sato 1995). The displacement of the lip is then defined by the single variable θ.

The lip mass m is taken to be concentrated at the end of a swinging rod of fixed length l and negligible mass, as illustrated in Fig. 3.22. The direction of F_{io} is always perpendicular to the inner and outer lip surfaces and therefore rotates with the lip. Assuming that F_{io} acts at the centre of the lip surface, the torque about the pivot point due to the pressure difference across the lips is

$$\mathcal{T}_{\text{io}} = F_{\text{io}}\, l/2 = (p_{\text{m}} - p) S_{\text{io}}\, l/2. \tag{3.7}$$

The moment of inertia of the mass about the pivot point is $I = ml^2$. The embouchure muscles are modelled as a spring providing a restoring torque

$$\mathcal{T}_{\text{emb}} = -kl^2\theta = -ml^2\omega_l^2\theta, \tag{3.8}$$

where ω_l is the lip resonance frequency (see Eq. 3.2).

The equation of angular motion for the swinging door model can now be written as

$$\frac{d^2\theta(t)}{dt^2} + \frac{\omega_l}{Q_l}\frac{d\theta(t)}{dt} + \omega_l^2\theta = \frac{S_{\text{io}}}{2ml}(p_{\text{m}} - p(t)). \tag{3.9}$$

If the mouth pressure p_{m} is below the threshold for self-sustained oscillation, a static equilibrium state exists for a lip angle

$$\theta_{\mathrm{eq}} = \frac{S_{\mathrm{io}} p_{\mathrm{m}}}{2ml\omega_l^2} \tag{3.10}$$

at which the restoring torque is equal to the torque due to the pressure difference across the lips $\Delta p = p_{\mathrm{m}}$. In contrast to the case of the sliding door model, the equilibrium position of the swinging door model increases with increasing mouth pressure.

When the mouth pressure is raised just above the playing threshold, the lip will start to oscillate about the threshold equilibrium angle. The relation between the angle θ and the lip opening height is

$$h = h_0 + l(1 - \cos\theta). \tag{3.11}$$

Since this equation is nonlinear, simple harmonic motion of the swinging door does not result in sinusoidal modulation of the lip opening. The behaviour of the lips near the oscillation threshold can however be represented by assuming that around the equilibrium position the movement of the mass in Fig. 3.22 is along the tangent to its actual circular path. The small amplitude equation of motion for the swinging door model can then be written in terms of the lip opening height as

$$\frac{d^2h(t)}{dt^2} + \frac{\omega_l}{Q_l}\frac{dh(t)}{dt} + \omega_l^2(h(t) - h_{\mathrm{eq}}) = -\frac{S_{\mathrm{io}} \sin\theta_{\mathrm{eq}} p(t)}{2m} = -\frac{p(t)}{\mu}, \tag{3.12}$$

where the effective mass per unit area of the vibrating lips is in this case $\mu = 2m/(S_{\mathrm{io}} \sin\theta_{\mathrm{eq}})$.

3.2.4 *Inward-Striking and Outward-Striking Reeds*

For a valve effect source to be capable of generating a continuous sound in a musical wind instrument, it must be capable of supplying energy to sustain the standing waves in the instrument's air column. Before continuing with the development of the elementary brass model, it is worth making a brief digression to examine the different ways in which this energy supply requirement is met in reed woodwind and lip-excited brass instruments.

The modelling of reed wind instruments was first put on a firm footing by the great nineteenth century German physicist Helmholtz (1877). He understood that the reed behaved as a pressure-controlled valve, modulating the flow of air from the player's mouth into the instrument. Helmholtz considered that there were two distinct types of reed mechanism, depending on the geometry of the reed structure relative to the direction of flow. He defined an 'inward-striking' (*einschlagende*) reed as one in which the motion of the reed blade in the direction of flow resulted in a decrease in the reed opening. An 'outward-striking' (*aussschlagende*) reed was defined as one in which the motion of the reed blade in the direction of flow resulted

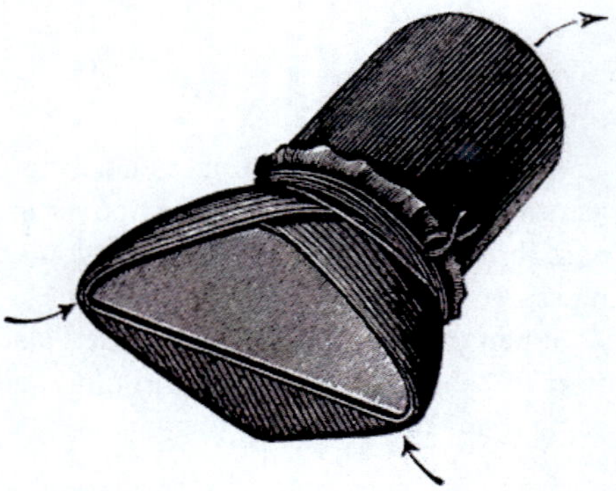

Fig. 3.23 Helmholtz's model of an inward-striking reed. Arrows indicate the direction of air flow. From Helmholtz (1877)

in an increase in the reed opening. This classification was also adopted by Bouasse (1929) and many subsequent researchers.

Figure 3.23 is a sketch of the rubber model which Helmholtz used to investigate the behaviour of an inward-striking reed. Air enters the model through the narrow slit on the left-hand side; as the angled flaps swing towards each other, the area of the entrance channel diminishes. This 'inward-swinging door' type of motion is found in the single reeds of the clarinet and saxophone and in the double reeds of the oboe and bassoon (Wilson and Beavers 1974; Fletcher 1993; Dalmont et al. 1995; Nederveen 1998a). Helmholtz noted that the rubber model reed could also be sounded by blowing from the other end; the flaps then swing outward in the direction of air flow, demonstrating the behaviour of an outward-striking reed. Helmholtz unambiguously identified the brass player's lips as 'membranous tongues which strike outwards', and the outward-swinging lips shown in Fig. 3.21 clearly fall into Helmholtz's outward-striking category.

Outward-striking and inward-striking reeds can also be distinguished by the way in which their behaviour depends on the pressure difference Δp across the reed. When Δp is slowly increased, an outward-striking reed tends to open, while an inward-striking reed tends to close. The specification of a slow pressure increase is necessary, since when an oscillating pressure difference is applied across the reed, the phase difference between pressure and reed opening depends on frequency. When an outward-striking reed is subjected to a pressure difference well above its natural resonance frequency, it closes as the pressure difference increases.

Equation 3.6 describes the dynamics of the 1DOF sliding door lip model. In considering the effect of a very slow change in the mouthpiece pressure $p(t)$, the first two terms on the left-hand side of this equation can be neglected, yielding the approximate equation:

$$\omega_l^2(h(t) - h_{\mathrm{eq}}) \simeq \frac{p(t)}{\mu}. \tag{3.13}$$

Inspection of this equation shows that the behaviour of the sliding door lip valve has the character of an inward-striking reed: a decrease in the outlet pressure $p(t)$, corresponding to an increase in Δp, results in an decrease in the lip opening height $h(t)$.

The small amplitude vibration behaviour of the outward-swinging door lip model is described by Eq. 3.12. For very slow pressure changes, this equation can be written as

$$\omega_l^2(h(t) - h_{\text{eq}}) \simeq -\frac{p(t)}{\mu}. \tag{3.14}$$

The negative sign of the term on the left-hand side of Eq. 3.14 confirms that the swinging door model has the character of an outward-striking reed: a decrease in the outlet pressure $p(t)$, corresponding to an increase in Δp, results in an increase in the lip opening height $h(t)$.

The swinging door model is not the only possible basis for an outward-striking pressure-controlled valve, although it is probably the most realistic picture of the behaviour of the brass player's lip. Figure 3.24 illustrates a notation introduced by Neville Fletcher which classifies reeds according to their response to upstream or downstream pressure changes. Figure 3.24a is a diagrammatic representation of an inward-striking reed: the notation $(-, +)$ indicates that the reed opens in response to a drop in upstream pressure (p_0) or a rise in downstream pressure (p). The outward-striking reed shown schematically in Fig. 3.24b is notated $(+, -)$ since a rise in upstream pressure or a drop in downstream pressure will open it. The symmetry of

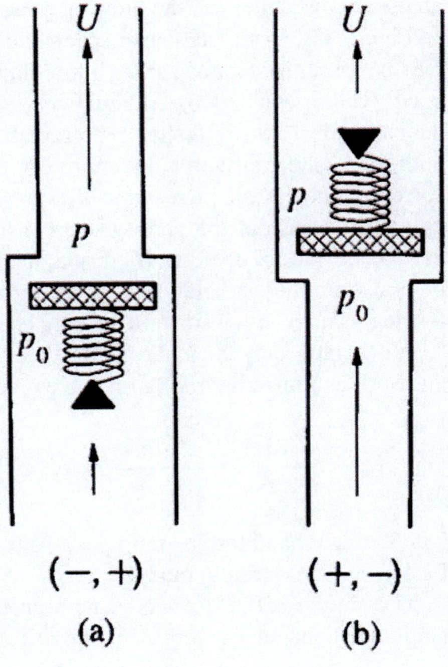

Fig. 3.24 Schematic diagrams of (**a**) an inward-striking reed and (**b**) an outward-striking reed. From Fletcher (1979)

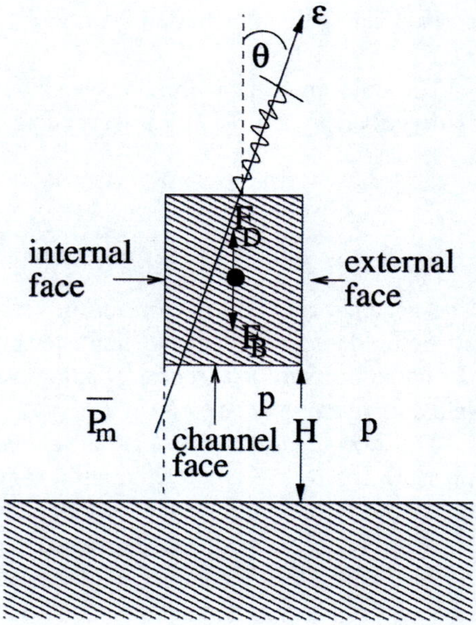

Fig. 3.25 A schematic diagram of a one-mass lip model exhibiting linear simple harmonic motion along transverse and axial directions. From Cullen et al. (2000)

these two diagrams explains why some designs of inward-striking reed behave like outward-striking reeds when the direction of flow is reversed. The free metal reeds in accordions and harmonicas can function as either (−, +) or (+, −) valves depending on the configuration and the blowing pressure (Fletcher and Rossing 1998).

Figure 3.25 is an alternative schematic representation of the lip valve, in which the lip is assumed to move along a straight line in the (x, z) plane at an angle θ to the y axis (Cullen et al. 2000). Although this is a more abstract picture than the swinging door model, it allows for the representation of linear simple harmonic motion in both axial and transverse directions. A decrease in the mouthpiece pressure p, corresponding to an increase in the pressure difference $\bar{P}_m - p$, drives the lip upwards because of the geometrical constraint; the corresponding force F_D tends to increase the lip opening H, displaying an outward-striking character. The drop in p also implies an additional downward force F_B on the lip surface; this is the Bernoulli force with its inward-striking character, tending to reduce the lip opening.

A generalised dynamical equation describing the small amplitude motion of both outward- and inward-striking lip valve models can be written as

$$\frac{d^2h(t)}{dt^2} + \frac{\omega_l}{Q_l}\frac{dh(t)}{dt} + \omega_r^2(h(t) - h_{eq}) = \pm\frac{p(t)}{\mu}, \tag{3.15}$$

with the right hand forcing term positive for the inward-striking model and negative for the outward-striking model.

The simple 1DOF lip model with an outward-striking character, whose small amplitude behaviour is represented by Eq. 3.12, corresponds to the scientific observations of low-frequency lip motion described in Sect. 3.1.5 and is in accord with the musician's experience that the lips bulge outward when buzzed. Nevertheless, experimental measurements of the mechanical response of artificial and human lips presented in Sect. 3.3 show evidence of resonances with both outward- and inward-striking behaviour. We therefore retain at this stage the more general Eq. 3.15, which can represent either outward- or inward-striking lip valves depending on the choice of sign in the forcing term.

3.3 The Mechanical Response of the Vibrating Lips

The equation of motion for the 1DOF lip model with no external force is found by setting the right-hand side of Eq. 3.3 to 0. There is a stationary solution to this equation for $h = h_{\mathrm{eq}}$ which represents the lips at rest with their equilibrium separation. When a sinusoidally varying external force $F(t) = A\cos\omega t$ is applied to the lips, they are driven into oscillation at the forcing frequency $f = \omega/2\pi$. The strength of the response depends on the relationship between the forcing frequency f and the lip resonance frequency $f_l = \omega_l/2\pi$. If the two frequencies are far apart, the response of the lips will be small, but when $f \simeq f_l$, the amplitude of the lip vibration can be many times larger.

The mechanical resonance behaviour of vibrating lips can be investigated experimentally by generating a small sinusoidally varying pressure difference $\Delta p(\omega)$ between the mouth and the mouthpiece and examining the resultant modulation of the lip opening height $h(\omega)$. The mechanical response H_{mr} at the angular frequency ω is defined as

$$H_{\mathrm{mr}}(\omega) = \frac{h(\omega)}{\Delta p(\omega)}. \tag{3.16}$$

If the lip valve did behave as a simple 1DOF oscillator, a plot of the mechanical response as a function of frequency would have a single peak at the natural resonance frequency of the lips. In reality the lips are complex structures with multiple resonances, and this complexity is revealed by experimentally measured mechanical response curves. It is nevertheless often possible to identify a peak corresponding to the mechanical resonance principally involved in the lip valve vibration, and analysis of the peak properties can provide information about the parameters of the corresponding one-mass model.

3.3.1 *Resonances of Artificial Lips*

An experimental arrangement for measuring the mechanical response of a pair of artificial lips is shown in Fig. 3.26. The artificial mouth illustrated in Fig. 3.15, which

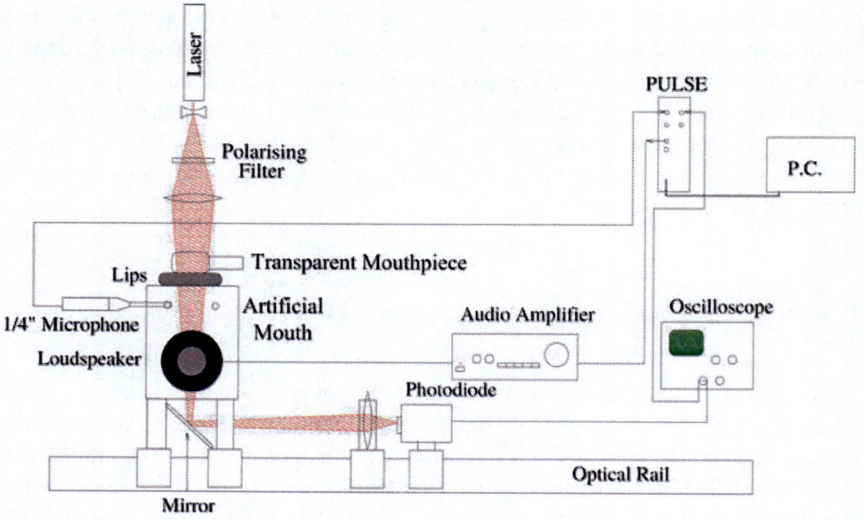

Fig. 3.26 Apparatus for measuring the mechanical response of artificial lips. From Richards et al. (2003)

included two cylindrical latex rubber lips filled with water, was mounted with the z axis vertical. The lips were pressed against the rim of the transparent trombone mouthpiece illustrated in Fig. 3.3. A loudspeaker driver mounted on the mouth was fed a computer-generated sine sweep signal, and the resulting acoustic pressure was measured by a microphone inside the mouth. A sinusoidal pressure difference was thus created between the interior of the mouth and the mouthpiece. The expanded light beam from a He-Ne laser illuminated the lip aperture from above. The light which passed through the aperture was refocused on to a photodiode which monitored the intensity of the beam.

An example of a typical measured mechanical response of the artificial lips is shown in Fig. 3.27. In this experiment the mouthpiece was not connected to a trombone, so that well below the mouthpiece resonance frequency (500 Hz), the mouthpiece pressure was close to atmospheric pressure. The magnitude of the microphone signal at angular frequency ω was therefore taken to represent the magnitude and phase of the pressure drop $\Delta p(\omega)$ across the lips at that frequency. Making the simplifying assumption that the lip opening was a rectangle of constant width, the magnitude of the alternating component of the photodiode signal was taken to be proportional to the magnitude of the lip opening height $h(\omega)$. The ratio of these quantities gave the magnitude of the mechanical response H_{mr} defined by Eq. 3.16, while the phase difference between the two signals yielded the phase of H_{mr}.

As expected, the mechanical response of these flexible, continuous structures is more complicated than the single resonant peak predicted by the 1DOF model. At

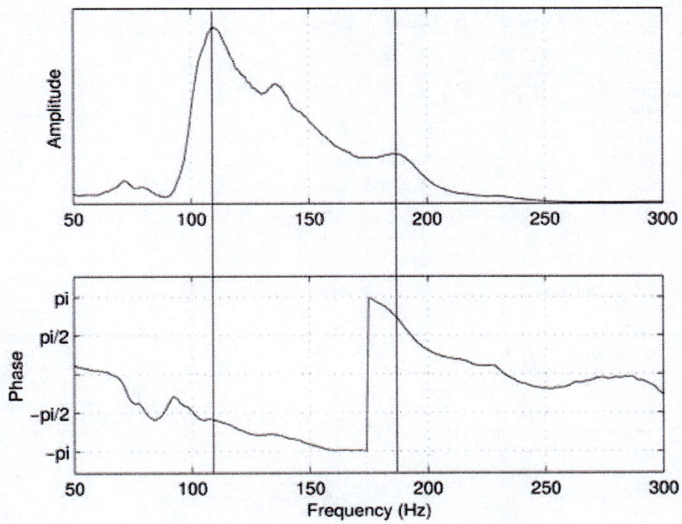

Fig. 3.27 Measured mechanical response of artificial lips. From Richards et al. (2003)

least three distinct peaks are evident in Fig. 3.27, each corresponding to a different mode of vibration of the lips. The peaks at 109 and 186 Hz, marked by vertical lines, have phase angles close to $-\pi/2$ and $+\pi/2$, while a third peak at 136 Hz has a phase angle intermediate between $-\pi/2$ and $-\pi$. The importance of these phase angles in explaining the operation of the valve effect source will be discussed further in Sect. 5.2.1 and Sect. 6.4.

3.3.2 Resonances of Human Lips

The mechanical response measurement technique illustrated in Fig. 3.26 could not be used for measurements on the lips of human players, since it requires the focusing optics and photodiode to be mounted behind the lips. An alternative technique using a high-speed video camera to record the lip motion of a brass player in synchronism with acoustical measurements of the pressure difference across the lips is illustrated in Fig. 3.28.

In order to ensure that the embouchures studied were capable of generating musically useful sounds, a special double-stemmed mouthpiece was constructed. In the normal playing configuration, shown in Fig. 3.29, the mouthpiece coupled the player to the instrument via the right-hand stem, with the left-hand stem remaining closed. Upon depression of the control valve, the right-hand stem was closed off and the left-hand stem opened, coupling the player to a cavity driven by the loudspeaker. This allowed the player to form an embouchure designed to play a specific note,

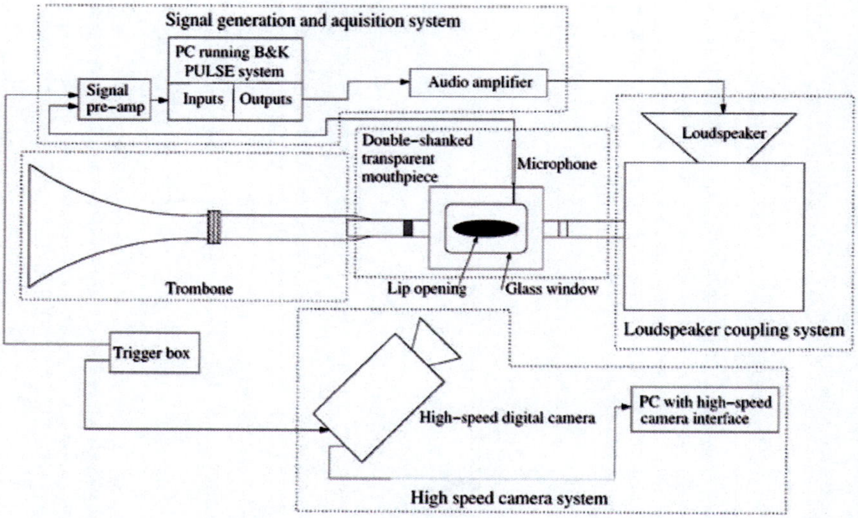

Fig. 3.28 Apparatus for measuring the mechanical response of a human trombonist's lips. Reproduced from Newton et al. (2008) with the permission of the Acoustical Society of America

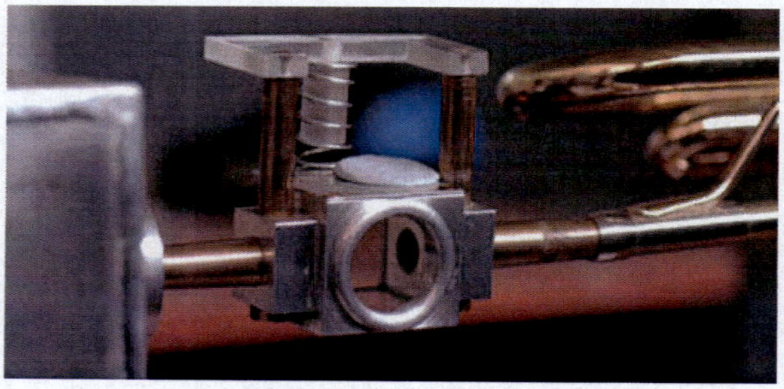

Fig. 3.29 Double-stemmed mouthpiece allowing a brass player to switch quickly between playing the instrument and connecting to the mechanical response excitation system. Reproduced from Newton et al. (2008) with the permission of the Acoustical Society of America

before subjecting the embouchure to forcing by the calibrated sine sweep. The front face of the rectangular mouthpiece shows the circular rim, with a diameter similar to that of a conventional tenor trombone mouthpiece, against which the player's lips were pressed. The opposite face was an optical quality glass window.

Figure 3.30 shows mechanical response curves for four embouchures formed by a trombonist to play the notes $B^\flat 1$, F2, $B^\flat 2$ and F3 on a tenor trombone (Newton et al. 2008). In each case there is one distinct resonance below the played frequency with a phase that is close to $-90°$ $(\pi/2)$. The peaks in the magnitude curves are much broader and flatter than those found in mechanical response curves of artificial lips, suggesting that the internal damping in the lip tissue is much stronger than in the water-filled latex lip.

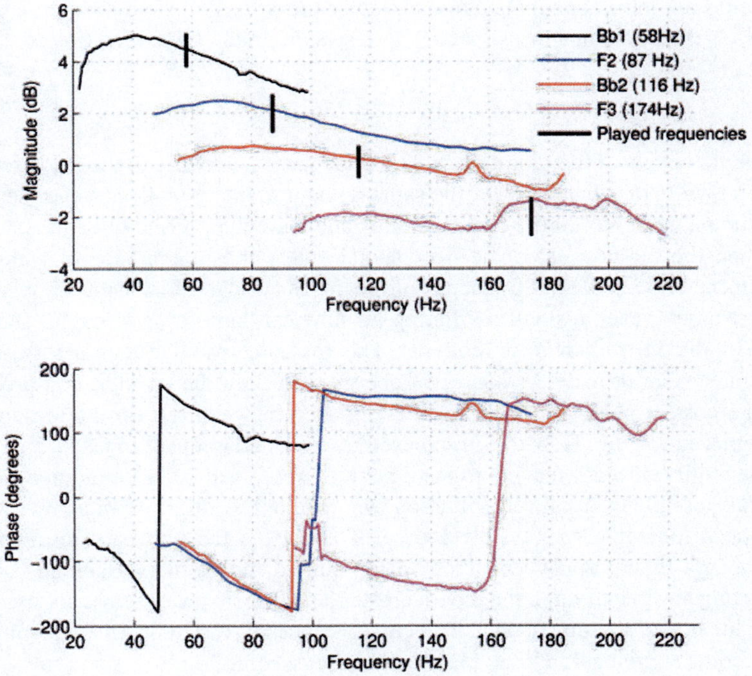

Fig. 3.30 A plot of four mechanical response curves obtained with human lips, using the video method. The played frequency for each note is marked by a vertical line. The magnitude curves have been offset by a few dB for display purposes. Reproduced from Newton et al. (2008) with the permission of the Acoustical Society of America

3.4 Why Do the Lips Buzz?

In Sect. 3.2 a model of the lips of a brass player as a 1DOF mechanical oscillator was presented. The equation of motion for this oscillator was given in Eq. 3.3 as

$$\frac{d^2h}{dt^2} + \frac{\omega_l}{q_l}\frac{dh}{dt} + \omega_l^2(h - h_{\text{eq}}) = \frac{F}{m}.$$

The second term on the left-hand side of this equation represents the damping of the oscillator, with the quality factor q_l determining the rate at which energy of motion is drained from the oscillator by dissipative processes. In Sect. 3.3 we have seen that the human lip is a strongly damped oscillator, which implies that a vibration started in the lips will die out very quickly unless the external force F is capable of supplying energy to the mass to compensate for the damping. When a player starts to buzz the lips, where does the energy come from to sustain the vibration?

A constant force cannot play this role, since the work W done by a force F acting in the $+y$ direction on a mass moving with speed v in the same direction is

$$W = Fv. \tag{3.17}$$

For the half cycle of the vibration in which the mass is moving in the $+y$ direction, the force will contribute energy; the same amount of energy will however be taken away in the other half of the cycle, when F and v are in opposite directions.

When the air flow from the lip channel enters the mouthpiece of a musical instrument, and the natural frequency of the lips is close to an acoustic resonance of the instrument, even a small oscillating air flow can build up a powerful standing wave in the instrument's air column. The resulting oscillating pressure in the mouthpiece can provide the source of energy to sustain the lip vibration provided that the correct phase relationship is established between the lip motion and the mouthpiece pressure. How this is achieved is explained in detail in Chap. 5.

The ability of brass players to buzz the lips freely without a mouthpiece shows that it is possible to sustain lip vibration even without feedback from a downstream acoustic resonator. One possible source of energy is the upstream resonance in the player's mouth and vocal tract. In the simple model, it is assumed that the fluctuating air flow through the mouth creates a negligible pressure oscillation at the upstream side of the lip channel. The circumstances in which upstream resonances can become significant in brass playing are discussed in Sect. 6.3.

There are several other possible processes by which energy could be supplied to sustain lip vibration without either downstream or upstream acoustic feedback, but these take us beyond the simple 1DOF lip model. One such mechanism is a periodic change in the profile of the lip channel. In Fig. 3.32 the lip channel is assumed to have a height which at any given time in the vibration cycle does not vary along the z axis (the flow direction). It is also assumed that the flow is uniform along the lip channel. A consequence of these assumptions is that if there is no downstream resonator, the lip channel pressure is also uniform and equal to atmospheric pressure. This will not change during the vibration cycle, and there will therefore be no fluctuating force on the lip which could supply energy to compensate for internal damping.

If, however, the lip channel has a converging profile, as shown in Fig. 3.31a, the pressure will decrease along the channel in the $+z$ direction, reaching atmospheric pressure only at the exit. There will then be a net upward force on the upper lip. If the channel remained convergent throughout the vibration cycle, the direction of the force would remain upward, and there would be no net transfer of energy to the lip. If however the channel became divergent as the lip started to move downward, as shown in Fig. 3.31b, the pressure would drop below atmospheric at the lip channel entrance, and there would be a net downward force in the same direction as the lip velocity. Energy would then be transferred to the lip in both halves of the cycle, giving the positive feedback necessary to sustain the vibration. It is not necessary that the switch from convergent to divergent should occur exactly at the change of

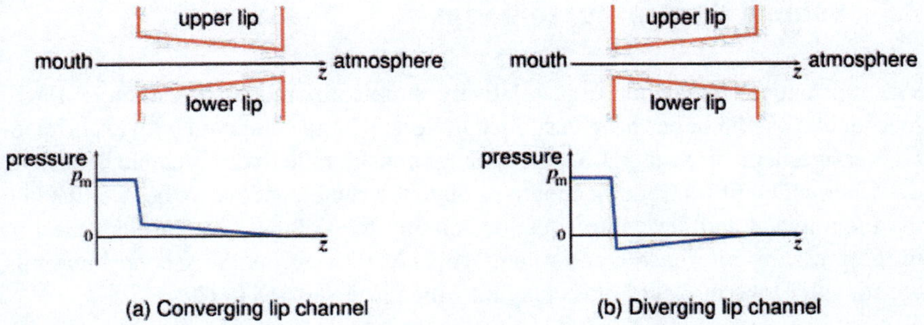

Fig. 3.31 Pressure distributions in converging (**a**) and diverging (**b**) lip channels

lip movement direction, as long as there is a net positive transfer of energy over the whole cycle.

A similar problem has been treated in great detail in modelling of vocal folds in phonation. Experimental studies have shown that the glottal channel separating the two vocal folds can change from a diverging to a converging profile, and this behaviour has been modelled either by a two-mass vocal fold model (Ishizaka and Flanagan 1972; Pelorson et al. 1994; Lous et al. 1999) or by a mucosal wave propagating along the vocal fold surface (Titze 1988). The aeroelastic coupling of two mechanical modes, known as the 'flutter effect' (Holmes 1977), has also been proposed as a model of the soft palate vibrations responsible for snoring (Auregan and Depollier 1995). It is also the commonly accepted oscillation mechanism for the vocal folds in voiced sound production. The energy transfer between the flow through the lip channel is the result of an asymmetry between the opening and closing phases of the lip movement.

This is analogous to the asymmetry between the opening and closing phases of the swimmer's breast stroke, resulting in a net propulsion. This process involves the oscillation of two mechanical structural modes in order to obtain different geometries during the opening and closing phases of the oscillation. With rigid arms and legs (one degree of freedom), we could not swim (Fabre et al. 2018).

Mechanical response measurements of artificial brass-playing lips (Cullen 2000; Cullen et al. 2000) have shown twin pairs of mechanical resonances with different phase behaviours (see Sect. 6.4). This lends support to the idea that lips can display self-sustained (autonomous) oscillations due to the coupling of two mechanical modes of vibration with the flow.

Flow separation, reattachment and vortex generation in the lip channel could perhaps also contribute to the positive feedback of energy to the vibrating lips (Hirschberg et al. 1995). In a diverging lip channel, the flow is likely to separate at a point well upstream from the exit; the lips could be destabilised by an aerodynamic force due to hysteresis in the flow separation (Hirschberg et al. 1990). Such subtleties can also be crucial in explaining self-sustained oscillation in reed instruments without major acoustic feedback such as accordions, harmonicas (Ricot et al. 2005) and lingual organ pipes without resonators (Miklos et al. 2003).

3.5 Volume Flow in Buzzing Lips

The mechanical behaviour of a relatively simple lip model based on a 1DOF mechanical oscillator has been described in Sect. 3.2 and compared with experimental measurements in Sect. 3.3. In Sect. 3.5 we consider the hydrodynamic behaviour of the air which flows from the mouth through the channel between the two lips into the mouthpiece and derive an equation relating the volume flow through the lips to the pressure difference across them. This equation forms the second important constituent of the model of brass playing which is developed in Chap. 5.

3.5.1 Acoustic Volume Flow Through the Lip Aperture

Figures 3.32 and 3.33 are schematic illustrations of the situation in which a steady flow of air from the player's lungs is passing through the mouth with speed $v_{\mathrm{m}}\,\mathrm{m\,s^{-1}}$. The pressure in the mouth is p_{m} pascals above atmospheric pressure. The lips are assumed to be stationary, with an opening between them in the form of a channel of fixed length l, width w and height h through which the air flows from the mouth into the mouthpiece. For simplicity the flow from the mouth into the lip channel is assumed to be incompressible and frictionless, following streamlines which are indicated schematically in Fig. 3.32.

The fundamental physical law of mass conservation requires that the mass of air leaving the mouth in 1 second must be equal to the mass of air flowing into the lip channel in the same time interval. The assumption of incompressibility is equivalent to considering that the air density is the same everywhere in the flow, so that the volume flow rate is also constant across any cross-section of the flow. If the cross-sectional area of the mouth cavity behind the lips is S_{m} and the mean

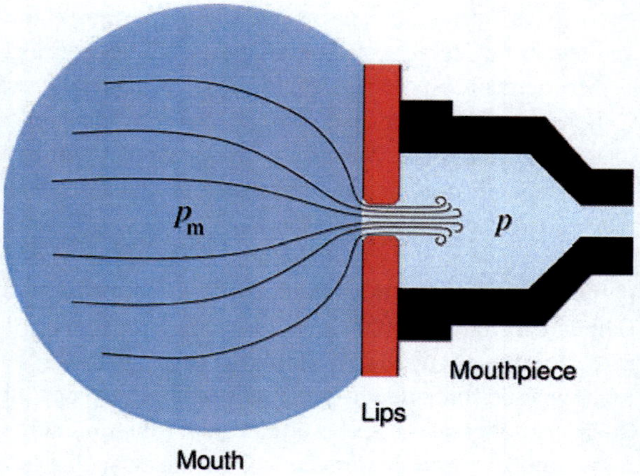

Fig. 3.32 Schematic view of the flow from the mouth through the lip channel into the mouthpiece of a brass instrument. The lips are shown in red and the air in blue. Air pressure amplitudes are indicated by the depth of colour. Black lines indicate streamlines (Color figure online)

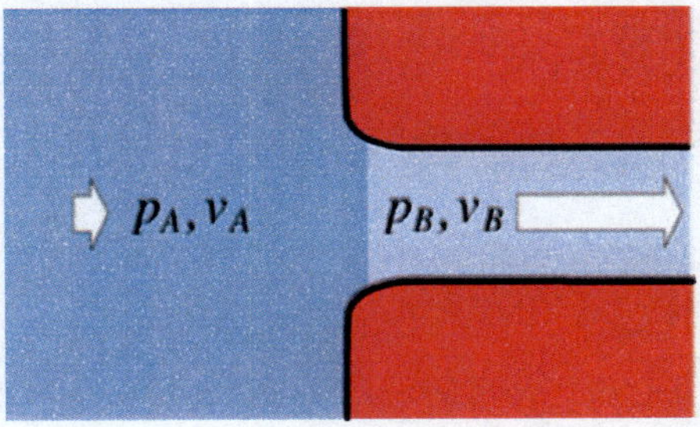

Fig. 3.33 Schematic view of the region around the entrance to the lip channel. The lips are shown in red and the air in blue. Air pressure amplitudes are indicated by the depth of colour and air particle velocities by the arrow lengths (Color figure online)

velocity of the air across this surface is $v_{\rm m}$, the volume flow rate through the mouth is $u = S_{\rm m}\, v_{\rm m}$. As the flow enters the lip channel, whose area $S_{\rm lc} = wh$ is many times smaller than $S_{\rm m}$, it accelerates to a new mean velocity $v_{\rm lc}$ which maintains the constancy of the volume flow rate:

$$u = S_{\rm m}\, v_{\rm m} = S_{\rm lc}\, v_{\rm lc}. \tag{3.18}$$

The assumption that the flow is frictionless implies that the sum of kinetic and potential energies is constant for a volume element of fixed mass travelling in the flow. This conservation law is often expressed in the form known as the Bernoulli equation, which relates the pressure p and particle velocity v at two different points A and B on the same streamline:

$$p_{\rm A} + \frac{1}{2}\rho v_{\rm A}^2 = p_{\rm B} + \frac{1}{2}\rho v_{\rm B}^2, \tag{3.19}$$

where ρ is the density of air.

Equations 3.18 and 3.19 can be combined to explain the relationship between the pressure in the mouth of a brass player and the pressure field which exists in the channel between the lips. Equation 3.18 shows that the ratio of mean upstream to downstream velocities is equal to the ratio of downstream to upstream cross-sectional areas:

$$\frac{v_{\rm m}}{v_{\rm lc}} = \frac{S_{\rm lc}}{S_{\rm m}}. \tag{3.20}$$

Equation 3.19 shows that as the velocity of the air entering the lip channel increases, there must be a decrease in the pressure exerted by the air on the lip surfaces which form the channel walls. The pressure drop between the mouth and the lip channel is given by

$$p_{\rm m} - p_{\rm lc} = \frac{1}{2}\rho(v_{\rm lc}^2 - v_{\rm m}^2) \tag{3.21}$$

$$= \frac{1}{2}\rho v_{\mathrm{lc}}^2 \left(1 - \frac{v_{\mathrm{m}}^2}{v_{\mathrm{lc}}^2}\right) \tag{3.22}$$

$$= \frac{1}{2}\rho v_{\mathrm{lc}}^2 \left(1 - \frac{S_{\mathrm{lc}}^2}{S_{\mathrm{m}}^2}\right). \tag{3.23}$$

The open area S_{m} of the mouth just behind the lips depends on the position of the tongue. There is evidence that some players deliberately constrict this area to assist high register playing (see Sect. 6.3). For low notes Farkas (1962) recommends a low tongue position corresponding to an 'oh' vowel, corresponding to an area $S_{\mathrm{m}} \simeq 500\,\mathrm{mm}^2$ (Story et al. 1996). The lip opening area S_{lc} in trombone playing is typically less than $50\,\mathrm{mm}^2$ (see Fig. 3.7); in this case $S_{\mathrm{lc}}^2/S_{\mathrm{m}}^2 \ll 1$, and

$$p_{\mathrm{m}} - p_{\mathrm{lc}} \simeq \frac{1}{2}\rho v_{\mathrm{lc}}^2. \tag{3.24}$$

.

3.5.2 *Acoustic Volume Flow Equation*

The left-hand side of Fig. 3.32 illustrates the mouth air flow pattern described in Sect. 3.5.1. As the streamlines converge into the narrow lip channel, the air particle velocity increases, and the pressure drops. If the flow emerging from the lip channel into the mouthpiece followed the same pattern, the streamlines would diverge smoothly from the channel exit, resulting in an increase of pressure and a decrease in the particle velocity. The final mouthpiece pressure p and particle velocity v would depend on the cross-sectional area of the mouthpiece cup, but not on the dimensions of the lip channel. If the lips started to oscillate, the changing cross-sectional area of the lip channel would have no effect on the downstream pressure and flow unless the amplitude of oscillation were enough to completely close the channel.

The flow control which makes the oscillating lips an effective valve effect source relies on subtle fluid dynamic effects localised in the viscothermal boundary layers close to the walls of the lip channel. Even when air is flowing through the lip channel with a speed greater than $10\,\mathrm{m\,s^{-1}}$, the air in contact with the lip surface remains at rest; there is a rapid velocity gradient across the boundary layer, which is typically much less than a millimetre thick. At a certain point along the lip channel, whose location depends on the channel profile and the air speed, instabilities in the boundary layer cause the flow to separate from the wall and become a free jet. The flow separation point occurs earlier in a diverging channel. Since the profile of the lip channel changes during the course of one cycle of lip vibration, the separation point is also likely to vary during the vibration cycle. There may also be a secondary separation point at the upstream entrance to the lip channel, giving rise to a reduction in the effective height of the lip channel known as the vena contracta (Hirschberg et al. 1996a).

In the simplest model, the separation point is fixed at the downstream edge for any values of the parameters. Such a jet flow is very unstable and becomes turbulent, as illustrated schematically in Sect. 3.32. The turbulence implies a rapid mixing of the jet flow with the surrounding stagnant air, leading to a deceleration of the jet and a transfer of momentum. Some distance downstream of the lip channel exit, the air particle velocity can be considered approximately constant across the mouthpiece cross-section. Since the conservation of volume flow is still applicable in this case, the volume flow rate in the mouthpiece is equal to that in the lip channel. However the dissipative forces acting during the turbulent mixing render the Bernoulli equation invalid in the mouthpiece cup, and the reduction in particle velocity is not accompanied by a recovery of the pressure. The kinetic energy given up by the decelerating air particles is not stored as potential energy, but converted through turbulence into heat.

The absence of pressure recovery in the mouthpiece means that the pressure difference Δp between the mouth pressure p_{m} and the mouthpiece pressure p is given by

$$\Delta p = p_{\mathrm{m}} - p = p_{\mathrm{m}} - p_{\mathrm{lc}}. \tag{3.25}$$

The acoustic volume flow rate $u(t)$ into the mouthpiece can then be derived using Eqs. 3.18 and 3.24:

$$\begin{aligned} u(t) &= S_{\mathrm{lc}}(t)\, v_{\mathrm{lc}}(t) \\ &= S_{\mathrm{lc}}(t)\sqrt{\frac{2\Delta p(t)}{\rho}} && (3.26) \\ &= S_{\mathrm{lc}}(t)\sqrt{\frac{2(p_{\mathrm{m}} - p(t))}{\rho}}. && (3.27) \end{aligned}$$

Based on the simplified model of the brass player's mouth and lips described in Sect. 3.5.1, Eq. 3.27 predicts that for a fixed mouth pressure p_{m}, the rate of flow of air into the mouthpiece depends on two variables: the cross-sectional area of the lip channel S_{lc} and the mouthpiece pressure p. When a player buzzes the lips without a mouthpiece, the air just in front of the lip channel exit is at atmospheric pressure, corresponding to $p = 0$. The flow rate equation then simplifies to

$$u(t)_{\mathrm{freebuzz}} = S_{\mathrm{lc}}(t)\sqrt{\frac{2p_{\mathrm{m}}}{\rho}}. \tag{3.28}$$

The flow rate is directly proportional to the cross-sectional area of the lip channel. When the lips vibrate in such a way that the area varies sinusoidally, the flow leaving the lips will also be modulated sinusoidally. If the amplitude of lip vibration is high enough that the lip channel closes for part of the vibration cycle, the flow leaving the lips will not be sinusoidal but will have additional harmonic components.

From Eq. 3.28, the flow rate from a lip buzz into open air is also proportional to the square root of the mouth pressure. As a quantitative example, if the player buzzes with a mouth pressure $p_{\mathrm{m}} = 1\,\mathrm{kPa}$, and the lip open area varies sinusoidally between 0 and $20\,\mathrm{mm}^2$, Eq. 3.28 predicts that the flow rate leaving the lips will also vary sinusoidally between 0 and $0.8\,\mathrm{L\,s}^{-1}$. The maximum particle velocity of the air at the exit to the lip channel will be just over $40\,\mathrm{m\,s}^{-1}$.

When the fluctuating air flow emerges from the lip channel into the mouthpiece of a brass instrument, the pressure p in Eq. 3.27 is no longer equal to 0, but instead fluctuates with an amplitude and phase which depend on the acoustical properties of the instrument. In Chap. 4 these properties are described in detail. The construction of an elementary global brass instrument model which includes the feedback loop which couples the buzzing lips to the acoustic resonances of the instrument is presented in Chap. 5.

Chapter 4
After the Lips: Acoustic Resonances and Radiation

In almost all instruments of the brass family, the fluctuating air stream from the buzzing lips enters a tube whose length is much greater than its largest diameter. The form and scale of the tube determine the resonant properties of the air column contained within the instrument and the nature of the sound radiated from it. In Sect. 4.1 the relationship between standing waves and acoustic resonances is explained. The distinction is made between time domain and frequency domain descriptions of the acoustic processes in brass instruments, and the related concepts of impulse response and input impedance are introduced. Experimental techniques for measuring the input impedance of a brass instrument are described in Sect. 4.2, and approximate methods for calculating this important quantity are discussed in Sect. 4.7. Section 4.3 includes a review of the different types of bore profile which are found in brass instruments, and a discussion of the influence of the mouthpiece, mouthpipe, cylindrical and conical sections of tubing and flaring bell on intonation and timbre. Section 4.4 provides a brief introduction to tonehole labrosones, and mutes for brass instruments are described in Sect. 4.5. The nature of sound radiation from brass instruments is reviewed in Sect. 4.6.

4.1 Internal Sounds in Brass Instruments

The sounding pitches available on a brass instrument are strongly influenced by the way in which sound waves travel through the air inside the instrument. Any hollow object with at least one hole in its wall has the property that a sound wave with the right frequency can elicit a strong response in the enclosed air. This is the characteristic behaviour of an acoustic resonator: the frequency of maximum response is called the resonance frequency, and the pattern of pressure variation at resonance is described as an acoustic mode. Almost all members of the brass

M. Campbell et al., *The Science of Brass Instruments*, Modern Acoustics and Signal Processing, https://doi.org/10.1007/978-3-030-55686-0_4

instrument family have resonators with many acoustic modes, each with its own characteristic resonance frequency.

4.1.1 Lumped and Distributed Resonators

One of the simplest examples of an acoustic resonator is the empty beer bottle shown in Fig. 4.1a. The volume of the bottle and the size of the neck determine the characteristic resonance frequency. An air-filled cavity behaving like this is known technically as a Helmholtz resonator, and we will return in Sect. 4.3.5 to discuss the properties of Helmholtz resonators in connection with the behaviour of brass instrument mouthpieces.

The resonance frequency of the bottle can be measured by generating a sinusoidal pressure signal near the neck and recording the response with an internal microphone. It can also be estimated by blowing a jet of air across the neck, since the nonlinear coupling of the air jet and the internal cavity results in a strong oscillation at a frequency close to that of the acoustic resonance. For the beer bottle shown in Fig. 4.1a, the pitch of this sound was very close to G3. The effect of the resonance on the response of the bottle to external sounds can be experienced by performing the *Song for a beer bottle in G* (Fig. 4.1b) close to the neck, using the vowel sound 'oo' (as in 'moon') to imitate a sine wave. The syllables sung to the note G3 should sound louder, and decay more slowly, than those at the other pitches. A similar demonstration can be made with other shapes and sizes of bottle if the song is appropriately transposed.

Slapping a hand on the neck of the bottle forces extra air into it, resulting in a sudden increase in the pressure in the neck. This pressure pulse travels through the

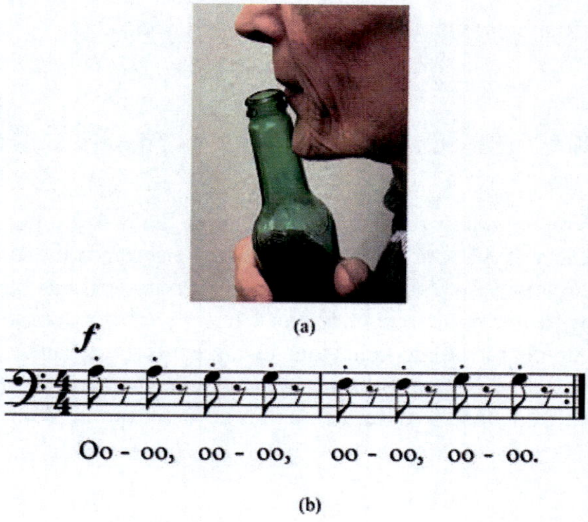

Fig. 4.1 (**a**) The Helmholtz resonant mode of the beer bottle can be excited by singing near the neck, or by blowing air across the neck. (**b**) *Song for a beer bottle in G*, illustrating the effect of the resonant mode at pitch G3

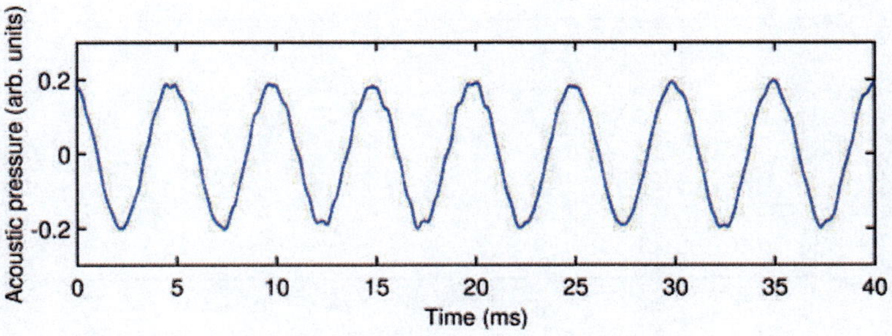

Fig. 4.2 Waveform of the sound generated by blowing across the neck of the bottle

air in the main volume of the bottle at the speed of sound c, which is 345 ms^{-1} at a temperature of 22 °C. In the small bottle shown in Fig. 4.1a, whose overall height was 22 cm, the pulse reaches the bottom in 0.64 ms. The waveform of the sound made by blowing across the bottle top is shown in Fig. 4.2. The resonant frequency of this bottle is 200 Hz, so one period of the oscillation is 5 ms. Since pressure changes propagate throughout the air volume in a time of the order of a tenth of the period, it is reasonable to assume that the pressure at a given stage in the oscillation is approximately the same everywhere in the volume. A hollow vessel with this behaviour is described as a lumped resonator.

It is also helpful to compare the largest linear dimension of the resonating volume L_{max} with the wavelength λ of the sound wave of interest. The system behaves like a lumped resonator if $\lambda \gg L_{\text{max}}$. Using the relationship $\lambda = c/f$, we find that a frequency of 200 Hz corresponds to a wavelength of 1.7 m, confirming that a bottle with maximum length 22 cm can be treated as a lumped resonator at this frequency.

The situation is different in the long thin tubes characteristic of most brass instruments. A tenor trombone, for example, has a total tube length from mouthpiece rim to bell of 2.8 m, and apart from the final rapid flare in the bell, the radius of the tube is less than 1 cm. Slapping a hand against the mouthpiece sends a pressure pulse travelling down the air column inside the tube. If the pulse travels at the speed of sound in open air, it will arrive at the bell after a time $T = (2.8/345) \times 1000 =$ 8.1 ms. This time interval is close to the period of the note $\text{B}^\flat 2$, which is 8.6 ms. It is not a coincidence that this is one of the playable notes on the instrument, but the factors determining the playable pitches on the trombone and other brass instruments will be discussed fully in Sect. 4.3. The important point to note here is that it is clearly not the case that the pressures at different points along the trombone tube rise and fall simultaneously when a note is being played on the instrument: the time taken for pressure changes to travel the length of the air column is comparable to the period of the oscillation.

The frequency of the note $\text{B}^\flat 2$ is 116.5 Hz, and the wavelength of a sound wave at this frequency is 2.96 m. The criterion $\lambda \gg L_{\text{max}}$ is clearly not satisfied for a 2.8-m-long trombone tube, which cannot therefore be considered as a lumped resonator. Instead it is treated as a distributed resonator or acoustic waveguide, the latter term

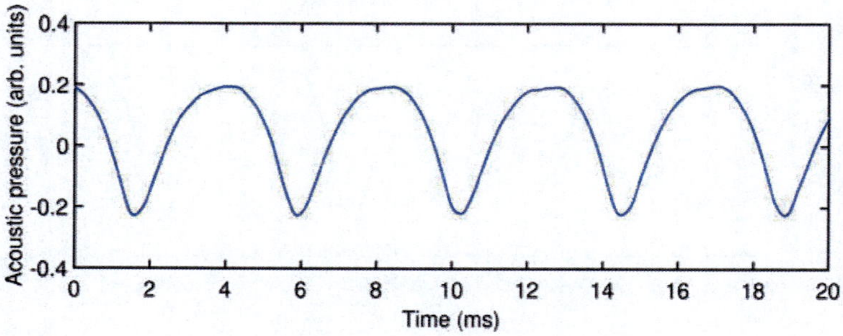

Fig. 4.3 Pressure signal in the mouthpiece of a trombone playing the note $B^{\flat}4$ piano

hinting that its acoustic behaviour can be fruitfully discussed in terms of travelling and standing waves.

4.1.2 Travelling Waves

Figure 4.3 shows the waveform of the pressure recorded in the mouthpiece of a trombone when the note $B^{\flat}4$ was played quietly by one of the authors.

The mouthpiece waveform is a smoothly varying periodic curve. It bears some resemblance to a sinusoid with a frequency of 466 Hz, but the broadened peaks and narrowed dips signify that there are several additional harmonic components in the mouthpiece pressure signal. Ignoring for the present these additional frequency components, we will consider the simple case of a pure sine wave generated in the mouthpiece. To further simplify the discussion, we will replace the trombone by a tube of constant radius $a = 5\,\text{mm}$ and length $L = 2.77\,\text{m}$, on which it is also possible to sound the note $B^{\flat}4$ using a trombone mouthpiece. The variable x represents distance along the axis of the tube. Let us follow one crest of the sine wave as it travels down the tube from the input ($x = 0$) and consider what happens to it at the open end ($x = L$).

We start the clock ($t = 0\,\text{s}$) when the crest we are following (marked by an arrow) leaves the entrance. As the curve in Fig. 4.4a shows, a succession of crests and troughs of pressure is already travelling down the tube ahead of the marked crest. The curve in Fig. 4.4b shows the situation after the marked crest has travelled one wavelength ($\lambda = 0.74\,\text{m}$) from the mouthpiece. It has taken one period ($T = 2.15\,\text{ms}$) for this journey, and another crest is just emerging from the mouthpiece. The curve in Fig. 4.4c illustrates the situation at $t = 8.05\,\text{ms}$, when the marked crest has just arrived at the open end. Since the tube length is equal to 3.75 wavelengths, the pressure at the mouthpiece is passing through zero at this moment.

The propagation of any sound wave whose pressure amplitude is small compared with atmospheric pressure is described by the linear acoustic wave equation (Pierce

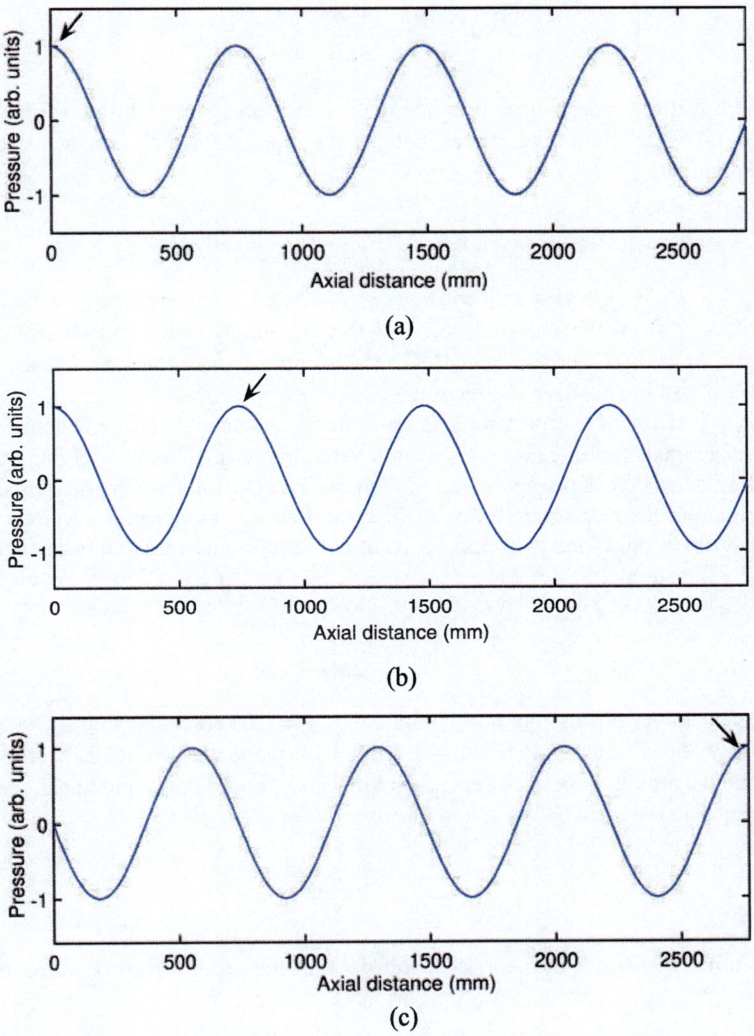

Fig. 4.4 Sinusoidal sound wave travelling down a cylindrical tube. (**a**) Time $t = 0\,\text{ms}$. (**b**) Time $t = 2.15\,\text{ms}$. (**c**) Time $t = 8.05\,\text{ms}$

1989). At this stage we neglect energy losses due to the viscosity of air (see Sect. 4.7). The linear acoustic wave equation then takes the form

$$\Delta p = \frac{1}{c^2}\frac{\partial^2 p}{\partial t^2}, \tag{4.1}$$

where Δ is the Laplacian operator. In a three-dimensional Cartesian coordinate system x, y, z, Eq. 4.1 can be written as

$$\frac{\partial^2 p}{\partial x^2} + \frac{\partial^2 p}{\partial y^2} + \frac{\partial^2 p}{\partial z^2} = \frac{1}{c^2}\frac{\partial^2 p}{\partial t^2}. \tag{4.2}$$

The travelling wave whose pressure dependence on space and time is illustrated in Fig. 4.4 is a solution of the acoustic wave equation which can be described mathematically as

$$p_+ = A\mathrm{e}^{j(\omega t - kx)}. \tag{4.3}$$

In Eq. 4.3 p_+ is the instantaneous value at time t of the acoustic pressure at a distance x from the mouthpiece, A is the maximum value (amplitude) of the pressure, $\omega = 2\pi f$ and $k = 2\pi/\lambda$. The subscript on p_+ indicates that the wave is travelling in the positive x direction.

The wavefront of a wave propagating in three-dimensional space is defined as a surface of constant phase. The only spatial variable in Eq. 4.3 is x, implying that the wavefronts are planes perpendicular to the tube axis. A full mathematical treatment (Chaigne and Kergomard 2016, p. 352) shows that this assumption of plane wave propagation is valid for a cylindrical tube of constant radius a up to a frequency f_{lim} given by the equation

$$f_{\mathrm{lim}} = \frac{1.84c}{2\pi a}. \tag{4.4}$$

In the case under discussion, $a = 5$ mm, so $f_{\mathrm{lim}} = 20$ kHz and a sound wave with frequency 466 Hz can safely be considered to have plane wavefronts inside the tube.

Since the pressure does not depend on the y and z coordinates, the linear acoustic wave Equation 4.1 can be written in the simpler form:

$$\frac{\partial^2 p}{\partial x^2} = \frac{1}{c^2}\frac{\partial^2 p}{\partial t^2}. \tag{4.5}$$

Substitution of Eq. 4.3 into Eq. 4.5 confirms that the speed of propagation of the wave is

$$c = \omega/k = f\lambda. \tag{4.6}$$

The complex exponential notation used in Eq. 4.3 is a convention which frequently simplifies the discussion of wave motion. Making use of the mathematical identity

$$\mathrm{e}^{j\theta} = \cos\theta + j\sin\theta, \tag{4.7}$$

Equation 4.3 can be rewritten as

$$p_+ = A\cos(\omega t - kx) + jA\sin(\omega t - kx). \tag{4.8}$$

If A is a real number, the first term on the right-hand side of Eq. 4.8 is purely real, and the second term is purely imaginary. Only the real part represents the physically measurable value of the variable. The pressure in the forward-going wave, which is of course a real quantity, is

$$p_+ = A\cos(\omega t - kx). \tag{4.9}$$

At time $t = 0$ the dependence of the acoustic pressure on distance along the tube is, from Eq. 4.9,

$$p_+ = A\cos(-kx) = A\cos(kx). \tag{4.10}$$

Comparison of Eq. 4.10 and Fig. 4.4a shows that Eq. 4.3 does indeed represent the illustrated travelling wave, with $A = 1$.

An expression for the acoustic particle velocity v_+ in the wave can be obtained from Eq. 4.3 by making use of the Euler equation :

$$\rho\frac{\partial v}{\partial t} = -\nabla p, \tag{4.11}$$

where ρ is the density of air. This equation is essentially Newton's second law of motion applied to a fluid particle: the left-hand side is proportional to the mass of the particle times its acceleration, while the right-hand side is proportional to the net external force. In general particle velocity and pressure gradient are vector quantities, but in the present case, both vectors are directed along the x axis, and Eq. 4.11 can be rewritten as

$$\rho\frac{\partial v}{\partial t} = -\frac{\partial p}{\partial x}. \tag{4.12}$$

The spatial pressure gradient can be found by differentiating Eq. 4.3 with respect to x. Substituting this into Eq. 4.12 and integrating both sides of the resulting equation with respect to t yields an expression for the particle velocity in the forward-going wave:

$$v_+ = \frac{A}{\rho c}\mathrm{e}^{j(\omega t - kx)} = \frac{1}{\rho c}p_+. \tag{4.13}$$

The ratio $p/v = \rho c$ is called the specific acoustic impedance of the wave. The fact that the specific acoustic impedance is real means that there is no phase difference between the plane travelling pressure wave and the associated particle velocity wave.

The amplitude A in Eq. 4.9 is written as a constant. In reality the amplitude of the wave decreases as it travels down the tube because some sound energy is lost through friction between the air and the wall and transfer of thermal energy to the wall. For the time being, we will ignore these viscothermal losses.

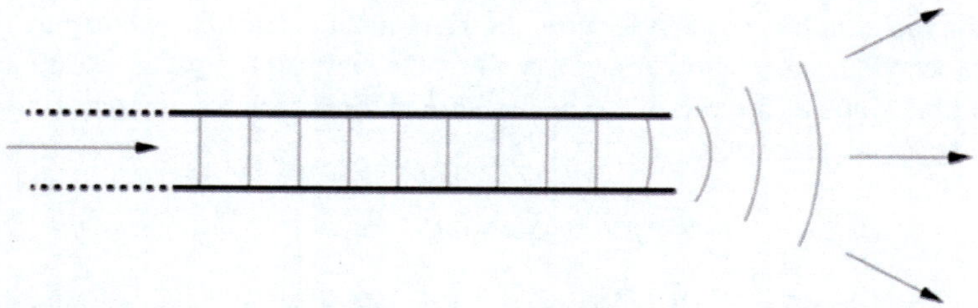

Fig. 4.5 Sound radiation from the end of a cylindrical tube

Equation 4.4 relates the behaviour of the sound wave to the boundary condition imposed by the solid wall of the confining cylinder. This boundary condition changes suddenly at the open end of the tube: the wave is now free to propagate in any direction in an effectively infinite space. Mathematical solution of the wave equation shows that outside the tube the wavefronts are no longer plane, but are close to being spherical, as illustrated in Fig. 4.5.

The nature of the sound waves radiated from different types of tube termination will be discussed in detail in Sect. 4.6. Here we are interested in the effect of the opening on the internal sound in the tube. An abrupt change in boundary conditions is usually accompanied by a strong reflection of the wave. As discussed in Sect. 2.2.5, most of the sound energy arriving at the open end is not radiated in the spherical wave, but reflected back up the tube as a plane wave p_-, described mathematically as

$$p_- = Be^{j(\omega t+kx)}. \tag{4.14}$$

The forward and backward travelling waves exist simultaneously in the tube, and at any give time t and place x, the total acoustic pressure is simply the sum of the pressures in the two waves:

$$p(x,t) = p_+(x,t) + p_-(x,t) = Ae^{j(\omega t-kx)} + Be^{j(\omega t+kx)}. \tag{4.15}$$

We chose to start the clock measuring t in Fig. 4.4 when a crest of the pressure wave was passing the point $x = 0$. This choice determined that the amplitude A of the forward-going wave was a real positive number. The amplitude of the backward going wave is determined by the boundary condition at $x = L$. Just outside the open end, the acoustic pressure is very close to zero, since any rise or fall is immediately compensated by air flow from the surrounding atmosphere. As a first approximation, we assume that $p(L,t) = 0$ for all values of t. Equation 4.15 shows that this is true if

$$Be^{jkL} = -Ae^{-jkL}$$

or

$$B = -Ae^{-2jkL} = -A(\cos 2kL - j \sin 2kL). \tag{4.16}$$

When a wave of unit pressure amplitude and zero phase travels down the tube, the ratio

$$R(\omega) = B/A \tag{4.17}$$

is a complex number describing the magnitude and phase of the reflected wave which returns to the input. This ratio is called the reflection coefficient of the tube. $R(\omega)$ is written as a function of angular frequency because for most types of tube, including the flaring tubes characteristic of brass instruments, the magnitude and phase of the reflected wave depend on the frequency of the wave. For the open-ended cylindrical tube described above, however,

$$R(\omega) = -e^{-2jkL}, \tag{4.18}$$

and its magnitude is independent of frequency. The use of the reflection coefficient in wind instrument modelling is discussed further in Sect. 4.1.5.

It is evident from Eq. 4.16 that B is in general a complex number. It is purely real if $2kL = n\pi$, where n is an integer, since in these cases, $\sin 2kL = 0$. If n is an even integer, $\cos 2kL = 1$, and $B = -A$. This means that at $x = 0$, the total pressure is

$$p(0,t) = p_+(0,t) + p_-(0,t) = (A + B)e^{j\omega t} = 0. \tag{4.19}$$

If n is an odd integer, $\cos 2kL = -1$, and $B = A$. The total pressure at $x = 0$ is then

$$p(0,t) = (A + B)e^{j\omega t} = 2Ae^{j\omega t}. \tag{4.20}$$

For the tube illustrated in Fig. 4.4, $2kL = 15\pi$, so the total pressure at the input of the tube is described by Eq. 4.20.

4.1.3 Standing Waves

The addition of two travelling waves going in opposite directions gives rise to a different type of wave, described as a standing wave. The reason for this description is clear when we consider the behaviour of crests and troughs in the standing wave setup by the forward and backward travelling waves discussed in Sect. 4.1.2. Figure 4.6 illustrates how the waves add together at four different stages in one cycle of oscillation. Forward travelling waves are shown by dashed red lines, backward travelling waves by dotted blue lines, and the standing wave sums by solid green lines.

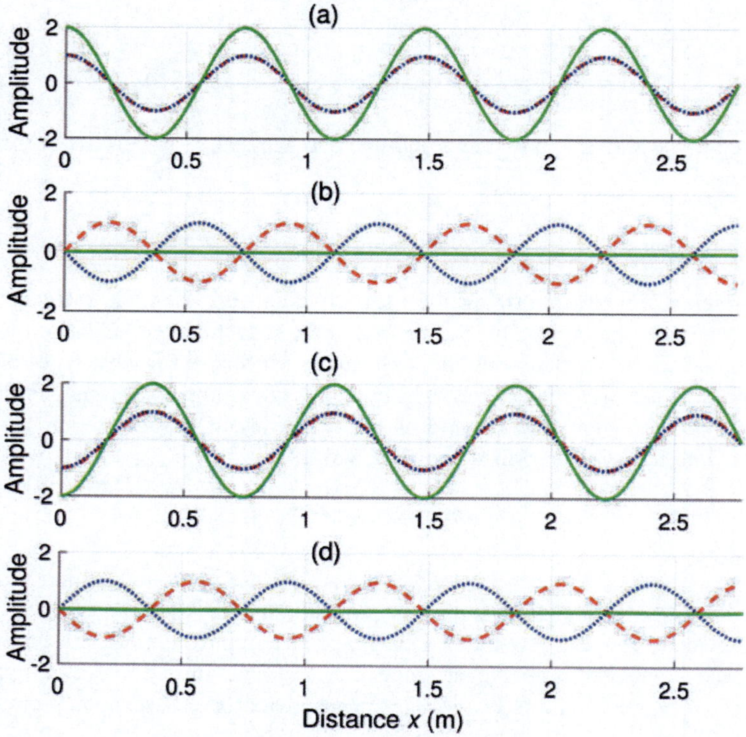

Fig. 4.6 Travelling and standing waves in the cylindrical tube. Red dashed line, forward-going wave p_+; blue dotted line, backward going wave p_-; green solid line, standing wave $p = p_+ + p_-$. **(a)** $t = 0$. **(b)** $t = T/4$. **(c)** $t = T/2$. **(d)** $t = 3T/4$ (Color figure online)

In Fig. 4.6a, at time $t = 0$, a crest of the forward wave (amplitude 1 Pa) is just leaving the tube entrance, while a crest of the returning wave with the same amplitude is arriving at the entrance. The two pressures add together to give a total pressure of 2 Pa. Half a wavelength along the tube, at $x = 0.37$ m, the two troughs add together to give a pressure of -2 Pa.

A quarter of the oscillation period later the two waves add to give a very different result, as shown in Fig. 4.6b. The crest of the forward-going wave has moved a quarter wavelength to the right; it now coincides with a trough of the returning wave, which has moved a quarter wavelength to the left. The two waves cancel to give zero pressure, not only at this point but everywhere along the tube.

The situation after a further quarter period is shown in Fig. 4.6c. The waves have moved a further quarter wavelength in opposite directions, and now crests and troughs coincide to give a maximum negative pressure of -2 Pa at the entrance.

For $t = 3T/4$, as shown in Fig. 4.6d, the two waves again cancel everywhere in the tube. At $t = T$ the situation is once more as shown in Fig. 4.6a.

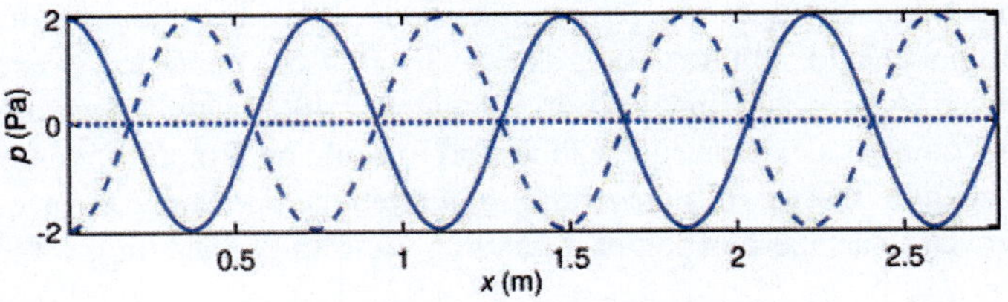

Fig. 4.7 The standing wave at different times during the oscillation cycle. Solid line, $t = 0$ and $t = T$; dotted line, $t = T/4$ and $t = 3T/4$; dashed line, $t = T/2$

The curves shown as solid green lines in Fig. 4.6 represent the total acoustic pressure in the tube at different times in the oscillation cycle. These four curves are superimposed in Fig. 4.7. As its name implies, there is no impression of movement to right or left in the vibration pattern of this standing wave. There are eight points along the tube, at distances $x = \lambda/4$, $3\lambda/4$, $5\lambda/4$, $7\lambda/4$, $9\lambda/4$, $11\lambda/4$, $13\lambda/4$, and $15\lambda/4$, at which the acoustic pressure remains zero throughout the period. These points are called pressure nodes. Intermediate between the nodes are points at which the pressure swing from positive to negative has its maximum value; these are called pressure antinodes. The node to antinode spacing is a quarter wavelength, and two adjacent nodes are half a wavelength apart. The general expression for the total pressure in a standing wave created by a forward-going wave with amplitude A and a backward going wave with amplitude B was given in Eq. 4.15. For the case discussed here, $B = A$, and

$$p(x,t) = p_+(x,t) + p_-(x,t) = A\left(e^{-jkx} + e^{jkx}\right)e^{j\omega t} = A(2\cos kx)e^{j\omega t}. \tag{4.21}$$

This form of the equation makes it obvious that at every point along the tube, the pressure variable has the same time dependence $e^{j\omega t}$. The amplitude of the pressure variation is $2\cos kx$, and the solid curve in Fig. 4.7 shows how this amplitude changes along the tube length.

In the previous discussion, it has been assumed that the open end of the tube is a pressure node for any standing wave. This is only an approximation, since it neglects the radiated sound wave illustrated qualitatively in Fig. 4.5. The detailed treatment of radiation impedance in Sect. 4.7 shows that the effective pressure node is displaced beyond the geometrical end of the cylinder by a distance $L_e \simeq 0.61a$, where a is the cylinder radius. The tube used as an example in this section has length $L = 2.77$ m, and radius $a = 5$ mm, so the end correction $L_e \simeq 3$ mm. This is small enough to be neglected for the purposes of the present discussion.

The length of the tube whose acoustical behaviour we have been discussing was chosen to be an odd number of quarter wavelengths ($L = 15 \times \lambda/4$) for the frequency of 466 Hz, so there must be a pressure antinode at the input. Figure 4.7 confirms this. It was noted in Sect. 2.2.2 that it is a necessary condition for successful sounding of a note on a lip-excited instrument that there should be strong feedback from the instrument to the lips of the player; the strength of this feedback is maximised when the input is a pressure antinode. If we had chosen a tube length which was not an odd number of quarter wavelengths, this condition would not have been satisfied. A pure standing wave with frequency 466 Hz and a pressure antinode at the input could not exist in such a tube, and it would be difficult for a brass player to sound the note $B^{\flat}4$ on it. In general, for an arbitrary frequency, the acoustic field can be viewed as a superposition of a standing wave and a travelling wave.

4.1.4 Frequency Domain and Time Domain

In discussing the acoustical behaviour of a brass instrument tube, there are two different but complementary approaches which can be taken. One is the time domain approach, which typically focuses on the way in which pressure changes in the air column develop over time. This method is particularly useful in dealing with transient phenomena, such as the starting of a new note or transitions between notes. The other is the frequency domain approach, which examines the behaviour of the instrument in response to different sine wave frequency components. Frequency domain techniques are very well adapted to investigating and describing continuously sounded notes.

Although brass players may not talk about time domain and frequency domain approaches, they are very aware of the corresponding aspects of musical performance. When presented with a new instrument for evaluation, a player will probably assess its intonation by playing octaves and slow arpeggios over the instrument's compass; this is an experimental investigation of its behaviour in the frequency domain. It is likely that the transient behaviour will be assessed by playing staccato notes, double-tonguings and trills; this is a typical time domain experiment. The instrument must not only play in tune, but it must be capable of responding flexibly to the rapid succession of commands sent to the instrument by the player in a virtuoso performance (see Sect. 1.2.10).

The picture in Fig. 4.7 showing nodes and antinodes of a standing wave in a cylindrical tube illustrates one aspect of a frequency domain view of the acoustical behaviour of the tube. It shows how the tube responds to excitation by a continuous sinusoidal disturbance at a specific frequency. To complete the frequency domain picture, it would be necessary to repeat the calculation for many different values of the frequency.

We have already described a simple but illuminating time domain experiment which can be carried out on a brass instrument without elaborate scientific equipment. Slapping the palm of the hand on to the open surface of the mouthpiece sends a pulse down the air column, and a microphone just outside the bell records the sound radiated by the instrument.

Figure 4.8 was obtained by gently slapping a palm against the mouthpiece rim on a Conn 8H tenor trombone with the slide in first position. The radiated sound was recorded using a small lapel microphone whose output was connected to the external microphone input of a laptop running the free software program Audacity.

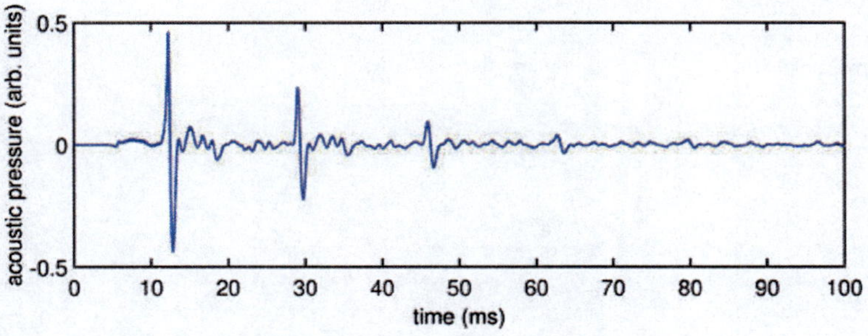

Fig. 4.8 Sound radiated from a tenor trombone after a slap closing the mouthpiece

The zero of time in Fig. 4.8 is approximately 4 ms before the slap was administered. The first large positive peak at $t = 12.3$ ms corresponds to the arrival at the microphone of the pulse which has travelled directly down the air column. What is the cause of the second large positive peak at 29.1 ms? To understand this we must recall that only a small fraction of the sound energy arriving in the direct pulse is transmitted through the bell. A large reflected pulse travels back up the tube, where it is again reflected by the palm still closing the mouthpiece. Once again it makes the journey down the tube. We saw in Sect. 4.1.1 that a pulse travelling at the speed of sound in free air would take 8.1 ms to travel the 2.8 m from mouthpiece to bell in a tenor trombone. The time delay expected between the first (direct) and second (doubly reflected) pulses is thus $2 \times 8.1 = 16.2$ ms. In fact the second pulse arrives 16.8 ms after the first. The third pulse, which has been reflected twice at each end of the tube, arrives 16.8 ms after the second, while the small fourth pulse which has undergone three double reflections is delayed by a further 16.8 ms. It thus appears that the pulses are taking around 0.3 ms longer than expected to make the journey from mouthpiece to bell. Reasons for this small discrepancy will be discussed in Sect. 4.3.

Figure 4.8 shows that a single pulse generated in the mouthpiece of a trombone is reflected internally many times before its energy is lost to either sound radiation from the bell or internal viscothermal losses. The nature of these reflections depends strongly on the boundary condition at the mouthpiece entrance. In the measurement discussed above, the mouthpiece remained closed after the pulse was generated. How different would the result be if the mouthpiece was not firmly closed?

Another simple experiment can help to answer this question. One possible approach would be to slap the mouthpiece and immediately withdraw the hand, but the result of this experiment suggests that the palm remains in contact with the mouthpiece for at least the first two reflections. Figure 4.9 illustrates an alternative method of creating a pressure impulse in the mouthpiece. A membrane (here a party balloon) is stretched tightly across the mouthpiece rim and then punctured by a cocktail stick. The sound is again recorded by a lapel microphone near the bell and illustrated in Fig. 4.10.

Fig. 4.9 Puncturing membrane to generate mouthpiece pulse

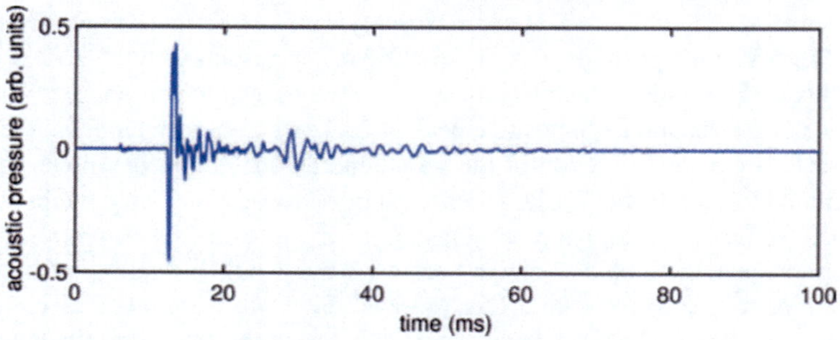

Fig. 4.10 Sound radiated from a tenor trombone after a membrane across the mouthpiece is punctured

The first pulse in Fig. 4.10 arrives at the microphone at $t = 12.7$ ms. It is a negative pulse, showing that the rupturing of the membrane creates a sudden drop in pressure in the mouthpiece. This negative pulse is followed within 0.8 ms by two positive pulses. The most obvious difference between Figs. 4.8 and 4.10 is that in the absence of a strongly reflecting closure at the mouthpiece rim, the internal reflections are less well defined and die out faster. If it were possible to create a perfectly absorbing surface closing the mouthpiece just behind the pulse source, we would expect to see only the direct wave, apart from some very weak reflections generated as the returning wave passes small irregularities in the internal bore of the tube.

4.1.5 *Impulse Response and Reflection Function*

In Sect. 4.1.4 we saw that the multiple reflections of a pulse inside the tube of a trombone could be investigated by recording the sound radiated from the bell. When discussing the way in which a brass player's lips interact with the acoustic

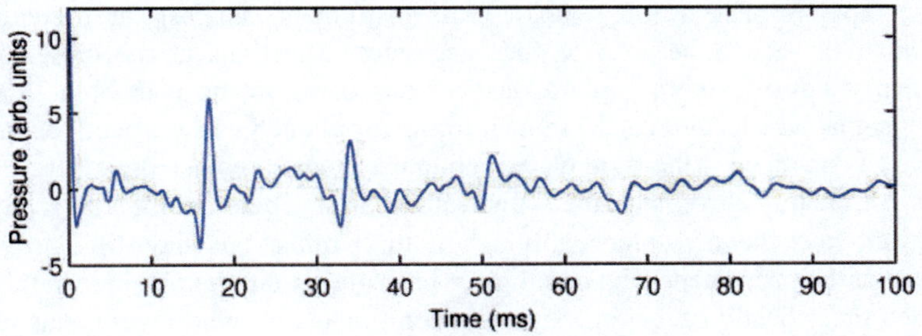

Fig. 4.11 Input impulse response $g(t)$ for a Conn 8H tenor trombone, first position

resonances of the instrument, it is useful to consider the pressure changes which occur just in front of the lips after a single pressure pulse leaves the mouthpiece. Since the player's lips almost completely seal the mouthpiece entrance, the relevant boundary condition is that the entrance should remain closed during the reflections which follow the initial impulse. A record of the variation of mouthpiece pressure following an idealised pressure impulse is described as the input impulse response of the instrument. Figure 4.11 illustrates the measured input impulse response $g(t)$ for a Conn 8H trombone.

Because the input plane of the mouthpiece is closed during the recording of the input impulse, the condition is similar to that applied in obtaining Fig. 4.8. An important difference is that the sound is now being recorded just inside the mouthpiece. The pulse generated in front of the input plane results in two pulses, one travelling forward and one backward. The backward pulse is immediately reflected by the rigid plate closing the mouthpiece and appears as a positive pulse at the start of the input impulse response. After a time $\tau \simeq 16$ ms, corresponding to the time taken to travel to the bell and back, the pulse returns as a negative spike, closely followed by a positive peak. Further pulses arising from multiple reflections from the closed input can be seen at integer multiples of τ.

When a plane wave travels down a cylindrical tube of constant radius, none of its energy is reflected until it reaches the end. The tubes of brass instruments usually include a cylindrical section, but there are many changes in radius along the bore profile (see Sect. 4.3). At any abrupt change in radius, for example, at the start or finish of a tuning slide, some of the sound energy of the forward-going wave will be reflected back towards the input. In Sect. 4.7 it is shown that a conical or flaring section of tube can be successfully approximated by a stepped bore corresponding to a sequence of very short cylinders with different radii. When the wave passes through such a section of tubing, there is a continuous reflection of sound energy. The input impulse response records all of these reflections: it is in a sense an acoustical map of the bore. The function $g(t)$ thus contains all the information about linear acoustical behaviour of the tube needed by a time domain model of the instrument.

The input impulse response of a brass instrument can last for a significant fraction of a second because of the successive reflections at the closed input. In effect it consists of multiple acoustic scans of the instrument bore. A much shorter signal which contains the same information about the instrument's acoustical response is obtained if the rigid plate sealing the input is replaced by a termination which completely absorbs all the sound falling on it. There are then no secondary reflections from the input plane, although multiple reflections may still exist within the bore of the instrument. The mouthpiece pressure signal $r(t)$ obtained in this way is called the reflection function. Time domain models of wind instruments can be made much more computationally efficient by using the reflection function rather than the input impulse response to represent the linear acoustic behaviour of the instrument (Schumacher 1981).

It is not experimentally feasible to provide a completely sound absorbing termination at the entrance to a brass instrument mouthpiece, although a long cylindrical tube with the same diameter as the mouthpiece rim is an approximate solution (see Sect. 4.2.4). The reflection function can be calculated from a knowledge of the bore profile (Sharp 1996). It can also be derived from the frequency domain measurements and calculations discussed in Sects. 4.2 and 4.7.

4.1.6 *Input Impedance*

While the reflection function is a useful representation of the acoustical behaviour of an instrument for time domain calculations, the frequency domain input impedance $Z(\omega)$ displays essentially the same information in a way which is much more intuitively related to the musical properties of the instrument. The input impedance can be defined directly by considering an experiment in which the mouthpiece excitation signal is not a pressure impulse but a continuous sinusoidally varying pressure

$$p(\omega) = p_{\mathrm{a}}(\omega)\mathrm{e}^{j\omega t} \tag{4.22}$$

generated by a sinusoidal volume air flow

$$u(\omega) = u_{\mathrm{a}}(\omega)\mathrm{e}^{j\omega t-\theta}. \tag{4.23}$$

The amplitude of the mouthpiece pressure signal is $p_{\mathrm{a}}(\omega)$ Pa, and the amplitude of the volume air flow is $u_{\mathrm{a}}(\omega)\ \mathrm{m}^3\,\mathrm{s}^{-1}$. The phase difference between the volume flow and the resulting pressure is $-\theta$ radians. The input impedance $Z(\omega)$ is then simply defined as the ratio of pressure to volume velocity, both quantities being evaluated at the entrance plane of the tube:

$$Z(\omega) = \frac{p(\omega)}{u(\omega)} = \left|\frac{p_{\mathrm{a}}(\omega)}{u_{\mathrm{a}}(\omega)}\right| \mathrm{e}^{j\theta}. \tag{4.24}$$

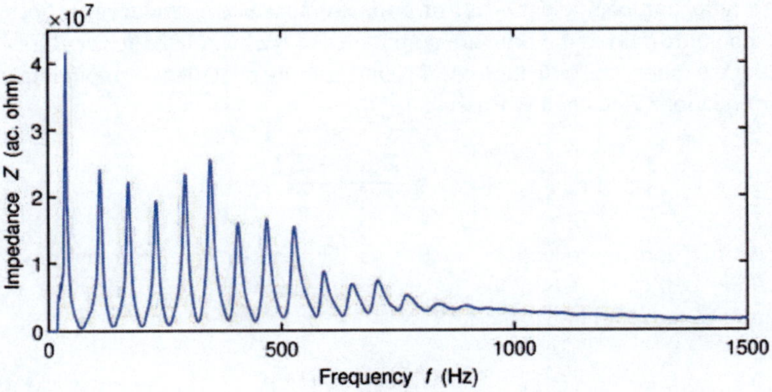

Fig. 4.12 Input impedance for a Conn 8H tenor trombone, first position

The magnitude of $Z(\omega)$ is the ratio of amplitudes of mouthpiece pressure and volume velocity, and the unit of input impedance is therefore Pa s m^{-3}. This unit is frequently described as the acoustic ohm.

For an infinitely long cylindrical tube, there is only a forward travelling wave, for which the relationship between pressure and particle velocity is given by Eq. 4.13. The input impedance is

$$Z_c = \frac{\rho_0 c}{S}, \tag{4.25}$$

where S is the input cross-sectional area. Z_c is defined as the characteristic impedance.

Figure 4.12 illustrates the magnitude of the measured input impedance for the Conn 8H trombone whose input impulse response is shown in Fig. 4.11. These two figures contain essentially the same information about the linear acoustic properties of the instrument air column, Fig. 4.12 expressed in the frequency domain and Fig. 4.11 in the time domain.

Equation 4.24 shows that the frequency spectrum $p(\omega)$ of the mouthpiece pressure in a brass instrument can be derived from the frequency spectrum $u(\omega)$ of the volume flow into the mouthpiece simply through multiplication by the input impedance:

$$p(\omega) = Z(\omega)u(\omega). \tag{4.26}$$

The equivalent calculation in the time domain makes use of the input impulse response $g(t)$:

$$p(t) = g(t) * u(t), \tag{4.27}$$

where the asterisk denotes the mathematical process of convolution.

The reflection coefficient $R(\omega)$ of an instrument was introduced in Sect. 4.1.2 as the complex ratio of the pressure amplitudes of backward and forward travelling waves at the input. The reflection coefficient is simply related to the input impedance by the equation (Fletcher and Rossing 1998)

$$R(\omega) = \frac{Z(\omega) - Z_c}{Z(\omega) + Z_c}. \tag{4.28}$$

The time domain reflection function is the inverse Fourier transform of the reflection coefficient, while the input impulse response is the inverse Fourier transform of the input impedance:

$$IFT[R(\omega)] = r(t); \tag{4.29}$$

$$IFT[Z(\omega)] = g(t). \tag{4.30}$$

The input impedance curve, which is a plot of input impedance magnitude as a function of frequency, has become a standard and invaluable tool in displaying and studying the linear acoustic properties of wind instruments. It is particularly useful in the case of brass instruments, since it shows explicitly the frequencies and amplitudes of the many acoustic resonances on which the played notes are based (see Sect. 1.2). In Fig. 4.12 it can be seen immediately that peaks 2–12 have the approximately equal frequency spacing corresponding to a set of quasi-harmonically related played notes. The first peak does not fit with this series, suggesting that there is something acoustically different about the pedal note. The peaks effectively disappear above 800 Hz, explaining the musical experience that above the 15th natural note, continuous glissandi are possible.

Figure 4.12 displays only the magnitude of the input impedance Z. Information about the phase of Z can also be enlightening, since it corresponds to the phase difference between the acoustic pressure and the volume velocity. Figure 4.14 shows the magnitude and phase of the input impedances of two similar brass instruments, the tenor and bass trombones illustrated in Fig. 4.13.

Fig. 4.13 Two trombones by Courtois. Above: bass trombone. Below: tenor trombone

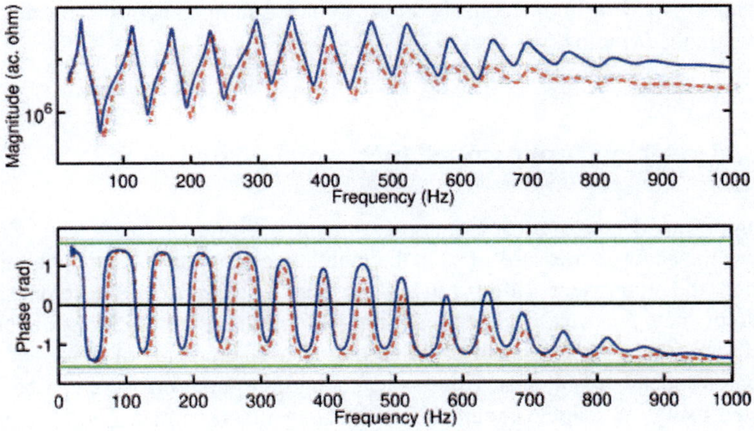

Fig. 4.14 Input impedance curves for the two trombones illustrated in Fig. 4.13. Top: magnitude, logarithmic scale. Bottom: phase. Horizontal green lines $\pm\pi/2$. Red dashed line: bass trombone. Solid blue line: tenor trombone (Color figure online)

In Fig. 4.14 the magnitude of the input impedance is shown on a logarithmic scale, revealing that the peaks representing resonances in the instrument (with pressure antinodes at the input) are mirrored by similar dips representing the antiresonances (with pressure nodes at the input). Comparing the magnitude and phase curves reveals that the phase of the impedance falls through zero at each resonance and rises through zero at each antiresonance. If there were no energy losses in the tube, the returning wave would have the same amplitude as the forward-going wave, and the phase would oscillate between $\pm\pi/2$. Below 100 Hz this is approximately true, since both wall losses and radiation from the open end are relatively small at low frequencies.

Above 500 Hz the swings in both magnitude and phase of the impedance become progressively weaker with rising frequency; this effect is more pronounced in the bass trombone than in the narrower bored tenor trombone, which has the musical consequence that playing in the high register is a little easier on the tenor than on the bass (see also Sect. 5.2.2). By 900 Hz the resonances and antiresonances have almost disappeared for both instruments, since practically no sound is being reflected back into the instrument at the bell. In this situation there is only a forward-going plane wave in the instrument, and the input impedance is the characteristic impedance of the cylindrical section of the bore:

$$Z_c = \frac{\rho_o c}{S} \tag{4.31}$$

with S being the bore cross-sectional area. Since S is greater for the bass than for the tenor, the input impedance magnitude of the bass asymptotically approaches a lower value of Z_c as the frequency rises. The phase of Z for both instruments approaches $-\pi/2$ at high frequency because of the lumped acoustic behaviour of the mouthpiece volume (see Sect. 4.3.5).

4.2 Measuring Input Impedance

The input impedance $Z(\omega)$ of a brass instrument is defined in Eq. 4.24 as the ratio between the acoustic pressure $p(\omega)$ in the mouthpiece and the acoustic volume flow $u(\omega)$ into the mouthpiece. Both $p(\omega)$ and $u(\omega)$ are assumed to be sinusoidal signals with frequency $f = \omega/2\pi$. The most direct way to measure the input impedance would therefore appear to involve simultaneous measurements of $p(\omega)$ and $u(\omega)$ for each frequency of interest. However, although the input pressure can be easily recorded using an adapted mouthpiece with a small microphone in the cup wall (see Sect. 2.1.2), the direct measurement of the acoustic flow velocity inside the mouthpiece is much more difficult. Studies of trombone input impedance have been carried out using a hot wire probe in the mouthpiece throat to measure the flow speed (Pratt et al. 1977; Elliott and Bowsher 1982), but this method is technically difficult and insensitive to the direction of flow.

The Euler equation (Eq. 4.11) shows a basic fluid dynamic relationship between the time derivative of the velocity and the spatial gradient of the pressure. It is therefore possible in principle to replace the measurement of flow velocity by pressure measurements from two or more spatially separated microphones. Most of the techniques currently employed to measure the input impedance of wind instruments employ at least two microphones. Depending on the method, calibration is necessary using a number of acoustical systems with known impedance. The calibrating systems used in practice range from cylindrical cavities a few millimetres deep to an impressive 95 m long plastic tube used as an anechoic termination by Joe Wolfe's research group at UNSW, Sydney. It is even possible to carry out input impedance measurements using a single microphone by an appropriate choice of calibrating systems (Sharp et al. 2011).

Valuable and comprehensive reviews of input impedance measurement techniques and calibration methods have been published by Benade and Ibisi (1987), Dalmont (2001a,b) and Dickens et al. (2007). Here we describe briefly three different approaches which have proved useful in the study of brass instrument acoustics.

4.2.1 *Capillary-Based Methods*

Many of the devices in current use for measuring input impedance rely on coupling a cavity in front of a loudspeaker to the mouthpiece of the instrument through a narrow diameter tube known as a capillary. The earliest mention of this idea in the scientific literature is an article written by J. C. Webster (1947), then working in the research

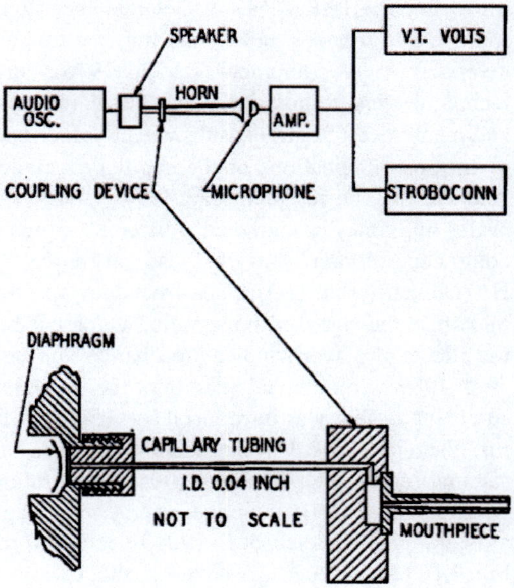

Fig. 4.15 Capillary-based system for measuring brass instruments. Reproduced from Webster (1947) with the permission of the Acoustical Society of America

laboratory of the brass instrument manufacturer C. G. Conn in Elkhart, Indiana. The purpose of the capillary is to deliver an oscillating air flow into the mouthpiece while decoupling the resonances of the instrument from those of the driving speaker and cavity. If the tube is sufficiently long and its diameter is sufficiently small, its acoustic impedance will be very much larger than even the peak impedance of the instrument under study. In that case the volume flow through the tube for a given pressure in the driving cavity will be almost independent of the acoustical properties of the instrument.

The sketch of Webster's apparatus, reproduced in Fig. 4.15, shows that the microphone recording the pressure response was outside the bell of the instrument. The device was therefore capable only of an uncalibrated measurement of the pressure transfer function (the ratio of radiated pressure to mouthpiece pressure as a function of frequency), rather than the input impedance. It is known, however, that the Conn research laboratory did develop a more sophisticated version of this apparatus including a microphone embedded in the mouthpiece. A feedback loop from a monitoring microphone in the cavity maintained the driving pressure amplitude constant over a wide frequency range, resulting in an effectively constant volume flow source (Benade and Ibisi 1987). For reasons of commercial confidentiality, details of the Conn input impedance measuring apparatus were not published, although its use was described by Earle Kent (1956), the leader of the Conn research team. Equipment based on the Conn design was used by Arthur Benade in pioneering studies of brass instrument input impedance (Benade 1973, 1976).

In 1974 John Backus described an impedance measuring apparatus based on a novel design of capillary in which a hexagonal rod was forced into a surrounding cylinder (Backus 1974). Detailed measurements of the input impedances of trumpets, trombones and french horns were subsequently made by Backus using a version of this equipment (Backus 1976). Further improvements to the capillary technique were proposed by Caussé et al. (1984) and Kergomard and Caussé (1986), and many research groups subsequently developed equipment of this type.

In most applications of the capillary method, the injected pressure signal is a sine sweep. The research team led by Joe Wolfe at the University of New South Wales in Sydney has adopted a different approach, in which the exciting signal is a computer-generated sum of sinusoidal frequency components at intervals of a few Hz (Dickens et al. 2007). The frequency spectrum of the signal can be tailored to maximise the signal-to-noise ratio, which has been a valuable feature in studies of the effects of player windway resonances during performance (see Sect. 6.3).

In 1989 a commercial version of the two microphone capillary impedance measurement system was introduced by Grigor Widholm and colleagues at the Institut für Wiener Klangstil in the Vienna University of Music and the Performing Arts (Ossman et al. 1989; Widholm 1995). In the following decades, the Brass Instrument Analysis System (BIAS) underwent several stages of development and refinement; a diagrammatic sketch of the 2019 version of the impedance head is illustrated in Fig. 4.16 (Artim 2020). A cavity in the lower part of the cylindrical head is driven by a small loudspeaker. Three separate bundles of capillary tubes connect the cavity to the upper face of the head, at the centre of which the response microphone is mounted flush with the surface. The driving and recording electronics is housed within the head. The BIAS equipment with its extensive accompanying analysis

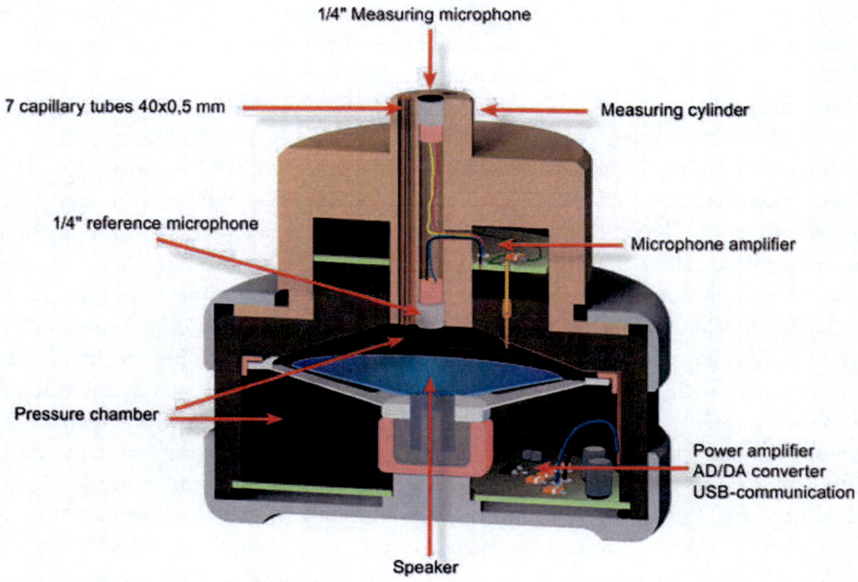

Fig. 4.16 The BIAS acoustic input impedance measurement head. Courtesy of Wilfried Kausel

software package has been adopted by many brass instrument manufacturers and has also been widely used in brass instrument research.

4.2.2 Complementary Cavity Methods

A number of input impedance measurement systems have been developed using a rigid piston as a driver (Benade and Ibisi 1987). An attractively simple idea is to couple the back of the driving piston to a closed cavity containing a microphone. We use the term 'complementary cavity' to describe this approach because a volume flow into the test system is complemented by an equal and opposite volume flow into the cavity. From a knowledge of the acoustical properties of the cavity, it is possible to deduce the volume flow from the measured pressure. This is particularly straightforward at low frequencies, since the cavity behaves as a lumped impedance (Dalmont 2001a).

A complementary cavity system using a loudspeaker driver described by Singh and Schary (1978) suffered from problems associated with the lack of rigidity of the driver and imperfect acoustical isolation between the cavity and the measurement system. A successful complementary cavity device using a piezoelectric disc as a driving piston has been developed and marketed by the Centre de Transfer de Technologie du Mans (CTTM) in association with the Laboratoire d'Acoustique de l'Université du Mans (LAUM) (Dalmont et al. 2012). The principle of the system is illustrated in Fig. 4.17, and Fig. 4.18 shows Jean-Pierre Dalmont using the CTTM system in a study of trumpet bore optimisation (Macaluso and Dalmont 2011).

4.2.3 Wave Separation Methods

A disadvantage of using a capillary tube of very high acoustic resistance to drive the flow in an impedance measuring device is that the flow amplitude is typically very much smaller than the values generated in human performance. This is not in

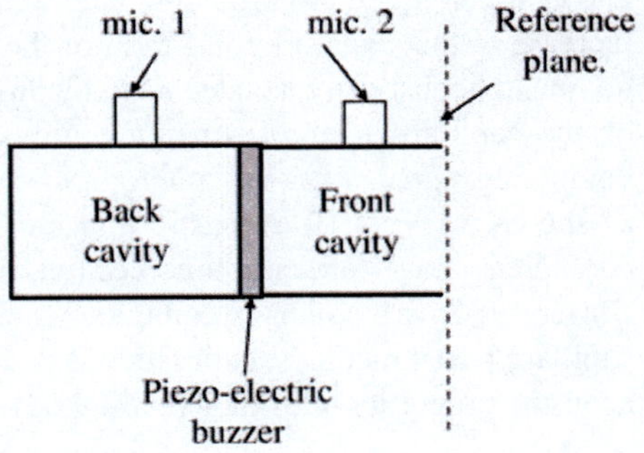

Fig. 4.17 Schematic diagram of the CTTM impedance measuring system. Courtesy of Jean-Pierre Dalmont

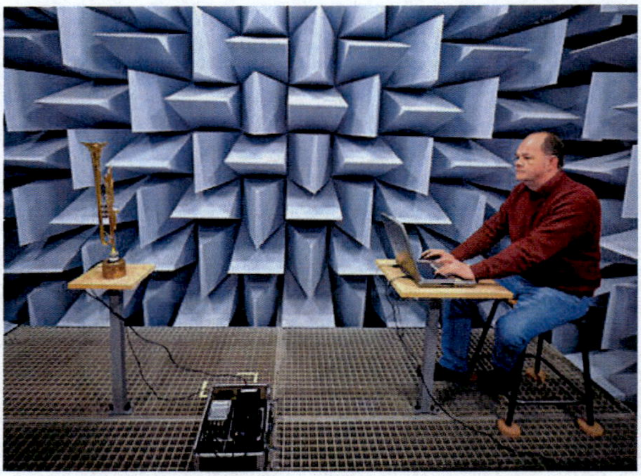

Fig. 4.18 Jean-Pierre Dalmont using the CTTM input impedance system in the LAUM anechoic chamber

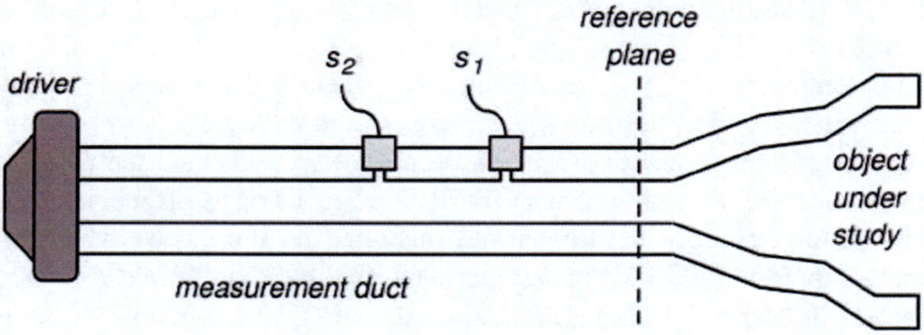

Fig. 4.19 Two microphone apparatus for wave separation (van Walstijn et al. 2005)

itself a problem, since the input impedance is a linear property of the instrument independent of the driving amplitude. However the correspondingly low amplitude of the pressure signal recorded by the response microphone makes it challenging to achieve a good signal-to-noise ratio. The susceptibility of the system to corrupting background sounds in noisy environments is exacerbated when a brass instrument is the object of study, since the bell funnels external sounds efficiently to the mouthpiece in the manner of an ear trumpet (Barbieri 2013).

Increasing the diameter of the delivery tube to match the input of the instrument allows much higher volume flow amplitudes to be achieved, with a corresponding increase in the signal-to-noise ratio of the response. However the driving cavity is no longer acoustically isolated from the instrument, invalidating a basic assumption of the capillary method. An alternative approach circumvents this problem by comparing signals from two microphones at different locations along the tube axis, as shown in Fig. 4.19. Acoustic impedance measurements of material absorption coefficients have for some time been carried out using this technique (Chung and Blaser 1980), which has become known as the two-microphone method. As with capillary based methods, calibration is necessary using a number of systems whose acoustic properties are known. Methods involving two microphones have been described with three calibrations (Gibiat and Laloë 1990) and four calibrations (van Walstijn et al. 2005).

Two-microphone wave separation breaks down at frequencies for which the microphone spacing is equal to an integer number of half wavelengths, since the two microphones then give essentially the same information. This problem can be partly overcome by using more than two microphones, and the 'two-microphone' apparatus described by Gibiat and Laloë (1990) in practice included four microphones with different spacings. Issues involved in the expansion of the frequency bandwidth using multiple microphones are discussed by Kemp et al. (2010).

4.2.4 Acoustic Pulse Reflectometry

The technique known as acoustic pulse reflectometry (APR) uses an experimental design, shown in Fig. 4.20, which is similar to that of wave separation. There is however only one microphone in the APR entrance tube, rather than the minimum of two required by wave separation, and the exciting signal is not a continuous sinusoidal wave but a very short impulse. The impulse travels down the launching tube from the driving loudspeaker; after travelling the distance l_1 shown in Fig. 4.20, it is recorded by the microphone. It then travels a further distance l_2 and enters the instrument under study. Reflections from changes in the instrument bore diameter travel back up the tube and are recorded by the microphone as they pass. A suitable choice of the distances l_1 and l_2 ensures that the last reflection from the instrument is recorded before the arrival at the microphone of further reflections from the loudspeaker. The history and practice of APR is reviewed by Sharp (1996).

During the time window within which the initial impulse and its reflections are recorded, the launching tube can be viewed from the instrument input as an anechoic termination. If the impulse were an ideal Dirac delta function with infinitesimal duration, the recorded reflection signal would be the reflection function $r(t)$ defined

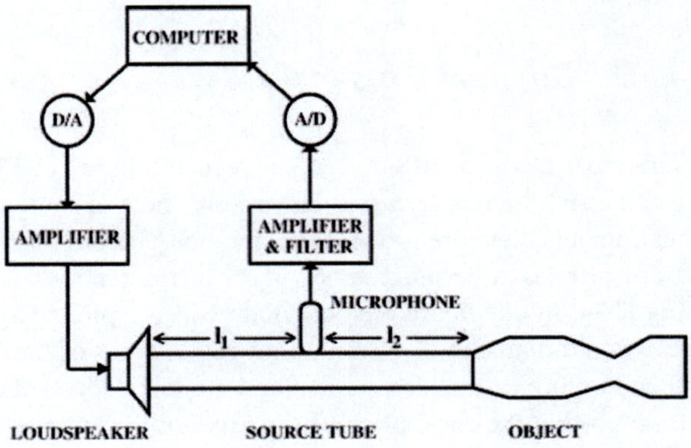

Fig. 4.20 Schematic diagram of an acoustic pulse reflectometer (Sharp 1996)

in Sect. 4.1.5. In practice $r(t)$ is obtained by deconvolving the reflected signal with the recorded impulse. The frequency domain reflection coefficient $R(\omega)$ of the instrument is found by Fourier transforming $r(t)$, and the input impedance $Z(\omega)$ can be deduced using Eq. 4.28.

Since the time domain reflection function measured at the entrance of a tube records the reflections arising from every change in diameter, it is possible to reconstruct the bore profile of a duct of varying cross-section from an APR measurement (Amir et al. 1995; Sharp 1996; Li et al. 2005). Bore reconstruction using APR has found many industrial applications and has been successfully used to study the bore profiles of musical wind instruments (Campbell and MacGillivray 1999; Buick et al. 2002; Gray 2005; Hendrie 2007; Dalmont et al. 2012).

4.3 Bore Profiles of Brass Instruments

The four brass instruments illustrated in Fig. 4.21a differ significantly in their overall appearance and also in their playing properties. In timbral terms the two instruments on the left could be classed as 'mellow' instruments, while the two on the right fall into the 'bright' instrument category (see Sect. 6.1). The major difference in tone colour is not however related to the proportion of straight to coiled tubing, or to the difference between valves and slides, but rather to the differences between the internal bore profiles of the instruments as shown in Fig. 4.21b.

In the following sections, we examine in some detail the relationship between the shape of the bore profile on a brass instrument and the frequencies and strengths of the acoustic resonances in the contained air column. The nature of these resonances plays a large role in determining the musical behaviour of the instrument at relatively low dynamic levels (generally speaking below forte). In loud playing another effect, known as nonlinear sound propagation, becomes increasingly important, in some cases giving rise to the very brilliant timbre described as 'brassy'. Although nonlinear sound propagation also depends strongly on the nature of the bore profile, we will defer a discussion of this effect until Sect. 6.1.

4.3.1 Different Parts of the Bore

The most obvious differences between the bores in Fig. 4.21b is that in the two trombones, the bore remains narrow for most of its length, while in the other two instruments, the bore grows more gradually before reaching the rapidly flaring bell. Bore profiles for a number of brass instruments of similar length are shown in Fig. 7.34. In this figure the bore diameters are plotted against axial distance up to a maximum diameter of 67 mm, allowing the part of the bore nearest the input to be seen in more detail. The serpent and ophicleide have an almost completely conical bore profile; the sackbut and bass trombone are cylindrical for the first 1.5 m and

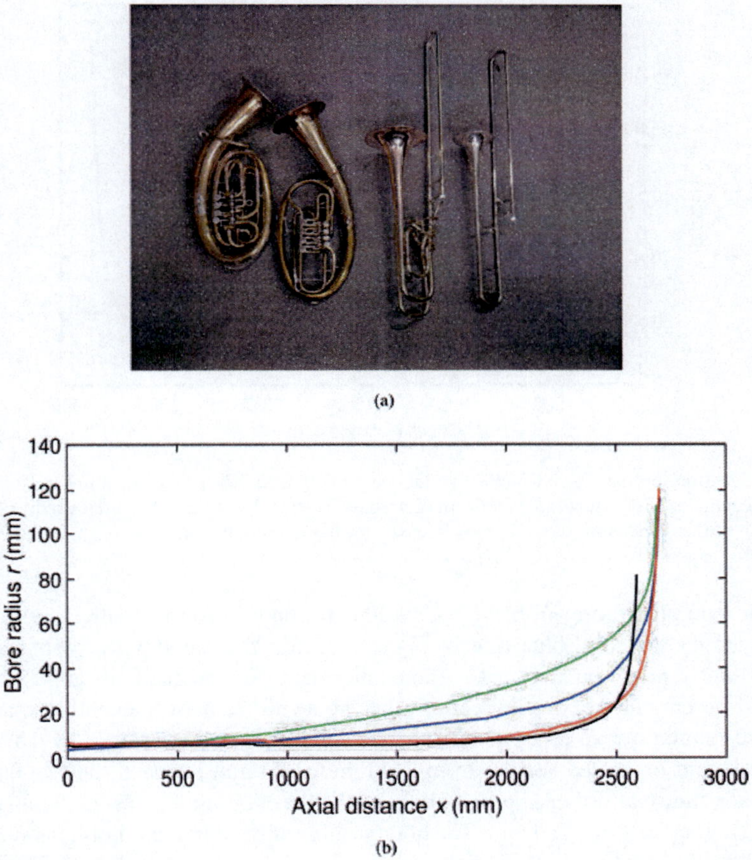

Fig. 4.21 (**a**) Photograph of four brass instruments. Left to right: Kaiserbaryton, Wagner tuba, bass trombone, tenor trombone. (**b**) Bore profiles of Kaiserbaryton (green), Wagner tuba (blue), bass trombone (red), tenor trombone (black) (Color figure online)

then expand more and more rapidly towards the bell; the french horn starts with a conical section, then becomes cylindrical and ends in a flaring bell almost identical to the bass trombone in the region plotted.

The nature of the bore profile is clearly determined to some extent by the practical way in which the instrument operates. On the ophicleide, side holes are opened to shorten the effective sounding length, and the rapidly expanding bore makes it possible to have large holes which radiate low frequencies more efficiently (see Sect. 4.4). On the trombone, the use of a long slide makes it essential that a substantial fraction of the tube length is of constant diameter. On the french horn, the cylindrical part corresponds to the valve section in which alternative lengths of tubing are switched in or out of the main bore: in this case a large section is switched out to raise the playing pitch from F to $B^{\flat}$.

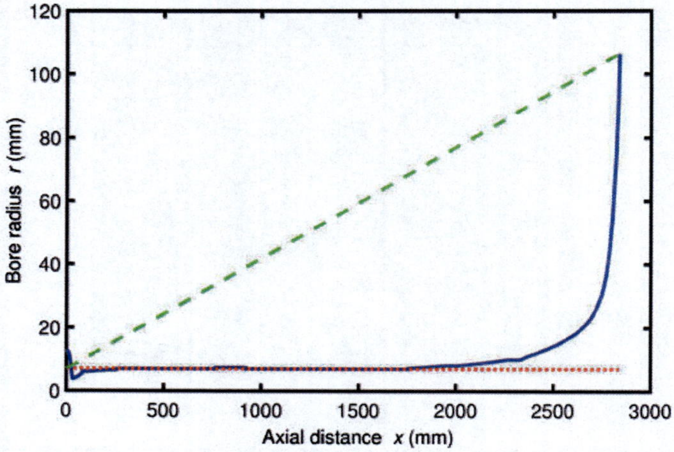

Fig. 4.22 Bore profiles for a Conn 8H trombone with a Denis Wick 5AL mouthpiece (solid blue line), a cylinder of constant radius 6.6 mm (dotted red line) and a cone with entrance radius 6.9mm and exit radius 106.5 mm (dashed green line) (Color figure online)

The complete bore profile of a modern trombone, from mouthpiece to bell, is shown by the solid blue line in Fig. 4.22. After the mouthpiece, with its cup, throat and rapidly expanding backbore, the main bore includes examples of the three different types of profile sections that we identified above: a gently expanding conical mouthpipe up to $x = 256$ mm, a cylindrical section up to $x = 1694$ mm (with a short expanded section from 761 mm to 951 mm between the two ends of the inner slide), a further approximately conical section up to $x = 2269$ mm and a flaring bell up to $x = 2764$ mm. The bore profiles of most modern brass instruments can be well approximated by combinations of cylinders, cones and flaring sections (Braden 2006). We start by considering how sound waves propagate and resonate in simple cylinders and cones, like those shown by the dashed and dotted lines in Fig. 4.22, before going on to examine the effects of adding a mouthpiece and a flaring bell to make a complete instrument such as a trombone, trumpet or horn.

4.3.2 *Cylindrical Tubes*

Many brass players like to amuse their friends by demonstrating that a piece of classical horn or trumpet music can be played on a length of plastic hosepipe like the example shown on the left of Fig. 6.6. The hosepipes in this figure are provided with brass instrument mouthpieces and plastic funnels, and these additions make the 'hosepipe horn' much easier to play for reasons which will be described in Sects. 4.3.5 and 4.3.7. It is however possible to play the excerpt from the Mozart D Major Horn Concerto shown in Fig. 4.23 simply by buzzing the lips against the

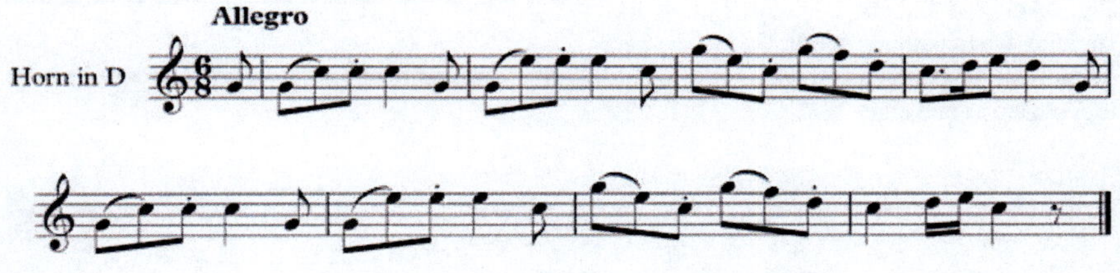

Fig. 4.23 Excerpt from the 2nd Movement of Mozart's Horn Concerto in D Major, K.412

open tube end if it is cut to give a reasonably flat rim. A comparison of Fig. 4.23 with Fig. 1.12 shows that the notes required are all found between the 6th and 12th members of a harmonic series based on C2; however the horn in D uses transposed notation in which the sounding pitches are a minor 7th (10 semitones) below the written pitches. It will be found that a length of hosepipe about 4.25 m long is necessary for the excerpt to sound at the correct pitch. This is roughly the same length as the bore of the D horn, which is described technically as being in '14 ft D' (Sect. 7.2.7).

The hosepipe horn is an amusing musical curiosity, but it is also a very interesting scientific demonstration that even a simple cylinder of constant diameter has a set of acoustic resonances which seem to be reasonably close to a harmonic series. However a musical exploration of the lower register of the hosepipe horn shows that below the sixth natural note, the pitches are increasingly flattened relative to the expected harmonics, the discrepancy reaching several semitones for the second natural note.

The explanation of the apparently strange behaviour of the cylindrical tube resonance frequencies lies in the nature of the reflections which create the standing waves in the tube. We saw in Sect. 4.1.2 that a travelling wave is reflected with a change of sign at the open end of the tube, creating a pressure node just outside the end; at the input end, which is effectively closed by the player's lips, the wave is reflected without a change of sign, making this point a pressure antinode. These end conditions can only be satisfied by waves for which the length L of the tube is an odd number of quarter wavelengths. Neglecting wall losses and radiation from the open end, the permitted wavelengths are

$$\lambda_n = \frac{4L}{2n-1}; \tag{4.32}$$

the corresponding frequencies form a series given by

$$f_n = \frac{c}{\lambda_n} = \frac{(2n-1)c}{4L} = f_1,\ 3f_1,\ 5f_1 \ldots \tag{4.33}$$

with $f_1 = c/4L$.

Fig. 4.24 The first six standing wave patterns for the cylindrical tube shown at the top, closed at one end

The first six standing wave patterns for a cylindrical tube closed at one end are shown in Fig. 4.24.

Equation 4.33 reveals that the expected resonances of a cylindrical tube closed and sounded by the lips do not form a complete harmonic series, but only the odd members of the series. The musical significance of this is brought out in Fig. 4.25, showing the first twelve pitches in an odd-member-only harmonic series based on C1. The number under each note is the multiple of the fundamental frequency, while the rank number of the note is given in brackets. Although this set of notes consists of odd harmonics of C1, the notes from the sixth upwards are fairly close to a complete set of harmonics based on B1. The 6th note, for example, is only slightly below the 6th harmonic of B1, but could be lipped up; the 8th, 9th and 10th notes are good approximations to the 8th, 9th and 10th harmonics of B1; the 12th harmonic is slightly sharper than an exact harmonic of B1, but could be lipped down. Playing in this range thus gives the impression that the notes are part of a (slightly mistuned) complete harmonic series based on B1, rather than an odd-member-only series based on C1.

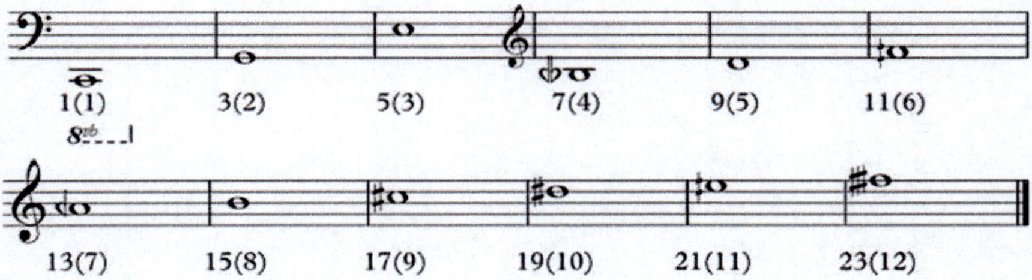

Fig. 4.25 The odd members of a harmonic series with fundamental C1

As explained in Sect. 4.1.6, the input impedance curve $Z(f)$ for an instrument gives a valuable frequency domain picture of its standing wave resonances. Figure 4.26a shows the calculated input impedance curve for the cylindrical tube whose bore profile is shown in Fig. 4.22. This 'hosepipe horn' is the same length as the Conn 8H trombone, and has a radius equal to that of the slide section of the trombone. Each peak in Fig. 4.26a corresponds to a resonance in the air column with a pressure antinode at the entrance. The broken green line which joins the points at the top of each peak is called the 'peak envelope' (Lurton 1981). For this simple cylinder, the peak envelope is a monotonically decreasing function of the frequency.

The dips correspond to antiresonances with pressure nodes at the entrance; they are less important in the functioning of brass instruments, since the valve effect source requires a pressure antinode in front of the lips. In contrast, the air jet excitation mechanism of the flute demands a pressure node at the entrance, so the playable note on a flute corresponds to the dips in the impedance curve (Nederveen 1998a).

The symmetrical nature of the peaks and dips is brought out when the impedance curve is plotted on a logarithmic scale (Fig. 4.26b). The musical intervals between the resonances are also more evident in this plot, since equal horizontal distances on the logarithmic frequency scale correspond to equal pitch intervals. A further striking feature of the plot in Fig. 4.26b is the linear decrease in the vertical distance between peaks and dips as the frequency increases. This distance is determined by the strength of the resonances, which in turn depends on the viscothermal losses mentioned in Sect. 4.1.2 and discussed fully in Sect. 4.7.3. The effect of these losses is to reduce the amplitude of a pressure wave travelling a distance x in a tube by a factor $e^{-\alpha x}$. The fact that the impedance peak height decreases linearly with frequency on a log-log plot confirms that the relationship between the decay constant α and the frequency f (Eq. 4.110) has the form of a power law:

$$\alpha = Af^n, \tag{4.34}$$

where A and n are constants. From the slope of the straight line representing the peak envelope in Fig. 4.26b, it can be deduced that $n = 0.5$: in other words, the losses increase with the square root of the frequency.

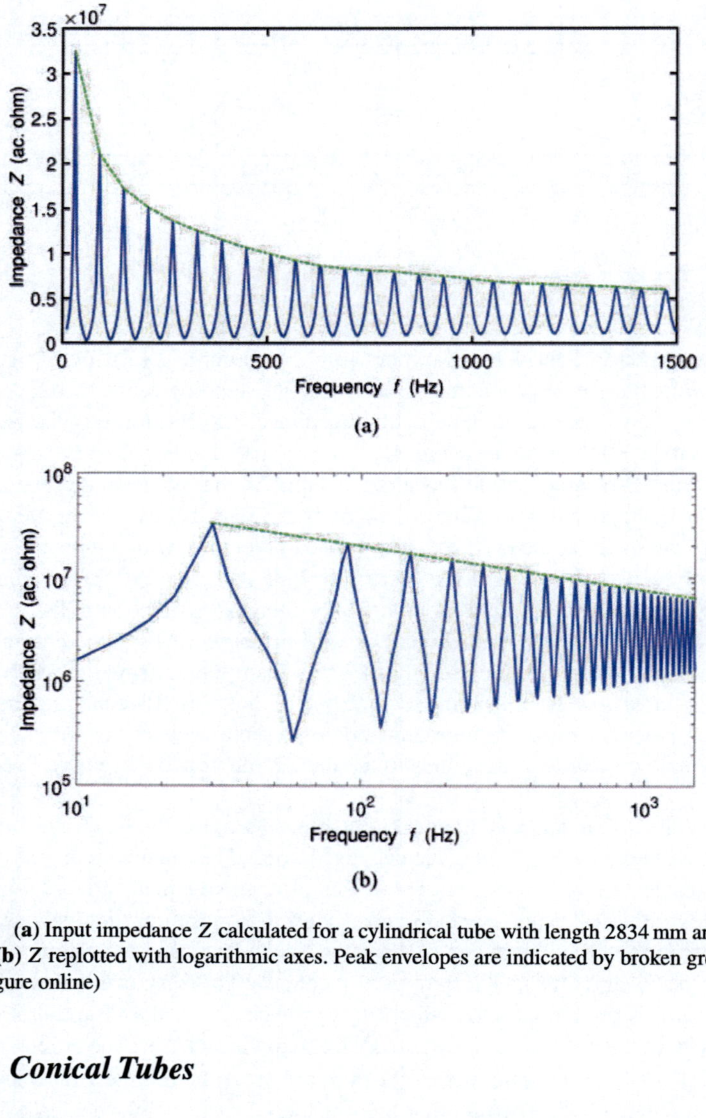

Fig. 4.26 (**a**) Input impedance Z calculated for a cylindrical tube with length 2834 mm and radius 6.9 mm. (**b**) Z replotted with logarithmic axes. Peak envelopes are indicated by broken green lines (Color figure online)

4.3.3 Conical Tubes

The propagation of sound waves in a conical tube is different from that in a cylindrical tube in one important aspect: the waves travel not with plane wavefronts but with spherical wavefronts. The nature of the wavefronts in a conical tube like that shown by the dotted line in Fig. 4.22 can be visualised by drawing a small circle on the surface of a partially inflated spherical balloon. The area of the balloon surface inside the circle is not plane (flat), but bulges outwards in the centre. As the balloon is inflated further, the area of the enclosed surface grows, just as the area of the spherical wavefront grows during its progress along a conical tube.

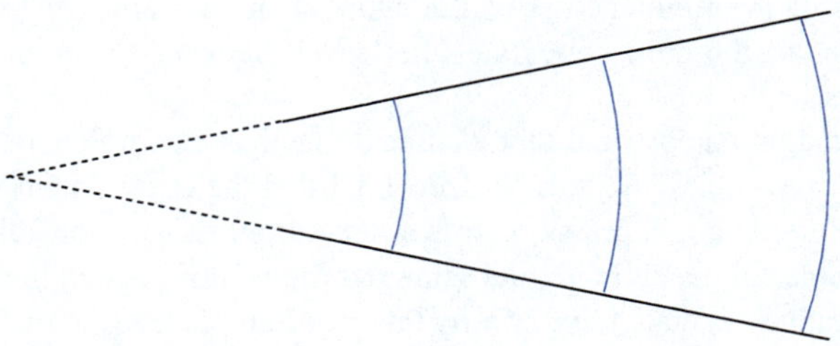

Fig. 4.27 Spherical wavefronts in a conical tube

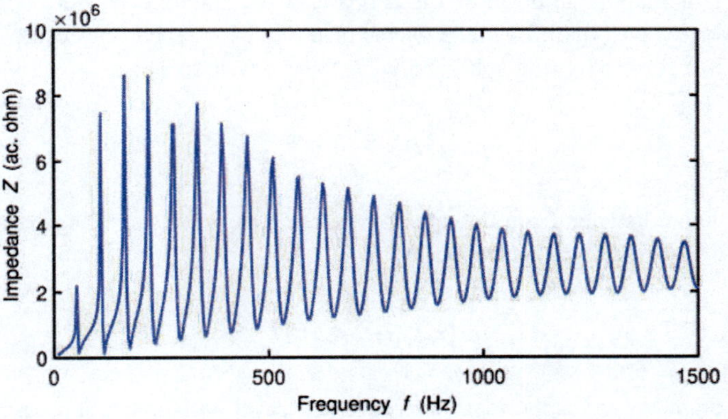

Fig. 4.28 Input impedance Z calculated for a conical tube with length 2834 mm, entrance radius 6.9 mm and output radius 106.5 mm

Examples of spherical wavefronts in a conical tube are shown in Fig. 4.27 as thin blue lines. In each case the wavefront is a section of the surface of a sphere centred at the vertex of the cone. In a playable instrument, the cone must be truncated by the removal of a section like that shown dotted in Fig. 4.27 to allow for the insertion of a mouthpiece.

Figure 4.22 showed an example of a truncated cone. In this hypothetical instrument the length, input diameter and final bell diameter were chosen to be the same as those of the Conn 8H trombone, but the tube is a cone of constant taper. This bore profile is in fact very similar to that of the B^b ophicleide illustrated in Fig. 7.28b. The calculated input impedance is shown in Fig. 4.28. Comparison with the input impedance curve for a cylinder of the same length in Fig. 4.26a shows that while the first impedance peak is the largest in the cylindrical tube, the first peak in the cone's impedance curve has remarkably low amplitude, and the third peak is the largest.

Another distinction between the properties of the impedance peaks in cones and cylinders, which is even more important from the musician's perspective, is that the frequencies of the peaks in the cone impedance curve are very close to a complete harmonic series with fundamental frequency 54.7 Hz. This is almost twice the frequency of the fundamental of the cylinder of the same length, and in fact the frequencies of the cone series are interleaved between the frequencies of the cylinder series.

It is at first sight paradoxical that a conical tube has a complete harmonic series of resonance frequencies. We saw in Sect. 4.3.2 that the requirement for a pressure antinode at the input and a pressure node at the output was responsible for the lack of even harmonics in the cylinder; the same requirement is present in the case of the cone. A qualitative understanding of why this requirement leads to different patterns of standing waves in cylindrical and conical tubes can be obtained by considering the nature of a travelling spherical wave.

The mathematical solution of the acoustic wave equation is more complicated for spherical waves than for plane waves (see Sect. 4.6.2), but the result is that the pressure in a forward-going spherical wave can be written as

$$p_{+}(r,t) = \frac{A}{r}\mathrm{e}^{j(\omega t - kr)}, \tag{4.35}$$

where r is the distance from the apex of the cone. A wave travelling back towards the apex can similarly be written as

$$p_{-}(r,t) = \frac{B}{r}\mathrm{e}^{j(\omega t + kr)}. \tag{4.36}$$

Unlike a plane wave, a spherical wave does not travel with constant pressure amplitude: the amplitude $\hat{p} = A/r$ decreases as the radius increases and conversely. This dependence can be seen as a consequence of energy conservation (remembering that here we are ignoring losses). The energy carried across the wavefront surface S by the wave is proportional to $\hat{p}^2 S$; since $S \propto r^2$, $\hat{p}r$ must remain constant throughout the cone to conserve energy.

Assuming for simplicity that reflections occur without loss of energy, we can set $B = A$. The standing wave in the cone is the sum of forward- and backward-going waves:

$$p = p_{+} + p_{-} = \frac{A}{r}\left(\mathrm{e}^{j(\omega t - kr)} + \mathrm{e}^{j(\omega t + kr)}\right). \tag{4.37}$$

We can gain understanding of the nature of this standing wave by considering a new variable $\psi = rp$. Replacing p by ψ/r in Eq. 4.37 and multiplying through by r gives the following equation:

$$\psi = rp = A\left(\mathrm{e}^{-jkr} + \mathrm{e}^{jkr}\right)\mathrm{e}^{j\omega t} = (2A\cos kr)\mathrm{e}^{j\omega t}. \tag{4.38}$$

Comparing the form of Eqs. 4.38 and 4.21 shows that the quantity ψ has the same mathematical behaviour as the quantity p in a plane standing wave. The boundary condition at the open end of the cone is the same for ψ as it is for p, since if there is

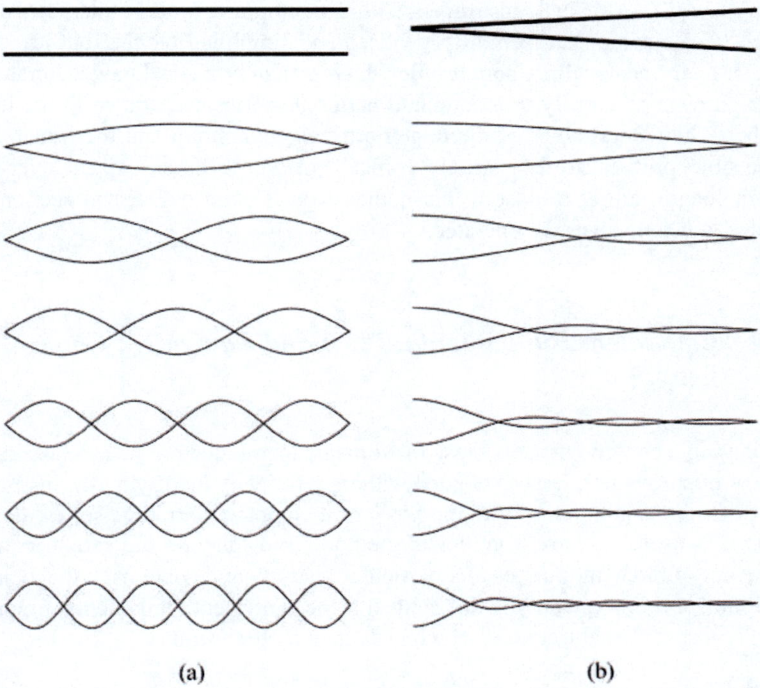

Fig. 4.29 (a) The first six standing wave patterns for the cylindrical tube shown at the top, open at both ends. (b) The first six standing wave patterns for the conical tube shown at the top, complete to the vertex

a pressure node there, the value of $\psi = rp$ must also be always zero. The condition at the input end is a little more subtle. As we approach the vertex of the cone, the value of $p = \psi/r$ becomes larger and larger; it will tend to infinity as $r \to 0$ unless ψ is also zero at $r = 0$. Thus the input end of the cone must also be a node for ψ.

What emerges from this burst of mathematics is that the standing waves in a cone have the same patterns when expressed in terms of ψ as do the standing waves in a cylinder open at both ends, since in both cases, there is a node at each end. The first six such patterns are shown in Fig. 4.29a. To derive from these the pressure standing wave patterns in the cone, we simply have to multiply the width of the ψ pattern by $1/r$: the result of this transformation is shown in Fig. 4.29b.

It is clear from the standing wave pressure distributions shown in Fig. 4.29b that they satisfy the conditions for sound generation using a valve effect source, since each has a pressure antinode at the input end. The distance between adjacent nodes is half a wavelength, so it is straightforward to confirm from the patterns that the resonance frequencies form a complete harmonic series given by the equation

$$f_n = \frac{c}{\lambda_n} = \frac{nc}{2L} = f_1,\ 2f_1,\ 3f_1 \ldots \tag{4.39}$$

with $f_1 = c/2L$. A realistic instrument with a mouthpiece, tuning slides, etc. will of course deviate somewhat from this idealised and simplified picture, but it remains true that instruments with a bore profile which is mainly conical have natural notes which correspond closely to a complete harmonic series even down to the lowest member. The natural notes of the flugelhorn, the euphonium and the french horn, whose bore profiles are not strictly conical but whose bores expand over most of their length, are also close to harmonic, even if extra cylindrical sections are introduced when valves are operated.

4.3.4 Equivalent Fundamental Pitch and Equivalent Cone Length

The playing technique of most brass instruments is founded on the availability of a range of pitches which can be sounded without physically modifying the instrument and which correspond closely to the notes of a complete harmonic series. For this reason it is useful to have a graphical method for displaying the extent to which the pitches of the natural notes of a particular instrument deviate from the expected harmonics. A useful quantity in this context is the equivalent fundamental frequency (EFF) of a specific natural note, which is defined by the formula

$$\mathrm{EFF} = \frac{f_n}{n}, \tag{4.40}$$

for a natural note with rank number n and frequency f. As an example we can take the frequencies of the first three impedance peaks in Fig. 4.28, which are 54.7 Hz, 109.8 Hz and 165.5 Hz. The corresponding EFF values are 54.7 Hz, 109.8/2 = 54.9 Hz and 165.5/3 = 55.2 Hz. The 'harmonicity' of the set of frequencies is measured by the correspondence between the EFF values: for a perfect harmonic series, the values would be identical.

For musical purposes it is often more useful to display the deviations of natural notes from perfect harmonicity in terms of pitch shifts rather than frequency changes. The equivalent fundamental pitch (EFP) is defined by the formula

$$\mathrm{EFP}(n) = \frac{1200}{\log(2)} \log\left(\frac{f_n}{nf_0}\right), \tag{4.41}$$

where f_0 is the frequency of a reference pitch; EFP(n) is the difference in cents between the pitch of the nth natural note and the nth harmonic of the reference pitch.

Figure 4.30 illustrates EFP plots derived from the calculated impedance peak frequencies of the cylinder and cone discussed above. Both tubes were the same length as a B$^\flat$ trombone, so the reference pitch was chosen to be B$^\flat$1 (f_0 =

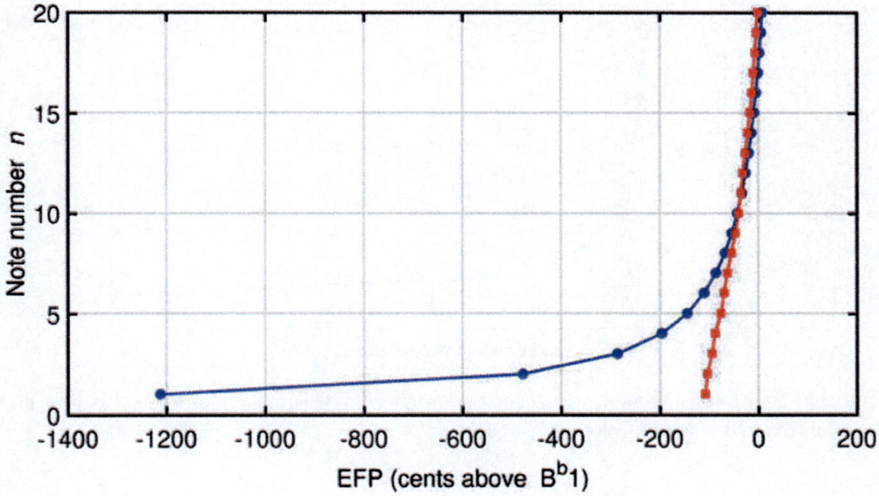

Fig. 4.30 EFP plots for the cylindrical tube (blue circles) and for the conical tube (red squares) (Color figure online)

58.3 Hz). The rank number of the impedance peak is plotted vertically, while the EFP value is plotted horizontally. Thin trend lines join the measured points.

The EFP values for a perfect set of harmonic peaks with fundamental frequency 58.3 Hz would all lie on the vertical line at EFP = 0. In agreement with the previous discussion, the points for the truncated cone are reasonably close to this line, with the first peak just over a semitone flat. The points for the cylinder show the expected large deviations for the lower peaks.

An idealised cone complete to the vertex (neglecting effects of losses and radiation to be discussed later) should have natural notes forming the complete harmonic series defined by Eq. 4.39. This has led to an alternative way of describing inharmonicity in terms of the equivalent cone length L_{ec} (Pyle 1975). Equation 4.39 can be rewritten in the form

$$L = \frac{nc}{2f_n}, \tag{4.42}$$

where L is the (constant) length of the ideal cone whose nth resonance has frequency f_n. For an instrument which is not a perfect cone, the ratio $nc/2f_n$ will not be constant; to reflect this L is replaced by a variable L_{ec}, defining the length of a perfect cone whose nth resonance frequency is f_n:

$$L_{ec}(n) = \frac{nc}{2f_n}. \tag{4.43}$$

Equivalent cone length plots for the cylinder and cone are shown in Fig. 4.31. These plots contain essentially the same information as the EFP plots, but display it

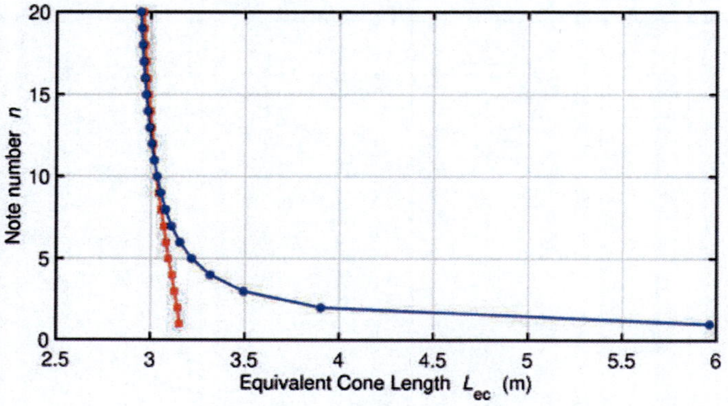

Fig. 4.31 Plots of equivalent cone length L_{ec} for the cylindrical tube (blue circles) and for the conical tube (red squares) (Color figure online)

in a way which focuses on the changes in effective length of the instruments rather than the deviations in pitch. The truncated cone is shown to change its effective length by only 19 cm over the first 20 natural notes, while L_{ec} for the cylinder is 3 m greater for the 1st natural note than for the 20th.

4.3.5 The Mouthpiece as a Helmholtz Resonator

The mouthpiece of a brass instrument is usually a separate item which is inserted into the instrument only when it is to be played. The choice of mouthpiece design is considered to be of great importance by brass players and teachers, and a change of mouthpiece can have a significant musical effect on the intonation, timbre and ease of playing of an instrument. The variety of mouthpiece shapes and sizes in common use is surveyed in Sect. 7.4. In the present section, we look at some of the fundamental principles which underlie the musical importance of mouthpiece design.

A typical modern trombone mouthpiece is illustrated in Fig. 4.32a. The internal profile of a mouthpiece of this type has the general form shown in Fig.4.32b. The width and profile of the external rim (not shown in Fig.4.32b) have an important influence on the player's embouchure (see Chap. 3), and some manufacturers offer a range of different rims which can be screwed on to the same mouthpiece body. From the acoustical point of view, however, the most important features are the volume of the cup, the radius of the throat and the length and taper of the backbore.

Not all brass instrument mouthpieces have such pronounced constriction as the throat in the trombone mouthpiece shown in Fig. 4.32b: french horn mouthpieces typically have a relatively deep conical shape with a gradually tapering profile (see Sect. 7.4), while some ancient brass instruments appear to have had mouthpieces

(a)

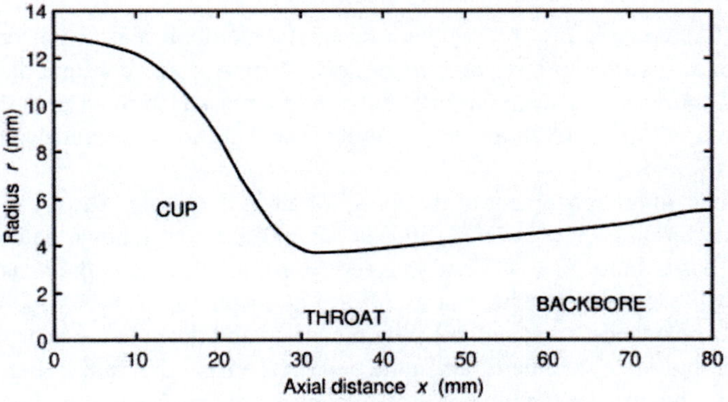

(b)

Fig. 4.32 **(a)** Photograph of a Denis Wick 5AL trombone mouthpiece. **(b)** The bore profile of a typical large bore tenor trombone mouthpiece

with almost no constriction (Sect. 9.1). The combination of a fairly large cup volume and a narrow throat does however result in a resonant behaviour of the mouthpiece which has several important musical consequences.

In Sect. 4.1.1 we saw that an empty bottle with a narrow neck could be treated as a lumped acoustic resonator, with a characteristic resonance frequency which could be excited by blowing across the neck. The description 'Helmholtz resonator' is given to this type of enclosure because its acoustical properties were first explained by the great nineteenth-century acoustician Hermann Helmholtz (1877). As can be seen from the diagrams in Fig. 4.33, a cup mouthpiece has two crucial properties in common with the bottle: a relatively large volume of air (closed at the rim by the lips in playing) and an exit through which air can enter and leave the volume. Since the model of the cup mouthpiece as a Helmholtz resonator sheds valuable light on its musical properties, we will derive the formula giving the frequency f_R of the resonator in terms of the geometrical properties of the mouthpiece.

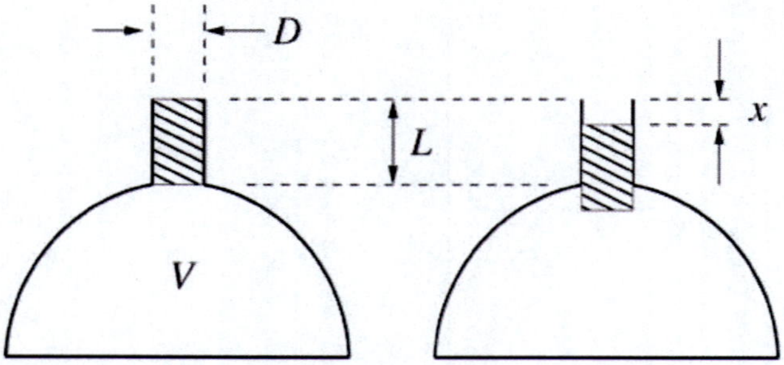

Fig. 4.33 Model of a mouthpiece cup and throat as a Helmholtz resonator. Cup volume V; throat length L and diameter D

The left diagram in Fig. 4.33 shows a schematic mouthpiece in cross-section. The volume of air in the cup is V, and the neck has length L and constant diameter D. This is of course a very oversimplified picture of a brass instrument mouthpiece, as comparison with Fig. 4.32b makes clear, but it will suffice to establish the general principles we need.

We focus on the movement of the 'plug' of air in the throat. The cross-sectional area of the throat is $S = \pi D^2/4$, and the air in the throat (shown hatched in the diagram) has volume $V = LS$ and mass $M = \rho LS$, where ρ is the density of the air. When the air is at rest, it is at a uniform atmospheric pressure p_{at}. When the Helmholtz resonance is excited, the plug moves up and down in the neck, bouncing on the compressible volume of air in the cup. In the right diagram in Fig. 4.33, the situation is shown after the plug of air has moved a distance x into the mouthpiece. Note that we are taking the positive direction of x downward. The volume of air which has entered the cup is Sx, so the air that was in the volume V on the left diagram is now compressed into a smaller volume $V - Sx$.

The compression of the air in the cup results in an increase in pressure from p_{at} to $p_{at} + p_{ac}$. As the air pressure rises and falls, the local temperature of the air also rises and falls. Since the period of an audible sound wave is always less than 0.05 s, there is insufficient time between an expansion and the next compression for isothermal conditions to be re-established. The relationship between pressure and volume therefore follows the adiabatic law

$$pV^{\gamma} = \text{constant}, \tag{4.44}$$

where γ is the ratio of the specific heats at constant volume and constant pressure ($\gamma = 1.4$ for air). The increase in pressure in the cup is therefore related to the decrease in volume by the equation

$$\frac{p_{at} + p_{ac}}{p_{at}} = \left(\frac{V - Sx}{V}\right)^{-\gamma}. \tag{4.45}$$

Using the Binomial Theorem, the right-hand side of Eq. 4.45 can be expanded in powers of Sx/V:

$$\left(\frac{V - Sx}{V}\right)^{-\gamma} = \left(1 - \frac{Sx}{V}\right)^{-\gamma} = 1+\gamma\left(\frac{Sx}{V}\right)+\frac{\gamma(\gamma+1)}{2}\left(\frac{Sx}{V}\right)^2 \ldots \quad (4.46)$$

If we assume that the fractional change in volume is always much less than 1, we can drop all the terms in the expansion involving powers of (Sx/V) higher than 1, and Eq. 4.45 simplifies to

$$p_{\mathrm{ac}} = p_{\mathrm{at}}\gamma\left(\frac{Sx}{V}\right). \quad (4.47)$$

In the situation illustrated in the right diagram in Fig. 4.33, there is a pressure difference of p_{ac} between the compressed air in the cup and the atmospheric pressure outside the neck. This pressure results in a net force $F = -p_{\mathrm{ac}}S$ on the plug of air; this upward force is written with a negative sign since it is in the opposite direction to the displacement x. Newton's second law applied to the plug then gives

$$F = -p_{\mathrm{ac}}S = -p_{\mathrm{at}}\gamma\left(\frac{S^2x}{V}\right) = M\frac{\mathrm{d}^2x}{\mathrm{d}t^2} = \rho LS\frac{\mathrm{d}^2x}{\mathrm{d}t^2}. \quad (4.48)$$

Rearranging Eq. 4.48 gives the equation of motion for the air plug:

$$\frac{\mathrm{d}^2x}{\mathrm{d}t^2} = -\left(\frac{\gamma p_{\mathrm{at}}S}{\rho VL}\right)x. \quad (4.49)$$

This equation has the standard form for Simple Harmonic Motion, in which the acceleration is equal to displacement multiplied by a negative constant $-\omega^2$:

$$\frac{\mathrm{d}^2x}{\mathrm{d}t^2} = -\omega^2x, \quad (4.50)$$

with

$$\omega = \left(\frac{\gamma p_{\mathrm{at}}S}{\rho VL}\right)^{1/2} = \left(\frac{c^2S}{VL}\right)^{1/2}. \quad (4.51)$$

In deriving Eq. 4.51, we used the relationship between the speed of sound c and the pressure and density of the air:

$$c = \left(\frac{\gamma p_{\mathrm{at}}}{\rho}\right)^{1/2}. \quad (4.52)$$

The solution of Eq. 4.50 is a sinusoidal displacement with frequency $f_R = \omega/2\pi$. The corresponding pressure change is also sinusoidal, as we found with the experiment in blowing across a bottle in Sect. 4.1.1. From Eq. 4.51 the resonance frequency is

$$f_R = \frac{c}{2\pi}\left(\frac{S}{VL}\right)^{1/2}. \tag{4.53}$$

4.3.6 Mouthpiece Effects on Intonation and Timbre

There are clearly serious limitations in applying this simplified model to a realistic brass instrument mouthpiece: since the throat is not a short cylinder but a relatively long tapering backbore, it is not obvious what value of L would be appropriate to insert into the formula for f_R. Equation 4.53 is nevertheless very useful in showing how the mouthpiece resonance frequency depends on the throat diameter and cup volume. f_R will be increased if the throat cross-sectional area S is increased, other variables remaining unchanged. Increasing only the volume V of the cup will reduce f_R.

The pitch of the Helmholtz resonance for a particular mouthpiece can be assessed experimentally by slapping the palm of the hand against the mouthpiece rim to close it. We already discussed this technique for sending an acoustic pulse down a complete instrument (Sect. 4.1.4). In the case of an isolated mouthpiece, it is fairly easy to hear the pitch of the rapidly decaying resonance in the sound radiated from the neck of the mouthpiece. The corresponding frequency is sometimes called the 'popping frequency' of the mouthpiece (Benade 1976).

Figure 4.34 shows the sound pressure signal recorded 20 cm from the end of the neck of a Denis Wick 5AL trombone mouthpiece when the rim was slapped and held shut. The successive peaks and dips in the pressure correspond to the plug of air in the throat oscillating towards and away from the cup, as discussed in the

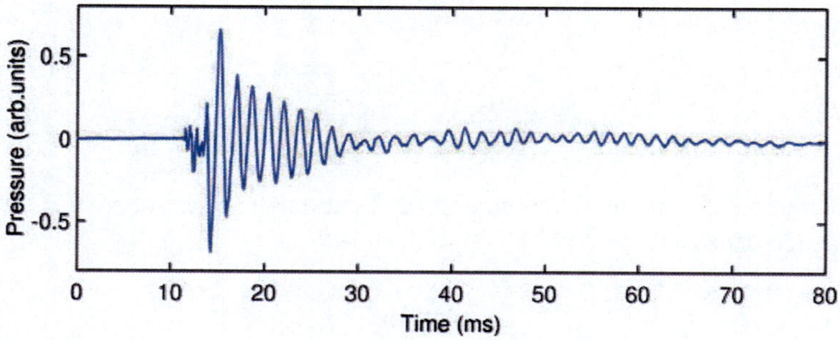

Fig. 4.34 Signal from a Denis Wick 5AL trombone mouthpiece when a palm is slapped against the rim

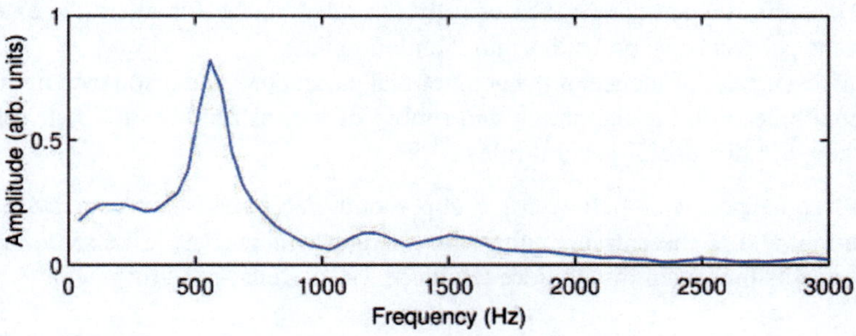

Fig. 4.35 Frequency spectrum of the signal shown in Fig. 4.34

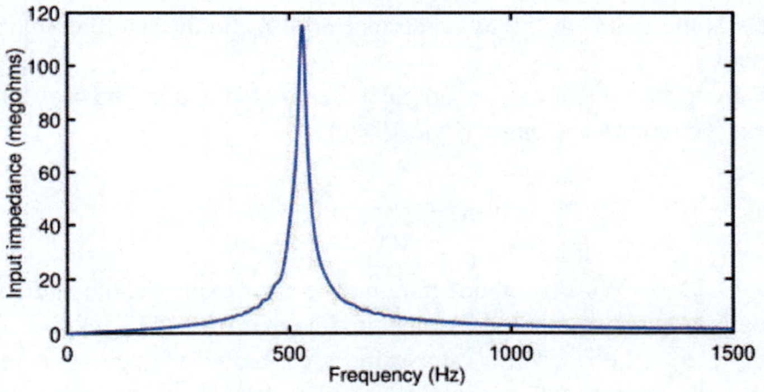

Fig. 4.36 Input impedance of a Denis Wick 5AL tenor trombone mouthpiece

previous section. The amplitude of the fluctuations in pressure diminishes rapidly as the energy is lost through sound radiation and frictional losses in the throat.

Comparing the sound from the slapped mouthpiece aurally with the sounds from a piano keyboard, several musicians agreed that the pitch class was D, although it was hard to judge the octave. A spectral analysis of the recorded sound was calculated using Audacity software; despite the relatively poor frequency resolution resulting from the short duration of the signal, the frequency spectrum in Fig. 4.35 shows that the pitch of the resonance was close to D5, with a frequency $f_R \simeq$ 580 Hz. A similar experiment with a Yamaha 11C4-7C trumpet mouthpiece gave a popping pitch of approximately G$^\sharp$5, with $f_R \simeq 830$ Hz.

The measured input impedance curve for the trombone mouthpiece is shown in Fig. 4.36. The Helmholtz resonance is a very high and narrow peak at 535 Hz. This is significantly lower than the 'popping frequency' quoted above because in the impedance measurement head, the mouthpiece is pressed against a flat rubber pad while in the slapping technique the surface of the palm bulges into the mouthpiece. This bulge reduces the mouthpiece volume, raising the resonance frequency. In fact the popping frequency of a brass instrument mouthpiece can be heard to change by around a semitone depending on whether the fingers of the hand against which the mouthpiece is slapped are relaxed or fully extended, since this alters the extent to which the palm surface protrudes into the mouthpiece.

In assessing the influence of the geometrical parameters and resonance frequency of a mouthpiece on the intonation and timbre of the instrument of which it forms part, there are two useful general rules:

1. At frequencies well below f_R, a cup mouthpiece behaves like a cylindrical continuation of the entrance tube with a length which gives it the same internal volume as the actual mouthpiece (Bouasse 1929; Benade 1976):

$$L_{eff} = \frac{V_{int}}{\pi r^2} \quad [f \ll f_R], \tag{4.54}$$

   where r is the radius at the tube entrance and V_{int} is the internal volume of the mouthpiece;
2. At the frequency f_R, a cup mouthpiece behaves like a cylindrical tube whose length is a quarter wavelength (Pyle 1975):

$$L_{eff} = \frac{c}{4f_R} \quad [f = f_R]. \tag{4.55}$$

Taking the Denis Wick trombone mouthpiece as an example, the internal volume of the mouthpiece was measured by blocking the exit from the backbore and filling the mouthpiece to the rim with water using a calibrated syringe. This took 12 ml of water, so the internal volume was $V = 12 \times 10^3\,\text{mm}^3$. The radius of the exit is 6.9 mm, so $L_{eff}(f \ll f_R) = 80\,\text{mm}$. This is just the external length of the mouthpiece, so in this case, the actual length and the low-frequency effective length coincide.

Assuming $f_R = 535\,\text{Hz}$, $L_{eff}(f = f_R) = 161\,\text{mm}$, suggesting that the effective length of the mouthpiece increases by 81 mm as the frequency rises from a low value to the resonance frequency. This is around 3% of the sounding length of a trombone, implying that the pitch difference between the first and tenth impedance peaks is about half a semitone lower than it would be if the mouthpiece were replaced by a cylindrical tube of the same length.

It is important to remember that the model illustrated in Fig. 4.33 is a very crude approximation to a realistic trombone mouthpiece with a tapering backbore, and its acoustical behaviour when attached to an instrument will be more complicated than the simple lumped impedance we assumed in deriving Eq. 4.53. Figure 4.37 shows the predicted effect of adding a mouthpiece to the cylindrical tube discussed in Sect. 4.3.2. In this case the complete bore including the mouthpiece was approximated by a series of cones and cylinders, and the wave equation was solved by a method which took into account the fact that the wavefronts in the mouthpiece are not plane (Braden 2006).

The blue circles in Fig. 4.37 show the calculated EFP values for a simple cylinder of length 2834 mm, while the red squares show the EFP values when the first 261 mm of the cylinder are replaced by the Denis Wick 5AL mouthpiece and a

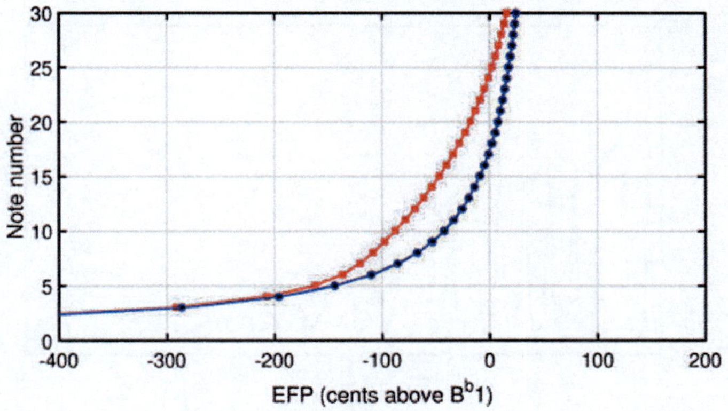

Fig. 4.37 EFP for two tubes of the same sounding length 2834 mm: a plain cylinder (blue circles) and a cylinder with a Denis Wick 5AL mouthpiece and mouthpipe (red squares) (Color figure online)

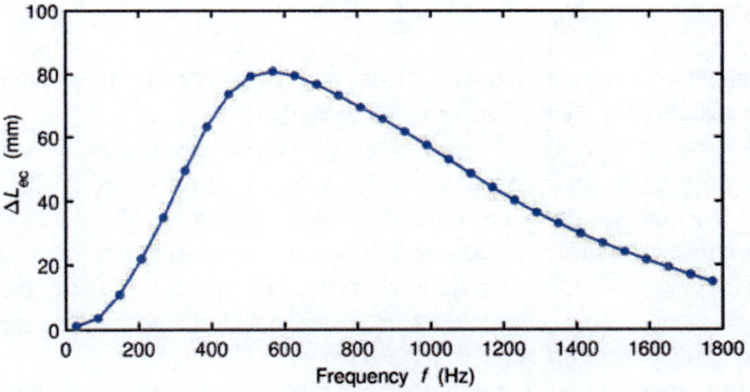

Fig. 4.38 Increase in equivalent cone length L_{ec} when the first section of the plain cylinder is replaced by a Denis Wick 5AL mouthpiece and mouthpipe

gently tapering mouthpipe. The introduction of the mouthpiece makes no significant difference to the frequencies of the first two impedance peaks (whose EFP values are not shown in Fig. 4.37); as expected, the mouthpiece is behaving in the low-frequency range like a cylindrical tube of the same length. As the peak number increases, the EFP difference also grows; at the tenth peak, whose frequency is close to f_R, the introduction of the mouthpiece has reduced the EFP by 45 cents. Since the pitch D5 is at the top of the normal playing range of the tenor trombone, we can summarise the effect of the mouthpiece on intonation by stating that it has little effect on the pitches of the low register but flattens the high register notes by an amount which increases with the number of the natural note.

In Fig. 4.38 the change in equivalent cone Length L_{ec} due to the mouthpiece is plotted as a function of frequency. This curve shows that the increase in effective

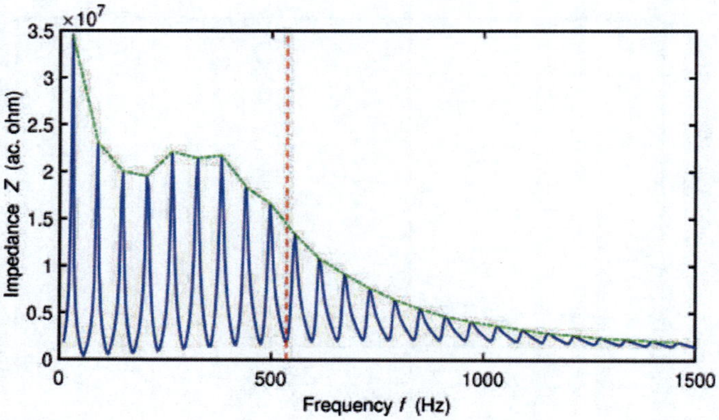

Fig. 4.39 Calculated input impedance for a cylindrical tube of length 2834 mm including a Denis Wick 5AL mouthpiece and mouthpipe. The resonance frequency of the trombone mouthpiece (535 Hz) is marked by a vertical dashed line and the peak envelope by a broken green line (Color figure online)

length reaches a maximum value of 80 mm just below 600 Hz, in broad agreement with the prediction of the lumped impedance model.

Another important effect of the addition of a mouthpiece to an instrument tube is illustrated by the peak envelope in Fig. 4.39. A comparison of this envelope with the corresponding peak envelope for a plain cylinder in Fig. 4.26a shows that the mouthpiece provides an additional boost to the amplitudes of a number of impedance peaks centred roughly on the sixth peak. Notes in this important range of the instrument's compass are therefore easier to sound, and the lower harmonics in their spectra are reinforced.

The impedance peaks of the cylindrical tube are reinforced by the trombone mouthpiece in a range significantly below the resonance frequency of the mouthpiece, as Fig. 4.39 demonstrates. The positioning of the reinforcement region relative to the mouthpiece resonance depends on the relative dimensions of mouthpiece and main tube diameter (Lurton 1981); (Chaigne and Kergomard 2016, pp. 312ff). Figure 4.40 shows that when the trombone mouthpiece is replaced by a trumpet mouthpiece, the peak reinforcement region is slightly above the mouthpiece resonance.

There is a general consensus among brass instrument players that a small mouthpiece cup volume favours a bright sound, while a deeper cup produces a darker sound. This view is borne out by statements in the catalogues of mouthpiece manufacturers. For a given throat diameter, reducing the cup volume increases the popping frequency and therefore raising the frequency range in which the impedance peaks are boosted. The extent to which such a change increases the predominance of upper harmonics and therefore brightness in the sound of a trumpet was investigated by Poirson et al. (2005), using a mouthpiece with a continuously adjustable cup depth. Objective studies involving input impedance measurements,

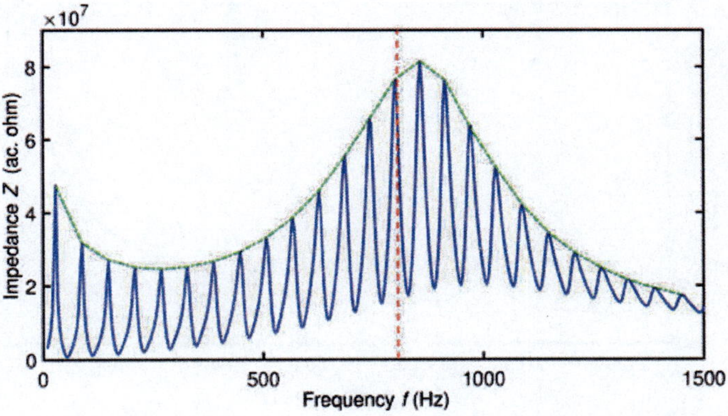

Fig. 4.40 Calculated input impedance curve for a cylindrical tube of length 2834 mm with a Vincent Bach 7C trumpet mouthpiece. The trumpet mouthpiece resonance frequency (806 Hz) is marked by a vertical dashed line and the peak envelope by a broken green line (Color figure online)

playing tests with an artificial mouth and physical modelling simulations were combined with a subjective study involving hearing tests and a panel of listeners. Some support was found for the idea that increased brightness was related to increases in the relative amplitudes of high-frequency impedance peaks, but it was noted that the variability in the timbre generated by human performers was of the same order as the variability related to changes in mouthpiece depth.

4.3.7 *Sound Waves in Flaring Bells*

The tuning effect of a mouthpiece on the resonances of a cylindrical tube below the mouthpiece resonance frequency is to reduce the inharmonicity, since the reduction in EFP increases with increasing note number. This brings the EFP curve of the cylinder a little closer to the vertical line corresponding to perfect harmonicity (Fig. 4.41). However the EFP curve for the hosepipe horn (cylindrical tube plus mouthpiece) is still far from a vertical line, and the mouthpiece effect is insignificant below the fifth resonant mode of the tube. To create a resonating tube with good harmonicity over most of its range, we need to find a means of preferentially raising the EFP values of the lower modes. Fortunately this can be done by replacing the last section of the cylindrical tube by a section in which the radius increases more and more rapidly, ending in a bell with a diameter several times the input radius. To understand the tuning effect of such a flaring bell, which is a characteristic feature of all musically useful trumpets and trombones, we need to consider how sound waves propagate in tubes with varying cross-sectional area.

Applying the linear acoustic wave equation (Eq. 4.1) to a segment of flaring pipe leads to the equation (Pierce 1989)

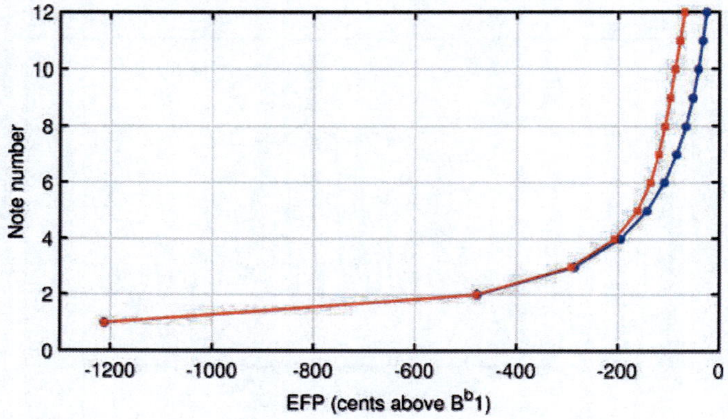

Fig. 4.41 EFP values for the first 12 resonant modes of the cylindrical tubes with (squares) and without (circles) a trombone mouthpiece

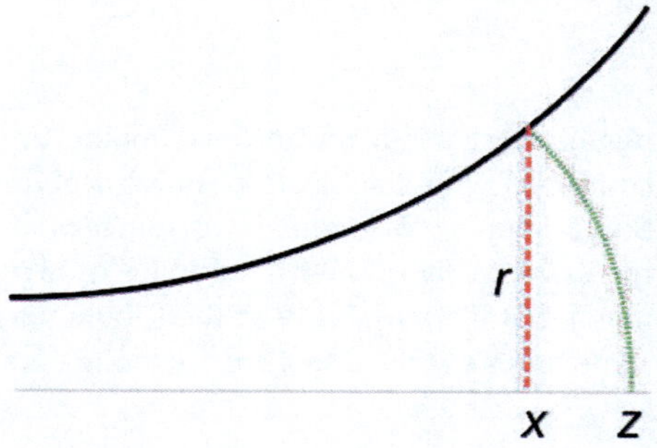

Fig. 4.42 Relationship between plane wavefront cutting the horn axis at x and quasi-spherical wavefront, perpendicular to the wall at x and cutting the horn axis at z

$$\frac{1}{S}\frac{\partial}{\partial x}\left(S\frac{\partial p}{\partial x}\right) = \frac{1}{c_0^2}\frac{\partial^2 p}{\partial t^2}, \tag{4.56}$$

where $S(x) = \pi r^2$ is the cross-sectional area of the pipe at position x along the bore and c_0 is the speed of sound propagation in open air. This is usually called the Webster horn equation. This formulation assumes plane wave propagation in the tube. As was already noted when discussing sound propagation in conical tubes (see Fig. 4.27), the wavefronts in a tube of varying radius are in general curved surfaces. It is possible to modify the plane wave treatment to take the curvature of the wavefronts into account by redefining S as the area of the wavefront surface which cuts the bore axis at z (see Fig. 4.42) (Benade and Jansson 1974; Fletcher and Rossing 1998). An effective bore radius can then be defined as

$$a = \left(\frac{S}{\pi}\right)^{\frac{1}{2}}. \tag{4.57}$$

When discussing spherical waves in conical tubes (Sect. 4.3.3), we found that the wave equation was simplified by changing the variable from p to $\psi = pr$. A similar substitution is helpful in the more general case of flaring tubes:

$$\psi = pS^{\frac{1}{2}}. \tag{4.58}$$

Assuming that the wave has a sinusoidal time dependence with angular frequency ω, the Webster equation becomes

$$\frac{\partial^2 \psi}{\partial z^2} + \left(\frac{\omega^2}{c_0^2} - U\right)\psi = 0, \tag{4.59}$$

where

$$U = \frac{1}{a}\frac{\partial^2 a}{\partial z^2} \tag{4.60}$$

is known as the horn function. In the plane wave approximation, $a = r$ and $z = x$; the horn function can then be calculated numerically from a known bore profile $r(x)$ using Eq. 4.60. In a more realistic treatment using curved wavefronts, it is first necessary to establish the functional relationships between (x, r) and (z, a) from the geometry of the assumed wavefront shape.

It has been pointed out by several authors that the form of the Webster equation 4.59 is formally identical to the time-independent Schrödinger Equation in quantum mechanics, with the horn function playing the role of the potential energy. It is a striking example of the power of the fundamental concepts of physics that the emission of light from a quantum well laser and the emission of sound from a trombone are governed by the same basic equation.

The constant c_0 which appears in Eqs. 4.56 and 4.59 was defined as the speed of sound in free air. The musically useful 'harmonicity-correcting' property of a flaring bell arises from the fact that the speed of sound is not constant everywhere in the tube. This is evident if we rewrite Eq. 4.59 as

$$\frac{\partial^2 \psi}{\partial z^2} + \left(\frac{\omega^2}{c^2}\right)\psi = 0, \tag{4.61}$$

with

$$c = \omega\left(\frac{\omega^2}{c_0^2} - U\right)^{-1/2}. \tag{4.62}$$

Equation 4.61 has a solution which can be written as

$$\psi = Ae^{j\omega(z/c-t)}, \tag{4.63}$$

representing a sinusoidal wave with angular frequency ω propagating with a speed c. For cylindrical or conical bores, $U = 0$ everywhere in the tube, and $c = c_0$. In a flaring bell, on the other hand, U increases with the rate of flare of the tube; Eq. 4.62 shows that this results in an increase in the local speed of sound propagation c. If the speed of propagation increases at constant frequency, the local wavelength must also increase: this means that nodes and antinodes in a standing wave are further apart in the flaring bell of a trombone than they are in the slide section.

If the maximum value U_{max} of the horn function is greater than ω^2/c_0, there will be a region of the tube in which the solution to Eq. 4.62 is imaginary. To understand what happens in this region, it is helpful to rewrite Eq. 4.63 in terms of a local wave number:

$$k = \omega/c = \left(k_0^2 - U\right)^{1/2}, \tag{4.64}$$

where $k_0 = \omega/c_0$. The solution to Eq. 4.61 then takes the form

$$\psi = Ae^{j(kz-\omega t)} = \left(Ae^{jkz}\right)e^{j\omega t}. \tag{4.65}$$

When both k and ω are real numbers, Eq. 4.65 represents a wave propagating in the positive z direction. When $U > k_0^2$, the local wave number k is imaginary, and the term in brackets on the right-hand side of Eq. 4.65 can be written as $A\exp(-kz)$. This solution represents an oscillating pressure change which is not propagating, but decaying exponentially; such a wave is described as evanescent.

It is now possible to understand why the lower modes in a flaring tube are higher in frequency than the corresponding modes of a cylindrical tube of the same length. At a frequency for which $U \gg k_0^2$, the region of the flaring tube in which a propagating wave can exist is much shorter than its sounding length, and the standing wave pattern is effectively squeezed into this shorter length. It should also be borne in mind that in the region of rapid flare, the distance between nodes in the standing wave pattern expands; the pressure distribution in a realistic instrument can be calculated numerically if the bore profile is known.

4.3.8 A Theoretical Example: The Bessel Horn

There is a family of theoretical horn profiles for which analytic solutions to the Webster equation can be found; because the solutions are expressed in terms of Bessel functions, these profiles are known as Bessel horns. The general expression for a Bessel horn bore profile is

$$r = \frac{B}{(x_0 - x)^{\alpha}}, \tag{4.66}$$

where B, x_0 and α are parameters which can be adjusted to provide an approximate fit to a measured bell profile. The exponent α determines how rapidly the bell flares: values of α between 0.5 and 0.65 have been found to give a similar rate of flaring to that found in the expanding sections of modern trumpets and trombones (Benade 1976; Chaigne and Kergomard 2016). Once α has been fixed, the coefficients B and x_0 determine overall shape and scale of the bore. With the input end of the horn at $x = 0$, there is a singularity at $x = x_0$ where $r \rightarrow \infty$; this is typically a few centimetres beyond the bell exit plane.

It should be emphasised that the Bessel horn is a mathematical abstraction rather than a practicable musical instrument. It would certainly not be possible to manufacture a playable Bessel trombone, since a slide section could not be incorporated into the continuously expanding bore. However its theoretical simplicity makes it a convenient vehicle for deriving and displaying some of the general principles of sound propagation in flaring tubes. In Sect. 4.3.9 these principles will be applied to a real brass instrument: the classic Conn 8H trombone.

The acoustical behaviour of the Bessel horn was explored in detail in a seminal paper by Arthur Benade and Erik Jansson (1974). Following their example, we consider a Bessel horn resembling the flaring section of a trombone. Unscrewing the slide section of the Conn 8H instrument leaves a 1.13 m tube with an input radius of 8×10^{-3} m, flaring out to the bell end with radius 0.108 m. We choose an exponent $\alpha = 0.65$, similar to that adopted by Benade and Jansson. With α fixed, substituting the values of $r(x)$ at the entrance and the bell plane yields the remaining two parameters in Eq. 4.66: $B = 8.77 \times 10^{-3}\,\mathrm{m}^{1.65}$ and $x_0 = 1.15$ m.

The resulting Bessel horn profile is shown as the solid curve in Fig. 4.43a. Comparison with the measured profile of the Conn 8H bell section, plotted as a dashed line in Fig. 4.43a, shows that the Bessel horn profile reproduces the overall shape of the bell well, but is not an exact fit. Bessel horn profiles have proved useful in modelling brass instrument bells, but it is usually necessary to divide the bell into several different sections, with a different set of parameters for each section. It is nevertheless enlightening to consider the nature of the standing wave patterns and mode frequencies in the idealised single Bessel horn shown in Fig. 4.43a.

Figure 4.43b shows the horn function U for the Bessel horn approximation to the trombone bell, calculated using Eq. 4.60 with the assumption that the wavefronts in the horn are spherical. The horn function is small in the first half of the horn, in which the flare rate is very low. In the more strongly flaring part near the bell exit, U increases rapidly with x, reaching a peak just under 5 cm inside the bell. It then drops steeply, approaching zero around 2 cm beyond the bell exit.

The calculated input impedance for the Bessel horn is shown by the solid line in Fig. 4.44. Each peak corresponds to a resonant mode of the contained air column with a pressure antinode at the entrance. The values of k_0^2 for the first four modes are marked by horizontal lines in Fig. 4.43b, and the pressure standing wave patterns for these modes are are illustrated in Fig. 4.43c. The first mode has $f(1) = 125.2$ Hz; the corresponding value of $k_0^2(1) = 2.28\,\mathrm{m}^{-2}$ is marked by the thin solid line in Fig. 4.43b. This line intersects the horn function curve at $x = 0.698$; at that point the

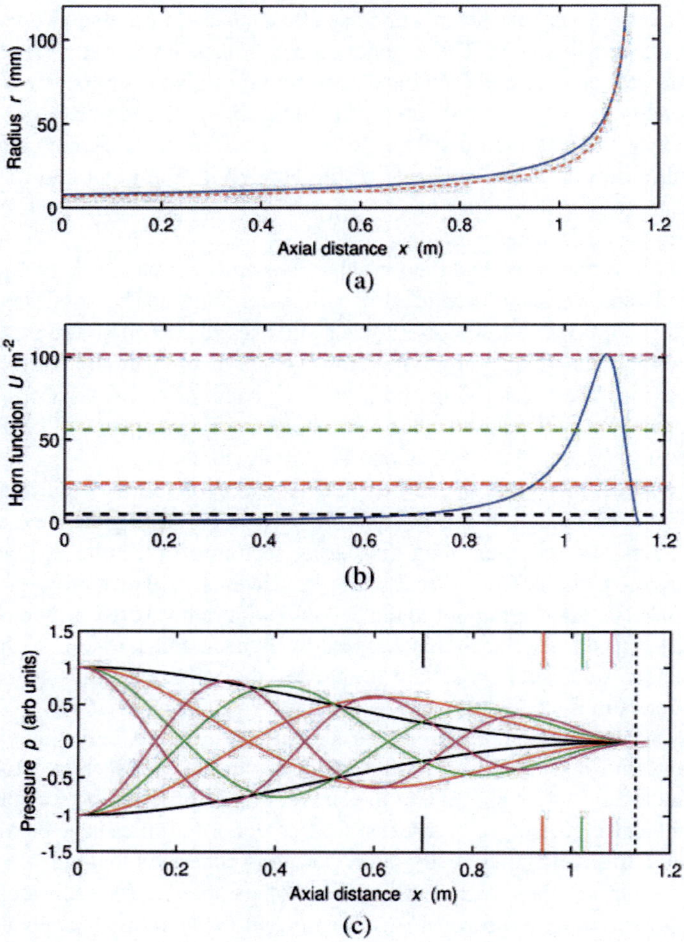

Fig. 4.43 Calculated horn function and internal pressure distributions for the first four acoustic modes of a Bessel horn approximation to a trombone bell. (**a**) Bore profiles for a Bessel horn trombone bell approximation (solid line) and a Conn 8H trombone bell section (dashed line). (**b**) Horn function calculated for the Bessel horn trombone bell approximation. Values of k^2 for the first four resonant modes are marked by horizontal lines. (**c**) Pressure distributions for resonant modes in the Bessel horn trombone bell approximation. Mode 1: black lines. Mode 2: red lines. Mode 3: green lines. Mode 4: magenta lines. The bell exit plane (at $x = 1.13$ m) is marked by a continuous vertical dashed line (Color figure online)

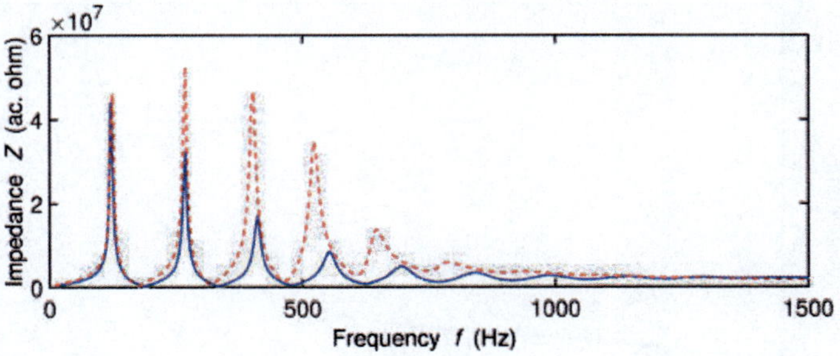

Fig. 4.44 Impedance curves calculated for the Bessel horn trombone bell approximation with no mouthpiece (solid curve) and with a Denis Wick 5AL mouthpiece inserted (dashed curve)

travelling wave is reflected, and for larger values of x, the solution is an evanescent (exponentially decaying) wave.

As the mode number n increases, the point of intersection of the $k_0^2(n)$ line with the U curve moves closer to the bell exit, and the length of the tube in which travelling and standing waves can exist increases. The limit of propagation for each of the first four modes is marked by an interrupted vertical line in Fig. 4.43c. For modes with $n > 4$, the wave can propagate over the entire length of the horn.

The pressure distributions in Fig. 4.43c were calculated for the simplified case in which radiation from the open end was neglected. In reality, even at low frequencies, some wave energy leaks through the barrier represented by the horn function and is radiated from the bell as a sound wave. The fraction of the incident energy which is transmitted rather than reflected by the horn function barrier depends strongly on the width of the barrier at the relevant value of k_0^2, and also on the magnitude of $U_{max} - k_0^2$. For $n = 1$ the barrier is wide and $U_{max} - k_0^2$ is large, so almost all the sound energy is reflected back into the horn: the internal standing wave is strong, as confirmed by the high first peak in the input impedance curve, but little sound energy is radiated. For $n = 4$ the barrier is thin, and $U_{max} - k_0^2$ is very small; the much reduced height of the fourth impedance peak shows that less energy is reflected back into the horn and more is transmitted.

The frequency

$$f_c = \frac{c}{2\pi} U_{max}^{1/2} \tag{4.67}$$

for which $k_0^2 = U_{max}$ is known as the cutoff frequency for the bell. Just above the cutoff frequency there is still some reflection of sound energy, but this diminishes rapidly as $k_0^2 - U_{max}$ increases. The effect of this on the input impedance of the Bessel horn, for which $f_c = 555$ Hz, can be seen in Fig. 4.44: the peaks for modes with $n > 4$ diminish progressively and by $n = 8$ have almost disappeared.

The effect of the variation of effective length with frequency on the inharmonicity of the modes of a flaring horn is demonstrated in Fig. 4.45. The EFP values for the

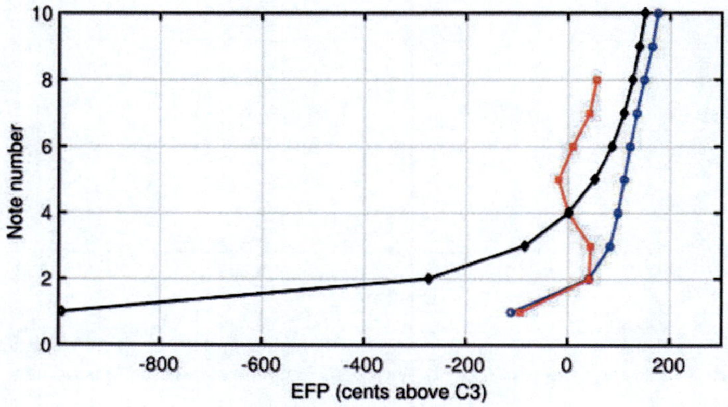

Fig. 4.45 EFP values calculated for a cylindrical tube 1.13 m long (black diamonds), a Bessel horn trombone bell approximation with no mouthpiece (blue circles) and the Bessel horn with a Denis Wick 5AL mouthpiece inserted (red squares) (Color figure online)

first ten modes of the Bessel horn are shown by circles, while the EFP values for a cylindrical tube of the same length are shown by diamonds. Replacing the cylinder by the flaring horn raises the pitch of the first mode by 880 cents; the second mode is raised by 313 cents, and by the sixth mode, the pitch increase is only 35 cents. The Bessel horn still displays a positive inharmonicity (the EFP values increasing with n), but the very large inharmonicities in the first few modes have been greatly reduced.

At the beginning of this section, we noted that the addition of a mouthpiece to a cylindrical tube reduced the EFP values of the resonances by an amount which increased with mode number n up to the mouthpiece resonance frequency. A similar effect is found when a mouthpiece is added to a flaring tube, as shown by the squares in Fig. 4.45 which show the EFP values for the Bessel horn with a 5AL trombone mouthpiece at the input. The mouthpiece has little effect on the first two mode frequencies, but lowers the EFP values in the region of the mouthpiece resonance by more than 100 cents. This has the effect of giving a set of resonances from $n = 2$ to $n = 8$ which do not deviate by more than ±38 cents from a harmonic series based on the pitch C3+19 cents.

The amplitude boosting effect of the mouthpiece on the input impedance curve, discussed in Sect. 4.3.6, is also strongly evident in Fig. 4.44. The peaks just below the mouthpiece resonance frequency $f_R = 535$ Hz are significantly amplified when the mouthpiece is present, while the peaks above f_R die away even more rapidly than those corresponding to the bell alone.

It must be remembered that the foregoing discussion has not related to a complete instrument, but to a Bessel horn approximation of a trombone bell section. It is possible, however, to verify qualitatively the predicted behaviour by carrying out a playing experiment on a real trombone bell which can be detached from the slide section. On the Conn 8H, for example, it is possible for an experienced player to sound the notes corresponding to the first few modes of the bell section by buzzing the lips against the tube entrance. The first mode has a pitch around $D^\flat 3$, but the higher modes are obviously too sharp to be true harmonics of this pitch. The trombone mouthpiece can be inserted in the bell section entrance if the stem is wrapped with paper or tape, and the notes sounded with the mouthpiece are audibly closer to true harmonics.

4.3.9 A Practical Example: The Complete Trombone

How well does the foregoing theoretical discussion describe the behaviour of a real brass instrument? As a practical example, we take the Conn 8H trombone whose separate components have been discussed earlier. An experimentally measured input impedance curve for this instrument, with the slide in first position (fully retracted), has already been presented in Fig. 4.12. This curve is reproduced in Fig. 4.46, together with an input impedance curve for the same instrument with the slide in seventh position (fully extended). Several of the features discussed above can be identified in these curves. In both the impedance boosting effect of the mouthpiece on the peaks around 330 Hz is noticeable. The drop in amplitude of the peaks above the bell cutoff frequency (around 550 Hz) is also evident.

It is interesting to observe that the peaks in the instrument with extended slide are systematically lower than the peaks in the same frequency region on the instrument with the slide retracted. The height of a peak is determined by the rate of energy loss in the air column due to viscothermal effects and sound radiation; for frequencies below the bell cutoff, the viscothermal effects are the dominant cause of energy loss. In a longer tube, the viscothermal losses are greater, and it is therefore expected that the impedance peaks will be lower.

The measured intonation for the particular instrument under study is shown by the EFP values in Fig. 4.47. It is clear that this instrument has been well designed to give a set of natural notes with the slide in first position which are very close to

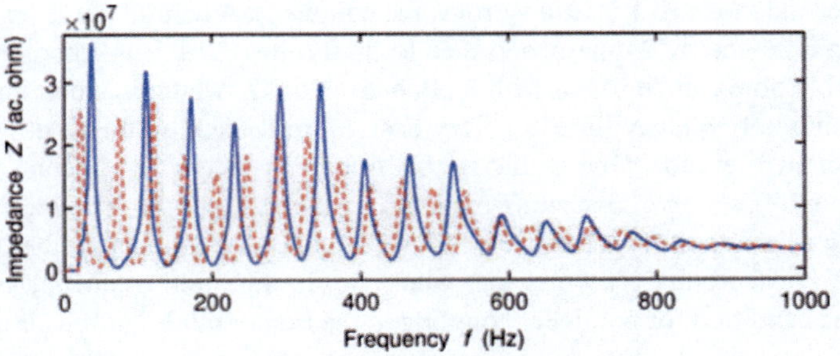

Fig. 4.46 Input impedance Z for a Conn 8H trombone with 5AL mouthpiece, slide in first position (solid line) and seventh position (dashed line)

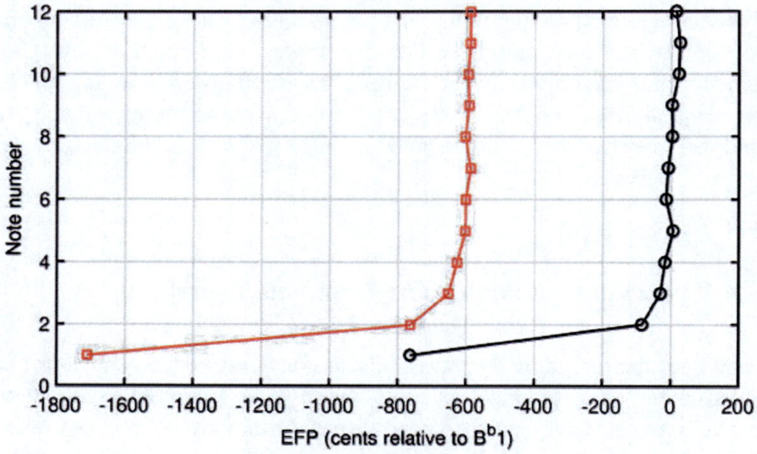

Fig. 4.47 EFP values for a Conn 8H trombone with 5AL mouthpiece, slide in first position (black circles) and seventh position (red squares) (Color figure online)

a harmonic series based on B$^\flat$1 from the third mode upwards. The EFP value for the second mode is -84 cents, while the first mode is more than a perfect fifth below the nominal pitch B$^\flat$1. These discrepancies are larger than those predicted for the bell section modes, but it must be remembered that the instrument now includes a large additional length of cylindrical tubing in the slide section. The frequency of the first impedance peak is 38 Hz, corresponding to $k_0^2 = 0.7\text{m}^{-2}$. This value is so low that the desired correction of effective length could only be achieved if the flaring section were longer than the existing bell section. A flaring slide is obviously impossible, so that the first mode on a trombone is always much too flat to be musically useful. It is still possible to play a note at the nominal pitch of B$^\flat$1 in first position, but this is the 'pedal note' which does not require to couple to a resonance at its fundamental frequency (see Sect. 5.4.4). Since there is no resonance in the vicinity of its first harmonic, the pedal note spectrum is characterised by a weak first harmonic amplitude (Sect. 2.2.3, Fig. 2.20).

The second mode EFP is also worth some comment. A helpful player experiment is to sound the note B$^\flat$2 quietly and then to lip it downward. It is commonly found that a stable note can be found with a pitch around A2, which is where the second resonant mode frequency lies. It is very easy to lip it back up to the desired pitch of B$^\flat$2 for two reasons. One is that in this frequency range, the lips are behaving predominantly as 'outward-striking reeds', for which the coupling to the air column resonance is strongest for a frequency above the natural frequency of the resonance (Sect. 5.2.1). The other reason is that when the lip vibration frequency is slightly raised, the condition for nonlinear coupling to the higher modes of the air column is enhanced.

As the slide is extended from the first position, the proportion of cylindrical to flaring tubing is increased, and the deviation of the lower mode frequencies from the nominal harmonic series becomes more serious. In seventh position the nominal pitch of the natural note series is lowered by six semitones to E1. The EFP values in Fig. 4.47 for the trombone with the slide in seventh position show that the modes from $n = 3$ to $n = 12$ are still very close to true harmonics of E1, but the second mode frequency (75 Hz) is 163 cents below a true second harmonic of E1. In practice the player can sound this note in tune by choosing the correct lip vibration frequency of 82.4 Hz; the mechanism by which energy is supplied to sustain the air column vibration is then primarily nonlinear coupling to the even-numbered higher modes. The first mode is almost an octave below the nominal pitch when the trombone slide is in seventh position; this has no musical significance, however, since the first mode plays an insignificant role in the regime of oscillation which sustains the pedal note (see Sect. 5.4.4).

4.3.10 Instruments with Predominantly Expanding Bore Profiles

The trombone is not the only instrument in which the first acoustic mode frequency is substantially lower than the nominal fundamental frequency: this behaviour is also characteristic of trumpets and other instruments in which a large proportion of the bore is approximately cylindrical. The impedance curves of instruments in which most of the tubing is steadily expanding do not show this dramatic deviation from harmonicity in the first peak. Instruments of the saxhorn family (see Chap. 7) fall into this category, although a substantial amount of approximately cylindrical tubing is inserted into a saxhorn when several valves are operated. The serpent (Sect. 7.8.2) and the ophicleide (Sect. 7.8.3) are instruments with mainly conical bore profiles, shown in Fig. 7.61. The lower resonances of such instruments can be close to exact harmonics of the nominal pitch without the necessity for a rapidly flaring bell (see Sect. 4.3.3). The EFP plots for the serpent in Fig. 7.63 demonstrate that when all the toneholes are closed, the EFP for the first acoustic resonance is almost exactly zero, meaning that the resonance frequency is very close to the nominal fundamental frequency. The EFP plots for the ophicleide in Fig. 7.65 show that with all holes closed, the first resonance frequency is in fact slightly above the nominal fundamental frequency. The EFP plots for the bass tuba in Fig. 7.66 show that the inharmonicity of the first acoustic resonance depends on the operation of the valves, which introduce approximately cylindrical tubing.

The effect of changes in the bore profile on the pitches of the lower natural notes can be observed in the comparison of three instruments of similar length illustrated in Fig. 4.48; Fig. 4.49 shows their bore profiles and Fig. 4.50 their EFP curves. The bore of the Russian bugle expands approximately exponentially and is close to cylindrical for the first 150 mm: its resonances from the second upwards are acceptably harmonic for an instrument pitched about 50 cents above A4 = 440 Hz. The cavalry trumpet is cylindrical for half its length and has a flared bell: although

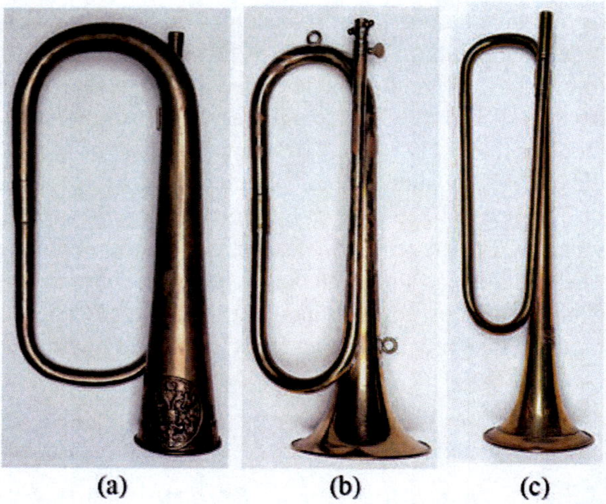

Fig. 4.48 Three instruments with contrasting bore profiles: (**a**) Russian bugle (EU3243); (**b**) Spanish bugle (EU2468); (**c**) cavalry trumpet (EU0224)

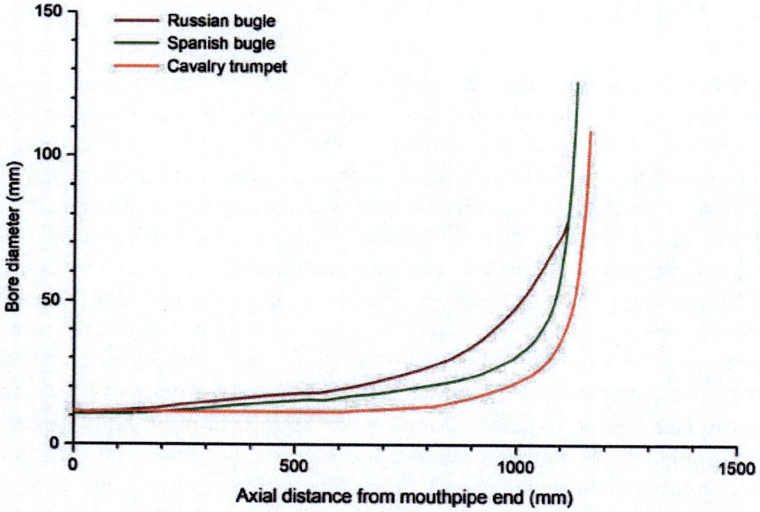

Fig. 4.49 Bore profiles of a Russian bugle (brown), a Spanish bugle (green) and a cavalry trumpet (red) (Color figure online)

the unused first resonance is very flat, from the second upwards, the resonances are sufficiently harmonic for the trumpet to respond effectively. The bore of the Spanish bugle, however, is approximately a shallow cone for half its length followed by a flared bell: the resonances are far from harmonic, and playing tests found the fourth, fifth and sixth natural notes to be a semitone sharper than the second and third natural notes. These instruments were made for signalling rather than concert use, and even the wayward Spanish bugle could issue coded commands using four or five distinguishable notes and thus be fit for purpose.

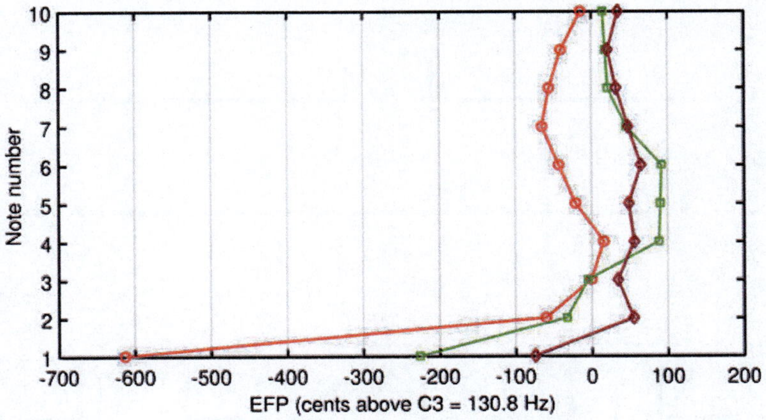

Fig. 4.50 EFP curves for a Russian bugle (brown diamonds), a Spanish bugle (green squares) and a cavalry trumpet (red circles) (Color figure online)

4.4 Toneholes

All brass instruments have an aperture at the entrance against which the lips are buzzed and another aperture at the exit from which sound is radiated. On modern orchestral brass instruments, the only other apertures are small side holes known as water keys, which are closed during performance but can be opened to release condensed water vapour which has accumulated inside the tubing.

Most woodwind instruments have a considerable number of side holes, which can be closed either directly by the fingers of the player or by padded discs controlled by key mechanisms. These apertures, which are used to change the effective sounding length and therefore the playing pitch of the instrument, are called toneholes. It was pointed out in Sect. 1.1 that many historically important labrosones, including the cornett, the serpent and the ophicleide, also employed toneholes; these instruments are discussed in detail in Chap. 7.

The acoustical behaviour of toneholes is reviewed by Benade (1960), Keefe (1982), Fletcher and Rossing (1998) and Chaigne and Kergomard (2016). A detailed discussion can also be found in Dubos et al. (1999). A summary of some aspects relevant to the performance of labrosones with toneholes is given here.

The idea that opening a tonehole is equivalent to cutting off the lower part of the tube just below the hole is attractive, but the reality is much more complicated. We consider first the theoretically simple case illustrated in Fig. 4.51a: an infinitely long cylindrical tube of radius a, with a single open tonehole of radius b. The tonehole is represented by a short cylinder of height h branching out from the side of the main tube. On some instruments, h is the thickness of the wall through which the side hole is bored; an example of this type of tonehole is illustrated in Fig. 4.52a. On others, the hole geometry includes a short chimney extending beyond the outer surface of the main tube, as shown in Fig. 4.52b.

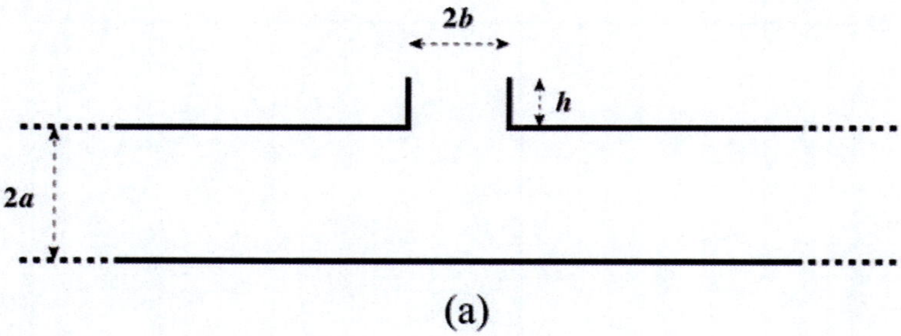

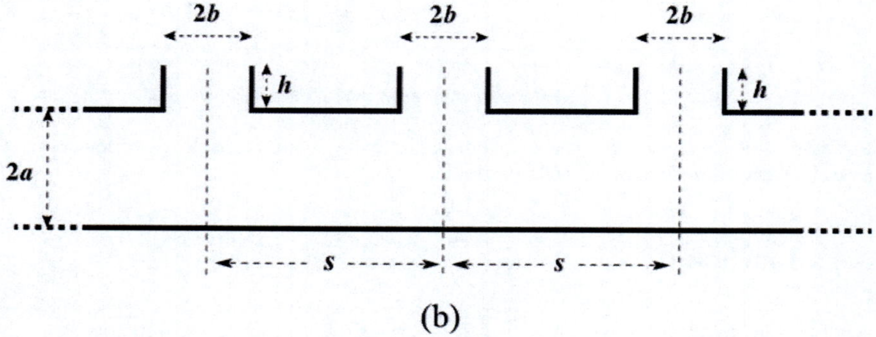

Fig. 4.51 (**a**) A single tonehole of radius b in an infinitely long cylinder of radius a. (**b**) Part of a lattice of toneholes with identical hole dimensions and equal inter-hole spacing s

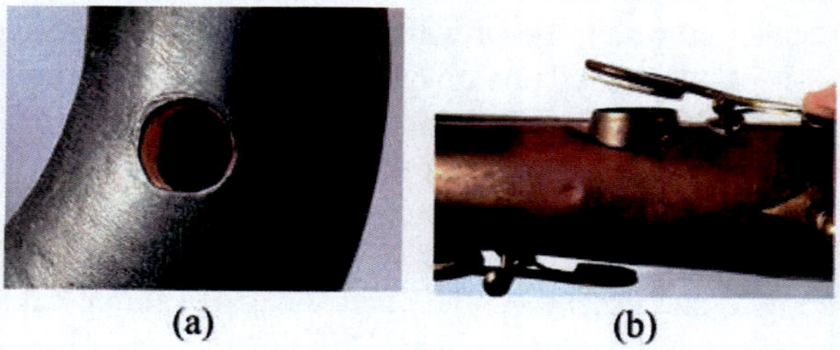

Fig. 4.52 (**a**) Tonehole on a serpent. (**b**) Tonehole on an ophicleide, with chimney, key and pad

The effective acoustic length of the side branch is $l_e \simeq h + 1.6b$; the second term on the right-hand side of this equation includes effects of the inertia of the air in the branch and the radiation of sound from the open end. When a sinusoidal sound wave travelling along the main tube arrives at the open tonehole, some of the energy is transmitted down the main branch, some is radiated from the tonehole, and the remainder is reflected back up the main tube. The sound power transmission coefficient T_m is the fraction of the sound power which is transmitted past the tonehole (Kinsler et al. 1999). For a tonehole of radius b and effective length l_e on an infinite cylinder of radius a,

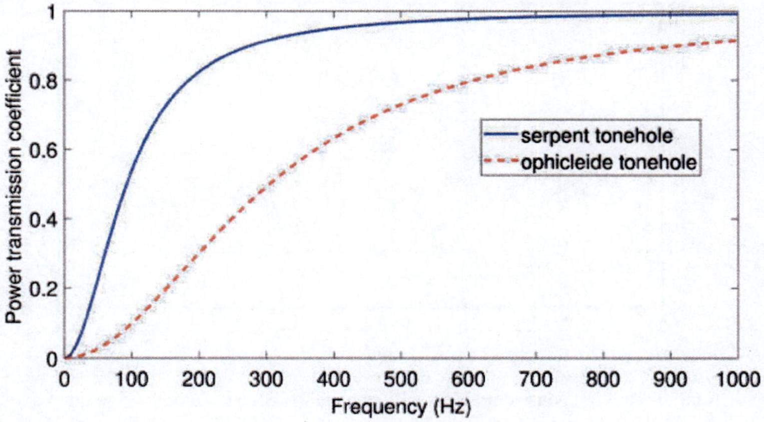

Fig. 4.53 Power transmission coefficients T_m for two different tonehole geometries

$$T_m = \frac{1}{1 + (cb^2/4\pi l_e a^2 f)^2}, \tag{4.68}$$

where c is the speed of sound.

Figure 4.53 shows how the transmission coefficient varies with frequency for the two toneholes illustrated in Fig. 4.52. In both cases the effect of the open tonehole can be described as a high-pass filter: very little sound energy gets past the tonehole at low frequencies, while at high frequencies, almost all the sound energy continues down the main tube. The cutoff frequency f_c is defined as the frequency for which $T_m = 0.5$; Eq. 4.68 shows that

$$f_c = \frac{c}{4\pi l_e}\left(\frac{b}{a}\right)^2. \tag{4.69}$$

Since the serpent tonehole is closed by the player's finger, its maximum diameter is determined by the size of the fingertip. Figure 4.52a shows a hole of 7 mm radius in a main tube of radius 30 mm; the relatively small value of b/a results in a low cutoff frequency $f_c = 93$ Hz. Figure 4.52b shows an ophicleide player's finger pressing a key which lifts the padded disc from the short chimney forming part of the tonehole. On a keyed instrument, the tonehole diameter can be many times the width of the player's finger: in the example shown in Fig. 4.52b the main tube radius is again 30 mm, but the tonehole radius is 20 mm, and the cutoff frequency is $f_c = 305$ Hz.

In the case of the infinite length cylinder discussed so far, the transmitted sound wave continues down the tube without any further reflection. In the realistic case of a wind instrument, the wave returning to the mouthpiece is a combination of the waves reflected by the tonehole and by the open end of the instrument. The effect of opening only the fifth tonehole on an ophicleide is illustrated by Fig. 4.54, which

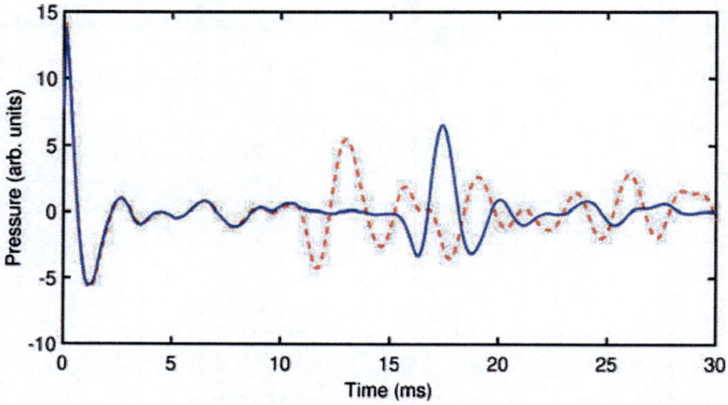

Fig. 4.54 Input impulse response curves for a B^b ophicleide. Blue solid line: fingered to play A (all toneholes closed). Red dashed line: fingered to play D (fifth tonehole open) (Color figure online)

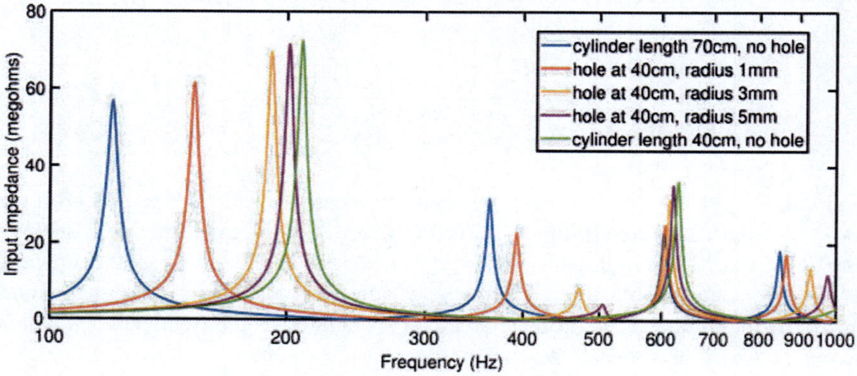

Fig. 4.55 Calculated input impedances for a cylindrical tube, length 70 cm, radius 1 cm, open at one end: effect of opening a side hole of varying diameter

shows the measured input impulse responses with the tonehole closed and open. The blue curve shows the response with the hole closed: the impulse generated at the input travels all the way to the end of the bell, where it is reflected. The round-trip time taken for the positive pulse to return to the input is 18 ms, which is the period of the 55 Hz note of pitch A1 corresponding to this fingering. The red curve shows the input impulse response with the hole open, which is a fingering for the note D2. Most of the energy in the wave travelling down the tube is now reflected near the open hole instead of continuing to the bell. The round-trip time for the impulse to return to the entrance is 13.5 ms; this period corresponds to a frequency of 74 Hz, close to the frequency of the desired pitch D2. Multiple reflections within the tube give rise to further peaks in the impulse response.

The effect of tonehole radius on the input impedance peak frequencies of an finite length tube is illustrated by Fig. 4.55, showing the calculated input impedance for a simple cylinder of length 70 cm and radius 1 cm, open at the output end and having a single tonehole centred at a point 40 cm from the input end. The blue curve shows the impedance in the absence of a hole. As expected for a tube of this length, the first resonance has a frequency close to 120 Hz, while the second peak has a frequency near 360 Hz, three times that of the first. Opening a tonehole with radius 0.5 cm, half that of the main bore, moves the first resonance up to 201 Hz: this is quite close to 208 Hz, the frequency of the first resonance of a cylinder of 40 cm shown by the green curve. The other curves show the input impedance for holes of smaller diameter at the same position. Opening a hole of radius 0.1 cm raises the first resonance peak to only 152 Hz. The higher-frequency resonance peaks display a much less regular behaviour; above the cutoff frequency, the sound wave continues to propagate past the hole, and resonances of the complete tube and the lower section are mixed with those of the upper section.

Most brass instruments with toneholes have bores which are closer to cones than cylinders. To illustrate the effect of opening a side hole in an approximately conical tube, input impedance curves have been calculated for the simplified serpent bore shown in Fig. 4.56. The blue curve in Fig. 4.57 shows that with no holes open, the frequencies of the resonances are close to harmonic. Opening the first hole results in the set of resonance frequencies shown in the red curve, which are far from evenly spaced at low frequencies. At frequencies above a few hundred hertz, the red and blue curves almost coincide, showing that opening the hole has little effect on the higher resonance frequencies.

The high-pass filtering effect of an open tonehole is further modified when several toneholes with similar geometries and approximately equal spacing are simultaneously opened (Benade 1960; Chaigne and Kergomard 2016). Part of a tonehole lattice on a cylindrical tube is illustrated in Fig. 4.51b. An approximate expression for the cutoff frequency of this lattice is

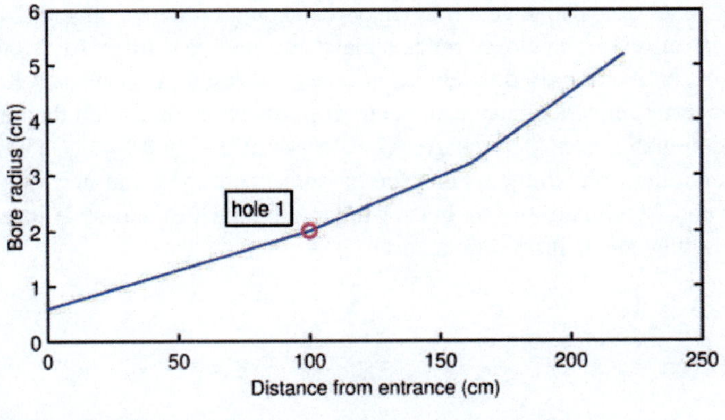

Fig. 4.56 Simplified model of a serpent bore with three conical sections

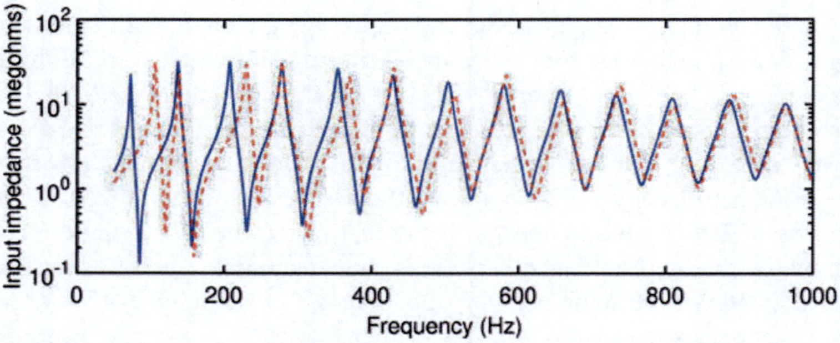

Fig. 4.57 Effect of opening the first hole on the input impedance of the model serpent. Solid blue line: hole closed. Dashed red line: hole open (Color figure online)

$$f_c \simeq \frac{c}{2\pi} \frac{b}{a} \frac{1}{(sl_e)^{1/2}}, \tag{4.70}$$

where s is the inter-hole spacing.

Equation 4.70 can also be used to estimate the cutoff frequency of a conical instrument provided that the ratio b/a is approximately the same for all the open holes. An example is provided by the ophicleide. The five holes nearest the bell have an average spacing $s \simeq 15$ cm; assuming equal values of h and b/a for all five holes give a predicted cutoff frequency, $f_c = 475$ Hz. This is considerably higher than the value of 305 Hz for a single tonehole calculated using Eq. 4.69. The tonehole lattice is effective at reflecting the wave arriving from the mouthpiece over a bigger-frequency range, leading to an increased number of high impedance peaks and a potential improvement in the centring and stability of the played notes in this frequency range. Opening all the five lowest holes on the instrument is in fact an often preferred alternative fingering for the note D2.

Closed toneholes also have a frequency-dependent influence on sound transmission, but in this case, the closed holes behave as a low-pass filter. Above the cutoff frequency, a lattice of closed toneholes attenuates the sound waves travelling down the bore. Strong reflections can also occur at frequencies for which the height h of the closed branch is equal to a quarter wavelength of the sound wave. These effects can be important in instruments like the bassoon, with long and narrow toneholes (Chaigne and Kergomard 2016), but for the relatively short and wide toneholes on labrosones, they are of little significance.

4.5 Mutes

In audio technology the term 'muting' refers to the silencing of a signal. In a musical score, the instruction 'mute', or the Italian equivalent 'con sordino', has a different meaning. Mutes used on musical instruments are not normally designed to suppress

Fig. 4.58 Excerpt from the first trumpet part in *Petrouchka* by Igor Stravinsky (1947 Version). Reproduced by permission of Boosey & Hawkes Music Publishers Ltd

the sound output completely, although the role of a 'practice mute' is to reduce the amplitude of the radiated sound to a level at which neighbours are not disturbed. All mutes modify both the loudness and the timbre of the sound, and the timbral change is usually the primary motivation for using a mute in performance. A muting instruction does not necessarily imply that the passage will be played quietly: the dramatic timbre of muted brass instruments played fortissimo has been exploited by many composers, including Igor Stravinsky (Fig. 4.58).

With the exception of the instruments with toneholes discussed in Sect. 4.4, on which mutes are not normally used, all the sound output of a brass instrument emerges through the bell. Brass instrument mutes are devices which modify the way in which sound waves are reflected and radiated when they reach the bell. In a comprehensive survey of the acoustical and musical properties of brass instrument mutes, Sluchin and Caussé (1991) draw a useful distinction between internal mutes which penetrate into the bell and external mutes which are held outside the bell. In the following sections, we discuss two types of internal mute, the straight mute and the Harmon mute, and also the external mute known as the plunger.

4.5.1 *Straight Mutes*

A typical trumpet straight mute is illustrated in Fig. 4.59. The mute is a thin-walled hollow metal vessel whose only opening is a circular hole at the narrow end. Three thin strips of cork, visible in Fig. 4.59a, act as spacers between the outer surface of the mute and the inner surface of the bell when the mute is inserted as shown in Fig. 4.59b.

The mute shown in Fig. 4.59 has a complex profile designed to match the inner profile of a trumpet bell. The cork spacers maintain a gap of a few millimetres between the outer surface of the mute and the inner surface of the bell. The sound is radiated through this narrow channel, which also allows the mean flow to escape.

The most basic design of straight mute is simply a truncated cone closed at the wide end, as illustrated in Fig. 4.60.

The effect of the straight mute shown in Fig. 4.60 on the input impedance of a Conn 8H tenor trombone is illustrated in Fig. 4.61. In Fig. 4.61a the slide is in first position (fully retracted). Comparison of the blue curve (without mute) and the red curve (with mute) shows that the insertion of the mute has almost no effect on the frequencies or amplitudes of the impedance maxima between 100 Hz and

(a) (b)

Fig. 4.59 A trumpet straight mute (**a**) out of the instrument and (**b**) inserted into the bell

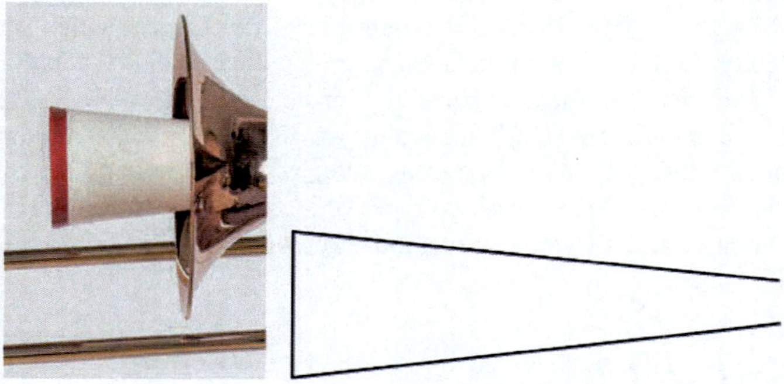

(a) (b)

Fig. 4.60 (**a**) A straight mute inserted into the bell of a tenor trombone. (**b**) Internal cross-section of the straight mute

600 Hz. Above 600 Hz, which is close to the bell cutoff frequency (see Sect. 4.3.9), the amplitudes of the impedance maxima are noticeably increased by the insertion of the mute. By 1000 Hz the impedance peaks without the mute have effectively disappeared, whereas with the mute inserted, they are still in evidence up to 1500 Hz.

The increase in the amplitude of the high-frequency impedance peaks when the mute is inserted arises because the partial closure of the bell opening results in greater reflection of energy back into the internal standing waves. The existence of significant peaks up to the 20th mode of the air column improves the centring and stability of the corresponding notes, making it easier for the performer to play in this very high register. On the other hand, the increase in the fraction of the energy which is reflected internally corresponds to a reduction in the fraction radiated as sound.

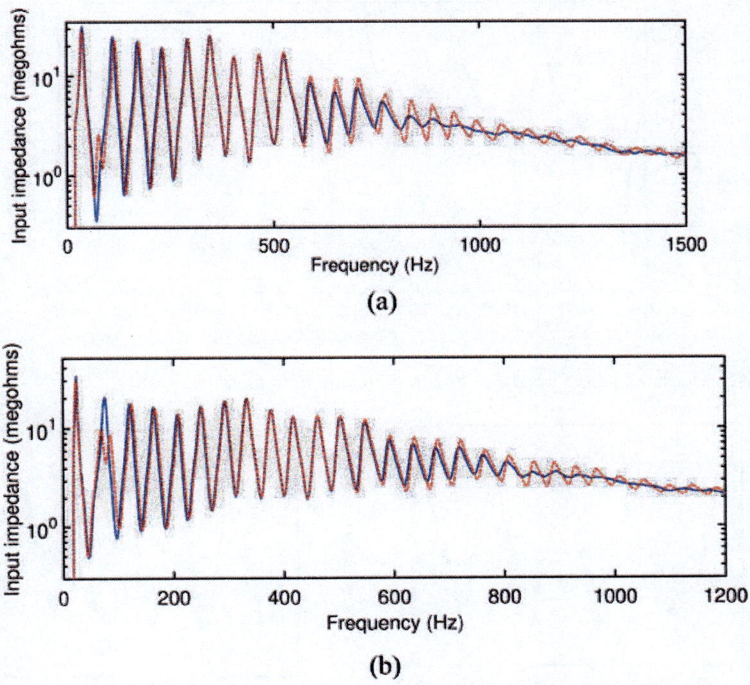

Fig. 4.61 Measured input impedance curves for a Conn 8H trombone. Solid blue line: no mute. Broken red line: straight mute inserted. (**a**) Slide in first position. (**b**) Slide in seventh position (Color figure online)

Figure 4.61b shows how the input impedance is changed by inserting the straight mute when the trombone slide is in the seventh position (fully extended). The effect on the impedance peaks is broadly the same as was observed with the slide in first position: the amplitudes of the input impedance peaks above 600 Hz are increased, and the range of visible peaks is extended to 1500 Hz.

4.5.2 *Effects of Internal Resonances in the Straight Mute*

The insertion of a straight mute does not only reduce the effective radiating area of the bell; it also introduces an additional resonant cavity formed by the internal volume of the mute. Figure 4.62 shows the input impedance curve measured at the open neck of the mute. When the mute is inserted into the bell, its internal resonances couple strongly to the air column of the instrument at frequencies corresponding to minima in the mute impedance, since a minimum in impedance implies a maximum rate of oscillating air flow into and out of the cavity. To make the

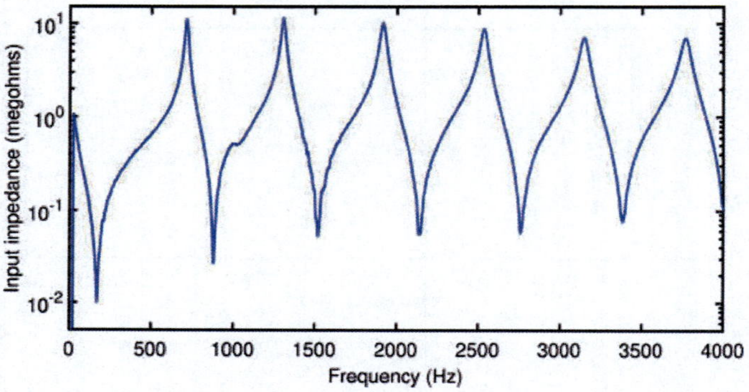

Fig. 4.62 Measured input impedance of a trombone straight mute

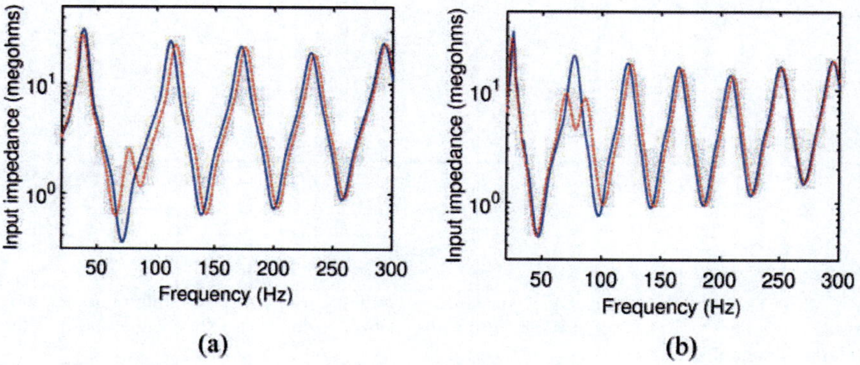

Fig. 4.63 Low-frequency impedance peaks for a trombone with (red curves) and without (blue curves) a straight mute. (**a**) Slide in first position. (**b**) Slide in seventh position (Color figure online)

impedance minima more visible in Fig. 4.62, the curve is plotted with a logarithmic impedance axis.

The first dip in the mute impedance curve occurs at 166 Hz. This is the Helmholtz resonance frequency of the mute (see Sect. 4.1.1); the corresponding pitch E3 can be sounded by blowing across its open neck. The effect of the Helmholtz resonance on the input impedance of the muted trombone is limited to the low-frequency region shown in Fig. 4.63. A small additional peak appears when the mute is inserted.

This extra peak is described by Sluchin and Caussé (1991) as 'parasitic', since it plays no useful role in the sound generation process and can in fact disrupt it. The frequency of the additional peak is lower than the Helmholtz resonance frequency of the mute by an amount which increases when the trombone slide is extended or the mute pushed further into the bell.

Figure 4.63a shows that with the trombone slide in first position, the parasitic peak appears at 77 Hz. The note B♭2 is played with this slide position, the player's lips coupling to the second main impedance peak at 116.5 Hz. The parasitic peak is far below this and causes no disruption. However the parasite has a more poisonous influence when the slide is fully extended, as shown in Fig. 4.63b. This is the position used to sound the note E2, at a frequency of 82.4 Hz. The parasitic peak frequency is now so close to this frequency that it splits the second main peak into two sub-peaks at 70 and 86 Hz. As a consequence the note E2 is unstable when played with the straight mute inserted.

The precise pitch at which the parasitic peak appears depends on the detailed profiles of the instrument bell and the mute. A novel design of trombone mute which uses active control technology to suppress the parasitic resonance is described in Sect. 9.2.2.

Several higher-frequency minima can also be seen in the mute impedance curve in Fig. 4.62. Each minimum corresponds to a standing wave inside the mute with a pressure antinode at the closed end and a pressure node at the open end. Near these frequencies the mute is more efficient at trapping and reflecting the sound energy arriving in the bell region, and the fraction of sound energy radiated externally is reduced. Figure 4.64 shows the frequency spectra of the note B$^{\flat}$2 played on a tenor trombone with and without a straight mute. In the spectrum of the muted sound, dips are evident in the amplitudes of the frequency components around 700 Hz, 1500 Hz, 2100 Hz and 2800 Hz. These dips are approximately correlated with the first four standing wave resonances of the mute.

In the playing experiment whose results are shown in Fig. 4.64, the dynamic levels of the unmuted and muted notes were chosen to equalise the peak waveform amplitudes. Comparing the two spectra, it is evident that inserting the mute significantly reduced the heights of the frequency components below 1000 Hz. Above 1500 Hz it appears that inserting the mute actually resulted in increasing the amount of radiated sound energy. This seems counter-intuitive, since the mute is a passive device incapable of generating additional energy. In fact, the additional energy came from the player, who had to increase the mouthpiece pressure amplitude when the mute was inserted to give the same peak waveform amplitude as the unmuted note. Nonlinear distortion, which increases with pressure amplitude, boosted the high-frequency part of the radiated spectrum (see Sect. 6.1).

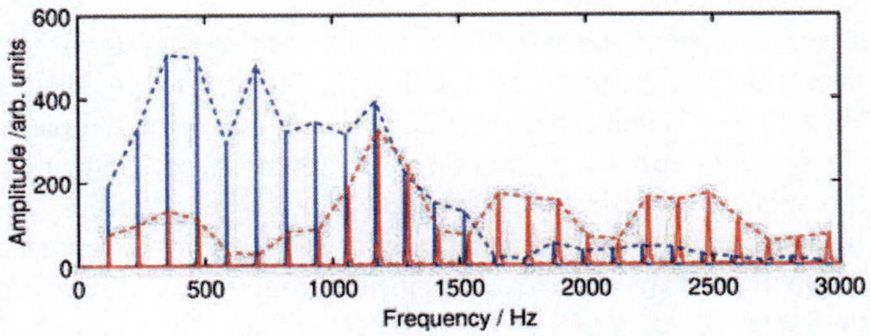

Fig. 4.64 Frequency spectrum of the sound radiated by a Conn 8H tenor trombone, slide in first position. Blue line: no mute. Red line: straight mute inserted. Dashed lines: peak envelopes (Color figure online)

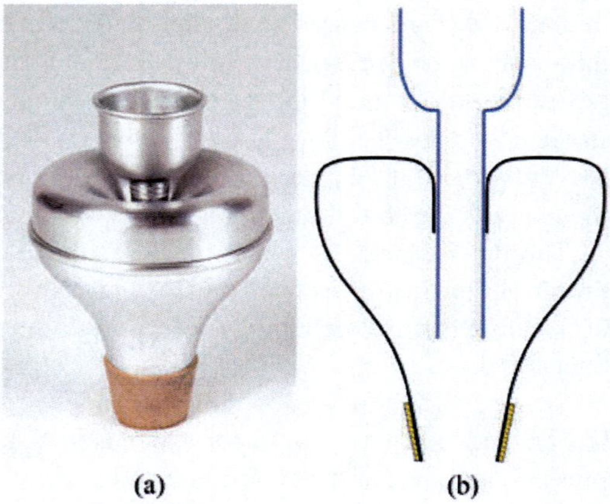

Fig. 4.65 (**a**) Trombone Harmon mute. (**b**) Cross-section of Harmon mute

4.5.3 Harmon Mute

Figure 4.65 illustrates a Harmon or 'wah-wah' mute. The acoustical behaviour of this type of mute is fundamentally different from that of the straight mute, since the cork spacer is a continuous strip around the neck which prevents any sound radiation through the gap between the mute and the bell. Instead, the sound emerges through a cylindrical channel which pierces the outer face of the mute. A second cylindrical tube with a small bell can be inserted into the exit channel, as shown in Fig. 4.65.

Impedance curves measured at the input aperture of the mute with and without the additional tube are shown in Fig. 4.66a. The effect of inserting the mute into a trombone is shown in the impedance curves in Fig. 4.66b. A sound wave arriving at the bell is most efficiently reflected when its frequency is near to a maximum in the mute impedance curve, since the mute entrance is then behaving almost as a closed surface. With the inner tube inserted, the first maximum in the impedance of the mute is at 72 Hz. A corresponding parasitic peak just below this frequency can be seen in the input impedance curve for the muted trombone; it is well below the frequency of the second main peak at 115 Hz, and therefore does not disrupt the sounding of the note $B^\flat 2$.

Several jazz trumpeters, notably Miles Davis, have made extensive use of the Harmon mute with the inner tube removed. It is important, however, that the mute is designed in such a way that removing the tube does not bring the parasitic peak too close to the frequency of a playable note. Figure 4.66a shows that the first impedance maximum of the trombone mute discussed here rises from 72 to 133 Hz when the tube is taken out. The effect on the input impedance of the muted trombone can be seen in Fig. 4.66b: the second main impedance peak is replaced by two peaks of similar amplitude at 104 and 127 Hz. This renders the note $B^\flat 2$ effectively unplayable.

The nickname 'wah-wah mute' has been given to the Harmon mute because of its ability to mimic this vocal effect. By almost closing the bell end of the inner tube

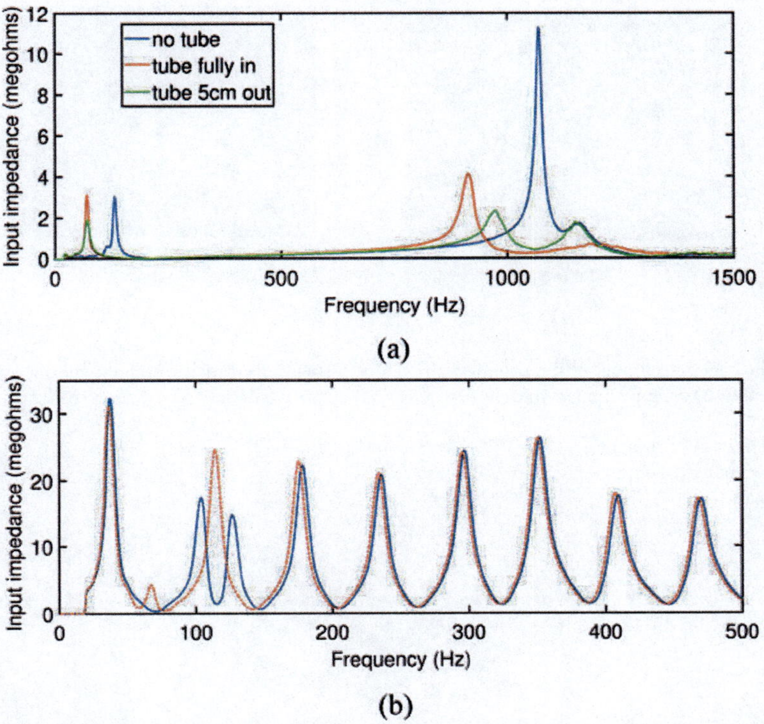

Fig. 4.66 (**a**) Input impedance of a trombone Harmon mute whose inner tube is removed (blue line), partially inserted (green line), and fully inserted (red line). (**b**) Input impedance of a Conn 8H trombone with a Harmon mute whose inner tube is removed (solid blue line) and fully inserted (broken red line) (Color figure online)

with the fingers, the player can filter out all but the lowest few harmonics from the radiated sound. A singer creates the vowel 'ooh' in a similar way, by almost closing the lips. The transition to an 'ah' vowel is made by lowering the jaw and opening the lips; a similar change in timbre is made on the Harmon mute by removing the fingers from the inner tube bell.

Figure 4.67a shows the spectrogram of the sound generated by a trombonist playing 'wah-wah-wah' using this technique; for comparison, Fig. 4.67b is the spectrum of the same phrase sung by a male vocalist. Although the spectra are significantly different below 1.5 kHz, the similarity in the higher-frequency bands is sufficient for the timbral changes in the trombone sound to be perceived as the same alternation of vowels as in the sung example.

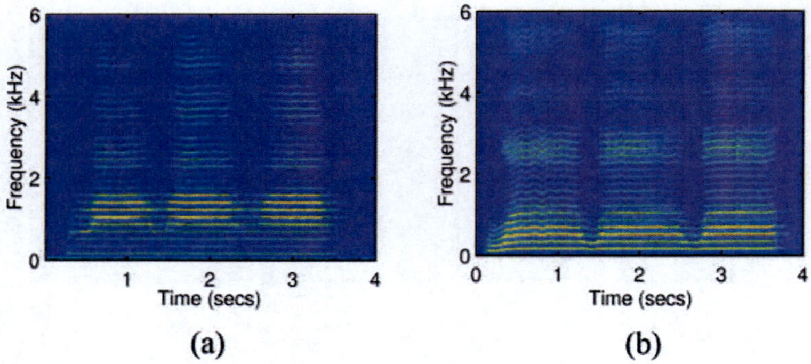

Fig. 4.67 Spectrograms of the sound 'wah-wah-wah' (**a**) played by a trombone with a Harmon mute, and (**b**) sung by a male voice

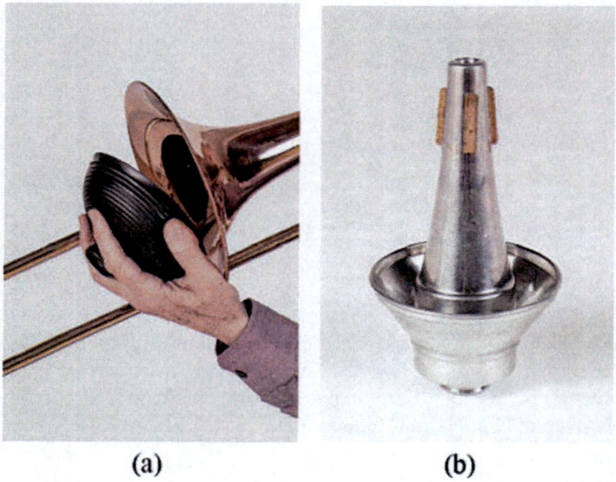

Fig. 4.68 (**a**) Trombone plunger mute. (**b**) Trombone cup mute

4.5.4 *Plunger and Cup Mutes*

The plunger mute belongs to the class of external mutes. It is in essence a shallow cup with a diameter a little smaller than the instrument bell. The name derives from the discovery by hard-up jazz musicians that suitable mutes could be obtained by removing the rubber caps from 'plungers' used to unblock drains. The plunger is held freely in the player's hand outside the bell, as shown in Fig. 4.68a. Since it does not penetrate into the bell, it has a negligible effect on the low-frequency resonances of the instrument, but when it is close to the bell, it strongly reduces the radiation of the higher harmonics of the played note.

Swing bands in the 1930s and 1940s made frequent use of plunger mutes in the trumpet and trombone sections. A notation was developed to indicate how the mute was to be employed, with a '+' sign above the note indicating that the plunger was to be close to the bell and an 'o' sign indicating that it should be removed. In the

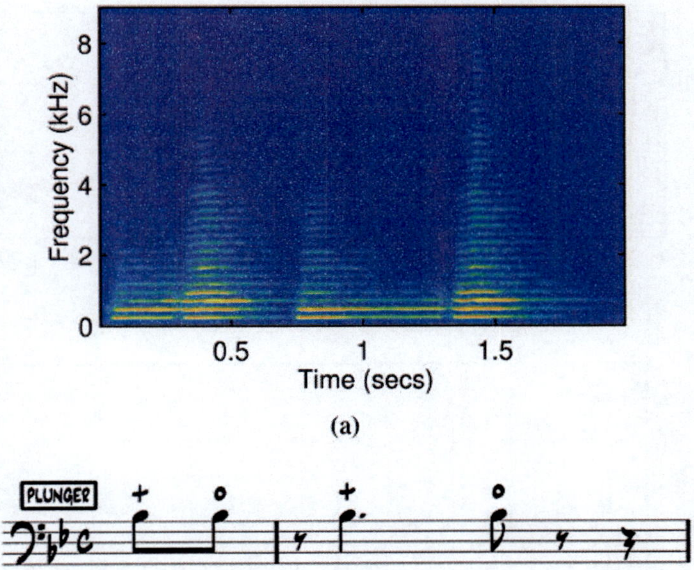

(a)

(b)

Fig. 4.69 Excerpt from the first trombone part of the Glenn Miller arrangement of Tuxedo Junction (Erskine Hawkins and Bill Johnson). (**a**) Spectrogram. (**b**) Notation

example of this notation in Fig. 4.69b, the trombonist is asked to alternate between closed and open positions of the plunger. The corresponding changes in timbre are illustrated by the spectrogram in Fig. 4.69a.

A cup mute is essentially a straight mute with the addition of a cup, similar to the plunger, surrounding the part which extends beyond the bell. It therefore shares features of both internal and external mute types. Figure 4.70a shows the input impedance of a Conn 8H trombone into which the cup mute illustrated in Fig. 4.68b has been inserted. The Helmholtz resonance of the internal cavity results in the appearance of a low-frequency parasitic resonance in the vicinity of the second main impedance peak of the muted instrument, similar to that observed with the straight mute described in Sect. 4.5.2.

When the cup is fitted, an additional cavity is introduced, defined by the internal walls of the bell and the cup. The resonant frequency of this volume is strongly dependent on the area of the annular gap between the bell and the cup rim, but is typically around 1000 Hz (Sluchin and Caussé 1991). Comparison of the input impedance curves with and without the cup in Fig.4.70a shows that for this particular trombone mute, the extra volume reduces the reflection of sound waves, and hence the amplitude of the impedance peaks, for frequencies around 600 Hz. A corresponding increase in the efficiency of sound radiation near this frequency is evident in the comparison of sound spectra in Fig. 4.70b. Around 1200 Hz, in

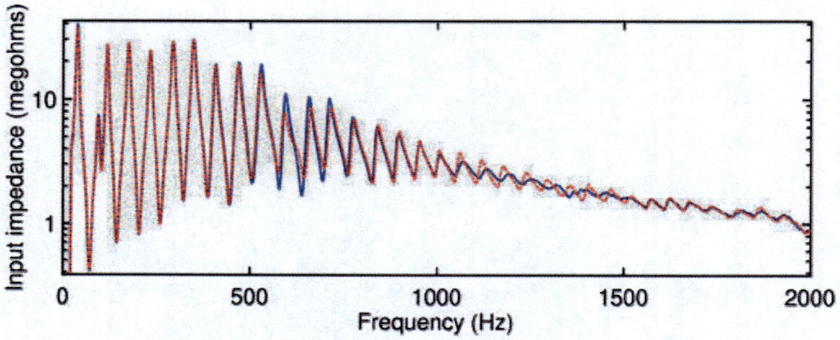

(a)

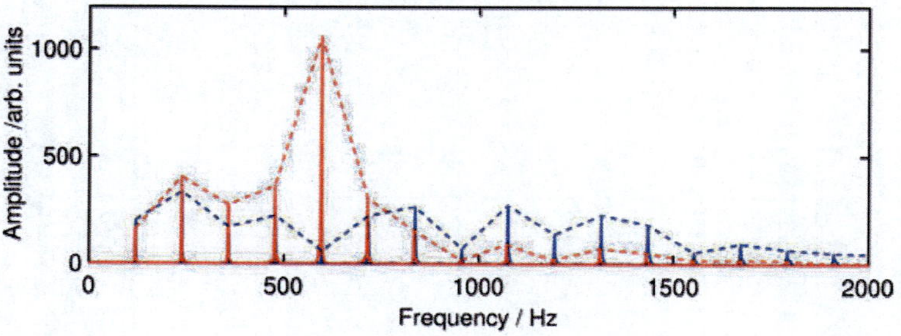

(b)

Fig. 4.70 (**a**) Input impedance of a tenor trombone with cup mute. (**b**) Spectra of the note B$^{\flat}$2 played on the trombone with cup mute. Blue lines: cup removed. Red lines: cup attached. Dashed lines: spectral envelopes (Color figure online)

contrast, the addition of the cup results in an increase in the input impedance peak amplitudes and a reduction in the amplitudes of the radiated spectral components. This effective transfer of sound energy from higher- to lower-frequency bands gives the cup mute a warmer and less strident sound than the straight mute.

4.5.5 Transposing Mutes

In the mid-sixteenth century the artist and architect Giorgio Vasari published his famous biographical encyclopedia '*Lives of the Most Excellent Painters, Sculptors, and Architects*' (Vasari 1568). This book includes a vivid description of a carnival procession in Florence in 1511, in which one of the floats caused such a sensation that it was still being talked of half a century later. This was the 'Chariot of Death', on which singers dressed as corpses emerged from tombs accompanied by 'trombe sorde e con suon roco e morto' (muted trumpets with a hoarse, dead sound). Vasari's text is one of the earliest references in literature to the use of mutes on brass instruments.

The first composer to specify brass instrument mutes in a score appears to have been Claudio Monteverdi (1567–1643). His opera *Orfeo*, first performed in 1607, begins with a Toccata played by a group of instruments including a trumpet ensemble. The original score, which is reproduced in Fig. 1.8b, includes an instruction that if the trumpets are muted, the other accompanying instruments must play a tone higher than the written notes. The implication is that the trumpet mute in Monteverdi's time was a transposing mute, raising the playing pitch of the instrument by a whole tone.

It is to be expected that inserting an object into the bell of a brass instrument will alter the resonant frequencies of the air column and therefore the playing pitches of the instrument. The modern mutes described in the previous sections are rendered non-transposing by the incorporation of a suitably chosen internal cavity, an invention usually ascribed to the eighteenth-century horn player Anton Joseph Hampel. The importance of this cavity can be readily verified by the simple experiment of closing the open end of a straight mute with a cork. The pitches of the natural notes from $B^\flat 4$ upwards are little affected by closing the internal cavity, but the notes from F4 downwards are flattened by approximately a semitone.

This musical experiment is confirmed by measurements of the input impedance of a Conn 8H trombone with a straight mute corked and uncorked. Figure 4.71 shows equivalent fundamental pitch curves (Sect. 4.3.4) derived from these measurements. From the 3rd to the 13th natural notes, the pitches are within ±30 cents of the ideal $B^\flat$ harmonic series when the cavity is open, but when it is closed, the lower notes are up to 140 cents too flat. The exception is the first resonance at 37 Hz, which is not significantly affected by the change in the mute since at this frequency, the wave is reflected before reaching the mute entrance (see Sect. 4.3.9).

It is thus easy to understand why an internal mute would lower the pitches of the natural notes if it did not have a large enough internal cavity. But how could it raise the pitches? The explanation of the apparently paradoxical behaviour of baroque

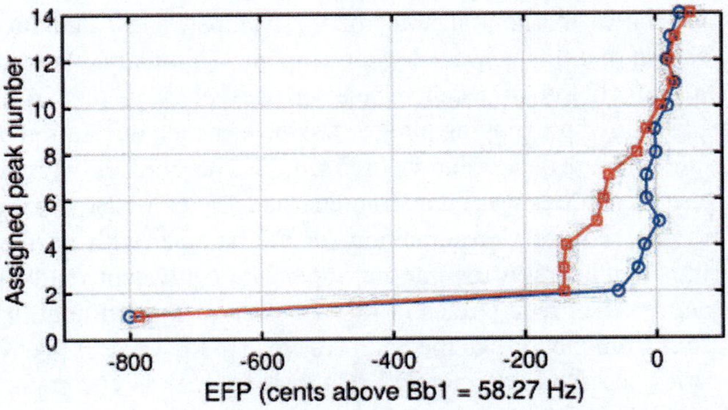

Fig. 4.71 Illustrating the effect of closing the cavity in a straight mute on the equivalent fundamental pitch curve for a muted Conn 8H tenor trombone, slide in first position. Blue line and circles: mute uncorked. Red line and squares: mute corked (Color figure online)

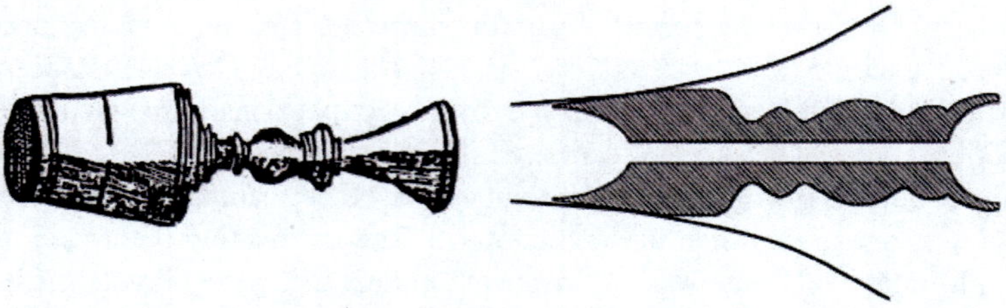

(a) (b)

Fig. 4.72 (**a**) Illustration of a baroque trumpet mute by Mersenne (1635). (**b**) Cross-section of a baroque pattern trumpet mute in the bell of a 1632 trumpet by Hainlein (Pyle 1991)

Fig. 4.73 Cross-section of the final 20 cm of a Hofmaster baroque trumpet in E^{b}. The bell diameter is 10.8 cm. A simplified solid mute with an open cylindrical channel along its axis is inserted into the bell

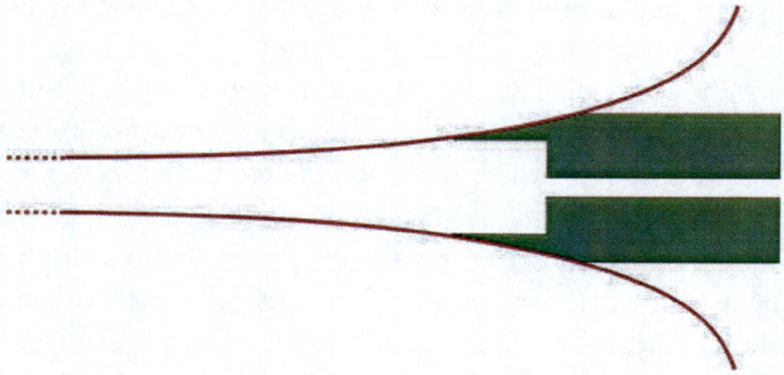

trumpet mutes emerges from an understanding of their acoustical functioning. No mutes are known to have survived from the seventeenth century, but Fig. 4.72a reproduces a drawing of a trumpet mute from Mersenne's *Harmonicorum libri XII* (Mersenne 1635). Mersenne gives no information about the internal geometry of the mute, but it was probably similar to the modern mute based on eighteenth-century designs shown in Fig. 4.72b. This mute is designed to fit closely into the bell, and the path for sound radiation is through the tubular opening along the axis.

It would indeed be surprising if a mostly solid mute of this type did not lower the pitches of each of the natural notes of the trumpet, and calculations by Pyle (1991) have shown that this is indeed what happens. To illustrate how a downward frequency shift of all the impedance peaks nevertheless results in an apparent upward transposition of the playing pitches, we have carried out similar calculations using the simplified mute design shown in Fig. 4.73. The trumpet whose bore profile was used in the calculations was an instrument made by Hofmaster in around 1760.

Input impedance curves were obtained for the trumpet with no mute inserted and with various diameters of the internal tube. The equivalent fundamental pitch values indicated by the green circles in Fig. 4.74a show that without the mute, the impedance peaks from the third to the tenth are close to the vertical line representing a harmonic series based on the pitch D2. The second peak is 169 cents flatter than the second harmonic of D2. As usual with trumpets and trombones, the first peak is many semitones too flat and does not appear on the diagram.

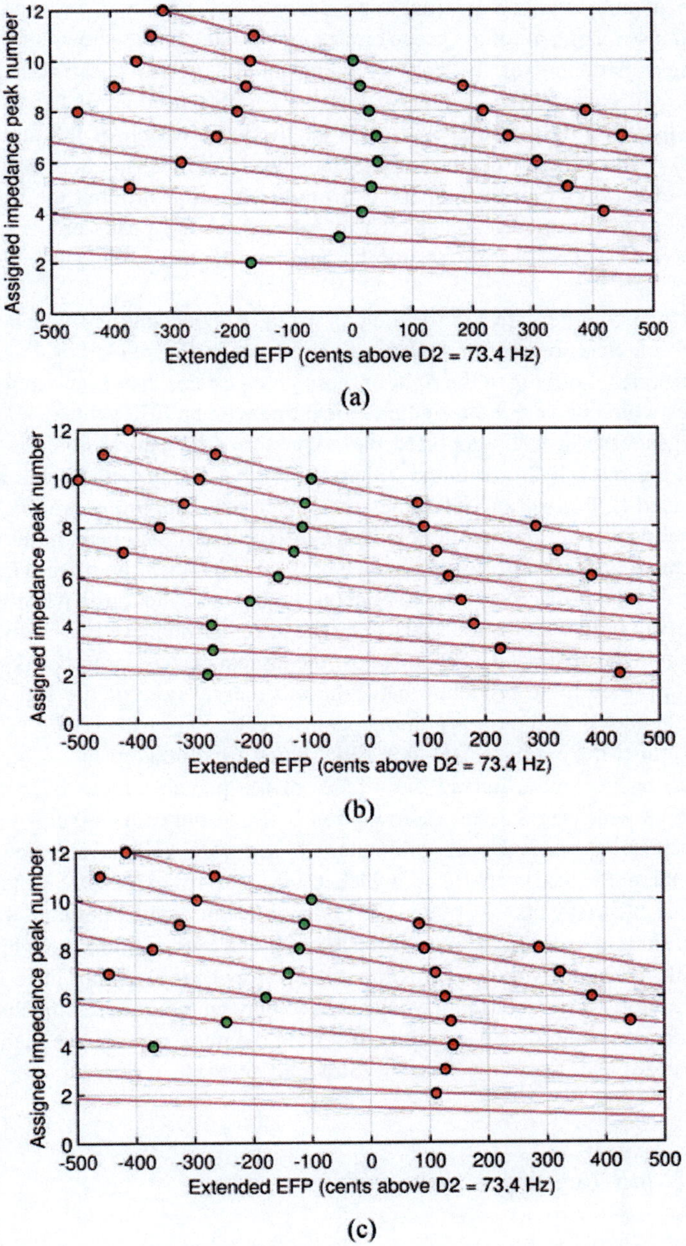

Fig. 4.74 Extended EFP of Hofmaster trumpet. (**a**) No mute. (**b**) Mute with internal tube diameter 12 mm. (**c**) Mute with internal tube diameter 6 mm

When a mute with an internal tube diameter of 12 mm is inserted into the modelled instrument, all of the green circles move to the left as shown in Fig. 4.74b, indicating a flattening of the corresponding natural notes. The fourth peak, for example, which was only 22 cents below the fourth harmonic of D2 without the mute, is flattened by a further 250 cents by the introduction of the mute. Figure 4.74c shows the effect of decreasing the diameter of the internal tube to 6 mm: this has little effect on the peak frequencies from the fifth upward, but the fourth peak is flattened by a further 100 cents. The third and second peaks are, respectively, 591 cents and 681 cents semitones too flat and have therefore disappeared from the diagram.

Since the flattening effect of the mute is strongly dependent on frequency, the set of peak frequencies represented by the green circles in Fig. 4.74c is very far from a harmonic series. Looking to the right of these green circles, however, we see a set of red circles which lie very close to the vertical line with an EFP value of +100 cents. This suggests that a well-supported harmonic series of natural notes is available based on the pitch E^{b}2, a semitone higher than that found without the mute. On the Extended EFP diagram, all circles on a given magenta line represent the same impedance peaks, but different circles have different assigned peak numbers. Thus the frequency of the fourth impedance peak is 237 Hz when the mute with a 6 mm diameter channel is inserted; this is 371 cents too flat to be the fourth harmonic of D2 (73.4 Hz), but only 26 cents too sharp to be the third harmonic of E^{b}2 (77.8 Hz). The red circles on the Extended EFP plot show the effects of reassigning peak numbers; for the simplified mute modelled here, the nth natural note of the E^{b}2 harmonic series is supported by the $(n - 1)$th impedance peak.

The trumpet mute whose extended EFP values are plotted in Fig. 4.74c behaves like a transposing mute, raising the pitches of the playable notes by a semitone. Experiments with original and reproduction baroque mutes have shown that they also transpose upwards by approximately a semitone (Klaus and Pyle 2015). No original mute has been found which transposes by a tone, as described by Monteverdi and many other seventeenth- and eighteenth-century authors. It has been suggested that, in order to avoid transposition by a semitone into keys which could be difficult for other woodwind instruments, trumpet players may have routinely removed a short section of tubing when asked to play with mutes, raising the pitch of the unmuted instrument by a semitone; inserting the mute would have then resulted in a total transposition of a whole tone (Stradner 2015).

4.5.6 Hand Technique on the Horn

Although mutes similar to those used on trumpets and trombones are employed on the french horn, the standard method of playing from the eighteenth century has involved the insertion of the hand of the player into the bell of the instrument (Fig. 4.75a). The pitch of each of the natural notes of the instrument can be modified by adjusting the position of the hand; increasing the degree of closure of the bell

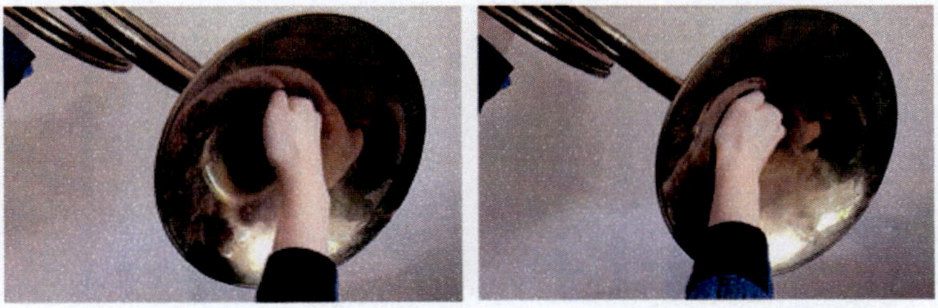

(a) (b)

Fig. 4.75 The player's hand inserted into the bell of a french horn. (**a**) Normal position; (**b**) stopped position. Courtesy of Lisa Norman

lowers the pitch of every natural note. Changes in hand position enable 'out-of-series' notes to be played and are used to control the intonation and timbre of the radiated sound. The basic acoustics of this hand technique is similar to that of the internal mutes described earlier in Sect. 4.5. The insertion of the hand increases the reflection of sound energy, extending the frequency range over which significant impedance peaks are available to support note production (Benade 1973; Yoshikawa and Nobara 2017).

When the hand of the player almost completely closes the bell (Fig. 4.75b), a series of 'stopped' notes can be sounded. On a horn whose nominal fundamental pitch is 12 ft F (Sect. 7.2.7), the stopped notes are a semitone above the natural notes of the unstopped horn. The player's hand is then performing the same role as the transposing baroque trumpet mute, lowering the frequencies of all the natural notes to such an extent that a new harmonic series can be found by reassigning the peak numbers to a higher nominal fundamental frequency (Campbell and Greated 1987).

On shorter horns with nominal fundamental pitch higher than 11 ft G, the stopped notes lie more than a semitone above the unstopped notes, and cannot therefore be used to supply in tune accidentals. The B^b side of a double horn often has an additional 'stopping valve', which allows the stopped notes to be played in tune without adjusting the main tuning slide.

4.6 Radiation of Sound from Brass Instruments

The previous sections of this chapter have been mainly concerned with the nature of the sound waves in the air inside the tubing of brass instruments. The behaviour of sound waves reaching the end of a flaring bell was discussed in Sect. 4.3.8, but the focus there was on the way in which the change in cross-section in the bell region influenced the standing waves set up in the air column of the instrument. The bell profile also has a strong influence on the nature of the sound field in the external space around the instrument, which is the topic of the present section.

4.6.1 Near Field and Far Field

In general, the sound field radiated by a musical instrument depends on the nature and shape of the emitting surface, the frequency of the radiation and the distance from the instrument. For example, the sound field measured at a point just above the bridge of a violin will be the sum of component waves radiated from many different parts of the instrument, arriving from different directions with different amplitudes and phases. A fairly small movement of the point of measurement will also result in significant changes in the directions of the component waves. There is no reason to expect that in such a case the wavefronts will be plane or spherical, or that there will be a simple relationship between the total measured pressure and the distance from the instrument.

The sound field becomes much simpler when the measurement point is moved away from the instrument by a distance

$$r \gg L, \tag{4.71}$$

where L is the largest spatial dimension of the instrument. It is then a good approximation to consider that the component waves have all arrived at the measurement point in the same direction, having travelled the same distance. The wavefronts can be expected to be spherical surfaces, although the amplitude of the wave may still have an angular dependence. The region of space for which Eq. 4.71 is valid is called the 'far field'. The region close to the instrument for which Eq. 4.71 is not valid is called the 'near field'.

Another criterion which is useful when discussing the nature of a radiated sound field is the ratio between the instrument dimension L and the wavelength λ. In the low-frequency region in which

$$kL = \frac{2\pi L}{\lambda} < 1, \tag{4.72}$$

the radiation is dominated by diffraction, and the far field radiation is approximately isotropic (independent of angle).

Radiation from a brass instrument without toneholes is relatively uncomplicated, since all the sound emerges from a single aperture at the end of the bell which is in most cases axisymmetric. For a trombone bell of diameter $D = 0.2$ m, the far field must begin at a radius $r_f \gg 0.2$ m to satisfy Eq. 4.71. The criterion for isotropic radiation in Eq. 4.72 is satisfied for frequencies below 275 Hz; in practice, the radiation field of a trombone remains effectively isotropic up to 400 Hz (see

Sect. 4.6.6). In this low-frequency region, the far field radiation has wavefronts which are concentric spherical surfaces, with pressure amplitude independent of angle and inversely proportional to r. This type of sound field is described as 'monopole radiation', since in theory it can be considered as emanating from a single point or monopole.

4.6.2 Monopole Radiation

The mathematical treatment of monopole radiation is most straightforward in the system of spherical polar coordinates r, θ, ϕ. The spherical symmetry of the radiation means that the pressure p depends only on r, not on θ or ϕ. In this case, the 3D linear acoustic wave equation (Eq. 4.1) can be rewritten in spherical polar coordinates as

$$\frac{\partial^2 p}{\partial r^2} + \frac{2}{r}\frac{\partial p}{\partial r} = \frac{1}{c^2}\frac{\partial^2 p}{\partial t^2}. \tag{4.73}$$

The expression

$$p_+ = \frac{A}{r}\mathrm{e}^{j(wt-kr)} \tag{4.74}$$

for the pressure in an outward travelling spherical wave was already presented in Sect. 4.3.3. Substitution in Eq. 4.73 confirms that it is indeed a solution of the 3D wave equation.

Since the only spatial variable in the expression for p_+ is r, the Euler equation takes the simple form

$$\rho\frac{\partial v_+}{\partial t} = -\nabla p_+ = -\frac{\partial p_+}{\partial r}. \tag{4.75}$$

Differentiation of Eq. 4.74 with respect to r, substitution in Eq. 4.75 and integrating with respect to time result in an expression for the acoustical particle velocity in a spherical wave:

$$v_+ = \frac{A}{\rho c r}\left(1 - \frac{j}{kr}\right)\mathrm{e}^{j(\omega t-kr)} = \frac{1}{\rho c}\left(1 - \frac{j}{kr}\right)p_+. \tag{4.76}$$

The specific acoustic impedance for the outward travelling spherical wave is

$$\frac{p_+}{v_+} = \rho c\left(\frac{jkr}{1+jkr}\right) = \rho_0 c\left(\frac{k^2r^2}{1+k^2r^2} + j\frac{kr}{1+k^2r^2}\right). \tag{4.77}$$

The impedance has both a real part and an imaginary part, showing that in general the particle velocity is not in phase with the pressure in a spherical wave. The phase difference θ is given by the expression

$$\tan\theta = \frac{1}{kr}. \tag{4.78}$$

When $kr \ll 1$, the phase difference between pressure and particle velocity is almost 90°. At a distance $r = \lambda/2\pi$ from the monopole source, $kr = 1$ and $\theta = 45°$. When $kr \gg 1$, the pressure and particle velocity are almost in phase, as they are for plane waves (see Sect. 4.1.2). This is to be expected, since at very large distances from the source, the spherical wavefronts are almost indistinguishable from plane waves.

A monopole is a mathematical abstraction, since any acoustic source must in reality have a finite spatial extent. A pulsating sphere, whose radius is expanding and contracting periodically as shown in Fig. 4.76, must by symmetry generate a spherical sound field in the region of space outside the surface of the sphere. The pressure and particle velocity in this external field are described by Eqs. 4.74 and 4.76, respectively. The monopole source can be viewed as the limiting case of the pulsating sphere as its equilibrium radius a tends to zero.

From Eq. 4.76, the particle velocity at $r = a$ in the spherical sound field is

$$v_+(a) = \frac{A}{\rho c a}\left(1 - \frac{j}{ka}\right) e^{j(\omega t - ka)}. \tag{4.79}$$

If the sphere is sufficiently small that $ka \ll 1$, Eq. 4.79 reduces to

$$v_+(a) = \frac{A}{\rho c a}\left(-\frac{j}{ka}\right) e^{j\omega t} = \left(\frac{-jA}{\rho c k a^2}\right) e^{j\omega t}. \tag{4.80}$$

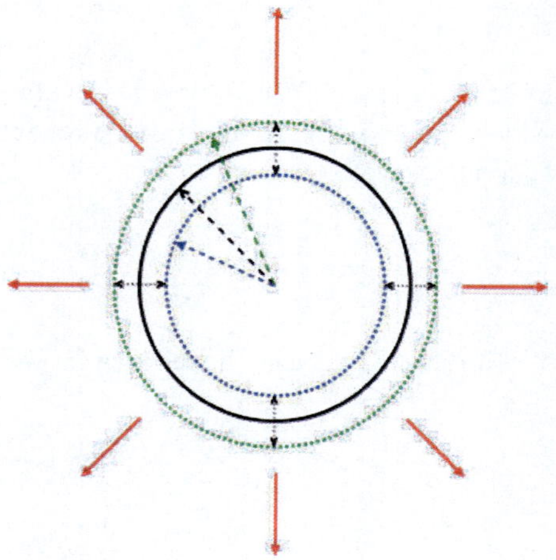

Fig. 4.76 A pulsating sphere. Solid black line: equilibrium radius (a). Dashed blue line: minimum radius. Dashed green line: maximum radius. Red arrows: isotropic sound radiation (Color figure online)

The velocity of a surface point on the sphere is along the radial direction and can be written as

$$v_s = \hat{v}_s e^{j(\omega t)}. \tag{4.81}$$

Since the surface velocity must match the sound field particle velocity at $r = a$,

$$\hat{v}_s = \left(\frac{-jA}{\rho cka^2}\right), \tag{4.82}$$

and

$$A = j\rho ck\hat{v}_s a^2 = \frac{j\rho ckQ}{4\pi}, \tag{4.83}$$

where $Q = 4\pi a^2 \hat{v}_s$ is the maximum volume velocity of the air displaced by the motion of the surface. The pressure in the radiated field can now be written as

$$p_+ = \frac{A}{r} e^{j(wt-kr)} = j\left(\frac{\rho ckQ}{4\pi r}\right) e^{j(wt-kr)}. \tag{4.84}$$

The pulsating sphere is described as a simple source, since it radiates a monopole field. The quantity Q is described as the source strength. The imaginary unit j in the expression on the right-hand side of Eq. 4.84 reflects the phase difference between the pressure and particle velocity at the surface of the source.

Any oscillating source whose dimensions are much smaller than the wavelength of the radiated sound can be treated as a simple source. The radiation from an object too large to satisfy this criterion can be calculated by considering it to be made up of a set of simple sources, although the interactions between the sources have to be taken into account (Chaigne and Kergomard 2016, p. 647).

4.6.3 Transition from Internal to External Sound Fields

In Sect. 4.1.2 it was explained that for low frequencies, only plane waves can propagate along a cylindrical tube. However the sound waves radiated from the open end of the cylinder are expected to have spherical surfaces when the wavelength is much larger than the tube diameter (see Sect. 4.6.2). There must clearly be a transition region in which the flat wavefronts approaching the cylinder exit gradually develop the curvature which is characteristic of the external spherical waves. Figure 4.5 indicates schematically the nature of the transition.

The assumption that the wall of the instrument is rigid requires that the acoustic velocity next to the wall is directed parallel to the surface, which in turn implies that the wavefront must be perpendicular to the wall. For a conical tube, this condition is

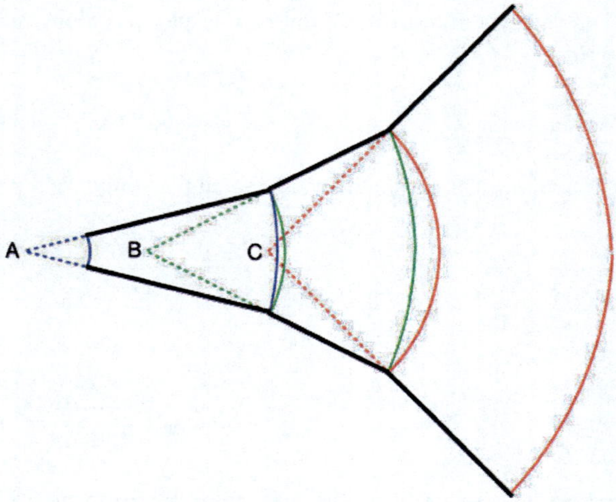

Fig. 4.77 Approximation of a flaring bell by three conical surfaces with different vertices and cone half-angles. The completion of each cone to its vertex is indicated by dashed lines. Coloured solid lines represent spherical wavefronts centred at A (blue), B (green) and C (red) (Color figure online)

met by wavefronts which are sections of spheres centred at the vertex of the cone. It is tempting to assume that the sound waves radiated from the mouth of the cone will also have spherical wavefronts, centred at the cone vertex. This is approximately true at low frequencies, but at higher frequencies, the sudden transition from rigid boundary wall to free space results in the transfer of some of the sound energy to nonspherical modes (Hall 1932).

Many brass instruments have a bore profile with a significant length of cylindrical tubing opening out into a widely flaring bell. In such tubes the wavefronts are planar in the cylindrical section, but develop a bulge well before they reach the open end of the bell (see Fig. 4.42). In a flaring bell, the internal wavefronts can still be described as 'quasi-spherical', but the apparent centre of the spherical surfaces is no longer a fixed point. The reason for this is illustrated in Fig. 4.77, in which a bore profile roughly resembling a brass instrument bell is assembled from three conical sections with increasing flare angle. The first section is a truncated cone with vertex at the point A; spherical wavefronts centred on this point are shown in blue. The vertex of the second conical section is at B, and spherical wavefronts in this section centred on B are shown in green. The wavefronts in the final section, part of a cone centred at C, are shown in red.

There are evident discontinuities in the wavefront curvature at the junctions between the three sections; in reality there would be a smooth transition across a finite region around each change in cone angle (Chaigne and Kergomard 2016, p. 334). By greatly increasing the number of conical sections, it is possible to construct a useful model of the continuously expanding flare of a trumpet bell, in which the apparent centre of the quasi-spherical waves moves steadily along the axis as the wavefront approaches the bell exit. If the last part of the bell is represented by a conical section with a projected vertex at a point C on the bell axis, as shown in Fig. 4.77, the wavefront approaching the bell will at low frequencies be approximately spherical and centred at C. The radiated sound waves do not necessarily radiate from the same point, however: the change in boundary conditions at the bell exit results in energy transfer from the spherical mode to higher (non-spherical) modes, altering the shape of the wavefronts (Benade and Jansson 1974; Jansson and Benade 1974).

In circumstances in which the radiated sound field can be well approximated by an isotropic spherical wave, the common centre of the spherical wavefronts can be described as the 'acoustic centre' of the instrument (López-Carromero 2018). The nature of the pressure fields radiated by real brass instrument bells, and the validity of the concept of an acoustic centre for brass instruments, will be explored further in Sects. 4.6.4–4.6.6. In Sect. 4.6.4 the near fields in front of brass instrument bells in the low-amplitude linear acoustic regime are discussed. High-speed photographs of shock waves in the near fields at amplitudes typical of 'brassy' playing are analysed in Sect. 4.6.5. In Sect. 4.6.6 the directional characteristics of brass instruments in the far field are reviewed.

4.6.4 Mapping the Radiation Fields of Brass Instruments

The way in which the nature of the radiated sound field near the bell of a trombone changes with frequency is illustrated by the pressure maps in Fig. 4.78 (Kemp et al. 2017). The maps in the left-hand column of this figure are derived from calculations using a multimodal theory. The input to the instrument was taken to be a sinusoidal signal with the specified frequency. The calculations were performed both inside and outside the bell, the last 10 cm of which is shown by the magenta curve in each map.

The right-hand column of Fig. 4.78 displays corresponding maps measured experimentally using a vertical linear array of 23 microphones in an anechoic chamber. The array was stepped along the horizontal instrument axis to record the acoustic pressure at different distances from the bell. A horn loudspeaker driver at the input to the instrument generated a swept sine pressure signal, which was repeated at each step of the microphone array. From these measurements it was possible to derive values of the magnitude and phase of the pressure at each measured point for any frequency in the range of the sine sweep (López-Carromero 2018).

The real amplitude of the acoustic pressure is represented by the colour scale in these maps, which are thus snapshots of the pressure distribution in the radiation field at a particular instant. A wave crest is represented by a colour towards the yellow (positive) end of the scale, while a trough is represented by a colour towards the blue (negative) end of the scale. The contours in the experimental maps are interpolated from discrete measurements on a relatively coarse grid, but broadly confirm the multimodal predictions.

Figure 4.78a and b illustrate the pressure distribution for the relatively low input frequency of 500 Hz. The free space wavelength at this frequency is 0.69 m, so the length of the scanned region is just under half a wavelength. The growing bulge as

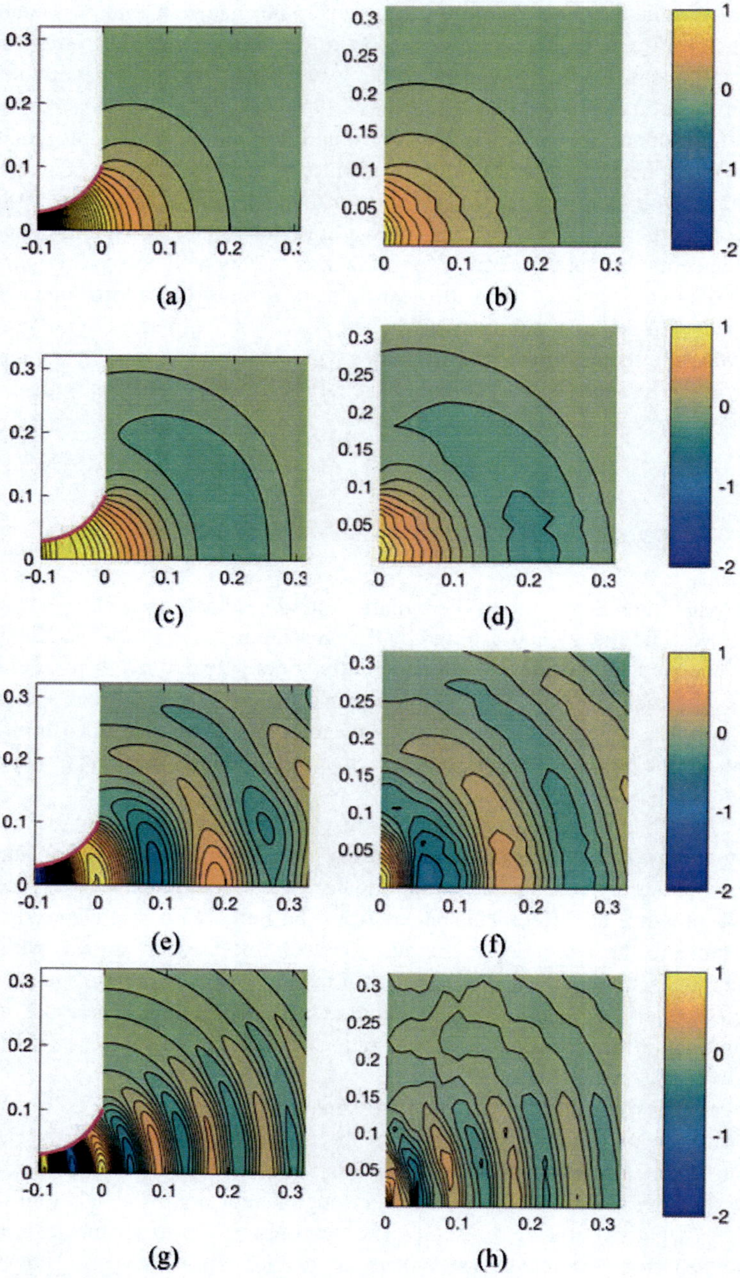

Fig. 4.78 Maps of the real component of the pressure in the sound field radiated by a sinusoidally excited trombone. (**a**) 500 Hz (theory); (**b**) 500 Hz (experiment); (**c**) 1000 Hz (theory); (**d**) 1000 Hz (experiment); (**e**) 2000 Hz (theory); (**f**) 2000 Hz (experiment). (**g**) 4000 Hz (theory); (**h**) 4000 Hz (experiment). Horizontal axes: axial distance x (m) from bell exit plane. Vertical axes: radial distance y (m) from bell axis. From Kemp et al. (2017)

the wavefront travels through the bell towards the exit plane is evident in Fig. 4.78a. The contour of equal real pressure at the bell exit is approximately a circular arc with its centre around 5 cm inside the bell. Once into free space, the contours are also close to circular, but the common centre is now near the ($x = 0$, $y = 0$) point where the bell exit plane cuts the axis. There is no evidence in either theoretical or experimental map of an angular dependence of the radiation pressure amplitude at 500 Hz. The (0,0) point could therefore be described as the acoustic centre at this frequency.

The radiation field looks a little different at a frequency of 1000 Hz, as illustrated in Fig. 4.78c and d. There is a slight suggestion of flattening of the contour lines and a reduction of the strength of the radiation at angles approaching 90°. The flattening of the wavefronts and the concentration of the radiated power in the forward direction is increasingly evident as the frequency is raised to 2000 Hz (Fig. 4.78e and f) and then to 4000 Hz (Fig. 4.78g and h). In Fig. 4.78g a wave crest is just passing the ($x = 0$, $y = 0$) point; a further three crests cut the bell axis at values of x which are multiples of the free space wavelength 0.086 m. Since the wavefronts are not spherical and the radiation is far from isotropic, the idea of an acoustic centre is of limited use at these higher frequencies.

4.6.5 Visualising Wavefronts with Schlieren Optics

When a brass instrument with a high proportion of narrow-bored tubing is played very loudly, shock waves can be generated within the air column of the instrument. This phenomenon, which is responsible for the 'brassy' timbre of fortissimo trumpets and trombones, is discussed in detail in Sect. 6.1. The crest of the wave which leaves the mouthpiece travels down the tube slightly faster than the trough; eventually the crest catches up with the trough in front of it, resulting in an almost instantaneous pressure rise. If the mouthpiece signal is a sine curve, the dependence of pressure on time inside the tube just after shock formation resembles the idealised 'N-wave' shown in the upper part of Fig. 4.79. When such a shock wave reaches the bell of the instrument, the rapid pressure rise is followed by a precipitous drop as the wave expands into free space; the pressure signal registered by a microphone in the radiation field is characterised by a sequence of sharp spikes as illustrated in the lower part of Fig. 4.79. The external pressure signal resembles the derivative of the internal N-wave.

The technique of schlieren photography registers gradients in the refractive index of a fluid as patterns of light and shade in an image (Settles 2001). Since the change in pressure due to the passage of a sound wave is accompanied by a change in the density and therefore refractive index of the air, it is in principle possible to make sound wavefronts visible by this technique. The density gradients in the radiation fields shown in Fig. 4.78 are too small to be visualised using this method: the distance between maximum and minimum pressure contours is half a wavelength, which even at 4000 Hz is 43 mm, and the pressure amplitude was restricted to a

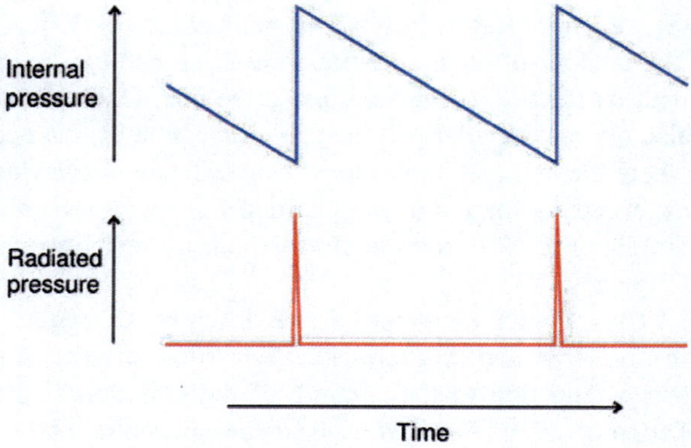

Fig. 4.79 Upper curve (blue): N-wave pressure signal inside an instrument tube. Lower curve (red): impulsive pressure signal in the external radiation field (Color figure online)

few Pa to avoid nonlinear propagation effects. In a shock wave, however, a pressure change of the order of several kPa takes place over a distance of the order of 1 mm, resulting in a very large density gradient suitable for visualisation using schlieren optics.

The first observations of shock waves from electrical sparks using the schlieren method were made by August Toepler in Riga in the 1860s, and some 20 years later, Ernst Mach recorded the first image of a shock wave on a photographic plate (Krehl and Engemann 1995). A schlieren photograph of a shock wave emitted by a loudly blown trombone was published by Hirschberg et al. (1996b). A spectacular image of the shock wave from a trumpet (Pandya et al. 2003) is reproduced in Fig. 6.5.

Figure 4.80 illustrates a schlieren optical setup including a high-speed video camera, which has been used to investigate shock waves radiated from several different types of brass instrument bell (López-Carromero et al. 2016). The apparatus includes a single spherical mirror with a radius of curvature of 3 m. Light originating from a 2 mm diameter source located at the centre of curvature of the mirror illuminates the field of study in front of the mirror. The reflected light is deviated by a beam splitter and enters the lens of the camera, which images the field of study. A knife edge intersects the beam at the point where the reflected waves converge. Rays which pass through a region of high refractive index gradient in the field of study are deviated and converge to a different point. The deviating region appears brighter or darker in the camera image, depending on whether the point of convergence moves away from or towards the knife edge.

Figure 4.81 shows four schlieren images of the progress of a shock wavefront emitted from a natural trumpet. The input signal was a 1000 Hz sine wave generated by a horn driver, with a pressure amplitude of approximately 10 kPa. The time interval between the images was 50 μs. The pressure signal measured by the microphone visible at the far side of each image had the form illustrated by the lower curve in Fig. 4.79, with the passage of each shock wave generating a spike with a width of a few μs.

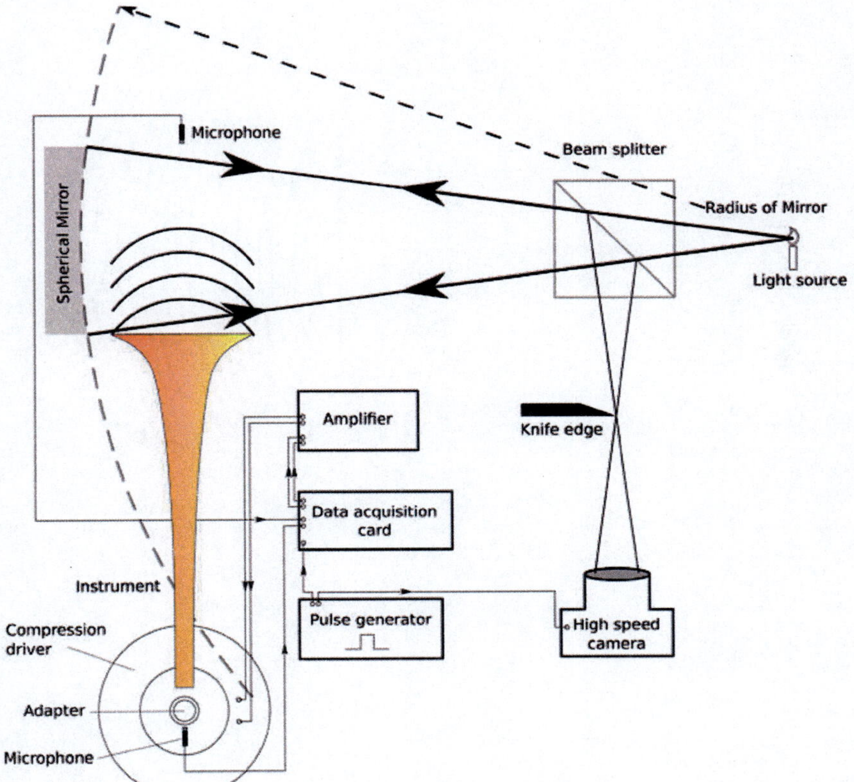

Fig. 4.80 Schlieren optical apparatus for high-speed imaging of shock waves from a brass instrument (López-Carromero et al. 2016)

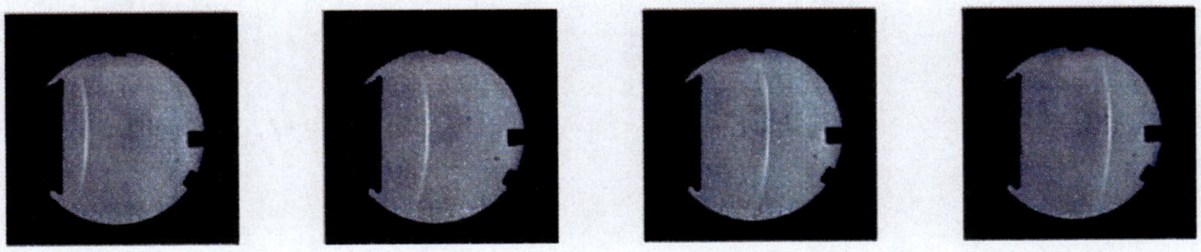

Fig. 4.81 Four successive schlieren images of a shock wave radiated from the bell of a natural trumpet driven by an electroacoustic transducer at the mouthpiece (López-Carromero et al. 2016)

The shock wavefronts are approximately circular arcs, and with appropriate processing, it is possible to fit circles to a set of expanding wavefronts in front of the bell (López-Carromero 2018). Figure 4.82 illustrates the fitted circles for two different shapes of bell profile. In Fig. 4.82a the ‘bell’ is a conical glass funnel. In this case the circles have a common centre very close to the apparent apex of the

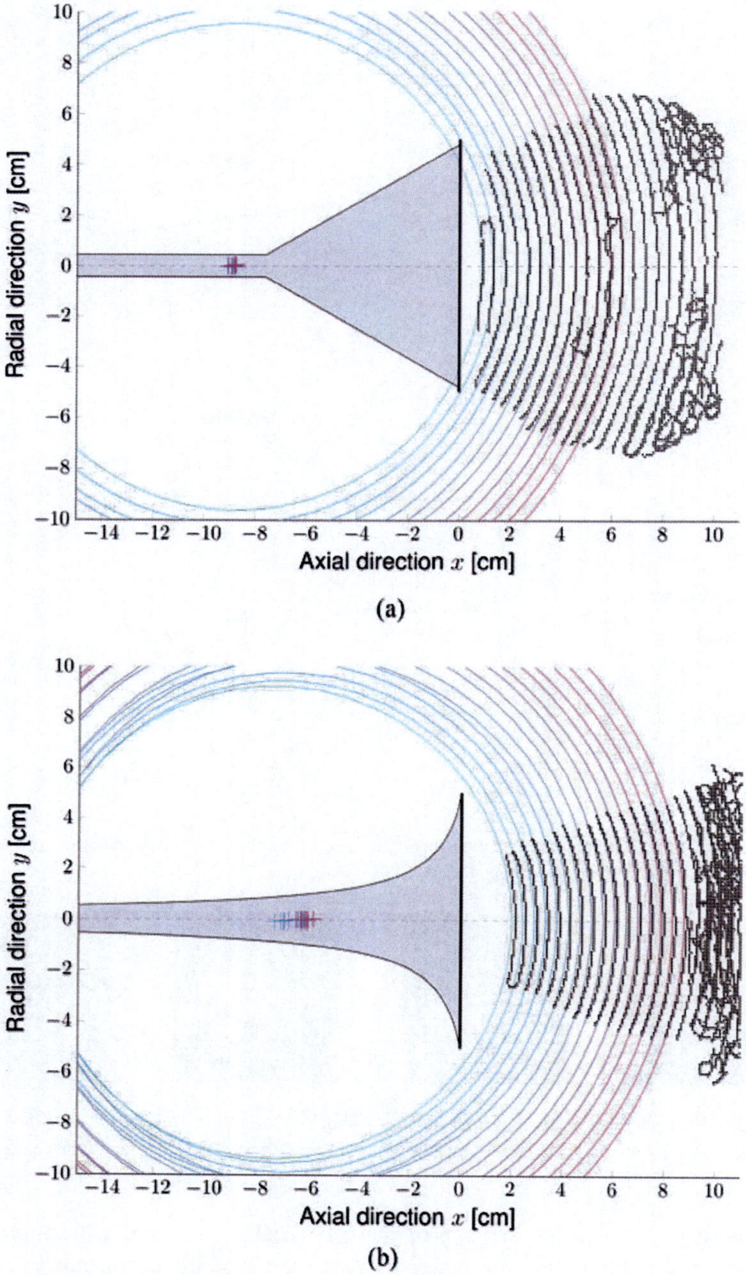

Fig. 4.82 Circles fitted to schlieren imaged shock waves emitted by (**a**) a conical glass funnel and (**b**) a piccolo trumpet bell (López-Carromero 2018)

internal conical surface. The pressure amplitude in the shock wave is greatest in the axial direction and falls off with increasing angle; the black lines recording the measured wavefronts end when the pressure has fallen so far that the trace is no longer detectable. The radiated sound intensity is largely concentrated in the solid angle defined by the internal surface of the cone.

The behaviour of the shock waves emanating from the piccolo trumpet bell is rather different. The apparent centre of curvature of the wavefronts is not fixed, but moves significantly as the wavefront expands. Although the conical funnel and the piccolo trumpet have similar exit diameters, the solid angle containing most of the radiated sound energy is much smaller in the case of the trumpet, confirming that for very loud playing, the flaring bell directs the radiation strongly in the forward direction.

4.6.6 Far Field Directivity in Brass Instruments

In the far field region, all acoustic sources are characterised by spherical wavefronts and a pressure amplitude which is inversely proportional to the radius (Sect. 4.6.1). The pressure amplitude is not necessarily the same at every point on a spherical wavefront, however, but may vary with the angle θ between the bell axis and the direction of radiation. In Fig. 4.83 the acoustic centre is shown in the plane of the bell, but the experimental measurements of López-Carromero et al. (2016) suggest that it is more likely to be approximately one bell diameter inside the bell.

The far field pressure from an axisymmetric acoustical source can be expressed as

$$p_{+} = \frac{A}{r} H(\theta, f) e^{j(wt-kr)} \tag{4.85}$$

where the directional factor $H(\theta, f)$ describes the angular dependence of the radiation for points on the spherical wavefront with radius r (Kinsler et al. 1999, p. 188). $H(0, f)$ is the pressure on the bell axis. For a monopole source, $H(\theta, f) =$

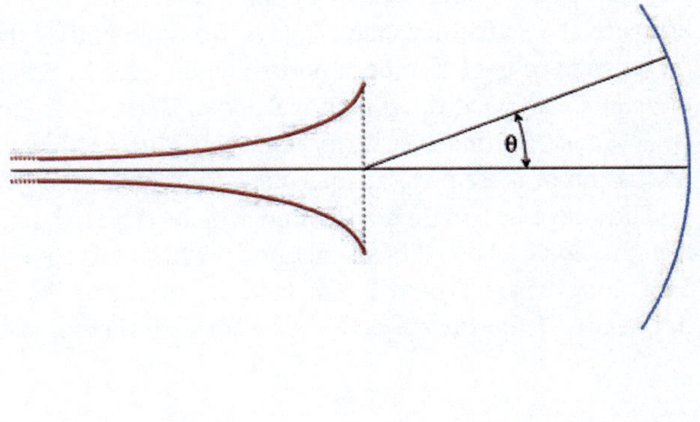

Fig. 4.83 Illustrating the definition of the directivity angle θ

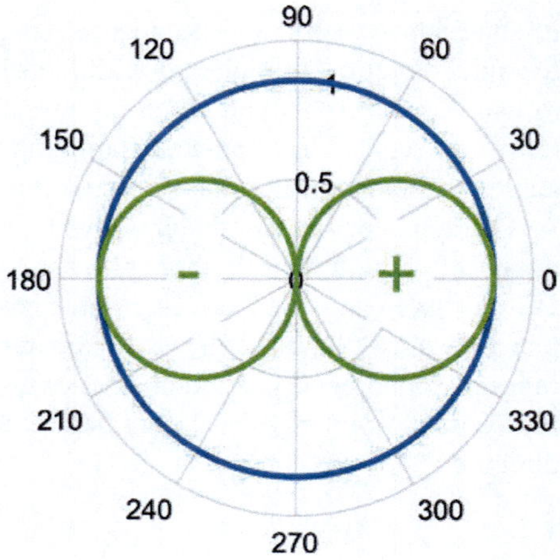

Fig. 4.84 Directivity plots for monopole (blue) and dipole (green) acoustic sources. Radial units: relative pressure amplitude. Angular units: deviation θ (degrees) from axis (Color figure online)

1 independently of angle and frequency. For an acoustic dipole, consisting of two closely spaced monopole sources of equal strength and opposite polarity (phases differing by 180°), $H(\theta, f) = \cos\theta$. The directivity plots for these two theoretical examples are shown in Fig. 4.84: the amplitude of the radiation at angle θ is proportional to length of the line drawn at this angle from the origin to its intersection with the curve.

The first experimental measurements on the directivity of brass instruments were carried out by Daniel Martin (1942). As part of his Ph.D. work at the University of Illinois, Martin built an outdoor measuring system on the roof of the Physics building which allowed him to study the free field radiation patterns of a cornet and a french horn. An electroacoustic driver fed pressure signals at various frequencies to the mouthpiece of the instrument under study. He confirmed that the radiation was effectively isotropic at low frequencies, but was more and more strongly directed along the bell axis as the frequency increased.

Some decades later, Jürgen Meyer and Klaus Wogram carried out an extensive series of directivity measurements on orchestral brass instruments in the anechoic chamber at the Physikalisch-Technische Bundesanstalt laboratory in Braunschweig (Meyer and Wogram 1969, 1970). In these measurements a sequence of notes was played on each instrument by a human performer, and the amplitude of the sound pressure at a radial distance of 3.5 m from the bell of the instrument was recorded for a range of angles in both horizontal and vertical planes. Frequency analysis of the signals allowed the directional factor $H(\theta, f)$ to be deduced. Meyer's classic book *Acoustics and the Performance of Music* (Meyer 2009) includes a detailed discussion of these measurements and their musical significance.

The ranges of frequency within which Meyer found the sound radiation from the common orchestral brass instruments to be effectively isotropic are indicated by the colour bars in Fig. 4.85. The lower limit of each bar represents the fundamental frequency of the lowest note in the normal playing range of the instrument. The

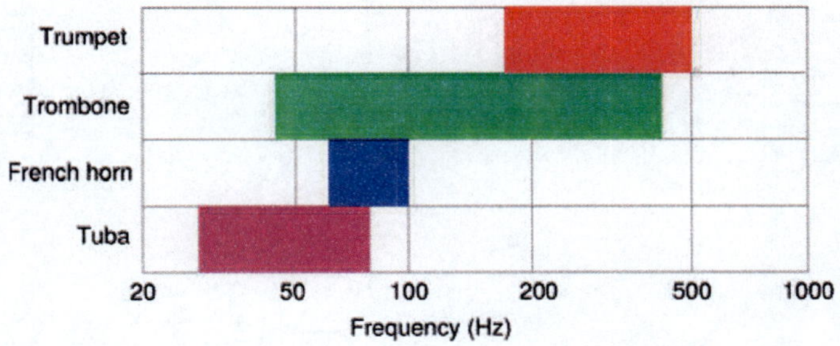

Fig. 4.85 Frequency regions in which radiation is approximately isotropic. Red: trumpet. Green: trombone. Blue: french horn. Magenta: tuba. After Meyer (2009) (Color figure online)

Table 4.1 Values (in dB) of on-axis directivity index $DI(0, f)$ for orchestral brass instruments

f	500 Hz	1000 Hz	3000 Hz
Trumpet	0.8	5.1	12.9
Trombone	4.1	6.4	13.1
Tuba	6.0	13.1	16.4
Horn	4.6	7.6	13.6

upper limit is around 500 Hz for the trumpet, decreasing to around 75 Hz for the tuba. These values are broadly in agreement with Eq. 4.72, which predicts that the monopole approximation will become unreliable when $kD \simeq 1$, with D the bell diameter. For a typical trumpet with $D = 0.12$ m, a frequency of 500 Hz corresponds to $kD = 1.1$; for a large tuba with $D = 0.48$ m, a frequency of 75 Hz corresponds to $kD = 0.66$.

Above the frequency limit of the monopole approximation, the radiated sound intensity is no longer uniform over a spherical surface in the far field. A useful measure of the extent to which the sound energy is beamed in a particular direction θ is the directivity index $DI(\theta, f)$, which is the difference (in dB) between the sound pressure level on the spherical surface at angle θ and the pressure which would be measured if the instrument were replaced by a simple source with the same acoustic power. The maximum value of the directivity index for an axisymmetric bell is usually the on-axis value $DI(0, f)$, although this can be modified by reflections from an external mute or the body of the player. Table 4.1 lists values of $DI(0, f)$ measured by Meyer for orchestral brass instruments at three different frequencies.

All of the instruments listed in Table 4.1 show very strong directivity at 3000 Hz, bearing in mind that the 16.4 dB value of $DI(0, 3000)$ measured for the tuba corresponds to a 44-fold increase in the sound intensity due to the concentration of the radiation in a narrow beam along the axis. This gain is of course compensated by a reduction in the efficiency of radiation at wider angles. Plots of the directivity index as a function of angle for trombone and trumpet are shown in Fig. 4.86a and b. It can be seen in Fig. 4.86a that for angles greater than 50°, the value of $DI(\theta, 3000)$

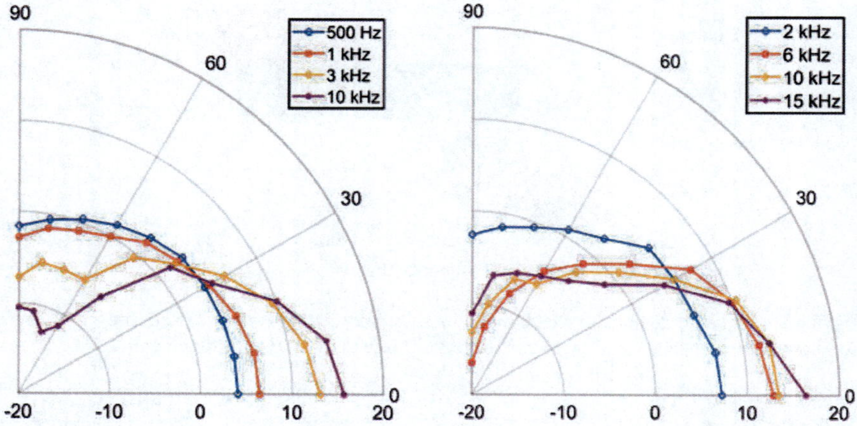

Fig. 4.86 Measured directivity index plots for (**a**) a trombone; (**b**) a trumpet. Axial units: degrees. Radial units: decibels relative to an isotropic radiator with equal acoustic power. Data from Meyer (2009)

for the trombone becomes negative, indicating that the intensity of radiation is lower than it would be from a simple source of the same total power.

An alternative method of graphing the beam patterns of radiation from brass instruments, devised by Meyer (2009), is illustrated in Fig. 4.87a. The shaded area represents the angular zone within which the sound intensity radiated by a trumpet does not fall below 3 dB of its maximum value. The view shown is taken from the player's right-hand side and shows the increasing concentration of energy in a narrow cone along the bell axis. An example of a view of the radiation pattern from a french horn, seen from above the player, is reproduced in Fig. 4.87b. Because of the way in which the horn is held by the player, the radiation pattern is more complicated, and it is notable that at high frequencies, the radiation is most strongly directed into the backward hemisphere.

Further studies of directivity patterns of symphonic brass instruments have been carried out by Otondo and Rindel (2004, 2005) and also by Pätynen and Lokki (2010). The approach taken by Pätynen and Lokki, using a dodecahedral microphone array in an anechoic chamber to record performances by professional players, is illustrated in Fig. 4.88. Four rings of five microphones encircled the players, at elevation angles of $\pm 11°$ and $\pm 53°$, with two additional microphones directly in front of and above the musician. From recordings of two-octave arpeggios played in registers appropriate to the instruments, maps of the directivity as a function of azimuthal angle and angle of elevation were derived. In the maps shown for a tuba in Fig. 4.88, a trumpet in Fig. 4.89a and a french horn in Fig. 4.89b, the frequency responses are averaged over third octave bands and normalised so that the highest pressure recorded in a given third octave band is recorded as 0 dB (white on the greyscale).

Fig. 4.87 Principal radiation directions for (**a**) trumpet in the vertical plane; (**b**) (on following page) horn in the horizontal plane. Shaded areas include directions for which the intensity is within 3db of its maximum value. Meyer (2009)

Maps such as those illustrated in Figs. 4.88 and 4.89 contain much detailed information about the complex radiation field patterns generated by brass instruments. The trumpet shows little variation of normalised pressure level over azimuth and elevation angles at frequencies below 400 Hz, confirming its quasi-monopole nature in this frequency range. By 4000 Hz the radiation is concentrated in a lobe in the forward direction which is most strongly registered by the microphone at elevation $-11°$, consistent with the normal playing direction. At high frequencies the tuba radiation is concentrated in a very narrow lobe at an azimuth angle of around 70° (to the player's left) and recorded by the microphone at an elevation of 53°,

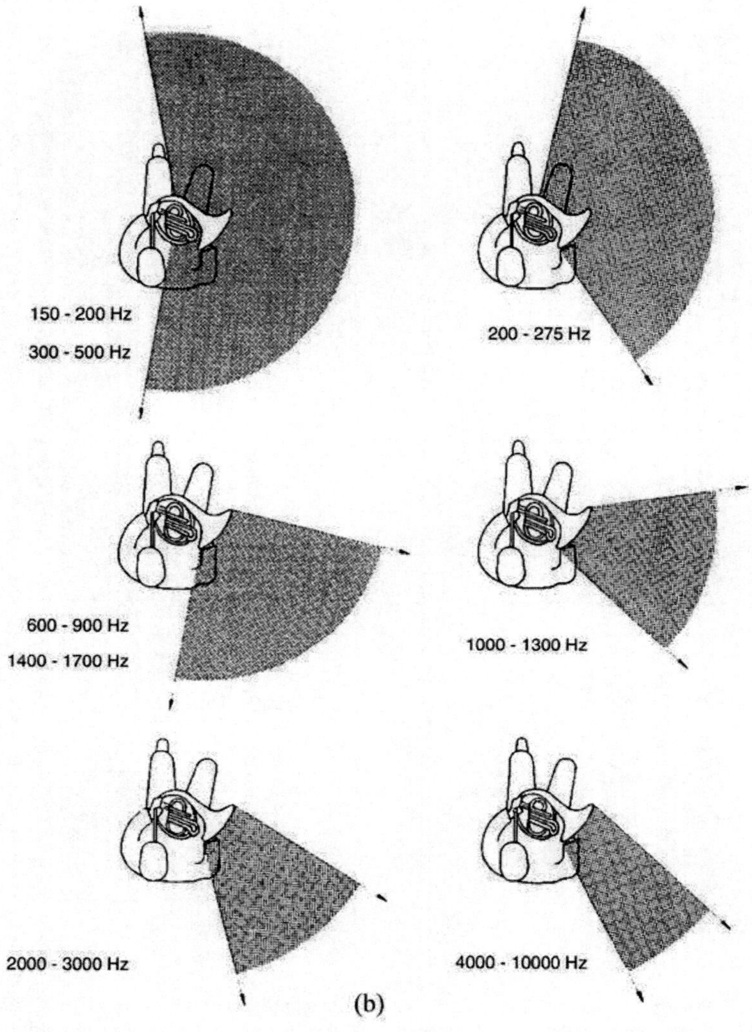

Fig. 4.87 (continued)

which is approximately on the axis of the instrument. The radiation from the french horn is strongly dependent on frequency, but consistently biased towards negative elevations and positive azimuth angles greater than 90°. This type of information is useful in determining the optimum microphone placements for recording sessions. It is also valuable in guiding the design and construction of loudspeaker arrays with directivities matching those of the instruments which they replicate in virtual

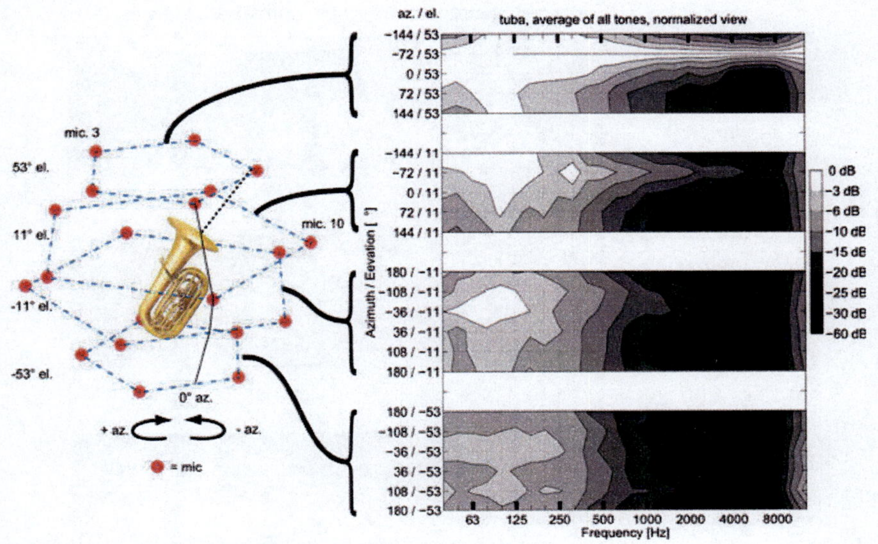

Fig. 4.88 Directivity patterns for a tuba Pätynen and Lokki (2010)

orchestras used in architectural auralisations (Pätynen 2011; Pelzer et al. 2012; Lokki 2014).

4.7 Going Further: Calculating Input Impedance

The importance of the input impedance as a convenient representation of the linear acoustical response of a brass instrument was explained in Sect. 4.1.6, and methods of measuring the input impedance of existing instruments were described in Sect. 4.2. In this section we review methods for calculating the input impedance of an instrument from knowledge of its bore profile. Very old instruments often suffer from leaks or other types of damage which make it impossible to use acoustical measurement techniques, but impedance calculation can provide estimates of the playing pitches and other properties of the instruments in their original state (see Sect. 9.1). Input impedance calculations are also the basis for reviewing and optimising projected designs for new instruments (Kausel 2001; Braden et al. 2009).

4.7.1 Analytical Calculations

In Sect. 4.1.2 a wave equation (Eq. 4.1) was introduced and used to describe the propagation of a sound wave in a cylindrical tube. It was shown that for a cylinder of length L open at the output end, the reflection coefficient is

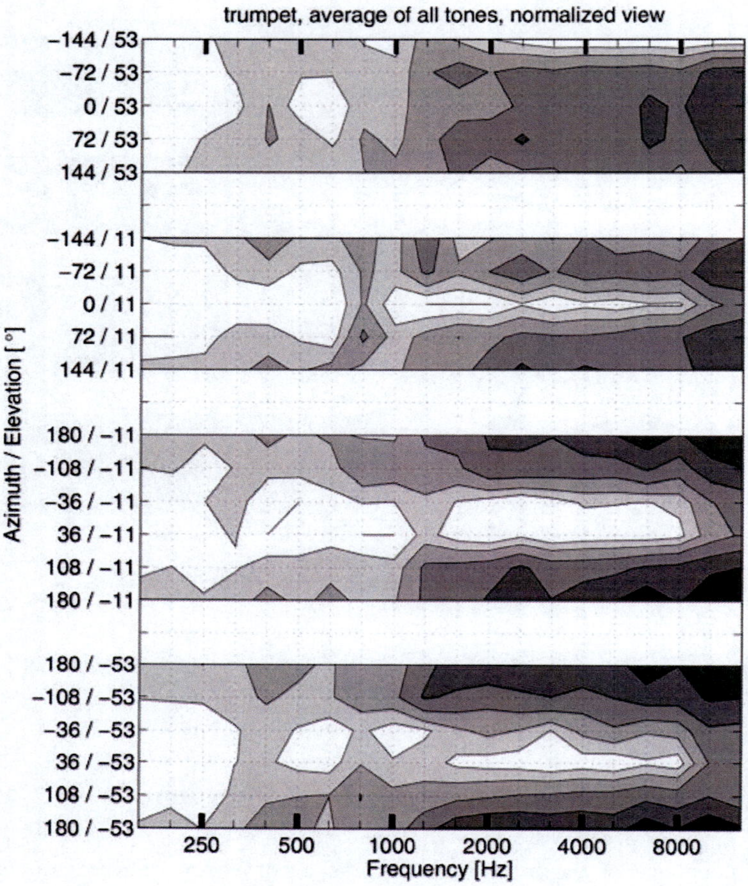

(a)

Fig. 4.89 Directivity patterns for **(a)** a trumpet; **(b)** (on following page) a french horn Pätynen and Lokki (2010)

$$R(\omega) = -e^{-2jkL}. \tag{4.18}$$

The reflection coefficient is related to the input impedance through Eq. 4.28, which can be rearranged to give

$$Z'(\omega) = \frac{1 + R(\omega)}{1 - R(\omega)}, \tag{4.86}$$

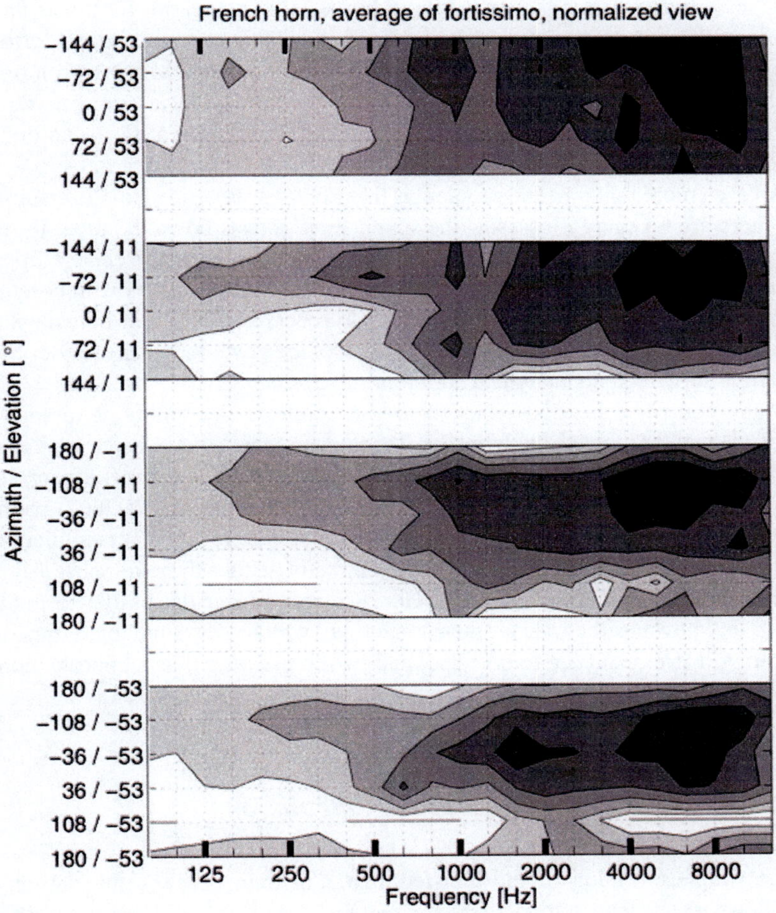

(b)

Fig. 4.89 (continued)

where

$$Z'(\omega) = \frac{Z(\omega)}{Z_c} = \frac{Z(\omega)S}{\rho c} \tag{4.87}$$

is the normalised input impedance. Substituting Eq. 4.18 in Eq. 4.86 yields

$$Z'(\omega) = j\tan(kL) = j\tan(\omega L/c). \tag{4.88}$$

Equation 4.88, which is an explicit expression for the input impedance of the tube, is described as an analytical solution for Z'. Analytical solutions can be found for a few geometries, including cylinders and cones, but not for the bores of realistic brass instruments such as those shown in Fig. 4.21b. One reason for this is that the shapes of these instruments cannot be described by simple mathematical formulae. Another problem is that the transition from the tube interior to the region outside the bell cannot be treated analytically. Even for the case of a cylinder, Eq. 4.88 was obtained by neglecting sound radiation and assuming that the open end was a pressure node.

An even more fundamental problem is that Eq. 4.1 was derived on the assumption that there are no energy losses as the sound wave travels down the tube. In reality there are various fluid dynamical loss mechanisms which drain energy from the sound wave during its passage (Pierce 1989; Keefe 1984). The most important are the viscothermal losses caused by frictional processes in the thin boundary layer near the internal wall of the tube. These processes are governed by the Navier-Stokes Equation, for which no analytical solutions are available.

Many numerical methods exist for finding approximate solutions to equations which cannot be solved analytically. Techniques which have been applied successfully to the study of wind instruments include finite element (Lefebre 2010), finite difference (Noreland 2002) and lattice Boltzmann (Kühnelt 2007) methods. Increasingly powerful computers have made it possible to carry out simulations in which the Navier-Stokes equation is solved directly using a three-dimensional finite difference method (Giordano 2017). Here we restrict our discussion to a simple approach, known as the transfer matrix method, which allows the input impedance of a brass instrument bore to be calculated by representing it as a concatenation of segments of simple geometry.

4.7.2 Lossless Plane Wave TMM Calculations

In the simplest version of the transfer matrix method (TMM), the flaring tube of a brass instrument is modelled as a sequence of short cylinders with different diameters, as shown in Fig. 4.90. It is assumed that only plane waves propagate in the cylinders. If, in addition, viscothermal losses are ignored, Eq. 4.5 can be used to derive an analytical espression for the input impedance Z_{in} of a cylindrical section in terms of the output impedance Z_{out}.

The total pressure at a distance x along the cylinder axis is the sum of the pressures in the forward and backward travelling waves, as shown in Eq. 4.15:

$$p(x,t) = p_+(x,t) + p_-(x,t) = A\mathrm{e}^{j(\omega t - kx)} + B\mathrm{e}^{j(\omega t + kx)}.$$

The acoustic particle velocity in the forward travelling wave was given in Eq. 4.13 as

$$v_+(x,t) = \frac{A}{\rho c}\mathrm{e}^{j(\omega t - kx)} = \frac{1}{\rho c}p_+(x,t).$$

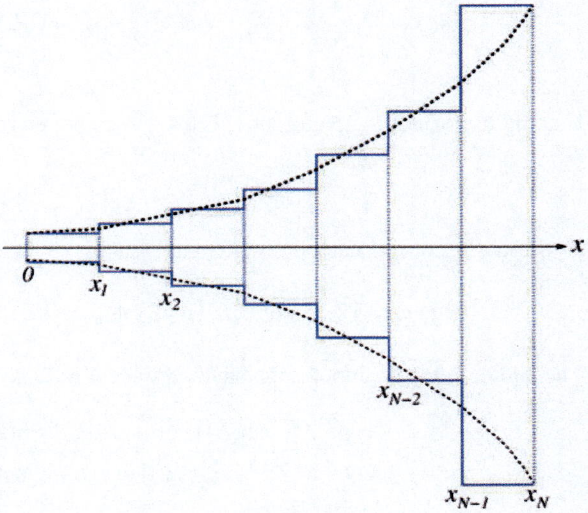

Fig. 4.90 Approximation of a flaring bell as series of cylindrical sections

For the backward travelling wave, the acoustic particle velocity (measured in the +x direction) changes sign, so

$$v_-(x,t) = -\frac{B}{\rho c}\mathrm{e}^{j(\omega t+kx)} = -\frac{1}{\rho c}p_-(x,t). \tag{4.89}$$

The total acoustic particle velocity at x is therefore

$$v(x,t) = \frac{1}{\rho c}(p_+(x,t) - p_-(x,t)) = \frac{1}{\rho c}\left(A\mathrm{e}^{j(\omega t-kx)} - B\mathrm{e}^{j(\omega t+kx)}\right). \tag{4.90}$$

Recalling that the volume velocity $u(x,t) = Sv(x,t)$, where S is the cross-sectional area of the cylinder, and the characteristic impedance of the cylinder is $Z_\mathrm{c} = \rho c/S$ (Eq. 4.31), the total volume velocity at x is

$$u(x,t) = \frac{1}{Z_\mathrm{c}}\left(A\mathrm{e}^{j(\omega t-kx)} - B\mathrm{e}^{j(\omega t+kx)}\right). \tag{4.91}$$

We consider the cylindrical section of length L starting at x_1 and ending at x_2 and derive expressions relating the pressure and volume velocity at the entrance to the corresponding quantities at the exit. To simplify the notation, the common factor $\exp(j\omega t)$ is omitted. From Eq. 4.15,

$$\begin{aligned} p(x_1) &= A\mathrm{e}^{-jk(x_2-L)} + B\mathrm{e}^{jk(x_2-L)} \\ &= A\mathrm{e}^{-jkx_2}(\cos(kL) + j\sin(kL)) + B\mathrm{e}^{jkx_2}(\cos(kL) - j\sin(kL)) \end{aligned}$$

$$= \cos(kL)\Big(Ae^{-jkx_2} + Be^{jkx_2}\Big) + j\sin(kL)\Big(Ae^{-jkx_2} - Be^{jkx_2}\Big). \tag{4.92}$$

Making use of Eqs. 4.15 and 4.91, Eq. 4.92 can be rewritten as

$$p(x_1) = [\cos(kL)]p(x_2) + [Z_c j\sin(kL)]u(x_2). \tag{4.93}$$

Similarly Eq.4.91 can be rewritten as

$$u(x_1) = [Z_c^{-1} j\sin(kL)]p(x_2) + [\cos(kL)]u(x_2). \tag{4.94}$$

The equation for the input impedance of the cylinder is

$$Z(x_1) = \frac{p(x_1)}{u(x_1)} = \frac{[\cos(kL)]p(x_2) + [Z_c j\sin(kL)]u(x_2)}{[Z_c^{-1} j\sin(kL)]p(x_2) + [\cos(kl)]u(x_2)}. \tag{4.95}$$

Dividing numerator and denominator of the fraction on the right-hand side of Eq. 4.95 by $u(x_2)$ and noting that $Z(x_2) = p(x_2)/u(x_2)$ yields an equation which allows the input impedance $Z(x_1)$ to be calculated if the output impedance $Z(x_2)$ is known:

$$Z(x_1) = \frac{Z(x_2)\cos(kL) + Z_c j\sin(kL)}{Z(x_2)Z_c^{-1} j\sin(kL) + \cos(kL)}. \tag{4.96}$$

Substituting $Z(x_2) = 0$ in Eq. 4.96 confirms that assuming a pressure node at x_2 leads to the expression for $Z(x_1)$ given in Eq. 4.88.

Equations 4.93 and 4.94 can be expressed in matrix form:

$$\begin{bmatrix} p(x_1) \\ u(x_1) \end{bmatrix} = \begin{bmatrix} \cos(kL) & Z_c j\sin(kL) \\ Z_c^{-1} j\sin(kL) & \cos(kL) \end{bmatrix} \begin{bmatrix} p(x_2) \\ u(x_2) \end{bmatrix}. \tag{4.97}$$

The two-element column vector $[p(x_2)\ u(x_2)]^{\mathrm{T}}$ is transformed to the vector $[p(x_1)\ u(x_1)]^{\mathrm{T}}$ by multiplication with the transfer matrix

$$\mathbb{T} = \begin{bmatrix} \cos(kL) & Z_c j\sin(kL) \\ Z_c^{-1} j\sin(kL) & \cos(kL) \end{bmatrix}. \tag{4.98}$$

The transition between two cylindrical sections with different diameters inevitably generates non-planar acoustic waves (see Sect. 4.7.6), but at sufficiently low frequencies, both pressure and volume flow are conserved across the boundary. The output impedance of one section is thus equal to the input impedance of the following section, although the characteristic impedance Z_c will change.

The transfer matrix calculation begins by approximating the instrument under study by a set of N cylinders. These are not necessarily of equal length, but for simplicity, we will assume here that each has a length L. The entrance to the first section is at $x = 0$, and the exit plane is at $x_N = NL$, as shown in Fig. 4.90. The radiation impedance $Z(x_N) = Z_{\text{rad}}$ is assumed known. For a unit volume flow $u(x_N) = 1$ at the exit, $p(x_N) = Z(x_N)$, so the pressure and volume flow at the output are represented by the vector $[Z_{\text{rad}}\ 1]^{\text{T}}$.

At the input to the Nth section,

$$\begin{bmatrix} P(x_{N-1}) \\ u(x_{N-1}) \end{bmatrix} = \mathbb{T}_N \begin{bmatrix} Z_{\text{rad}} \\ 1 \end{bmatrix}. \tag{4.99}$$

This then becomes the output vector for the preceding section, whose input is found from

$$\begin{bmatrix} P(x_{N-2}) \\ u(x_{N-2}) \end{bmatrix} = \mathbb{T}_{N-1}\mathbb{T}_N \begin{bmatrix} Z_{\text{rad}} \\ 1 \end{bmatrix}. \tag{4.100}$$

This process continues until the pressure and volume velocity at the entrance of the instrument are found:

$$\begin{bmatrix} p(0) \\ u(0)) \end{bmatrix} = \mathbb{T}_1\mathbb{T}_2 \ldots \mathbb{T}_{N-1}\mathbb{T}_N \begin{bmatrix} Z_{\text{rad}} \\ 1 \end{bmatrix}. \tag{4.101}$$

The input impedance of the instrument is then

$$Z(0) = p(0)/u(0). \tag{4.102}$$

4.7.3 Including Losses in TMM Calculations

The model of a wind instrument presented in Sect. 4.7.2 is described as lossless because it is assumed that waves propagate in the tube without viscothermal losses. To obtain agreement with experimentally measured input impedance curves, it is necessary to include these losses in the model.

The two principal energy loss processes take place near the tube wall, in a boundary layer which for musical instruments is typically around 0.1 mm thick (Benade 1968). The air is assumed at rest next to the wall, but moving with the speed of the bulk flow just outside the boundary layer. Frictional shearing forces exist within the boundary layer because of the viscosity of the air. The rise and fall of temperature due to the passage of a sound wave also results in flow of heat to and from the wall. These two processes depend on the frequency of the wave and also on the ratios of the radius of the tube to the viscous and thermal boundary layer thicknesses, which are slightly different (Zwicker and Kosten 1949; Bruneau 2006; Pierce 1989).

The acoustic pressure in a plane wave propagating without losses along the x axis was expressed in Sect. 4.1.2 as

$$p_+ = A\mathrm{e}^{j(\omega t - kx)}. \tag{4.3}$$

Since the argument of the exponential function on the right-hand side of Eq. 4.3 is purely imaginary, the pressure is an oscillating quantity of constant magnitude. The major effect of viscothermal losses is to add a negative real component $-\alpha$ to the argument, corresponding to a exponential decay of the pressure magnitude as the wave propagates. There is also a small reduction in the phase velocity v_p from its free space value c.

These viscothermal effects can be incorporated into the expression for a travelling wave by rewriting it in terms of the propagation constant Γ:

$$p_+ = A\mathrm{e}^{j\omega t}\mathrm{e}^{-\Gamma x}. \tag{4.103}$$

In the lossless case, $\Gamma = jk = j(\omega/c)$. When losses are included,

$$\Gamma = \alpha + j\left(\frac{\omega}{v_\mathrm{p}}\right). \tag{4.104}$$

For the tube diameters and playing frequencies typical of brass instruments, Keefe (1984) gives the following approximate expressions for α and v_p:

$$\alpha = \left(\frac{\omega}{c}\right)\left(1.045\, r_\mathrm{v}^{-1} + 1.080\, r_\mathrm{v}^{-2} + 0.303\, r_\mathrm{v}^{-3}\right) \tag{4.105}$$

$$v_\mathrm{p} = \frac{c}{1.045\, r_\mathrm{v}^{-1}}. \tag{4.106}$$

The parameter in Keefe's equations is

$$r_\mathrm{v} = \frac{a}{b_\mathrm{v}} = a\left(\frac{\omega\rho}{\eta}\right)^{1/2}, \tag{4.107}$$

where a is the tube radius, b_v is the viscous boundary layer thickness, ρ is the density of air and η is the coefficient of viscosity. Using the values $\rho = 1.1769\,\mathrm{Kg\,m^{-3}}$ and $\eta = 1.846.10^{-5}\,\mathrm{Kg\,m^{-1}\,s^{-1}}$ listed by Keefe for a temperature of 300 K,

$$r_\mathrm{v} = 632.8\, f^{1/2}\, a. \tag{4.108}$$

In a tube of radius, $a = 5$ mm r_v is 31.64 at 100 Hz and 100.05 at 1000 Hz. A good approximation to the decay constant α can therefore be made by retaining only the first term on the right-hand side of Eq. 4.105. Making use also of Eq. 4.106, the propagation constant can then be written:

$$\Gamma = \alpha + j\left(\frac{\omega}{v_{\mathrm{p}}}\right)$$
$$\simeq \left(\frac{\omega}{c}\right)\left[1.045\, r_{\mathrm{v}}^{-1} + j\left(1 + r_{\mathrm{v}}^{-1}\right)\right]. \tag{4.109}$$

The decay constant in a musical instrument tube

$$\alpha \simeq 3 \times 10^{-5} f^{1/2}/a \tag{4.110}$$

is thus to a good approximation proportional to the square root of the frequency and inversely proportional to the tube radius.

The introduction of losses also modifies the expression for the characteristic impedance Z_{c} which appears in the TMM matrix Eq. 4.98 because of the change in the propagation velocity v_{p} (Keefe 1984):

$$Z_{\mathrm{c}} = \left(\frac{\rho c}{\pi a^2}\right)\left[1 + 0.369\, r_{\mathrm{v}}^{-1} - j\left(0.369\, r_{\mathrm{v}}^{-1} + 1.149\, r_{\mathrm{v}}^{-2} + 0.303\, r_{\mathrm{v}}^{-3}\right)\right]$$
$$\simeq \left(\frac{\rho c}{\pi a^2}\right)\left[1 + 0.369\, r_{\mathrm{v}}^{-1} - j\left(0.369\, r_{\mathrm{v}}^{-1}\right)\right]. \tag{4.111}$$

With these changes, the matrix for lossy transmission though a cylindrical section of radius a and length L becomes (Amir et al. 1997; Braden 2006)

$$\mathbb{T} = \begin{bmatrix} \cosh(\Gamma L) & Z_{\mathrm{c}} \sinh(\Gamma L) \\ Z_{\mathrm{c}}^{-1} \sinh(\Gamma L) & \cosh(\Gamma L) \end{bmatrix} \tag{4.112}$$

4.7.4 TMM with Non-Cylindrical Elements

The transfer matrix method using cylindrical elements can in principle be used for any duct with cylindrical symmetry, but to achieve accuracy in the rapidly flaring bell sections of brass instruments, a very large number of elements may be required. The computational effort can be significantly reduced by modelling the tube as a sequence of cones (Keefe 1990; Caussé et al. 1984), as shown schematically in Fig. 4.91.

An analytical solution of the lossless wave equation in a conical tube was presented in Sect. 4.3.3 and discussed further in Sect. 4.6.2. The wavefronts of constant phase in the cone are sections of a sphere centred at the (normally virtual) apex of the cone, and a forward travelling pressure wave can be represented in spherical coordinates as

$$p_{+}(r,t) = \frac{A}{r}\mathrm{e}^{j(\omega t - kr)}. \tag{4.35}$$

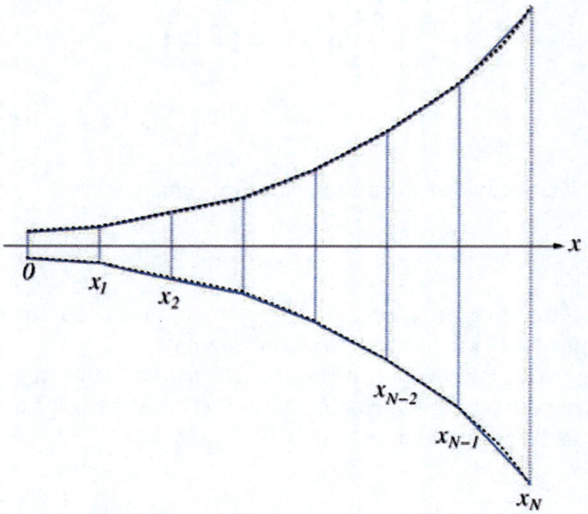

Fig. 4.91 Approximation of a flaring bell as series of conical sections

Here the spatial variable r is the radial distance from the apex. It is possible to develop the TMM approach using spherical wave solutions, but problems arise at junctions between two sections of different taper (see Sect. 4.6.3). It must also be borne in mind that the wavefronts in realistic brass instruments are only approximately spherical, particularly in a rapidly flaring bell (Benade and Jansson 1974). An alternative approach which takes into account the bulging nature of the wavefronts in non-cylindrical tubes is outlined in Sect. 4.7.6. Here the plane wave assumption is retained, which for conical elements can be done simply by replacing the radial variable r by the axial distance x in Eq. 4.35 and related equations (Caussé et al. 1984; Braden 2006).

The TMM calculation using conical elements follows the procedure of matrix multiplication described by Eq. 4.101. The inclusion of losses in the transfer matrix for a conical element is not as straightforward as in the cylindrical case since the tube radius, and therefore the value of the viscothermal loss parameter r_v, is not constant. If the tube is discretised into a very large number of short elements it can be assumed that the propagation constant, phase velocity and characteristic impedance are equal to the values obtained for a cylindrical element whose radius is the mean of the input and output radii of the conical element (Caussé et al. 1984). For real-time synthesis applications, in which it is desirable to minimise the number of separate tube elements to maximise calculation speed, an analytic expression for the propagation constant based on integration along the length of the element has been proposed by van Walstijn et al. (1997).

Using the expressions for Γ and Z_c given in Eqs. 4.109 and 4.111, the lossy transfer matrix for a conical element of length L with input and output planes at distances x_1 and x_2, respectively, from the virtual cone vertex is

$$\mathbb{T}_{\text{cone}} = \begin{bmatrix} A_{\text{cone}} & B_{\text{cone}} \\ C_{\text{cone}} & D_{\text{cone}} \end{bmatrix} \tag{4.113}$$

with

$$A_{\text{cone}} = \left(\frac{x_2}{x_1}\right)\cosh(\Gamma L) - \left(\frac{1}{\Gamma x_1}\right)\sinh(\Gamma L), \tag{4.114}$$

$$B_{\text{cone}} = Z_c\left(\frac{x_1}{x_2}\right)\sinh(\Gamma L) \tag{4.115}$$

$$C_{\text{cone}} = Z_c^{-1}\left\{\left(\frac{x_2}{x_1} - \frac{1}{\Gamma^2 x_1^2}\right)\sinh(\Gamma L) + \left(\frac{L}{\Gamma x_1^2}\right)\cosh(\Gamma L)\right\} \tag{4.116}$$

$$D_{\text{cone}} = \left(\frac{x_1}{x_2}\right)\left\{\cosh(\Gamma L) + \left(\frac{1}{\Gamma x_1}\right)\sinh(\Gamma L)\right\}. \tag{4.117}$$

In Sect. 4.3.8 it was pointed out that the theoretical bore profile known as the Bessel horn (Benade and Jansson 1974; Fletcher and Rossing 1998) provides a fairly close match to the bell sections of many brass instruments. The general equation for a Bessel horn profile is given in Eq. 4.66, and a comparison of the profile of a Conn 8H tenor trombone bell with a fitted Bessel horn bore is shown in Fig. 4.43a. Braden (2006) has provided explicit expressions for the elements of a transfer matrix representing a Bessel horn section and has included multiple Bessel horn sections in a TMM model used in trombone optimisation (Braden et al. 2009).

Instruments with toneholes can also be incorporated in the TMM framework, with a transfer matrix representing each open or closed hole along the bore. The nature and function of toneholes in historical instruments of the brass family, including cornetts, serpents and ophicleides, was reviewed in Sect. 4.4. Details of the required transfer matrices are given in Keefe (1990) and Dubos et al. (1999).

4.7.5 Radiation Impedance

The essence of the transfer matrix method for impedance calculations is that it begins with a known radiation impedance at the exit of the instrument and works backwards to the input. Finding a suitable expression for the radiation impedance at the bell of a brass instrument is however far from straightforward. The plane wave TMM approximation implies that the exit plane of the bell is a wavefront of uniform amplitude and phase. In this case the external radiation is the same as that from a flat vibrating piston. An analytical solution was found by Rayleigh (1894) for a piston surrounded by an infinite plane baffle. This is equivalent to a tube with a very large flange at the end, like the Danish lur (Fig. 1.7a), for wavelengths much greater than the flange radius. The real and imaginary components of the radiation impedance

$Z_{\text{flanged}} = R_{\text{flanged}} + jX_{\text{flanged}}$ are expressed as power series in the variable ka, where k is the wave number and a the radius of the opening:

$$R = Z_c \left\{ \frac{1}{2}(ka)^2 - \frac{1}{12}(ka)^4 + \frac{1}{144}(ka)^6 \ldots \right\} \tag{4.118}$$

$$X = Z_c \left\{ \frac{8}{3\pi}(ka) - \frac{32}{45\pi}(ka)^3 + \frac{128}{175\pi}(ka)^5 \ldots \right\}. \tag{4.119}$$

In the low-frequency limit $ka \ll 1$, which is valid for the playing frequencies of most brass instruments (although not the upper harmonics in the radiated sound),

$$Z_{\text{flanged}} \simeq Z_c\{0.5(ka)^2 + j[0.85(ka)]\}. \tag{4.120}$$

In the case of unflanged instruments without a rapidly flaring final section, such as the Russian horns illustrated in Fig. 7.51, it is more appropriate to make use of expressions for the plane wave radiation impedance of an unflanged cylinder derived by Levine and Schwinger (1948). The low-frequency expression for the radiation impedance of the unflanged pipe is

$$Z_{\text{unflanged}} \simeq Z_c\{0.25(ka)^2 + j[0.61(ka)]\}. \tag{4.121}$$

In both flanged and unflanged cases, it can be seen that when $ka \ll 1$, the radiation impedance is dominated by the imaginary component. The low-frequency normalised input impedance $Z' = Z/Z_c$ for an unflanged pipe is

$$Z' \approx j[0.61(ka)] \approx j\tan(kl_e), \tag{4.122}$$

where $l_e = 0.61a$. A comparison of Eqs. 4.122 and 4.88 shows that at low frequencies, the radiation impedance is approximately equal to the input impedance of a short open cylinder of length l_e with a pressure node at the open end. The effect of the radiation can be seen as displacing the pressure node beyond the exit plane by a distance l_e, which is often described as the open end correction (see Sect. 4.1.3).

The expression for the radiation impedance of an unflanged pipe is strictly valid only for an instrument with an infinitely thin wall. In some wooden brass instruments such as the cornett and the didgeridoo, the wall thickness can be a significant fraction of the pipe radius. Dalmont et al. (2001) give expressions for the radiation impedance of tubes for a range of values of the ratio of external and internal radii and also for a range of terminating geometries.

Accurate expressions of radiation impedance for instruments with rapidly flaring bells must be based on models which take account of the non-planar nature of the wavefronts. Hélie and Rodet (2003) have proposed an approach in which the radiating surface at the bell of a brass instrument is represented as a portion of a pulsating sphere. A comparison of experimentally measured input impedance curves for a trumpet and trombone with TMM calculations based on a plane wave model and on the model of Hélie and Rodet has shown that the pulsating sphere model, together with a spherical wave approach to propagation in the bell section, gives much closer agreement with experiment than a plane wave model (Eveno et al. 2012).

4.7.6 Multimodal Calculations

The acoustical disturbances which travel through a brass instrument tube and radiate from the bell cannot be exactly represented as either purely plane or purely spherical waves (Benade and Jansson 1974). To arrive at a more accurate representation of the bulging wavefronts, it is necessary to find more general solutions of the three-dimensional wave equation

$$\Delta p = \frac{1}{c^2}\frac{\partial^2 p}{\partial t^2}. \tag{4.1}$$

Since most brass instrument bores are to a good approximation cylindrically symmetric, it is convenient to describe the pressure and velocity fields in the tube using a cylindrical polar coordinate system in which x represents the distance along the tube axis, r the displacement perpendicular to the axis and θ the angle of rotation about the axis (Félix et al. 2012). In this system Eq. 4.1 can be written

$$\Delta_{\perp} p + \frac{\partial^2 p}{\partial^2 x} = \frac{1}{c^2}\frac{\partial^2 p}{\partial t^2}, \tag{4.123}$$

where

$$\Delta_{\perp} = \frac{1}{r}\frac{\partial}{\partial r}\left(r\frac{\partial}{\partial r}\right) + \frac{1}{r^2}\frac{\partial^2 p}{\partial \theta^2} \tag{4.124}$$

is the transverse Laplacian (Pagneux et al. 1996).

For a hard walled uniform cylindrical tube of radius a without losses, solutions of Eq. 4.123 must satisfy the boundary condition that the acoustic particle velocity (and therefore the radial pressure gradient) must be zero at $r = a$. An infinite set of such solutions exists, each representing a mode with its own characteristic patterns of pressure and flow velocity. These modes form a complete orthogonal set, and any type of acoustic disturbance can in principle be reproduced by adding together members of the set. This is the basis of the multimodal method of impedance calculation, and the process of identifying the neccessary modes is described as modal decomposition.

Multimodal treatments of brass instruments have been described by several authors (Pagneux et al. 1996; Amir et al. 1997; Kemp 2002; Braden 2006; Félix et al. 2012). The pressure in a forward travelling wave is derived as a sum of the

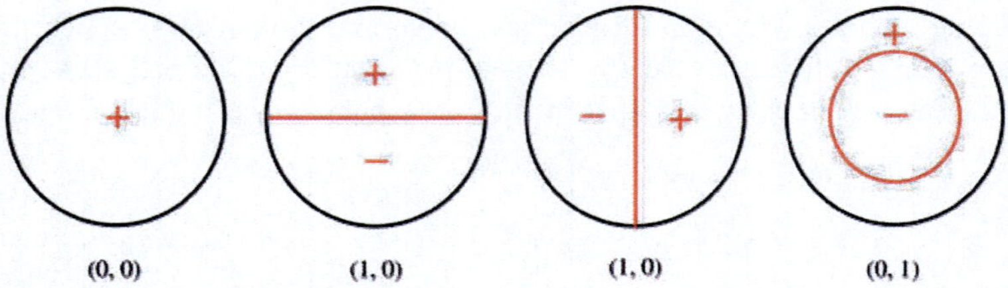

Fig. 4.92 Phase patterns for transverse modes (μ, ν) in a cylindrical tube

contributions from the component modes, each represented by an equation of the form

$$p_n(r, \theta, x) = A_n \phi_n(r, \theta) \mathrm{e}^{j(\omega t - k_n x)}. \tag{4.125}$$

The simplest possible solution is the familiar plane wave mode identified by the index $n = 0$, for which

$$\phi_0 = 1, \quad k_0 = \omega/c,$$
$$\text{and} \quad p_0(x) = A_0 \mathrm{e}^{j(\omega t - k_0 x)}. \tag{4.126}$$

This is the only mode for which the pressure distribution is uniform across the (r, θ) plane and which can exist as a propagating wave at all frequencies. The function ϕ_n, which is an eigenfunction of the transverse Laplacian operator, describes a pressure amplitude whose dependence on r and θ is determined by three indices (μ, ν, σ) (Félix et al. 2012):

$$\phi_n(r, \theta) = \Lambda_n J_\mu \left(\gamma_{\mu\nu} \frac{r}{a}\right) \sin\left(\mu\theta + \frac{\sigma\pi}{2}\right). \tag{4.127}$$

In Eq. 4.127 J_μ is the μth-order Bessel function of the first kind, and $\gamma_{\mu\nu}$ is the $(\nu+1)$th zero of the derivative J'_μ. Λ_n is a normalising constant. Solutions satisfying the boundary conditions correspond to integer values of μ and ν from 0 to ∞.

A transverse mode $\phi_{\mu\nu}$ has a phase pattern characterised by μ nodal diameters and ν nodal circles. Figure 4.92 illustrates some of these patterns. Each pattern corresponds to two orthogonal modes differentiated by the symmetry index σ, which takes the values 0 or 1. A change in σ corresponds to a rotation of the pattern by $\pi/2$ radians around the x axis, as shown by the two patterns for the (1,0) mode in Fig. 4.92. For axially symmetric tubes, these two modes are degenerate; in this case the index σ can be suppressed and μ set equal to zero (Kemp 2002). In the musically significant case of tubes with toroidal bends , the axial symmetry is lifted, and the full treatment must be used (Braden 2006; Félix et al. 2012).

The wave number k_n describing the pressure amplitude variation along the x axis in the nth mode is

Table 4.2 Mode indices ordered by increasing values of $\gamma_{\mu\nu}$ (adapted from Braden (2006))

n	μ	ν	$\gamma_{\mu\nu}$
0	0	0	0
1	1	0	1.84
2	2	0	3.05
3	0	1	3.83
4	3	0	4.20

$$k_n = \sqrt{k^2 - \left(\frac{\gamma_{\mu\nu}}{a}\right)^2}, \tag{4.128}$$

where $k = \omega/c$ is the free space wave number. The wave number is therefore smaller than the free space value for all modes except the plane wave. For each higher mode, there is a cutoff frequency f_c defined by

$$f_c = \frac{c\gamma_{\mu\nu}}{2\pi a} \tag{4.129}$$

below which k_n becomes imaginary. The index n orders the modes in increasing magnitude of the cutoff frequency, which for a fixed tube radius a is determined by $\gamma_{\mu\nu}$. The relationship between n, μ, ν and $\gamma_{\mu\nu}$ for the first few modes is shown in Table 4.2.

The x dependence of the pressure in the nth mode was written in Eq. 4.125 as

$$p_n(x) \propto e^{-jk_n x}. \tag{4.130}$$

For $f > f_c$, k_n is real: the propagation constant $\Gamma = jk_n$ is imaginary, and the pressure has the sinusoidal dependence on x characteristic of a travelling wave. For $f < f_c$, on the other hand, k_n is imaginary, and the propagation constant is $\Gamma = -|k_n|$, which describes an exponentially decreasing pressure amplitude as x increases. This is the behaviour of an evanescent wave. For a cylindrical tube with radius $a = 5$ mm, all non-planar modes are evanescent up to $f = 20.2$ kHz, which is the cutoff frequency for the $n = 1$ (1,0) mode. Since this mode is not axisymmetric (see Fig. 4.92), it can only be excited by a source which breaks cylindrical symmetry. The $n = 3$ (0,1) mode is the first axisymmetric mode to change from evanescent to propagating, at a frequency $f = 42.1$ kHz. It is evident that for frequencies of musical interest, only plane waves can propagate in the cylindrical or gently flaring sections of most brass instruments.

Impedance calculations have been carried out on brass instruments using the TMM method extended to include multiple modes (Pagneux et al. 1996; Kemp 2002; Braden 2006). These have employed short cylindrical elements, and losses have been incorporated in the manner described in Sect. 4.7.3. A complication in multimodal TMM calculations is that although the different modes are uncoupled in a uniform cylinder, they are coupled at a transition between two elements involving a change in radius. This is readily understandable, since even although the waves a few diameters on either side of the transition may be close to planar, the representation of the acoustic field in the immediate vicinity of an abrupt change in radius will require a large number of (probably evanescent) higher modes.

In the multimodal approach, the pressures and volume velocities p_n and u_n are the elements of two column vectors **P** and **U**, respectively. The impedance matrix $\mathbb{Z}$ is defined by

$$\mathbf{P} = \mathbb{Z}\mathbf{U}. \tag{4.131}$$

The diagonal elements $\mathbb{Z}_{ii}$ relate pressure and volume flow in the same mode and are called direct impedances. The off-diagonal elements $\mathbb{Z}_{ij}$, $i \neq j$, relate the pressure in one mode to volume flow in another mode and are called coupled impedances.

As in the TMM methods previously discussed, the calculation begins with a known radiation impedance, in this case represented by a matrix $\mathbb{Z}^{\mathrm{rad}}$. The multimodal radiation impedance derived by Zorumski (1973) for a cylindrical duct surrounded by an infinite baffle has been shown to provide an acceptable approximation to $\mathbb{Z}^{\mathrm{rad}}$ for the case of a brass instrument bell (Amir et al. 1997; Kemp 2002; Braden 2006), although it does not take into account the radiation field in the hemisphere behind the bell. Propagating the impedance from the bell to the input through a set of matrix multiplications yields the input impedance $\mathbb{Z}^{\mathrm{in}}$. The matrix equations describing propagation along a cylindrical element and across a discontinuity between two elements are presented in Pagneux et al. (1996).

For numerical treatment it is necessary to truncate the set of modes at a finite number $N = n_{\mathrm{max}} + 1$. It is not possible to exclude all evanescent modes, since the reflection of a propagating mode at a discontinuity is influenced by coupling to evanescent modes. The number of modes required depends on the shape and scale of the duct and in practice is found by carrying out successive calculations with increasing N until the results converge. Kemp (2002) showed that for a trumpet bell, acceptable convergence was achieved with seven modes. Since the process involves repeated multiplication of $N \times N$ matrices, the computational load is strongly dependent on the number of modes, and the time taken for a multimodal TMM input impedance calculation is typically several orders of magnitude greater than for the equivalent plane or spherical wave calculation.

4.7.7 Bends in Brass Instruments

For practical reasons almost all brass instruments include sections in which the tube axis is curved or bent rather than straight (see Sect. 1.2.11). Many of the instruments illustrated in Chap. 7 have tuning slides with tight 180° bends of the type shown in Fig. 4.93a, while sound waves are obliged to undergo several rapid changes of direction when travelling through valves. Alpine pastures offer ample space for the majestic length of the alphorn, but the more crowded environments of the orchestra pit and bandstand have led to the evolution of the folded and curved forms of the

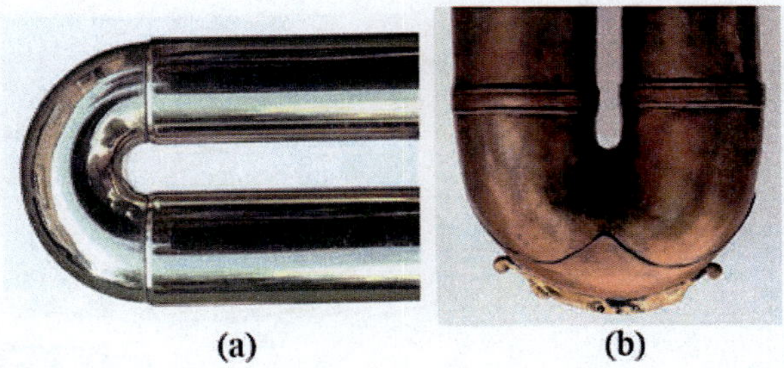

Fig. 4.93 (**a**) The third valve tuning slide on a B^b trumpet (tube diameter $\sim$ 12 mm). (**b**) The bend at the foot of a B^b ophicleide (tube diameter $\sim$ 50 mm)

serpent, ophicleide, horn and tuba. The limitation imposed by the length of the human arm makes the bent tube essential in slide instruments like the trombone.

The acoustical effect of a curvature in the axis of a lossless duct was considered briefly by Rayleigh (1894), who concluded that the influence of bending was unimportant in tubes of diameter much smaller than the wavelength. In the first edition of his influential textbook *Acoustical Aspects of Woodwind Instruments*, Kees Nederveen (1969) included a discussion of toroidal bends in tubes. On the assumption that the pressure was invariant across a tube cross-section, he derived a formula showing that in the low-frequency limit, the curvature of the tube reduces the effective length and increases the effective area by the same percentage.

An important parameter in discussing toroidal bends is the axis curvature κ, defined as the ratio of the internal radius a of the tube to the radius of curvature R_0 of the axis:

$$\kappa = a/R_0. \tag{4.132}$$

Nederveen's low-frequency prediction is that the effective length L_{eff} of the toroidal duct is related to the geometrical length L of the curved axis by the equation $L_{\text{eff}} = \alpha L$, where

$$\alpha = \frac{0.5\kappa^2}{1 - \sqrt{1-\kappa^2}}. \tag{4.133}$$

Since $\alpha < 1$ for all values of κ, the length correction derived from Eq. 4.133 is always negative. The prediction that at low frequencies the effect of a bend is to increase the resonance frequencies was confirmed by Keefe (1984) and in subsequent work by Nederveen (1998a,b) which also took into account additional length corrections due to transitions from straight to curved sections.

The assumption of plane wave propagation in a curved duct is only valid for frequencies satisfying the condition $f \ll f_{c(1,0)}$, where $f_{c(1,0)}$ is the cutoff frequency of the first higher mode (see Sect. 4.7.6). Even for frequencies well below $f_{c(1,0)}$, the effect of evanescent modes can be significant. A treatment of bends using a multimodal calculation was proposed by Félix and Pagneux (2001) and included in optimisation software developed by (Braden 2006; Braden et al. 2009).

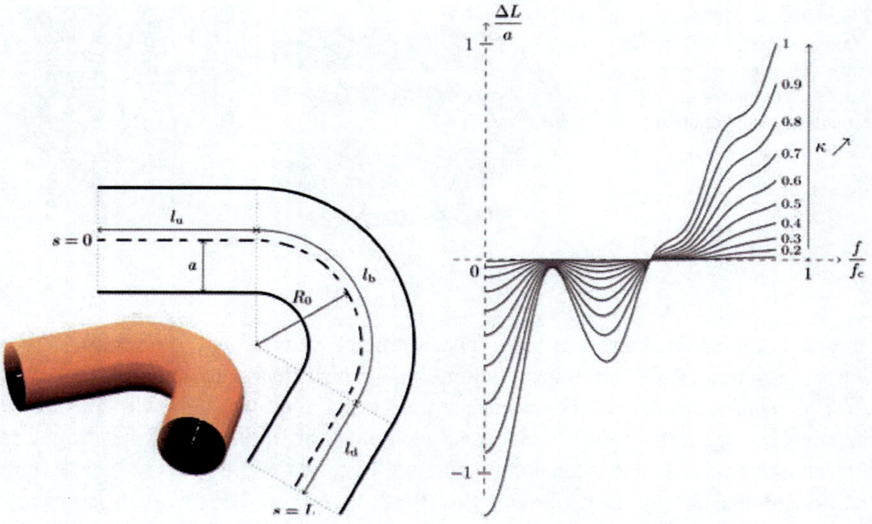

(a) (b)

Fig. 4.94 (**a**) Geometrical description of a toroidal bend. (**b**) Influence of the curvature parameter κ on the length correction Δ/n. Reproduced from Félix et al. (2012) with the permission of the Acoustical Society of America

In a comprehensive treatment of the effects of bends on acoustic resonances in wind instruments, Félix et al. (2012) compared multimodal calculations of curved sections of duct with finite difference calculations and experimental measurements. The calculated length correction $\Delta L = L_{\text{eff}} - L$ for the toroidal bend illustrated in Fig. 4.94a is shown as a function of $f/f_{c(1,0)}$ with κ as a parameter in Fig. 4.94b. A striking feature of these results is that at a frequency $f \simeq 0.5 f_{c(1,0)}$ the length correction changes its sign from negative to positive. This means that the bending of a pipe not only changes its effective length but also affects the inharmonicity of the resonances.

4.8 Going Further: The Wogram Sum Function

The close relationship between input impedance peak frequencies and natural note frequencies was introduced in Sect. 2.2.3. The sounding of a natural note involves a collaboration between several acoustic resonances, each of which is close in frequency to a harmonic of the played note. The pitch of the natural note therefore depends not only on the frequency of the impedance peak nearest to the note frequency but also on the frequencies of the other peaks involved in the collaboration. The Bouasse-Benade prescription, which recommends that the impedance peak frequencies should be harmonically related (Sect. 2.2.3), is designed to ensure that there is no conflict between the tuning messages sent to the lips by the different acoustic resonances.

In reality the acoustic resonances are never exactly harmonic, and mixed messages can arrive at the lips. If, for example, there is a strong impedance peak at a little more than twice the frequency of the fundamental of the played note, this could be expected to pull the repetition frequency upwards. Such a situation does indeed arise in the case of the pedal note, whose frequency is close to half the frequency of the second impedance peak but much higher than the frequency of the first peak (see Sect. 5.4.4).

The collaboration between several acoustic resonances and the mechanical resonance of the lips cannot be properly discussed without taking account of the nonlinear nature of the coupled system, which is the topic of Chap. 5. It is possible, however, to develop a function based on the input impedance $Z(f)$ of a brass instrument which attempts to capture the potential influence on a note played at frequency f due to the impedance at harmonic multiples of f. This was first proposed in the Ph.D. thesis of Klaus Wogram (1972). The Wogram sum function is defined as

$$SF(f) = \frac{1}{n}\sum_{n} \mathrm{Re}[Z(nf)]. \qquad (4.134)$$

The idea behind the sum function is that the real part of the input impedance is related to the transfer of energy to the air column; the sum is designed to represent the energy transfer to each of the standing waves present in the instrument at a given playing frequency.

Use of the sum function has been described by various authors. Elliott and Bowsher (1982) found that peaks in the sum function for a trombone gave a closer estimate of the actual playing frequencies than did the peaks in the raw input impedance curve. Caussé et al. (2013) confirmed that the sum function predicts correctly the playing frequency of the pedal note of the trumpet (see Fig. 4.95). They found, however, that the strict application of the sum function as defined in Eq. 4.134 resulted in pitches which were higher than measured playing frequencies.

Benade (1976) noted that the influence of higher impedance peaks was likely to increase with the amplitude of the played note. This can be taken into account by a weighted sum function

$$SF_W(f) = \frac{1}{n}\sum_{n} a(n)\mathrm{Re}[Z(nf)], \qquad (4.135)$$

with weighting factors $a(n)$ which depend on the playing dynamic. Figure 4.96 Illustrates a practical application to the investigation of level-dependent instability in the note E3 on a serpent in C. This note is played with the two lowest toneholes open, and serpents with this fingering tend to have very irregular input impedance curves (see Sect. 7.8.2). Figure 4.96a shows the detail of the impedance curve (in blue), and the sum function weighted heavily towards the first harmonic (in red). This could

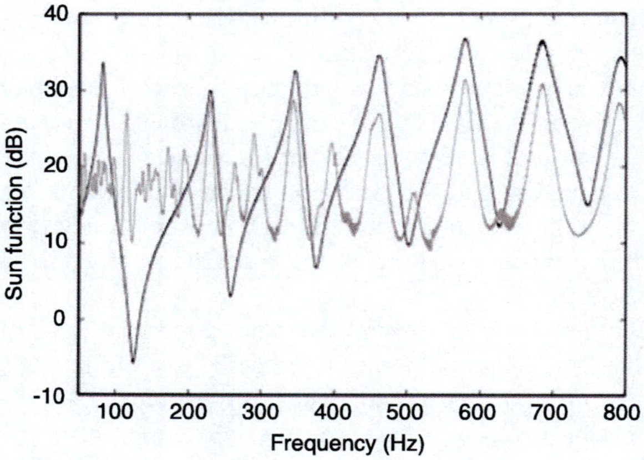

Fig. 4.95 Comparison between measured trumpet input impedance curve (black line) and calculated sum function (grey line). From Caussé et al. (2013) (Color figure online)

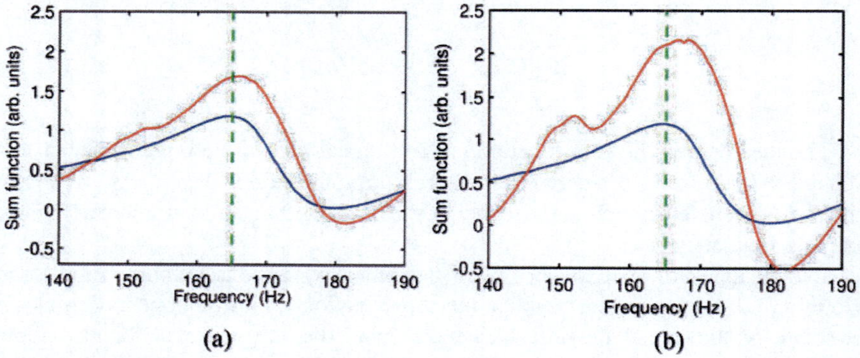

Fig. 4.96 Input impedance (blue curves) and weighted sum function (red curves) for a serpent in C with two toneholes open (fingering for E3, frequency 165 Hz). (**a**) Dynamic level piano: $a(1) = 1$, $a(2) = 0.5$, $a(3) = 0.2$, $a(4) = 0.1$. (**b**) Dynamic level forte: $a(1) = 1$, $a(2) = 1$, $a(3) = 0.7$, $a(4) = 0.4$ (Color figure online)

be considered appropriate for a piano dynamic level. Figure 4.96b shows the same detail of each curve, but this time with increased weighting of the second to fourth harmonics, which could be expected in a forte performance. The desired playing pitch is marked by a vertical dashed green line. It can be seen that a potentially disruptive second peak around 15 Hz below the desired frequency becomes more significant as the weighting of the upper harmonics increases. This accords with the musician's experience that a downward pitch shift of around a tone has to be avoided as the dynamic level rises.

Chapter 5
Blow That Horn: An Elementary Model of Brass Playing

A brass instrument comes to life when a musician picks it up and plays it. In this chapter we draw together the scientific investigations of buzzing lips discussed in Chap. 3, and the studies of the properties of brass instrument tubes reviewed in Chap. 4, to create an elementary model of brass instrument performance which includes both player and instrument. The aim of this task is to provide a scientific tool which can be used to improve our understanding of the ways in which sound is generated and controlled in a brass performance.

The model presented in the present chapter is classed as elementary because a number of major simplifications are made in deriving it. The vibrating lips are modelled as a one degree of freedom oscillator, and it is assumed that the lip vibration amplitude is small enough that the lips do not collide (or the collision model is basic). The upstream resonances of the player's windway are neglected, as is nonlinear propagation of sound in the air column of the instrument. Wall vibrations are also ignored. Despite these simplifications, the elementary model is capable of reproducing many of the important aspects of performance by human players which were described in Sect. 1.2. These include the musician's ability to select by embouchure control one of a number of pitches which form an approximate harmonic series and to pull the pitch up or down by the subtle embouchure modification described as 'lipping'. The model also explains why there is a minimum blowing pressure necessary to sound each note and why this threshold pressure increases as the played pitch rises.

The model is based on a set of three equations, which have to be solved simultaneously to predict the nature of the sound radiated by the instrument. These three constituent equations of the model are reviewed in Sect. 5.1. Low-amplitude self-sustained periodic solutions of the elementary model are explored in Sect. 5.2, which includes an introduction to the technique of linear stability analysis. In Sect. 5.3, the restriction to small amplitudes is relaxed: numerical simulation methods developed in physical modelling, including the use of bifurcation diagrams, are employed to illustrate a large range of playing situations. In Sect. 5.4 the

M. Campbell et al., *The Science of Brass Instruments*, Modern Acoustics and Signal Processing, https://doi.org/10.1007/978-3-030-55686-0_5

theoretical and computational techniques used to identify the oscillatory states of the elementary model are explained in more detail, and predictions of the model are compared with the behaviour experienced by players of real brass instruments.

5.1 The Three Equations of the Brass Instrument Model

Figure 5.1 is a schematic diagram of the structure of the model. The black box on the left-hand side represents the source of pressurised air which is created by the player's lung muscles prior to the start of a sound. The circle containing a cross represents the fact that the player's lips act as a valve controlled by the pressure difference between the mouth and the mouthpiece. The blue box labelled 'lip dynamics' represents the behaviour of the lips under the influence of this pressure difference: when the pressure in the player's mouth is raised above a threshold value the lips begin to vibrate. The way in which air then flows through the lips is represented by the red box labelled 'flow conditions of lips': the volume velocity of air depends on both the pressure difference across the lips and the size of the opening between them. The pressure variation in the mouthpiece due to the volume flow of air into it depends on the acoustic behaviour of the instrument, represented by the green box on the right-hand side. There is of course another output from the 'acoustic behaviour of the instrument' box, which is the radiation of sound by the instrument.

Each of the coloured boxes in Fig. 5.1 corresponds to one of the constituent equations of the model, which were identified and discussed in Chaps. 3 and 4. Before we proceed to show how the equations can be combined to explain what happens when a horn is blown, we pause briefly to recall some of the essential features of the brass playing process which our model must reproduce.

The sound in a brass instrument is the consequence of self-sustained mechanical oscillations of the lips, driven by an air flow from the player's lungs. A steady excess pressure in the mouth results in an oscillating pressure in the instrument when the mouth pressure is sufficiently high to destabilise the lip valve. The loss of stability

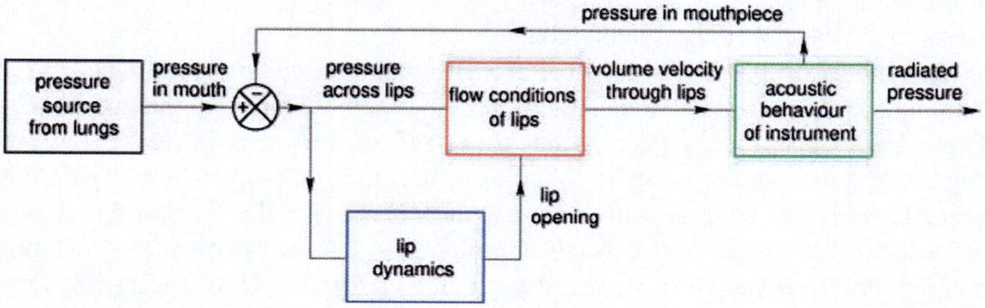

Fig. 5.1 Feedback loop as a block diagram of the interaction between player and instrument. Adapted from Elliott and Bowsher (1982)

of a mechanical oscillator interacting with a continuous flow is described as a flow-induced vibration phenomenon.

Similar oscillations occur in reed woodwind instruments because of the destabilisation of the valve formed by the mechanical reed. There is however an important difference between the roles played by the brass player's lips and the clarinet reed. The brass player is able to control and modify the lip natural resonance frequency, which is usually very close to the playing frequency. For that reason lip dynamics are an essential component of the brass playing model. In contrast, a cane single or double reed has a natural resonance frequency which is much higher than the highest note in the normal compass of the instrument and can be modified only to a limited extent by the player's lips. It is therefore possible to develop a two-equation model for a woodwind instrument based on a low-frequency approximation in which the reed is treated as a massless, lossless spring (McIntyre et al. 1983).

Normally the flow through the lips enters the mouthpiece of a brass instrument. The lip valve and the instrument are then coupled in an acoustic feedback loop, as shown in Fig. 5.1. The oscillations are amplified and controlled by the coupling of a localised nonlinear sound generating element (the source resulting from the valve-flow interaction) with a pipe in which acoustic energy can accumulate in resonant modes (standing waves). A small part of the acoustic energy is not trapped inside the pipe, but is radiated through the bell and any other open holes in the instrument.

When the mechanical system is destabilised, different permanent regimes of oscillation can be obtained. The sounding of a note of constant pitch corresponds to a periodic regime with a steady fundamental frequency. This frequency is often called simply the playing frequency, although the frequency spectrum of the sound typically contains many harmonic components. The playing frequency is usually close to one of the resonant mode frequencies of the air column in the pipe, but 'factitious notes' not satisfying this criterion can also be sounded. Quasi-periodic regimes which create multiphonics sounds, and even chaotic regimes with no definable periodicity, are also possible outcomes.

The generation of an oscillating output from a steady input is a characteristic feature of a nonlinear dynamical system such as that represented by Fig. 5.1. In reality, all three components represented by the coloured boxes display nonlinear behaviour, since neither the mechanical response of the lips (represented by the blue box) nor the acoustical response of the air column (represented by the green box) is strictly independent of amplitude. However the linear description of the lip dynamics in Sect. 5.1.1 provides a good approximation for moderate playing levels, as does the description of the acoustical response of the instrument by its linear input impedance in Sect. 5.1.3. The strongly nonlinear behaviour of the overall dynamical system is due to the nature of the pressure-flow relationship represented by the red box in Fig. 5.1 and discussed in Sect. 5.1.2.

5.1.1 The First Constituent Equation: Lip Dynamics

In order to describe the vibrating lips, the first of the three constituent equations of the brass instrument model was given in Sect. 3.2.4 as

$$\frac{d^2h(t)}{dt^2} + \frac{\omega_l}{q_l}\frac{dh(t)}{dt} + \omega_l^2(h(t) - h_{\mathrm{eq}}) = \pm\frac{p(t)}{\mu}. \tag{3.15}$$

In this equation, which describes the lips as a one degree of freedom (1DOF) mechanical oscillator, the symbols ω_l, q_l, h_{eq} and μ represent the angular lip resonance frequency, the quality factor of the lip resonance, the equilibrium value of the lip opening height and the effective mass per unit area of the lips, respectively. These quantities are parameters of the model, which are either constant (in a stable note) or changing slowly in a prescribed way (in a music performance). There are three variables in this equation: t, the time, $h(t)$, the lip opening height and $p(t)$, the pressure in the mouthpiece of the instrument. The variables representing height and pressure are written as $h(t)$ and $p(t)$ to emphasise that they are dependent variables, whose values are predicted by the equation for each value of the independent variable t.

Equation 3.15 can describe either inward-striking or outward-striking reed behaviour, depending on the sign of the term on the right-hand side. Evidence for both types of behaviour in the lips of brass players was presented in Chap. 3, but after a review of the phase relationships in the lip valve in Sect. 5.2.1, the development of the elementary model is implemented with the simplifying assumption that only outward-striking behaviour is present. The first constituent equation of the model then has the form

$$\frac{d^2h(t)}{dt^2} + \frac{\omega_l}{q_l}\frac{dh(t)}{dt} + \omega_l^2(h(t) - h_{\mathrm{eq}}) = -\frac{p(t)}{\mu}. \tag{5.1}$$

As was noted in Sect. 3.2, the equilibrium lip height h_{eq} depends on the mouth pressure p_{m}. At zero mouth pressure, the opening height for the outward-striking reed has its minimum opening $h_{\mathrm{eq}} = h_0$; as the mouth pressure is gradually raised, the lips swing outwards and h_{eq} increases. The force exerted on the lip by a static pressure p_{m} is

$$F = p_{\mathrm{m}}S_{\mathrm{eff}}, \tag{5.2}$$

where S_{eff} is the effective area on which the mouth pressure acts. The increase in h_{eff} due to this static force is

$$h_{\mathrm{eff}} - h_0 = \frac{F}{k} = \frac{p_{\mathrm{m}}S_{\mathrm{eff}}}{k}, \tag{5.3}$$

where k is the spring constant of the 1DOF oscillator. Multiplying both sides of Eq. 5.3 by $\omega^2 = k/m = k/(\mu S_{\text{eff}})$ leads to the relationship

$$\omega^2(h_{\text{eff}} - h_0) = \frac{p_{\text{m}}}{\mu}. \tag{5.4}$$

Equation 5.4 can be used to write an alternative form of Eq. 5.1 which explicitly includes the mouth pressure p_{m}:

$$\frac{d^2h(t)}{dt^2} + \frac{\omega_l}{q_l}\frac{dh(t)}{dt} + \omega_l^2(h(t) - h_0) = \frac{p_{\text{m}} - p(t)}{\mu}. \tag{5.5}$$

5.1.2 *The Second Constituent Equation: Flow Conditions*

The second constituent equation, given in Sect. 3.5.1, describes the relationship between pressure and flow in the lip channel:

$$u(t) = S_{\text{lc}}(t)\sqrt{\frac{2(p_{\text{m}} - p(t))}{\rho}}. \tag{3.27}$$

In this equation, the mouth pressure p_{m} and the air density ρ are parameters, while the volume flow rate $u(t)$, the cross-sectional area of the lip channel $S_{\text{lc}}(t)$ and the mouthpiece pressure $p(t)$ are dependent variables. If the pressure difference under the square root becomes negative, we use the absolute value of the pressure difference, with a negative sign in front of the square root. When the lips close ($S_{\text{lc}}(t) = 0$), the volume flow becomes zero regardless of the magnitude of the pressure difference.

In Sect. 3.1.4, a power law dependence of S_{lc} on h was postulated:

$$S_{\text{lc}}(t) = S_o\left(\frac{h(t)}{h_o}\right)^q. \tag{3.1}$$

For the purposes of the elementary model, a linear dependence can be chosen by setting $q = 1$, equivalent to the assumption that the open area is a rectangle of constant width w. Although this hypothesis is well grounded for single reed woodwinds, the measurements reported in Sect. 3.1.4 show that it is a considerable simplification when discussing the vibrating lips of a brass player. Accepting this simplification, we get

$$S_{\text{lc}}(t) = wh(t). \tag{5.6}$$

Substituting this relationship in Eq. 3.27 gives a new version of the second constituent equation containing only the three dependent variables $u(t)$, $h(t)$ and $p(t)$:

$$u(t) = wh(t)\sqrt{\frac{2(p_{\mathrm{m}} - p(t))}{\rho}}. \tag{5.7}$$

5.1.3 *The Third Constituent Equation: Instrument Acoustics*

The third constituent equation describes the relationship between flow and pressure in the instrument mouthpiece. Equations representing this relationship were given in Sect. 4.1.6 in terms of both the frequency domain input impedance $Z(\omega)$ and the time domain impulse response $g(t)$. In the development of our elementary model, it is most useful to begin with the frequency domain version

$$p(\omega) = Z(\omega)u(\omega). \tag{5.8}$$

This equation can be transformed into a linear relationship between the time domain variables $p(t)$ and $q(t)$ in a number of ways, for example, through the use of Eqs. 4.28 and 4.29. A transformation involving the superposition of acoustic modes is described in Sect. 5.4.2.

5.2 Crossing the Threshold: Small Amplitude Oscillating Solutions

In Sect. 5.1 three equations were presented linking the three time-dependent variables $p(t)$, $u(t)$ and $h(t)$ used in the formulation of our elementary model of brass instrument playing. Since the number of variables equals the number of equations, it is in principle possible to derive solutions in which the time dependence of each variable is separately expressed in terms of the parameters of the model which define the physical properties of the lips and the instrument. Unfortunately the nonlinear nature of the second constituent equation makes it impossible to derive straightforward analytical solutions even to this very simplified model. Instead, we describe a number of different approaches which allow us to use the model to predict how a brass instrument will perform. Section 5.2 looks at the model behaviour in regimes in which the blowing pressure is just above the threshold and the resulting acoustic pressure amplitude is very small. The near-threshold phase relationships which allow the lip valve to supply energy to sustain the oscillations of the air column are discussed in Sect. 5.2.1 for both inward- and outward-striking lip reed models. A review of the experimental evidence justifies the decision to retain only the outward-striking behaviour in the elementary model. In Sect. 5.2.2 the three equations are reformulated in a way which allows the stability of low-amplitude brass sounds to be explored using linear stability analysis.

5.2.1 Phase Relationships in the Lip Valve

The ultimate goal of a brass instrument model is to predict the nature of the sound radiated by the instrument from a knowledge of the control parameters such as the mouth pressure p_m and the lip resonance frequency f_r. The first step in this process is to be able to predict the pressure $p(t)$ generated in the instrument mouthpiece under the influence of these control parameters. The simplest possible solution, $p(t) = 0$, corresponds to the situation in which the mouth pressure is insufficient to induce lip vibrations. From the musical point of view, solutions which do not result in sound generation are not very interesting, but it is important to understand the factors which determine the threshold pressure for self-sustained lip vibration. Some insight into this question can be gained by examining the phase relationships between air flow and pressure in the two different modes of lip valve operation represented by the two different signs on the right-hand side of Eq. 3.15, in the limiting case in which the blowing pressure is just above the threshold.

Although the single and double reeds of woodwind instruments fall unambiguously into the inward-striking class, the brass player's lips have proved much harder to categorise. Helmholtz considered that the lips were outward-striking, and this view was also adopted by Fletcher (1979). In a seminal paper on regeneration in brass instruments, Elliott and Bowsher also argued that the lip valve was predominantly outward-striking, but noted that the Bernoulli force in the lip channel introduced an alternative flow control mechanism with an inward-striking character. This aspect of the model was further developed by Pelorson et al. (1994) and Hirschberg et al. (1995). More recent studies (Yoshikawa 1995; Adachi and Sato 1996; Chen and Weinreich 1996; Ayers 2001; Cullen et al. 2000; Campbell 2004; Boutin et al. 2015b) have confirmed that it is not possible to assign a single phase characteristic to the lip valve: both inward-striking and outward-striking behaviours have been observed in experiments and simulations.

For very low-amplitude oscillations, it can be assumed that the acoustic variables are to a good approximation sinusoidal functions of time. The phase relationship between mouthpiece pressure and valve opening height for the different valve classifications can be found by inserting the expressions $h(t) = \hat{h}\cos(\omega t + \phi_h)$, $p(t) = \hat{p}\cos(\omega t)$ in Eq. 3.15 for each of the two signs of the right-hand forcing term. The resulting variations of the phase ϕ_h with frequency are shown schematically in Fig. 5.2.

Figure 5.1 shows that the player and instrument are linked by a feedback loop (see also Sect. 2.2.1). The arrow from the pressure in the mouthpiece back to the control of the lip valve represents the feedback, which is similar in character to the feedback from the loudspeaker of a public address system to the microphone whose signal drives it. If the feedback is negative, an increase in the output results in a decrease in the input, so the signal dies away; if the feedback is positive, an increase in the output increases the input, and the buildup can result in an unpleasant howl of self-sustained oscillation from the PA system. In the case of the musical instrument,

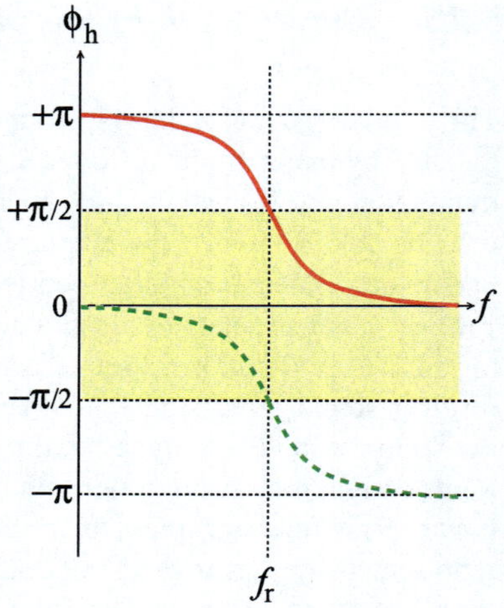

Fig. 5.2 Phase ϕ_h of the opening height $h(t)$ relative to the mouthpiece pressure $p(t)$ for an outward-striking reed (solid red line) and an inward-striking reed (dashed green line). f_r is the reed resonance frequency. The yellow shaded area indicates the range of phases for which the average energy flow into the instrument is positive (Color figure online)

of course, the sound of the self-sustained oscillation is what the instrument is designed to produce.

Feedback loop stability analysis has been used by several authors to analyse the threshold behaviour of reed and brass instruments (Worman 1971; Wilson and Beavers 1974; Fletcher 1979, 1993; Elliott and Bowsher 1982; Saneyoshi et al. 1987). These studies show that to maintain a note near the threshold of oscillation, an outward-striking $(+, -)$ reed must vibrate at a frequency which is above both the nearest acoustic resonance frequency and the natural resonance frequency of the reed; an inward-striking $(-, +)$ reed must vibrate at a frequency which is below both air column and reed resonances.

These conditions can be understood qualitatively by recalling that the oscillating component of the air flow through the valve constitutes a transfer of momentum equivalent to an oscillating force on the air in the mouthpiece. If the maximum of the air flow into the mouthpiece occurs during the positive half-cycle of the mouthpiece pressure, there will be a net transfer of energy in each cycle to sustain the standing wave in the air column. To achieve this, the phase difference $\phi_{ac} = \phi_u - \phi_p$ between the acoustic pressure p at the input of the instrument and the acoustic volume velocity u through the valve must satisfy the condition $-\pi/2 < \phi_{ac} < \pi/2$.

Equation 3.27 shows that the volume velocity $u(t)$ depends both on the valve open area $S(t)$ and on the pressure difference across the valve $\Delta p = p_m - p(t)$. For near-threshold playing at fairly high pitches, the mouth pressure p_m is considerably greater than the mouthpiece pressure $p(t)$. The fractional change in Δp is then relatively small, and the flow oscillation is mainly governed by the changing open area, which for constant width is proportional to $h(t)$. Under these circumstances $\phi_{ac} \simeq \phi_h$, and the yellow shaded area in Fig. 5.2 marks the region of the graph in which ϕ_h meets the condition for positive energy supply to the air column. The dashed green curve representing the inward-striking reed is within this area

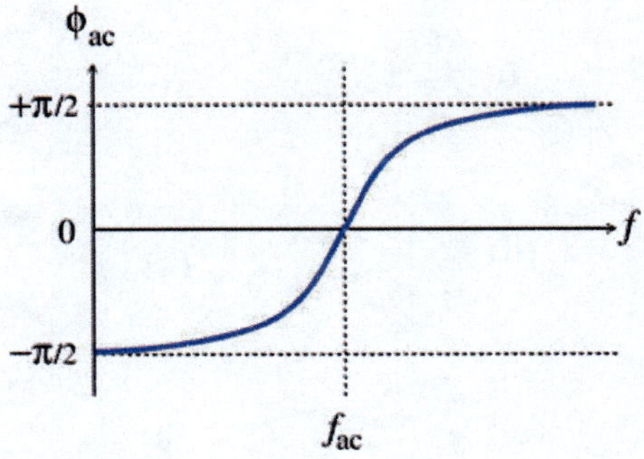

Fig. 5.3 Phase ϕ_{ac} of air flow relative to pressure in the frequency range around an air column resonance

only below the reed resonance, while the solid red curve representing the outward-striking reed is in the region only above the reed resonance.

The nature of the standing wave in the instrument air column imposes a further condition on ϕ_{ac}. Figure 4.14 in Sect. 4.1.6 showed that the phase θ of the input impedance of a trombone was close to zero at the frequency f_{ac} of an impedance peak, rising towards $\pi/2$ below f_{ac} and falling towards $-\pi/2$ above it. Since the phase of the air flow relative to the pressure is $\phi_{ac} = -\theta$ (Eq. 4.23), the variation of ϕ_{ac} with frequency in the vicinity of a resonance is as shown in Fig. 5.3. Below the resonance the phase falls towards $-\pi/2$, and above the resonance, it rises towards $+\pi/2$. This phase of the air flow entering the instrument must equal the phase of the air flow leaving the valve. A comparison of Figs. 5.2 and 5.3 shows that this match can only be made for a playing frequency which is above the air column resonance frequency for an outward-swinging valve and below the air column resonance frequency for an inward-swinging valve (or a sliding door valve).

The relationship between valve motion, air flow and mouthpiece pressure for inward-striking and outward-striking valves is illustrated schematically in Fig. 5.4. Five stages in the vibration cycle of an inward-striking valve, operating well below its natural resonance frequency f_r, are shown in Fig. 5.4a. In Stage (i) the double reed is half open; the pressure in the mouthpiece is at its mean value, and air is flowing into the mouthpiece from the valve. In Stage (ii) the pressure in the mouthpiece has fallen, and the rise in Δp has pushed the reed blades together, closing the valve and cutting off the air flow. In Stage (iii) mouthpiece pressure has risen again, half opening the valve and allowing air to flow. In Stage (iv) the mouthpiece pressure has reached its maximum, pushing the reed blades widely apart to give a large air flow. Finally in Stage (v) the mouthpiece pressure and air flow return to their median values with the valve half open, and a new cycle of vibration is about to commence.

The series of stages shown in Fig. 5.4a corresponds to an efficient transfer of energy to the air column of the instrument, since the maximum flow into the mouthpiece occurs at the phase of maximum pressure. This is of course a highly simplified view of the interaction between valve and instrument: it assumes that the flow is simply proportional to the opening area, and it also ignores the dynamics of the reed. In reality, as the frequency increases, a phase difference builds up

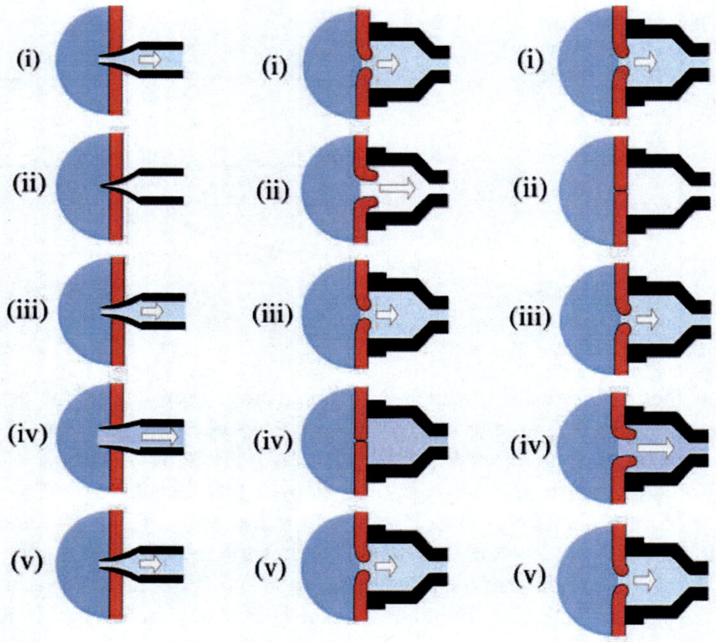

(a) Double reed, playing below reed resonance

(b) Outward-swinging lip valve, playing below lip resonance

(c) Outward-swinging lip valve, playing above lip resonance

Fig. 5.4 Air flow (represented by the length of the arrow) and pressure (represented by the depth of shading) in (**a**) a double reed operating well below its natural frequency; (**b**) an outward-striking lip valve operating well below its natural resonance frequency; (**c**) an outward-striking lip valve operating well above its natural resonance frequency. Adapted from Campbell (1999) with permission from Taylor & Francis Ltd, www.tandfonline.com

between the pressure on the reed and the reed displacement, reaching $\pi/2$ at the reed resonance frequency. However the playing frequencies of woodwind instruments are normally well below the reed resonance, which therefore does not play a crucial role in the performance of the valve.

The situation is quite different for the outward-striking lip valve. Figure 5.4b shows the same series of stages in the mouthpiece pressure cycle as Fig. 5.4a, again assuming that the playing frequency is well below the valve resonance. In this case, the drop in pressure in Stage (ii) sucks the valve open, allowing a strong air flow into the mouthpiece. In Stage (iv) the maximum of mouthpiece pressure has pushed the valve shut, cutting off the air supply. The air flow velocity is thus π out of phase with the pressure, draining energy from the air column oscillation instead of feeding energy into it.

Figure 5.4b represents graphically the phase difference of π between the mouthpiece pressure and the valve opening height shown by Fig. 5.2 for an outward-swinging valve driven well below its resonance frequency. This phase difference diminishes as the operating frequency rises, passing through $\pi/2$ at the resonance and approaching zero for frequencies well above the resonance. The latter situation is illustrated in Fig. 5.4c. Stage (ii) represents maximum outward suction on the lips from the low mouthpiece pressure, but in response the lips have swung shut. In Stage (iv) the mouthpiece pressure is exerting the maximum force on the outer surfaces of the lips, but they have opened outwards instead of closing inwards. As a consequence the pressure and air flow velocity are in phase, maximising the energy transfer to the air column oscillation.

Yoshikawa (1995) attempted to measure directly the phase difference between lip motion and mouthpiece pressure in horn and trumpet players. The experimental method involved attaching a small strain gauge to the upper lip of the player, and the interpretation of the results was complicated by the difficulty of establishing the relationship between strain gauge signal and lip movement. Yoshikawa concluded that the lip reed had an outward-striking character for the lowest played notes, but changed to an inward-striking behaviour at around the third natural note.

Chen and Weinreich (1996) used an ingenious single mode 'brass instrument' (Fig. 5.5) consisting of a Helmholtz resonator with a loudspeaker providing active feedback. By adjusting the parameters of the electronic circuit controlling the feedback, the resonance frequency and quality factor of the system could be varied through a sizeable range (see Sect. 9.2.2). The instrument was sounded by buzzing the lips against a short tube (the 'mouthpiece') inserted in the wall of

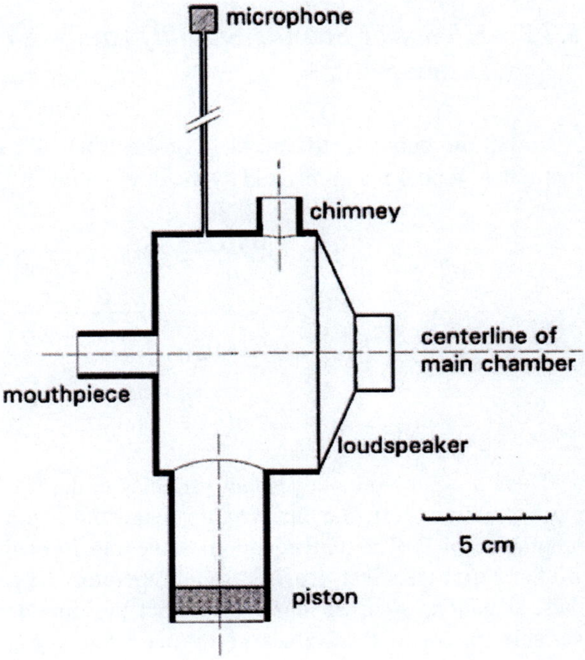

Fig. 5.5 The Helmholtz resonator single mode 'brass instrument' with active control. From Chen and Weinreich (1996)

the resonator. The vibrating lips were identified as inward- or outward-striking by noting whether the playing frequency was below or above the measured resonance frequency of the resonator. They found that with extreme effort, the player could sound notes either below or above the air cavity resonance, but that the most comfortable note was almost always above. They concluded that the lips could not be entirely modelled by a simple one degree of freedom model, but that the most characteristic behaviour resembled that of an outward-striking reed. Ayers (2001) used a different experimental procedure in order to compare the playing frequencies with the resonant frequencies and came to a different conclusion.

Although these experiments were carefully conducted and analysed, their interpretation is far from straightforward. A human brass performer is able to lip a played note over a wide range of frequencies, both above and below the acoustic resonance frequency. It must also be borne in mind that the distinction between inward- and outward-striking lip valve behaviour is based on approximations valid only for blowing pressures very near the threshold of oscillation and must therefore be treated with caution in discussing playing at levels above pianissimo. Nevertheless, a simplified lip model with only outward-striking behaviour is capable of reproducing the most important features of brass instrument behaviour, including the generation of sustained sounds, the dependence of loudness and timbre on blowing pressure and the dependence of sounding pitch on lip frequency (Petiot and Gilbert 2013). In the rest of this chapter, solutions of the elementary model equations will be explored on the assumption that the lips behave as an outward-striking reed.

5.2.2 *Silence or Sound? Stability Analysis of Brass Instruments*

Adopting the outward-striking view of the 1DOF lip valve, the elementary brass instrument model is summarised by the three following equations:

$$\frac{d^2h(t)}{dt^2} + \gamma\frac{dh(t)}{dt} + \omega_{\mathrm{r}}^2(h(t) - h_{\mathrm{eq}}) = -\frac{p(t)}{\mu} \tag{5.1}$$

$$u(t) = wh(t)\sqrt{\frac{2(p_{\mathrm{m}} - p(t))}{\rho}} \tag{5.7}$$

$$p(\omega) = Z(\omega)u(\omega). \tag{5.8}$$

There are three time-dependent variables in the first two equations: the mouthpiece pressure $p(t)$, the lip opening height $h(t)$ and the air flow rate into the mouthpiece $u(t)$. The third equation can be transformed into a relationship between $p(t)$ and $u(t)$ (see Sect. 4.1.6). It should therefore be possible in principle to solve these equations simultaneously to predict the time-dependent behaviour of each variable, provided that values of the parameters p_{m}, h_{eq}, w, γ, ω_{r} and μ are either known or can be estimated. It is also necessary to know the form of the input impedance Z of the instrument; in Chap. 4 methods were described for either calculating or measuring this function.

In reality finding the solution of the brass model equations is far from straightforward. It is possible to measure typical values of mouth pressure p_{m} during performance, and realistic estimates of the lip opening width w and equilibrium opening height h_{eq} can be extracted from video recordings of players using transparent mouthpieces. However the parameters γ, ω_{r} and μ depend on the effective mass, stiffness, surface area and internal damping of the vibrating section of lip; these quantities are not susceptible to direct measurement.

A more fundamental difficulty in finding solutions to the model equations arises from the nonlinear nature of the lip valve dynamics. In Sect. 5.2.1 the view of the player-instrument system as a feedback loop was discussed. This approach helps to explain the system behaviour very near its oscillation threshold, which is a useful if limited step on the way to a fuller understanding of brass playing. An alternative way of investigating near-threshold behaviour treats the player-instrument combination as a dissipative nonlinear dynamical system, which can be analysed using the mathematical technique of linear stability analysis (see, e.g. Cullen et al. (2000) and Velut et al. (2017a)). The principle of linear stability analysis (LSA) is described in the present section, and an example is given of its ability to explain some basic features of brass playing. A full mathematical treatment is reserved for the Going Further Sect. 5.4.

The linearisation of the model equations can be achieved by replacing each system variable $q_i(t)$ by $q_{i,\mathrm{eq}} + \tilde{q}_i(t)$, where $q_{i,\mathrm{eq}}$ is the threshold equilibrium value of q_i. Terms which are more than first order in one of the new variables $\tilde{q}_i(t)$ are then eliminated. The resulting linearised equations describe infinitesimal oscillations of the system variables about their equilibrium values. As a simple example, we consider the case in which the instrument has only one acoustic mode. The equations can then be rewritten in terms of the four variables $\tilde{h}$, $\frac{d\tilde{h}}{dt}$, $\tilde{p}$, and $\frac{d\tilde{p}}{dt}$, which can be combined into a 4×1 state vector

$$\mathbf{X} = \left[\tilde{h} \quad \frac{d\tilde{h}}{dt} \quad \tilde{p} \quad \frac{d\tilde{p}}{dt}\right]^{\mathrm{T}}. \tag{5.9}$$

The linearised model equations can be written as a single matrix equation

$$\frac{d\mathbf{X}}{dt} = \mathbb{M}\mathbf{X} \tag{5.10}$$

where $\mathbb{M}$ is a 4×4 matrix. This equation has solutions of the form

$$\mathbf{X} = \mathbf{W}e^{\lambda t}, \tag{5.11}$$

where **W** is the state vector describing the amplitudes of the four variables and λ is the eigenvalue of the state. The eigenvalue is in general complex; if λ has a positive real part, the state is unstable, since its amplitude grows exponentially with time. Linear stability analysis tracks the evolution of the eigenvalues of possible states, while a control variable (such as the quasi-static mouth pressure p_m) increases and identifies the point at which a given state becomes unstable.

At sufficiently low mouth pressure, the stable state of a brass instrument corresponds to the situation in which the lips are not oscillating. At the threshold value of the mouth pressure, this state becomes unstable, and the system switches to a new stable state in which the lips are oscillating. This change, which is described in the language of nonlinear dynamics as a Hopf bifurcation, is of great musical significance, since this marks the transition from silence to sound. Linear stability analysis allows the calculation of the pitch of the oscillating state and the threshold pressure required to reach it. Charting the variation of these two quantities as a function of the lip frequency provides information about the intonation and ease of playing of an instrument.

An example of the way in which LSA can be used to compare the behaviour of different brass instruments is shown in Fig. 5.6, which plots the threshold pressure and playing pitch as a function of lip frequency for the tenor and bass trombones illustrated in Fig. 4.13. The analysis is based on the elementary model, and the linear acoustical response of each instrument is represented as a set of acoustic modes derived from the measured impedance curves shown in Fig. 4.14. Looking first at the tenor trombone threshold frequency curve, shown in black in Fig. 5.6b, we see a series of almost horizontal segments separated by vertical gaps. Each line slopes upwards from a starting point just above one of the acoustic resonance frequencies. This is a graphical representation of a musical experience familiar to trombonists when playing the arpeggio of natural notes shown in Fig. 5.6c: starting with the pedal note and gradually increasing the lip frequency, the pitch stabilises at a value close to one of the natural notes of the instrument, bends gently upwards and then jumps to the next higher natural note. The heights, lengths and slopes of each segment of the curve in Fig. 5.6b contain information about the overall intonation of the instrument.

The pressure threshold curve for the tenor trombone, shown in black in Fig. 5.6a, contains further useful information about the playing properties of the instrument. Corresponding to each separate segment of the threshold frequency curve is a U-shaped valley in the threshold pressure curve. The bottom of the U represents the lowest threshold pressure at which the note can be sounded; the corresponding frequency is optimal in the sense that it requires the minimum effort to sound it. The value of the minimum in the pressure threshold curve increases as the natural note number rises: for the second natural note, the minimum is 0.1 Pa, while for the eighth natural note, it is 1.2 Pa. The LSA prediction that higher pitches require more effort to sound is borne out by musical experience.

Turning to the LSA results for the bass trombone, represented by red curves in Fig. 5.6, the same general features can be observed in the threshold frequency and pressure curves as for the tenor trombone. Tenor and bass trombones both

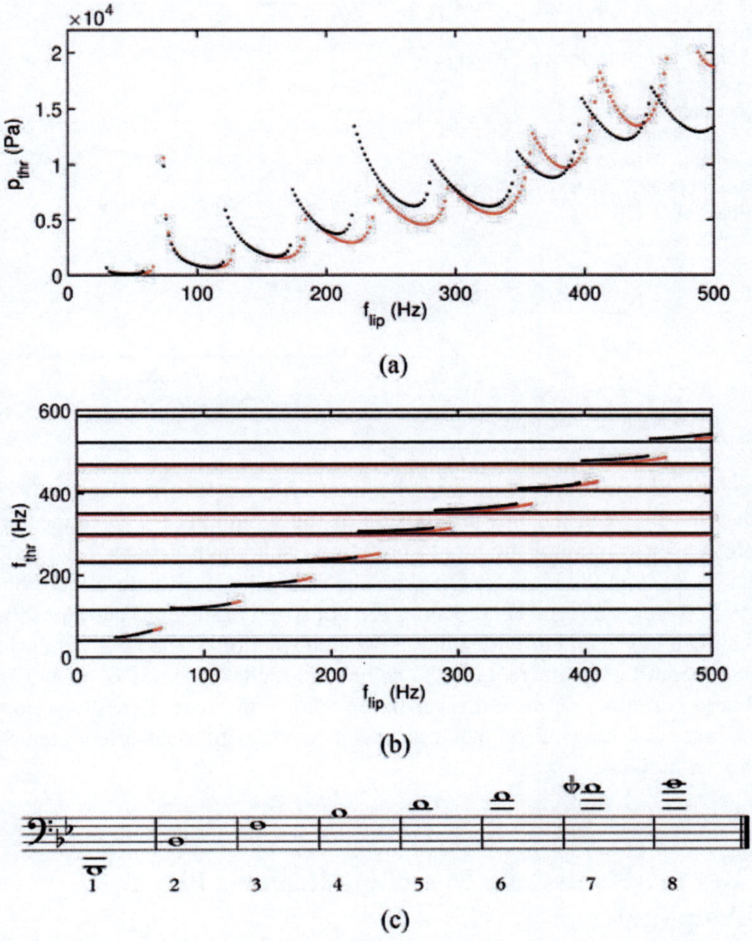

Fig. 5.6 Linear stability analysis applied to a tenor trombone and a bass trombone. Threshold mouth pressure p_{thr} (**a**) and threshold oscillation frequency f_{thr} (**b**) are plotted against lip frequency f_l. Horizontal lines represent acoustic resonance frequencies. Black: tenor trombone. Red: bass trombone. The corresponding musical pitches are shown in (**c**). Adapted from Gilbert et al. (2018) (Color figure online)

have a nominal fundamental pitch of B$^\flat$1, but differences in their input impedance curves are reflected in differences in threshold behaviour. The most striking of these differences is in the heights of the minima in the threshold pressure curves; from the second to the sixth natural note, the minima are lower for the bass trombone, but from the seventh natural note upwards the minima are lower for the tenor.

The LSA prediction that high register notes can be sounded with less effort on a tenor trombone than on a bass trombone was tested in an experiment by Gilbert et al. (2018). Four experienced trombonists were asked to play the first eight natural

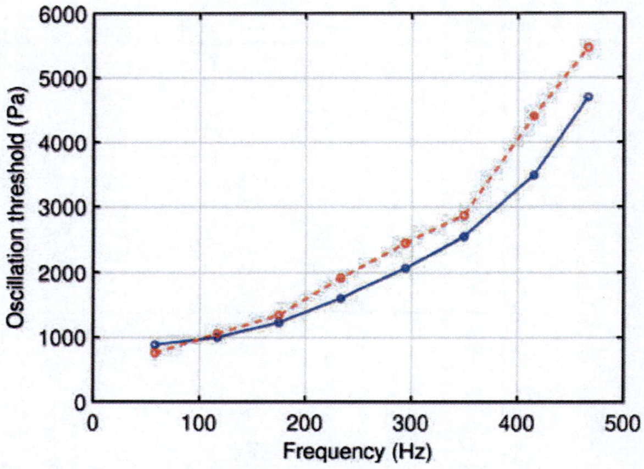

Fig. 5.7 Mouth pressures for sounding at piano dynamic level of the first eight regimes for the tenor and bass trombones displayed in Fig. 4.13. Solid blue line: tenor trombone. Dashed red line: bass trombone. Adapted from Gilbert et al. (2018) (Color figure online)

notes of the B^b trombone at a piano dynamic level on each instrument. A probe tube inserted between the lips at one side of the mouthpiece was connected to a pressure sensor to monitor the mouth pressure. Each player repeated the sequence of notes twice, and the eight mouth pressure values for each natural note on each instrument were averaged. These values are not strictly threshold pressures, but the results shown in Fig. 5.7 provide qualitative confirmation of the LSA result that the minimum mouth pressure required to sound the higher regimes is greater on the wide-bored bass than on the narrower-bored tenor. This accords with the common view of brass instrumentalists that a narrower-bored instrument offers greater ease in high note playing.

5.3 Beyond Pianissimo: Modelling Realistic Playing Amplitudes

The results presented in Sect. 5.2.2 demonstrate the usefulness of linear stability analysis in exploring the behaviour of the complete nonlinear model of a brass musical instrument. However these results are only valid near the threshold of sustained oscillations, and cannot be relied on to explain how brass instruments function even at medium playing levels. To progress to an understanding of brass performance at all dynamic levels, the equations of the global model must be solved without linearising them. Apart from a few very simplified cases, such as a clarinet-like model with a lossless cylindrical tube (Maganza et al. 1986; Hirschberg et al. 1995; Chaigne and Kergomard 2016), the equations are not tractable analytically. Section 5.3.1 presents results of numerical methods which have been used to simulate brass instrument performance in the nonlinear regime corresponding to realistic playing levels. Section 5.3.2 presents an introduction to bifurcation diagrams and shows that the concept of linear stability analysis can in fact be extended beyond the threshold regime using the mathematical technique of continuation. These ideas are more fully explained in the Going Further Sect. 5.4.

5.3.1 Analysis of Brass Performance Using Simulations

Many different numerical methods can be used to exhibit the behaviour of the nonlinear brass instrument model far from threshold. The harmonic balance method, for example, gives a Fourier series approximation to the steady state of periodic regimes, including unstable ones (Schumacher 1978; Gilbert et al. 1989; Farner et al. 2006; Cochelin and Vergez 2009). Following the pioneering work of Schumacher (1981), McIntyre et al. (1983), and Gazengel et al. (1995), it is possible to carry out time domain simulations based on reflection functions at moderate computational cost. These simulations permit the study of transients and also non-periodic solutions; with fast-enough processors, they can generate the real-time output necessary for synthesisers. The numerical methods available include the finite element and finite-difference methods which are widely used in many branches of physics and engineering (Bilbao and Chick 2013). Methods specifically adapted to the acoustics of ducts include digital waveguides, wave digital filters, impedance-based methods and those involving impulse responses and reflection functions. It is beyond the scope of the book to give an extensive view of all the different methods which have been used in wind instrument simulations; valuable reviews are provided by Beauchamp (2007), Bilbao (2009) and Smith (2010).

Sound synthesisers based on physical modelling are impressive in their ability to mimic musical phrasing, even when they are based on the elementary physical model discussed in this chapter. A critical requirement for musically convincing output is the control of the slow variation in time of the parameters of the physical model. One method of acquiring realistic data for the control parameters of a musician's embouchure, which has been successfully employed in clarinet and saxophone synthesis, is to use an instrumented mouthpiece (Guillemain et al. 2005).

Apart from their use in sound synthesis, time domain simulations can provide interesting and useful insights into particular aspects of wind instrument behaviour. For example, Velut et al. (2016) have simulated period-doubling and quasi-periodic oscillations. These unusual regimes are sometimes deliberately employed by performers to create exotic sounds such as multiphonics, but are also familiar traps for players with tired lips! Simulations have been used to study transients and unsteady sounds in clarinet-like instruments (Bergeot et al. 2013) and vibrato on the saxophone (Gilbert et al. 2005), but much work remains to be done on unsteady regimes in brass instruments.

We present here a few examples of simulations of brass performance using finite-difference time domain (FDTD) methods developed by Reginald Harrison-Harsley (Harrison et al. 2015, 2016). Figure 5.8 illustrates the attack transient of a brass instrument: a tenor trombone in first position playing the second natural note $B^{b}2$. The attack is controlled by the shape of the control parameter, in this case the

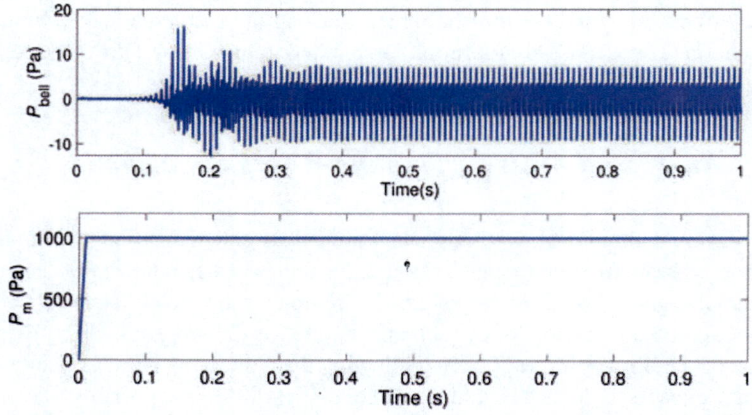

Fig. 5.8 Simulation of the attack transient of the note B^b2 played on a tenor trombone. Bell acoustic pressure (top) and control parameter p_m (bottom) as a function of time. Adapted from Harrison et al. (2015)

static mouth overpressure p_m. The duration of the acoustic attack transient can be modified by changing the time length of the p_m ramp: the longer the ramp, the longer the attack transient. In the example shown in Fig. 5.8, the mouth pressure reaches its maximum value in 10 ms, but the oscillation regime only starts to develop significantly after around 80 ms. At the beginning of the note, the envelope of the acoustic pressure is not smooth, illustrating the fact that the attack is not perfectly clean.

A brass player performs a crescendo by gradually increasing the mouth pressure p_m. This can be easily simulated by slowly ramping up the mouth pressure, the lip frequency and other control parameters being maintained constant. An illustration is given Fig. 5.9, where it can be seen that the amplitude of the envelope of the radiated signal increases with p_m.

During the crescendo, the playing frequency is not completely stable. In this simulation the decrease in pitch is very small (less than 2 cents). If this effect is more pronounced in practice, the brass player will have to compensate by slightly modifying the embouchure to increase the lip frequency in order to keep the pitch constant.

Sometimes a given note can be played with several valve fingerings or slide positions. For example, the note D5 can be played on a B^b trumpet using two fingerings: one with no valve depressed (fifth regime, the fundamental regime being B^b2), another with valves 1 and 2 depressed (sixth regime, the fundamental regime being G2). A comparison of simulations of D5 played with these two fingerings shows that the resulting bell acoustical pressures are slightly different (Fig. 5.10). The small differences are already visible in the mouthpiece pressure signal as a consequence of the self-sustained oscillations which are slightly different: self-

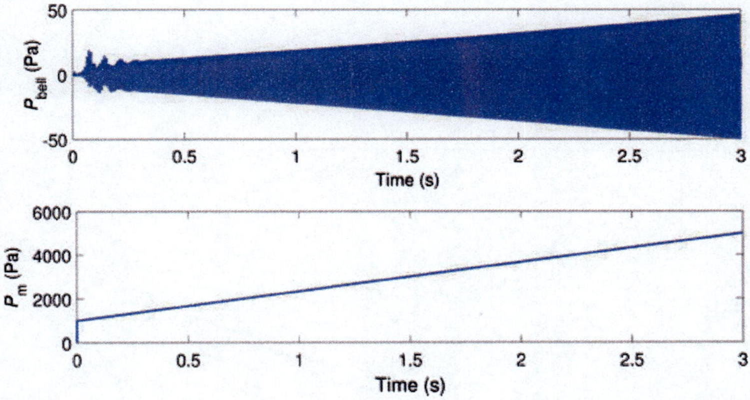

Fig. 5.9 Simulation of a crescendo on the note B^{b}2 played on a tenor trombone. Bell acoustic pressure (top) and control parameter p_{m} (bottom) as a function of time

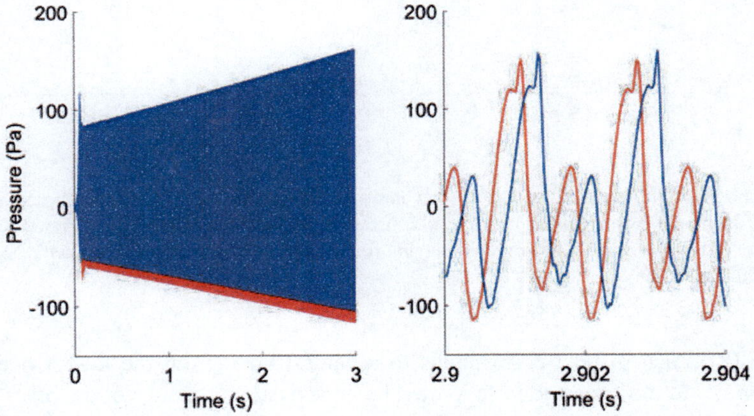

Fig. 5.10 Simulation of the bell acoustical pressure for a crescendo on the note D5 played on a trumpet with two different fingerings: no valves depressed (blue), and valves 1 and 2 depressed (red). Left: amplitude envelope. Right: zoom showing waveforms (Color figure online)

sustained regime 5 for the first fingering and self-sustained regime 6 for the second fingering. Nonlinear distortion is not taken into account in the simulation tool used here. If nonlinear propagation effects are included in the simulation (Maugeais and Gilbert 2017), the enrichment during a crescendo is much more pronounced, and the characteristic brassy effect is modelled (see Sect. 6.1).

In Sect. 5.2.2 it was shown that modelling of near-threshold oscillations can explain the existence of the natural notes of a brass instrument. By imposing a sweep of the lip frequency, linear stability analysis reveals the appearance of almost horizontal 'plateaus' in the threshold playing frequency (see Figs. 5.6b

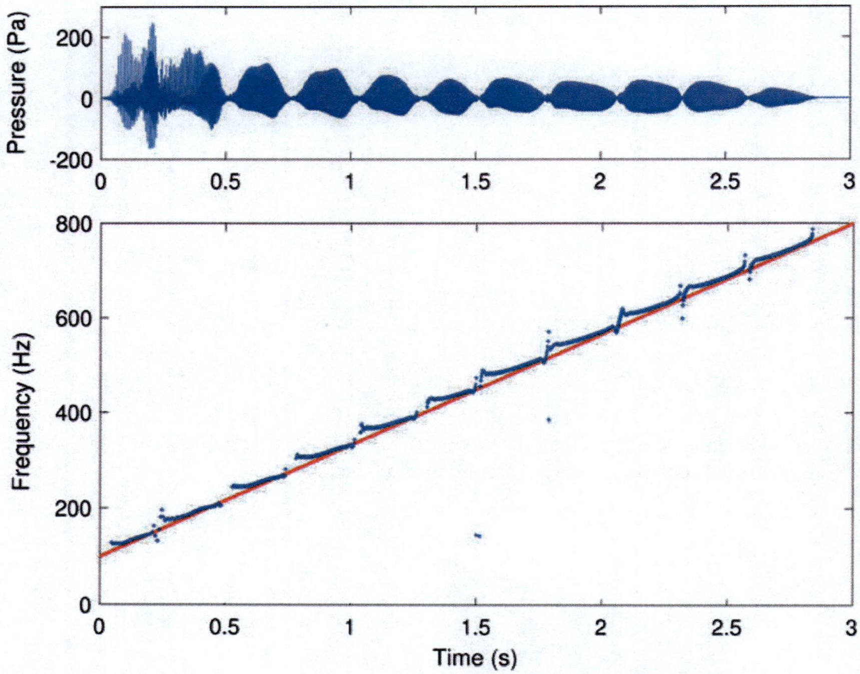

Fig. 5.11 Simulation of the set of natural notes of a trombone using a linear lip frequency sweep, the mouth pressure being maintained constant. Horizontal axis: time. Upper graph: acoustic pressure at bell exit plane. Lower graph: playing frequency (blue), and control parameter F_{lip} (red) (Color figure online)

and 5.18). Each plateau corresponds to a natural note, from the lowest one (the pedal note) to the highest one supported by an acoustic resonance of the brass bore. In performance, a human player will normally use mouth pressures well above the oscillation threshold values: Fig. 5.11 shows the result of an FDTD simulation using a realistic mouth pressure. The simulation is carried out using a lip frequency sweep, the mouth pressure and other control parameters being maintained constant. The plateaus in the playing frequency are clearly visible.

Although brass players are all able to get the lowest natural notes, they do not have the same ability to get the highest ones. A relatively unskilled player may have difficulty ascending beyond the eighth natural note, while the embouchure control developed by years of practice allows professional performers to play glissandos beyond the highest air column resonance (Fig. 2.20). This can be partially illustrated using simulations in which the lip resonance quality factor and other embouchure parameters are varied.

5.3.2 Bifurcation Diagrams

In Sect. 5.3.1 simulations in the time domain have illustrated how approximate solutions of an elementary model can mimic qualitatively the performance of a brass instrument in the nonlinear regime corresponding to realistic playing levels. To progress from qualitative to quantitative agreement between simulations and the sounds generated by live musicians requires accurate estimations of the parameters of the embouchure of the musician and how they are slowly evolving in time. This remains a significant challenge in the physical modelling of brass instruments.

The mouth pressure p_{m} is the most important control parameter in the elementary model. When p_{m} is raised above its threshold value p_{thr}, the equilibrium state is destabilised. For a fixed value of p_{m} above the threshold, a permanent periodic regime is sustained in which the state variables oscillate with a finite amplitude. The equations describing the system cannot be applied in their linearised form to this situation, as was done during the linear stability analysis described in Sect. 5.2.2; information about the permanent oscillatory regimes can only be extracted using the equations in their original nonlinear form. An example of a periodic solution (after a transient regime) obtained numerically using the FDTD method was displayed in Fig. 5.8. If this kind of simulation is repeated for many different values of p_{m} larger than p_{thr}, an overview of the dynamic behaviour of the nonlinear behaviour of the brass instrument can be obtained by collecting all the simulated periodic regimes in the same picture. This can be done by plotting the RMS value of the periodic solution as a function of the control parameter p_{m}. Such a plot is called a bifurcation diagram.

An example of a wind instrument bifurcation diagram is shown in the upper curve of Fig. 5.12. As p_{m} reaches and exceeds p_{thr}, a periodic solution is found with an oscillation amplitude which increases progressively from zero at $p_{\mathrm{m}} = p_{\mathrm{thr}}$. For very small amplitudes, the oscillation is quasi-sinusoidal. This kind of transition from a stable equilibrium position ($p_{\mathrm{m}} < p_{\mathrm{thr}}$: silence) to a stable periodic regime ($p_{\mathrm{m}} > p_{\mathrm{thr}}$: sound) is called a direct (or supercritical) Hopf bifurcation.

The lower curve in Fig. 5.12 shows the evolution of the playing frequency as the blowing pressure is increased. The horizontal dashed line indicates the threshold playing frequency f_{thr} derived from linear stability analysis. For blowing pressures below p_{thr}, there are no periodic vibration states predicted by the simulation. As the pressure rises through the threshold, a vibration state appears with a frequency f_{thr} when p_{m} reaches p_{thr}; in this case, frequency decreases slowly as the blowing pressure increases above the threshold. Only at the pianissimo level is the frequency value f_{thr} found by LSA a reasonable approximation to the playing frequency.

The evolution of the amplitude and frequency of the simulated sound in a slow crescendo followed by a diminuendo can be seen by following the green arrows in Fig. 5.12 as the blowing pressure is increased and the red arrows as it is reduced. For this direct Hopf bifurcation, the path traced as the pressure rises is simply retraced as it falls again, and the sound is extinguished when the blowing pressure falls below p_{thr}.

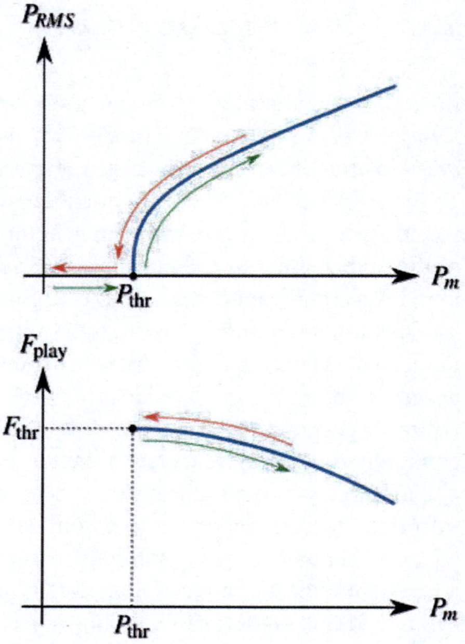

Fig. 5.12 Bifurcation diagram showing a direct Hopf bifurcation. Upper curve: RMS value of the periodic solution $p(t)$ as a function of the control parameter p_m. Lower curve: playing frequency as a function of p_m

While a direct Hopf bifurcation corresponds well with what might be expected intuitively, it is by no means the only way in which a wind instrument valve can destabilise. Another common process is described as an inverse (or subcritical) Hopf bifurcation. In this case, the amplitude of the oscillation does not increase progressively from zero when the control parameter exceeds the threshold value, but jumps discontinuously to a finite amplitude. This inverse Hopf bifurcation case is illustrated by the bifurcation diagram shown in Fig. 5.13.

The upper curve in Fig. 5.13 illustrates the dependence of the oscillation amplitude on the control parameter p_m. When p_m rises from zero and reaches p_{thr}, the equilibrium state becomes unstable, but there is no stable periodic regime with infinitely small amplitude which can replace it. Instead, there is a transition to a stable oscillation state with finite amplitude, indicated by the letter F on the upper curve in Fig. 5.13. This type of behaviour makes it very difficult to start a note pianissimo. As the blowing pressure is further increased, the amplitude grows, as shown by the branch from F to the right in Fig. 5.13.

The branch to the left from the threshold point $p_m = p_{thr}$ represents a set of periodic solutions which are unstable, and therefore not observable in practice. At the inflection point I, corresponding to a blowing pressure p_{subthr}, this branch turns to the right: it now represents stable periodic solutions and therefore playable sounds. This behaviour has two main consequences. The first is that periodic regimes exist for values of p_m lower than the threshold of oscillation found with a rising control parameter p_m: the minimum blowing pressure at which periodic oscillations are possible is p_{subthr}. The second is that there are two possible

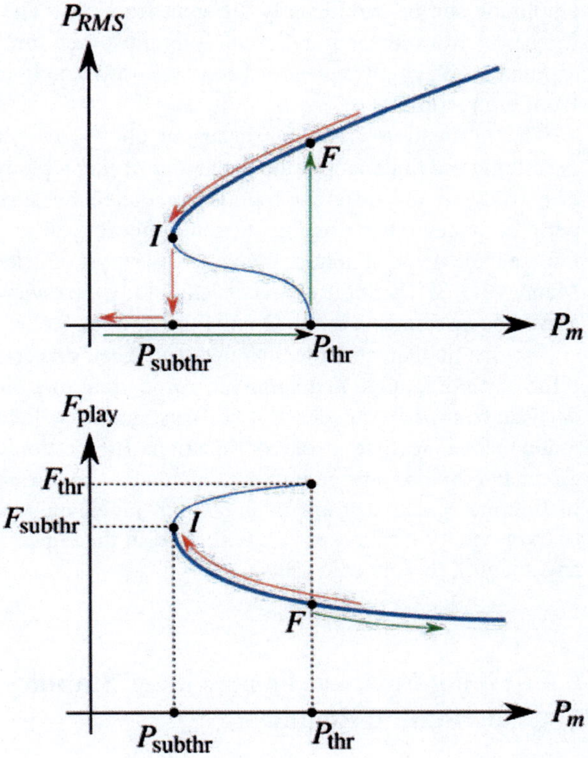

Fig. 5.13 Bifurcation diagram showing an inverse Hopf bifurcation. Upper curve: RMS value of the periodic solution with respect to the control parameter p_m. Thick (respectively thin) line corresponds to stable (respectively unstable) periodic solution. Lower curve: playing frequency with respect to p_m

observable regimes, silence and sound, for values of p_m between p_{subthr} and p_{thr}. Which one will be obtained in practice? It depends on the past history of the oscillation. With an inverse Hopf bifurcation, a crescendo (following the green arrows in Fig. 5.13) does not trace the same path as a decrescendo (following the red arrows). This is the phenomenon known as hysteresis. The stable states which exist for values of $p_m < p_{thr}$ can be reached only from above.

The lower curve of Fig. 5.13 is derived by aggregating the fundamental frequencies of the periodic regimes. The branch representing periodic solutions begins at the point corresponding to the threshold value $p_m = p_{thr}$, and this point does correspond to a regime whose frequency is equal to the threshold frequency f_{thr} calculated from LSA. Following the branch to the left, the frequency decreases along the branch up to the inflection point I, where its value is f_{subthr}. However all these points correspond to unstable periodic solutions, which are not observable in practice. Beyond the inflection point, the periodic solutions become stable and observable. Following the branch from I to the right, the playing frequency continues to decrease with increasing blowing pressure.

An important consequence of the behaviour of the inverse Hopf bifurcation is that the fundamental frequency f_{subthr} of the periodic oscillation with the lowest possible amplitude can be significantly different from $f_{\rm thr}$. The prediction of the threshold frequency from linear stability analysis must therefore be used with caution as an estimate of playing frequency even at very low amplitudes in the case of an inverse Hopf bifurcation.

The bifurcation diagrams discussed above are generic examples capable of describing the finite amplitude behaviour of many types of wind instrument model. The study of the different regimes predicted by such models, including quasi-periodic states corresponding to multiphonics, falls within the general scope of the area known as nonlinear dynamical systems (Guckenheimer and Holmes 1983; Manneville 2010). Some aspects of this study are reviewed and applied to both reed and brass instruments in the Going Further Sect. 5.4.

Various mathematical techniques have been developed in this field to calculate bifurcation diagrams. A diagram corresponding to a direct bifurcation is explicitly derived from the Van der Pol self-sustained oscillator in Sect. 5.4.1, using the mathematical technique of continuation. Bifurcation diagrams corresponding to direct and inverse bifurcations are calculated and discussed at the end of Sect. 5.4.5 in the context of an instrument air column having only two acoustic resonances close to harmonicity (Gilbert et al. 2020), and in the context of brass instrument design and analysis (Fréour et al. 2020).

5.4 Going Further: From Linear Stability Analysis to Oscillation Regimes

The generation of oscillations by the destabilisation of a nonlinear dynamical system is a major field of academic study, with practical applications ranging from the explanation of snoring (Auregan and Depollier 1995) to the design of liquid fuel valves in rocket engines (Chang 1994). Some useful mathematical techniques which have been developed to study the nature of these oscillations, including linear stability analysis and bifurcation diagrams, were outlined in Sects. 5.2 and 5.3. In this section the mathematical background to these techniques is explained, and some applications to the study of brass instruments are reviewed.

To introduce the mathematical treatment of self-sustained oscillations, we begin in Sect. 5.4.1 with a discussion of the Van der Pol oscillator. The analytical tools which will be used to study brass instrument behaviour are first demonstrated in this mathematically simple case. State-space representations of the elementary brass playing model are developed in Sect. 5.4.2 and used to carry out linear stability analysis of equilibrium states of the model in Sect. 5.4.3. The threshold results obtained from LSA, and the periodic regimes generated by numerical simulation above the oscillation threshold, are discussed and compared with the experience of brass instrument players. In Sect. 5.4.5 bifurcation diagrams of reed and brass instruments are obtained using the mathematical technique of continuation. Discussion of these results offers a theoretical insight into the Bouasse-Benade prescription for optimising playability. In Sect. 5.4.6 the more complicated oscillation regimes which correspond to multiphonics are also simulated and discussed.

5.4.1 Introduction: A Van der Pol Self-Sustained Oscillator

The equation for a variable $p(t)$ performing damped simple harmonic oscillations with equilibrium value $p_{eq} = 0$ and no external forcing is

$$\frac{d^2p(t)}{dt^2} + \frac{\omega_1}{q_1}\frac{dp(t)}{dt} + \omega_1^2 p(t) = 0. \tag{5.12}$$

As in the similar Eq. 3.3 representing the lip as a 1DOF oscillator, $f_1 = \omega_1/2\pi$ is the natural resonance frequency of the oscillator, and q_1 is the quality factor of the resonance. Since q_1 is positive, the second term on the left-hand side of the equation represents the loss of energy through damping.

For consistency with the following discussion of the Van der Pol oscillator, we choose to rewrite Eq. 5.12 as

$$\frac{d^2p(t)}{dt^2} - \gamma\omega_1\frac{dp(t)}{dt} + \omega_1^2 p(t) = 0, \tag{5.13}$$

where $\gamma = -1/q_1$. The general solution in free oscillation of the above equation is

$$p(t) = p_0 e^{\mathrm{Re}(\lambda)t}\cos(\mathrm{Im}(\lambda)t + \phi), \tag{5.14}$$

with

$$\lambda = \frac{\gamma\omega_1}{2} + j\omega_1\sqrt{1 - \frac{\gamma^2}{4}}, \tag{5.15}$$

p_0 and ϕ being determined by the initial conditions (p and $\frac{dp}{dt}$ at $t = 0$). Since γ is negative the amplitude is exponentially damped. If $1 - \gamma^2/4 > 0$, $p(t)$ is pseudo-periodic, with frequency $f_{pp} = f_1\sqrt{1 - \gamma^2/4}$. If $1 - \gamma^2/4 \geq 0$, λ is purely real, and $p(t)$ decays to zero without oscillation.

The Van der Pol equation can be written as

$$\frac{d^2p(t)}{dt^2} - \left[\gamma - \left(\frac{p(t)}{P_{\mathrm{ref}}}\right)^2\right]\omega_1\frac{dp(t)}{dt} + \omega_1^2 p(t) = 0. \tag{5.16}$$

Equation 5.16 resembles the single degree of freedom oscillator Eq. 5.13, with ω_1 being its natural resonance frequency. The important difference is in the nature of the second term on the left-hand side of each equation. For the simple harmonic oscillator, $\omega_1 \dfrac{dp(t)}{dt}$ is multiplied by the constant $-\gamma$; if γ is negative the amplitude is exponentially falling, while if γ is positive, it is exponentially rising. For the Van der Pol oscillator, on the other hand, $\omega_1 \dfrac{dp(t)}{dt}$ is multiplied by the function

$$\left[\gamma - \left(\frac{p(t)}{p_{\text{ref}}}\right)^2\right],$$

which contains a nonlinear term proportional to the square of the unknown variable $p(t)$. If γ is negative the amplitude is again falling. If γ is positive the amplitude increases for small values of $p(t)$, but once $(p(t)/p_{\text{ref}})^2$ is higher than γ, the term becomes dissipative, and the amplitude starts to fall again. As a consequence, stable periodic self-sustained oscillations can be obtained in a permanent regime.

In practice, solutions of the Van der Pol Eq. 5.16 are obtained by a numerical solver which integrates a system of ordinary differential equations (ODEs) of the form $y' = f(t, y)$ with known initial conditions. Periodically oscillating solutions of the Van der Pol equation with $f_1 = \omega_1/2\pi = 220$ Hz are illustrated in Fig. 5.14 for two values of γ. For $\gamma = 0.5$, it is evident that the oscillation is not a pure sine curve even though the Van der Pol equation is based on a single mode system. This distortion of the waveform is a consequence of the nonlinear term in the equation. The fundamental frequency of the oscillation is $f = 217$ Hz, which is intermediate between the natural mode frequency $f_1 = 220$ Hz and the frequency $f_{\text{pp}} = 213$ Hz of the pseudo-periodic decay which would take place in the absence of the nonlinear term. For $\gamma = 0.1$ the waveform is much closer to a pure sine curve.

Many numerical tools have been developed to study physical systems represented by sets of first-order ODE equations. To make use of these tools to investigate the behaviour of a brass instrument as a nonlinear dynamical system, it is fruitful to reformulate the equations of the elementary model of brass playing as an autonomous system of first-order ODE equations. Before carrying through this

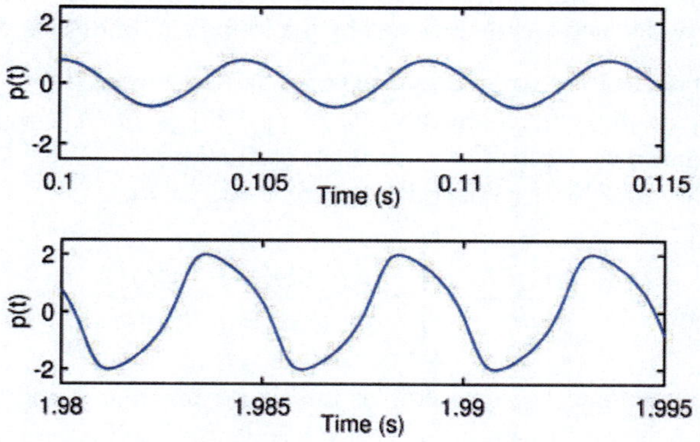

Fig. 5.14 Three periods of $p(t)/P_{\text{ref}}$ for a Van der Pol oscillation in permanent regime with $\omega_1/2\pi = 220$ Hz. Upper curve: $\gamma = 0.1$; lower curve: $\gamma = 0.5$

reformulation of the brass model in Sect. 5.4.2, we illustrate the procedure by reformulating the simpler system of the Van der Pol self-sustained oscillator as a dynamical system of dimension 2. We begin by defining a state vector **X**, with the following components:

$$\begin{bmatrix} X_1 = p(t) \\ X_2 = \dfrac{dp(t)}{dt} \end{bmatrix}. \tag{5.17}$$

The second-order Van der Pol Eq. 5.16 is then reformulated as two autonomous ODEs, the second of which is nonlinear:

$$\begin{cases} \dfrac{dX_1}{dt} = X_2 \\ \dfrac{dX_2}{dt} = -\omega_1^2 X_1 + \omega_1 \left(\gamma - \left(\dfrac{X_1}{P_{\text{ref}}} \right)^2 \right) X_2. \end{cases} \tag{5.18}$$

The Van der Pol equation can then be written in vector form as

$$\frac{d\mathbf{X}}{dt} = F_\gamma(\mathbf{X}), \tag{5.19}$$

with γ being the control variable of the Van der Pol dynamical system.

The next step in studying the Van der Pol oscillator as a dynamical system is to determine the equilibrium positions and their stability. There is only one equilibrium position of the system represented by Eq. 5.19: it is $\mathbf{X}_{\text{eq}} = [0\ 0]^{\text{T}}$, corresponding to $p(t) = 0$. To determine the stability of this equilibrium solution, it is necessary to linearise the equations by dropping the term X_1^2 in the equation for X_2. The linearised version of Eq. 5.19, which represents the small amplitude behaviour of **X** around its equlibrium state $\mathbf{X}_{\text{eq}}$ is described by the Jacobian matrix $\mathbb{M}$ of the system, defined by

$$\frac{d\mathbf{X}}{dt} = \mathbb{M}\mathbf{X} = \begin{bmatrix} 0 & 1 \\ -\omega_1^2 & +\omega_1 \gamma \end{bmatrix} \mathbf{X}. \tag{5.20}$$

The eigenvalues of an $n \times n$ matrix $\mathbb{M}$ are the values λ_n which satisfy the matrix equation

$$\mathbb{M}\mathbf{X} = \lambda_n \mathbf{X}. \tag{5.21}$$

For the Van der Pol oscillator, assuming that $\gamma < 2$, there are two eigenvalues of the Jacobian matrix, defined by

$$\lambda_{1,2} = \frac{\gamma \omega_1}{2} \pm j\omega_1 \sqrt{1 - \frac{\gamma^2}{4}}. \tag{5.22}$$

The stability of the equilibrium solution is determined by the sign of the real part of the eigenvalues of the Jacobian matrix: if and only if any of the eigenvalues has a positive real part, the solution is unstable. In the case of the Van der Pol oscillator, the two eigenvalues have equal real parts $\mathrm{Re}(\lambda_1) = \mathrm{Re}(\lambda_2) = \gamma\omega_1/2$, so the equilibrium state is stable when γ is negative and unstable when γ is positive. The threshold value of the control parameter is the value at which the transition from stability to instability takes place; for the Van der Pol oscillator $\gamma_{\mathrm{thr}} = 0$.

The imaginary parts of the eigenvalues are $\mathrm{Im}(\lambda_{1,2}) = \pm\omega_1\sqrt{1 - \gamma^2/4}$. At threshold the two eigenvalues are purely imaginary and equal to $\pm\omega_1$. The threshold frequency is defined as $f_{\mathrm{thr}} = \omega_1/(2\pi)$.

Although the equilibrium position is unstable when $\gamma > 0$, a different kind of stable permanent regime exists at positive values of γ. This is a periodic oscillation, in which the frequency, amplitude and waveform are determined by the control parameter. Two such regimes, for $\gamma = 0.1$ and $\gamma = 0.5$, are illustrated in Fig. 5.14.

The use of bifurcation diagrams to give an overview of the permanent regimes as a function of a control parameter was introduced in Sect. 5.3.2. Continuation methods, such as AUTO software (Doedel et al. 1997) or MANLAB software (Karkar et al. 2013), are capable of calculating bifurcation diagrams from autonomous system of first-order ODE equations. As an example of the application of such methods, AUTO has been used on Eq. 5.18 to get the bifurcation diagram for the Van der Pol oscillator displayed in Fig. 5.15. The upper graph of the figure shows a classical direct Hopf bifurcation behaviour: a stable periodic oscillation is born at the threshold of instability of the equilibrium position, and the amplitude of the oscillation increases progressively from zero as the control parameter γ increases from its threshold value $\gamma_{\mathrm{thr}} = 0$. The lower graph in Fig. 5.15 shows the fundamental frequency of the periodic oscillation decreasing slightly from the threshold frequency f_{thr} as the control parameter increases from its threshold value $\gamma_{\mathrm{thr}} = 0$.

Although the single mode Van der Pol oscillator is a much more straightforward dynamical problem than the multimode brass instrument model to be discussed in Sects. 5.4.2–5.4.6, it is already possible to draw a few illuminating comparisons between the behaviours of the two systems. In both cases the fundamental oscillation frequency is close to but not necessarily identical with the natural frequency of a mode, and distortion of the output signal is generated by a nonlinear term in the model equations. The qualitative parallel can be extended by simulating the playing of a crescendo on the Van der Pol oscillator, as illustrated in Fig. 5.16. Here γ is increased continuously from 0.1 to 1 over a time interval of 2 s. In the parallel case of the trombone crescendo simulated in Fig. 5.9, the control parameter is the pressure in the player's mouth. In both cases the increase in the control parameter results in a growth in amplitude, a downward deviation of the oscillation frequency and an increasing distortion of the waveform.

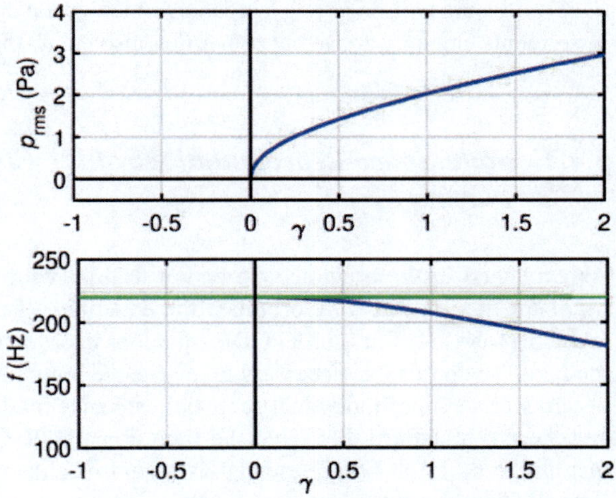

Fig. 5.15 Bifurcation diagram for the Van der Pol autonomous dynamical system as a function of the control parameter γ. Upper graph: RMS value of the periodic solution. Lower graph: fundamental frequency of the periodic oscillation (blue line); threshold frequency $f_{thr} = 220$ Hz (green line) (Color figure online)

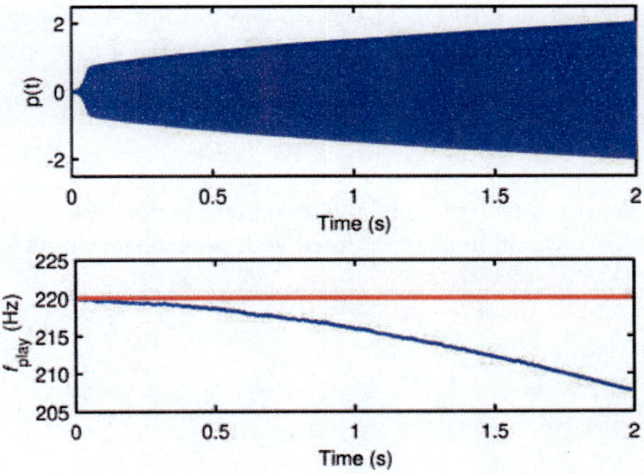

Fig. 5.16 Simulated solution of the Van der Pol equation with γ increasing continuously in time from 0.1 to 1, $\omega_1/(2\pi) = 220$ Hz remaining constant. Upper graph: the oscillating solution. Lower graph: playing frequency (blue line); $f_{thr} = 220$ Hz (red line) (Color figure online)

The study of the Van der Pol self-sustained single mode oscillator as an example of an autonomous dynamical system of dimension 2 has introduced several concepts and techniques which will be useful in the analysis of the elementary model of brass playing which follows in Sect. 5.4.2. The preliminary discussion of linear stability analysis in Sect. 5.2.2 is developed in a more formal way in Sect. 5.4.3, making use of the matrix representation of the set of autonomous first-order ODEs. Examples are given to demonstrate that LSA is a fruitful tool in analysing threshold mouth pressures and playing frequencies, leading to a clearer understanding of the intonation and ease of playing of brass instruments. The introductory treatment of bifurcation diagrams in Sect. 5.3.2 illustrated the way in which these diagrams allow the discussion of playing behaviour to be extended beyond the threshold region. In Sect. 5.4.5 the continuation methods for calculating bifurcation diagrams are shown to be promising tools in giving a broad overview of the dynamical behaviour of brass instruments, not only above but sometimes also below the mouth pressure threshold.

5.4.2 State-Space Representations of the Elementary Brass Playing Model

As explained in the previous subsection dealing with Van der Pol self-sustained oscillator, it is useful to reformulate the equations of the brass instrument model (Sect. 5.1) as a set of first-order ODE equations in order to use LSA and continuation methods. To do this it is necessary to rewrite the input impedance equation (Eq. 5.8) in terms of a sum of individual acoustic resonance modes in the frequency domain and then to transform them into the time domain. The acoustical response of the instrument tube can be represented as a sum of either real modes, as in Eq. 5.23 (see, e.g. Debut and Kergomard (2004)), or complex modes, as in Eq. 5.33 (see, e.g. Silva et al. (2014)). These two ways of approximating the input impedance in the frequency domain lead to two different sets of first-order vector equations

$$\frac{d\mathbf{X}}{dt} = F(\mathbf{X}).$$

We begin with the 'real mode' representation of the input impedance Z.

The input impedance fitted with N resonance modes is written as

$$Z(\omega) = \sum_{n=1}^{N} Z_n \frac{j\left(\dfrac{\omega_n}{q_n}\right)\omega}{\omega_n^2 + j\left(\dfrac{\omega_n}{q_n}\right)\omega - \omega^2}. \tag{5.23}$$

where the nth resonance is defined by three real constants, the amplitude Z_n, the dimensionless damping coefficient q_n and the angular frequency ω_n.

Transformation of Eq. 5.23 into the time domain, and decomposition of $p(t)$ into its real modal components p_n defined by

$$p(t) = \sum_{n=1}^{N} p_n(t), \tag{5.24}$$

results in a second-order ODE for each p_n:

$$\frac{d^2 p_n}{dt^2} + \frac{\omega_n}{q_n}\frac{dp_n}{dt} + \omega_n^2 p_n(t) = Z_n \frac{\omega_n}{q_n}\frac{du(t)}{dt}, \tag{5.25}$$

with $u(t)$ being the total volume flow rate.

Adding the equation describing the mechanical 1DOF oscillator (Eq. 5.1) gives the following set of $N+1$ second-order ODEs:

$$\begin{cases} \dfrac{d^2h(t)}{dt^2} = -\omega_l^2 h(t) - \dfrac{\omega_l}{Q_l}\dfrac{dh(t)}{dt} - \dfrac{p(t)}{\mu} + \omega_l^2 h_0 + \dfrac{p_m}{\mu} \\ \dfrac{d^2 p_n}{dt^2} = -\dfrac{\omega_n}{q_n}\dfrac{dp_n}{dt} - \omega_n^2 p_n(t) + Z_n \dfrac{\omega_n}{q_n}\dfrac{du}{dt} \quad \text{for } n \in [1:N]. \end{cases} \tag{5.26}$$

The derivative of the volume flow nonlinear equation (Eq. 5.7) is

$$\frac{du}{dt} = w\frac{dh}{dt}\left[\frac{2}{\rho}(p_m - p(t))\right]^{1/2} - \frac{wh}{\rho}\frac{dp}{dt}\left[\frac{2}{\rho}(p_m - p(t))\right]^{-1/2}. \tag{5.27}$$

Assuming that $p_m - p(t) > 0$ and $h(t) > 0$ for all t, Eq. 5.27 can be rewritten using Eq. 5.24 as

$$\frac{du}{dt} = w\frac{dh}{dt}\left[\frac{2}{\rho}\left(p_m - \sum_{n=1}^{N} p_n(t)\right)\right]^{1/2} - \frac{wh}{\rho}\frac{d}{dt}\left(\sum_{n=1}^{N} p_n(t)\right) \times \left[\frac{2}{\rho}\left(p_m - \sum_{n=1}^{N} p_n(t)\right)\right]^{-1/2}. \tag{5.28}$$

Substitution of Eq. 5.28 into Eq. 5.26 yields a set of $2(N+1)$ equations in the variables h, dh/dt, p_n, dp_n/dt. The equations of the brass model can now be put into a state-space representation

$$\frac{d\mathbf{X}}{dt} = F(\mathbf{X}), \tag{5.29}$$

where F is a nonlinear vector function and $\mathbf{X}$ is the state vector having $2(N+1)$ real components defined as follows:

$$\mathbf{X} = \left[h \quad \frac{dh}{dt} \quad p_1 \quad \dots \quad p_N \quad \frac{dp_1}{dt} \quad \dots \quad \frac{dp_N}{dt}\right]^{\mathrm{T}}. \tag{5.30}$$

The nonlinear vector function $F(\mathbf{X})$ can be written as

$$F(\mathbf{X}) = \frac{d\mathbf{X}}{dt} = \begin{pmatrix} \frac{dX(1)}{dt} \\ \frac{dX(2)}{dt} \\ \frac{dX(3)}{dt} \\ \vdots \\ \frac{dX(2+N)}{dt} \\ \frac{dX(2+N+1)}{dt} \\ \vdots \\ \frac{dX(2+N+N)}{dt} \end{pmatrix} = \begin{pmatrix} \frac{dh}{dt} \\ \frac{d^2h}{dt^2} \\ \frac{dp_1}{dt} \\ \vdots \\ \frac{dp_N}{dt} \\ \frac{d^2p_1}{dt^2} \\ \vdots \\ \frac{d^2p_N}{dt^2} \end{pmatrix} \tag{5.31}$$

$$= \begin{pmatrix} X(2) \\ -\omega_l^2 X(1) - \frac{\omega_l}{Q_l} X(2) - \frac{\sum_{n=3}^{2+N} X(n)}{\mu} + \omega_l^2 h_0 + \frac{p_m}{\mu} \\ X(2+N+1) \\ \vdots \\ X(2+N+N) \\ -\frac{\omega_1}{q_1} X(2+N+1) - \omega_1^2 X(3) + Z_1 \frac{\omega_1}{q_1} \frac{du}{dt} \\ \vdots \\ -\frac{\omega_N}{q_N} X(2+N+N) - \omega_N^2 X(2+N) + Z_N \frac{\omega_N}{q_N} \frac{du}{dt} \end{pmatrix}, \tag{5.32}$$

with

$$\frac{du}{dt} = wX(2)\left(\frac{2}{\rho}\left(p_m - \sum_{n=3}^{2+N} X(n)\right)\right)^{1/2}$$
$$-\frac{w}{\rho}X(1) \sum_{n=2+N+1}^{2+N+N} X(n) \left(\frac{2}{\rho}\left(p_m - \sum_{n=3}^{2+N} X(n)\right)\right)^{-1/2}.$$

The same exercise of reformulation is now presented using the 'complex mode' representation of the input impedance Z (see, e.g. Silva et al. (2014)). The modal-fitted input impedance with N resonance modes is written as follows:

$$Z(\omega) = Z_c \sum_{n=1}^{N} \frac{C_n}{j\omega - s_n} + \frac{\bar{C}_n}{j\omega - \bar{s}_n}, \tag{5.33}$$

where s_n and C_n are the complex poles and the complex residues of the nth complex mode, and $\bar{s}_n$ and $\bar{C}_n$ their respective complex conjugates. The characteristic impedance of the resonator is $Z_c = \rho c/S$, where S is the input cross-section of the bore at the mouthpiece rim.

As an example, a comparison between the measured input impedance of a trombone and its value derived from a superposition of 18 complex modes is given Fig. 5.17.

Transforming Eq. 5.33 into the time domain and decomposing $p(t)$ into its complex modal components $p_n(t)$ such that

$$p(t) = \sum_{n=1}^{N} 2\mathrm{Re}(p_n(t)), \tag{5.34}$$

results in a first-order ODE for each p_n:

$$\frac{dp_n}{dt} = Z_c C_n u(t) + s_n p_n(t) \quad \text{for} \quad n \in [1:N]. \tag{5.35}$$

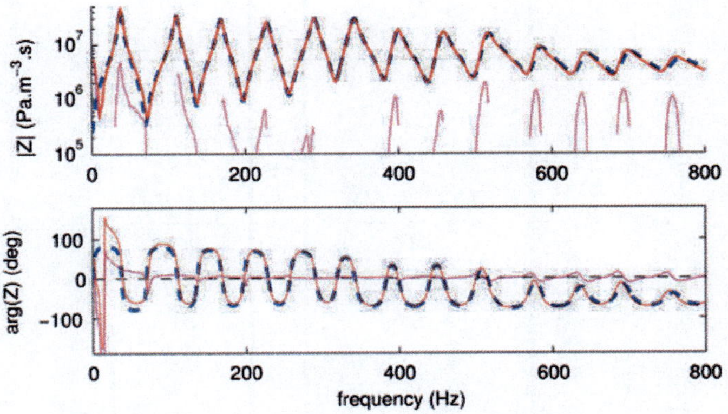

Fig. 5.17 Magnitude (top) and phase (bottom) of the input impedance of a tenor trombone with slide in first position. Dashed blue lines: measured impedance. Solid red lines: fitted curves with 18 complex modes. Adapted from Velut et al. (2017a) (Color figure online)

Having found a time domain representation of the third constituent equation of the elementary model in terms of complex mode parameters, we follow the same procedure as we used with the real mode representation. The two other equations of the elementary model, describing the 1DOF lip oscillator (Eq. 5.1) and the volume flow through the lips (Eq. 5.7), are employed to obtain the following set of second- and first-order ODEs:

$$\begin{cases} \dfrac{d^2h(t)}{dt^2} &= -\omega_l^2 h(t) - \dfrac{\omega_l}{q_l}\dfrac{dh(t)}{dt} - \dfrac{p(t)}{\mu} + \omega_l^2 h_0 + \dfrac{p_m}{\mu} \\ \dfrac{dp_n}{dt} &= s_n p_n(t) + Z_c C_n \sqrt{\frac{2}{\rho}} w h(t) \sqrt{p_m - p(t)} \text{ for } n \in [1:N], \end{cases} \tag{5.36}$$

where as before we assume that at each time t, $p_m - p(t) > 0$ and $h(t) > 0$.

In the state-space representation (Eq. 5.29), the state vector $\mathbf{X}$ has $2(N+1)$ real components defined as follows (see, e.g. Velut et al. (2017a)):

$$\mathbf{X} = \begin{bmatrix} h & \dfrac{dh}{dt} & \text{Re}(p_1) & \ldots & \text{Re}(p_N) & \text{Im}(p_1) & \ldots & \text{Im}(p_N) \end{bmatrix}^{\mathrm{T}}, \tag{5.37}$$

where each complex component p_n has been split into its real part $\text{Re}(p_n)$ and its imaginary part $\text{Im}(p_n)$.

The nonlinear vector function F can be written as:

$$\frac{d\mathbf{X}}{dt} = \begin{pmatrix} \dfrac{dX(1)}{dt} \\ \dfrac{dX(2)}{dt} \\ \dfrac{dX(3)}{dt} \\ \vdots \\ \dfrac{dX(2+N)}{dt} \\ \dfrac{dX(2+N+1)}{dt} \\ \vdots \\ \dfrac{dX(2+N+N)}{dt} \end{pmatrix} = \begin{pmatrix} \dfrac{dh}{dt} \\ \dfrac{d^2h}{dt^2} \\ \dfrac{d\text{Re}(p_1)}{dt} \\ \vdots \\ \dfrac{d\text{Re}(p_N)}{dt} \\ \dfrac{d\text{Im}(p_1)}{dt} \\ \vdots \\ \dfrac{d\text{Im}(p_N)}{dt} \end{pmatrix} \tag{5.38}$$

$$
= \begin{pmatrix}
X(2) \\
-\omega_l^2 X(1) - \frac{\omega_l}{q_l} X(2) - \frac{p(t)}{\mu} + \omega_l^2 h_0 + \frac{p_m}{\mu} \\
\mathrm{Re}\left[s_1 (X(3) + jX(2+N+1)) + C_1.Z_\mathrm{c}.\sqrt{\frac{2}{\rho}} wX(1)\sqrt{p_m - p(t)} \right] \\
\vdots \\
\mathrm{Re}\left[s_N (X(2+N) + jX(2+N+N)) + C_N.Z_\mathrm{c}.\sqrt{\frac{2}{\rho}} wX(1)\sqrt{p_m - p(t)} \right] \\
\mathrm{Im}\left[s_1 (X(3) + jX(2+N+1)) + C_1.Z_\mathrm{c}.\sqrt{\frac{2}{\rho}} wX(1)\sqrt{p_m - p(t)} \right] \\
\vdots \\
\mathrm{Im}\left[s_N (X(2+N) + jX(2+N+N)) + C_N.Z_\mathrm{c}.\sqrt{\frac{2}{\rho}} wX(1)\sqrt{p_m - p(t)} \right]
\end{pmatrix}, \tag{5.39}
$$

with

$$
p(t) = 2 \sum_{k=3}^{N+2} X(k).
$$

Many computational tools which have been developed for investigating the behaviour of general dynamical systems can be applied to brass instruments once the underlying physical equations are expressed in the state-space formalism, using either the ‘real mode’ or the ‘complex mode’ representation. Section 5.2.2 included an overview of one of the most important of these tools, linear stability analysis. In Sect. 5.4.3 the relationship between the state-space representation and LSA is explained and examples given of its use in investigating different aspects of brass instrument behaviour.

5.4.3 Linear Stability Analysis Applied to Brass Instruments

In the discussion of the elementary model of brass playing in Sect. 5.2.2, an equilibrium state was defined as one in which the lip opening has a constant value h_eq. The simplest equilibrium state is the one in which the mouth pressure $p_\mathrm{m} = 0$: no excess pressure is generated by the player’s lungs, and the equilibrium lip opening has its lowest value $h_\mathrm{eq} = h_0$. In this state, described mathematically as the ‘trivial’ solution, both the mouthpiece pressure $p(t)$ and the volume flow rate $u(t)$ are equal to zero.

Equilibrium states also exist for values of the mouth pressure $p_{\rm m} > 0$. Since in these states there is a pressure gradient across the lip opening, there will be a steady volume flow $u_{\rm eq}$. We assume that $p(t) = 0$, which is equivalent to setting $Z = 0$ in the third constituent equation (Eq. 5.8). Setting all the terms involving time derivatives equal to zero in the first constituent equation (as expressed in Eq. 5.5) and substituting the result in the second constituent equation (Eq. 5.7) gives the following expressions for $h_{\rm eq}$ and $u_{\rm eq}$ as functions of the control parameter $p_{\rm m}$:

$$\begin{cases} h_{\rm eq} = h_o + \dfrac{p_{\rm m}}{\mu\omega_l^2} \\ u_{\rm eq} = w\left(h_o + \dfrac{p_{\rm m}}{\mu\omega_l^2}\right)\sqrt{\dfrac{2p_{\rm m}}{\rho}}. \end{cases} \tag{5.40}$$

These equilibrium states theoretically exist for all positive values of $p_{\rm m}$. They can however only be observed in practice if $p_{\rm m}$ is below the threshold $p_{\rm thr}$ for self-sustained oscillation. When $p_{\rm m}$ is raised slowly through the threshold, the equilibrium state becomes unstable, and the system bifurcates towards a stable oscillating regime. It is the aim of linear stability analysis to derive these threshold values from the equations of the global model.

The analysis of the stability of a state corresponding to a given value of the control parameter $p_{\rm m}$ follows the method outlined in Sect. 5.4.1 for the example of the Van der Pol oscillator. The nonlinear dynamical system represented in the state-space formalism by Eq. 5.31 or Eq. 5.38 is linearised around the equilibrium position defined by Eq. 5.40. After linearisation the $2(N + 1)$ equations are coupled by the linearised air flow equation (for details see Velut et al. (2017a) using the ‘complex mode’ representation). The resulting set of $2(N + 1)$ simultaneous first-order linear ordinary differential equations can be written as the vector equation

$$\frac{d\mathbf{X}}{dt} = \mathbb{M}\mathbf{X}, \tag{5.41}$$

where $\mathbb{M}$ is the $2(N + 1) \times 2(N + 1)$ Jacobian matrix.

Once the Jacobian matrix for the system is known, its stability is examined by determining the eigenvalues of the matrix (Thomsen 1997). As explained in Sect. 5.4.1, linear stability is preserved provided that the real parts of all eigenvalues of the square matrix $\mathbb{M}$ are negative. If the control parameter $p_{\rm m}$ is slowly increased from zero, the equilibrium solution of the system becomes unstable when the real part of one of the eigenvalues becomes positive. The threshold mouth pressure $p_{\rm thr}$ at which this transition occurs can thus be determined by studying the eigenvalue behaviour as a function of $p_{\rm m}$. In a direct Hopf bifurcation, a stable oscillatory state exists for $p_{\rm m} > p_{\rm thr}$ (see Sect. 5.3.2). The threshold frequency $f_{\rm thr}$, defined as the limiting value of the oscillation frequency as $p_{\rm m} - p_{\rm thr} \to 0$, is determined by the imaginary part of the eigenvalue λ:

$$f_{\mathrm{thr}} = \frac{\mathrm{Im}\lambda}{2\pi}. \tag{5.42}$$

Although by its nature linear stability analysis can only yield information about the behaviour of oscillating states for mouth pressures very near p_{thr}, its application to the elementary model offers valuable insights into several aspects of realistic brass playing. A good example is provided by a study of trombone playing using LSA by Velut et al. (2017a). The elementary model was expressed in the state-space formalism described in Sect. 5.4.2, with the input impedance represented as a sum of complex modes. The study was carried out for a range of lip frequencies $f_l \in [20 : 500]$ Hz. For each value of f_l, the eigenvalues of the Jacobian matrix were computed for increasing values of p_{m} until the first instability occurred.

The results are shown in Fig. 5.18. For each value of f_l, the top graph represents the lowest mouth pressure $p_{\mathrm{m}} = p_{\mathrm{thr}}$ for which the equilibrium solution is unstable. The lower graph shows the corresponding values of the threshold frequency f_{thr}. The horizontal blue lines on the lower graph show the acoustic resonance frequencies of the trombone, given by the maxima of the input impedance amplitude. It should be noted that, for p_{m} values higher than p_{thr}, other pairs of conjugate eigenvalues may have a positive real part. In this case the system will have multiple instabilities for the same lip frequency f_l, each with its own threshold pressure and frequency. If different oscillating solutions with the same parameter values are stable, the system would be able to start oscillating in different registers, depending on the initial conditions. In Fig. 5.18 only the lowest value of p_{thr} and the corresponding f_{thr} are plotted for each f_l value.

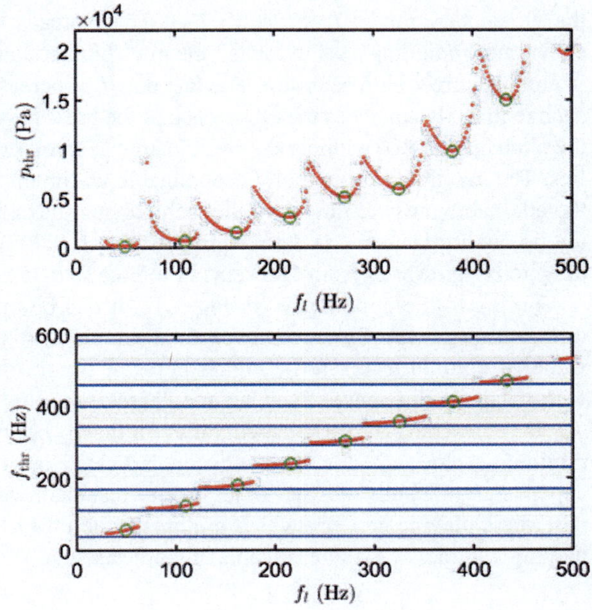

Fig. 5.18 Results of linear stability analysis applied to a trombone (adapted from Velut et al. (2017a)). The abscissa represents lip resonance frequency f_l. Upper graph: threshold mouth pressure p_{thr}. Lower graph: threshold frequency f_{thr}. Solid lines: values of acoustic resonance frequencies. Green circles: values of the threshold pressure and frequency at the 'optimal' lip frequency f_l^{opt} corresponding to the lowest local p_{thr} (Color figure online)

Figure 5.18 can be viewed as simulating a musical performance in which the trombonist attempts to play a note with the slide in fixed position while gradually increasing the lip frequency f_l from 20 to 500 Hz, continually adjusting the mouth pressure to the minimum value necessary to get the note to sound. The lower curve, which effectively traces the playing frequency for a mouth pressure just above the sounding threshold, replicates well the musician's experience that this exercise results in an arpeggio of notes which are close to the natural notes of the instrument (see Sect. 1.2.2).

The span of lip frequencies displayed in Fig. 5.18 can be divided into nine distinct ranges based on the discontinuities in the threshold frequency curve. Each range corresponds to one regime or register of the instrument. The first regime spans the frequency range [30 : 62 Hz], within which the pedal note B$^\flat$1 is played; this is a somewhat anomalous regime, which will be discussed later. The second to eighth regimes correspond to the frequency ranges [72 : 123 Hz], [124 : 179 Hz], [180 : 234 Hz], [235 : 288 Hz], [289 : 352 Hz], [353 : 404 Hz] and [408 : 454 Hz], respectively. The ninth regime begins at 488 Hz and continues beyond the 500 Hz limit of the calculation. No stable oscillatory state was found for lip frequencies below 30 Hz or between 63 and 72 Hz; there are also an interval between 454 and 488 Hz in which the equilibrium state remained stable for the tested values of p_{m}.

The shape of the lower curve in Fig. 5.18 is characterised from $f_l = 72$ Hz upwards by a series of plateaus of almost constant height, each plateau representing the variation of the threshold oscillation frequency as a function of the lip frequency within one of the regimes. In the nth regime f_{thr} increases gently with increasing f_l from a value just above the frequency $f_{\mathrm{ac},n}$ of the nth acoustic resonance and is also always above the lip resonance frequency f_l. This behaviour is the expected consequence of our choice of the outward-striking valve to represent the lips in the elementary model (see Sect. 3.2.4). The plateaus correspond to the player's experience in finding a set of 'slots', the pitch being relatively insensitive to changes in embouchure within each slot. The fact that f_{thr} increases slightly with rising f_l in each regime also matches the experience of the brass player, who can slightly 'bend' the pitch up or down within a slot by adjusting f_l through the muscular tension of the lips. The maximum change in f_{thr} obtainable within one regime, which determines the attainable musical range within each register, has analytic limits depending on the lip quality factor q_l as detailed in Silva et al. (2007). For the model used here, the pitch range varies from 186 cents in regime 2 to 45 cents in regime 8.

Turning to the upper curve in Fig. 5.18, it can be observed that the oscillation threshold p_{thr} globally increases with the rank n of the register. A greater p_{m} value is required to reach the higher notes of the instrument, which again agrees with musical experience. Each regime characterised by a plateau in the curve for f_{thr} is represented by a U-shaped valley in the p_{thr} curve (Silva et al. 2007). The minimum value $p_{\mathrm{thr}}^{\mathrm{opt}}$ for each register, indicated by a green circle in Fig. 5.18, depends significantly on the losses in the resonator. Since musicians commonly describe a playing strategy which minimises the effort to produce a sound in each playing regime, it is a reasonable hypothesis that $p_{\mathrm{thr},n}^{\mathrm{opt}}$ and the associated lip resonance frequency $f_{\mathrm{thr},n}^{\mathrm{opt}}$ are the optimal values for a human performer playing the nth natural note. The musical 'centre' of the note (see Sect. 1.2.2) can then be identified as the pitch whose frequency is $f_{\mathrm{thr},n}^{\mathrm{opt}}$. The experience of musicians broadly supports this hypothesis, although the pitch centres marked by circles in Fig. 5.18 are a little sharper than the corresponding reference pitches of the equally tempered scale. These pitch discrepancies, which are related to the restriction of the lip valve to purely outward-striking behaviour (Velut et al. 2017a), are less than 50 cents for registers 3–8.

It is noticeable, particularly in regimes $n = 2 - 5$, that $f_{l,n}^{\mathrm{opt}}$ is close to the upper frequency limit of the regime and that the pressure threshold $p_{l,n}^{\mathrm{thr}}$ increases faster when $f_{l,n}$ is above $f_{l,n}^{\mathrm{opt}}$ than when it is below. These results are compatible with the brasswind playing experience that it requires less effort for a musician to bend a note down than to bend it up it up. The downward limit of pitch bending is lower in practice because of several factors not considered in the elementary model, including windway resonances (Sect. 6.3) and multiple lip resonances (Sect. 6.4).

The ease with which the onset of pianissimo sounds can be controlled is an important factor in the player's judgement of the playability of a brass instrument. Many wind instrumentalists describe the sensation of overcoming resistance when starting a note, and it was suggested in Sect. 1.2.9 that this 'sounding resistance' might be related to the threshold pressure p_{thr} of the note in question. If this suggestion is adopted, the upper curve in Fig. 5.18 could be viewed as a plot of sounding resistance, the value of $p_{\mathrm{thr},n}^{\mathrm{opt}}$ corresponding to the resistance experienced when the nth natural note is sounded.

5.4.4 *The Trombone Pedal Note Regime*

We return now to consider the first regime of the trombone. The notes played in this regime are called pedal notes; with the slide in first position, the pedal note is B$^\flat$1, which is the nominal pitch of the instrument. Pedal notes are frequently employed in trombone playing and appear in many orchestral scores (see Sect. 7.8.5). A remarkable feature of the pedal B$^\flat$1 note is that its sounding frequency is far from any acoustic resonance frequency. The first resonance of the trombone is at a frequency $f_{\mathrm{ac},1} = 38$ Hz, while the frequency of B$^\flat$1 is 58 Hz. The reason for this discrepancy is explained in Sect. 4.3.9. The fact that the pedal note can be sounded at the desired pitch without the support of a nearby acoustic resonance was commented on by Bouasse (1929) and has been explained as the consequence of nonlinear coupling of higher acoustic modes whose frequencies are close to harmonics of 58 Hz (Benade 1973).

The LSA study of the trombone reported by Velut et al. (2017a) offers a fresh perspective on the nature of the pedal note. In the first regime, illustrated in Fig. 5.19, oscillating states exist with threshold frequencies ranging from $f_{\mathrm{thr}} = 47.1$ Hz at $f_l = 30$ Hz to $f_{\mathrm{thr}} = 65.4$ Hz at $f_l = 62$ Hz. The minimum threshold pressure $p_{\mathrm{thr}}^{\mathrm{opt}}$

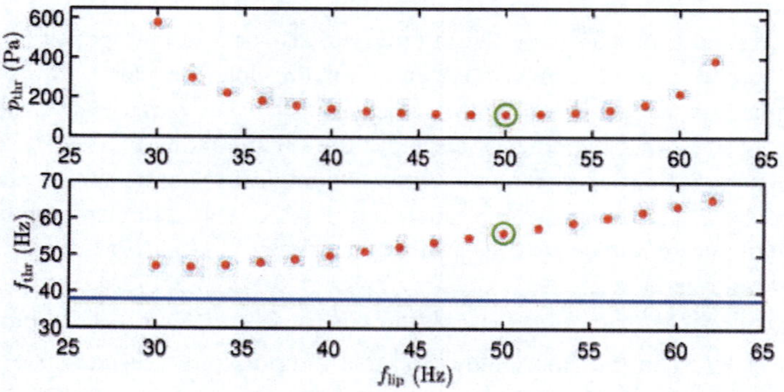

Fig. 5.19 Detail of Fig. 5.18 showing the variation of threshold pressure (upper graph) and threshold oscillation frequency (lower graph) as a function of lip resonance frequency for the first regime of a tenor trombone

is found at a lip frequency $f_l^{\text{opt}} = 50\,\text{Hz}$, corresponding to a threshold oscillation frequency $f_{\text{thr}}^{\text{opt}} = 56.3\,\text{Hz}$. This LSA prediction is very close to the frequency of 58 Hz at which the pedal note is sounded in musical performance, although it is 48% higher than the first acoustic resonance frequency $f_{\text{ac},1} = 38\,\text{Hz}$.

Comparison of the predicted optimum threshold frequencies with playing experience shows that, despite the simplifications accepted in framing the elementary model, LSA is able to give a good account of the pitches of the natural notes of the trombone. Threshold frequencies which correspond to minimum values of threshold pressure are found to agree with the musically defined frequencies of the first five natural notes to within 5%. That this level of agreement is found even for the first natural note is perhaps surprising, given that the conventional explanation for the existence of the pedal note in terms of nonlinear coupling seems at odds with the linearisation of the model equations required in the LSA treatment. The linearisation does not remove all the off-diagonal terms in the Jacobian matrix, which means that the behaviour of the model at the destabilisation threshold of one mode is still subject to some influence by the other modes. However a similar LSA calculation for the case in which all the higher modes are suppressed, leaving only the first acoustic resonance of the trombone at 38 Hz, predicted essentially the same playing behaviour for the first regime, with an optimum threshold frequency of 61.1 Hz.

Nonlinear coupling to higher modes is not therefore essential to the pianissimo playing of the pedal note at 58 Hz. The first acoustic mode, at 38 Hz, is too far below to exert a significant influence on the sounding pitch, but it has the correct phase relationship to gain energy from an outward-striking valve with a lip resonance at 50 Hz in a feedback loop (see Sect. 5.2.1). The threshold oscillation frequency at 56 Hz is 12% above the lip resonance frequency, which is broadly comparable to the outward-striking valve behaviour when playing the higher register notes. Raising the lip frequency by 2 Hz is enough to ‘lip’ the pitch up to the desired pitch of $\text{B}^\flat 1$.

Would a pedal note at the nominal fundamental pitch of the instrument still be playable if the outward-striking lip valve was replaced by an inward-striking reed? To answer this question, the ingenious acoustician Henri Bouasse devised an experiment in which a B$^{\flat}$ tenor trombone was played with a saxophone mouthpiece (Bouasse 1929). The result was an instrument whose first natural note was $E^{\flat}1$, with an oscillation frequency just under $f_{ac,1} = 38$ Hz. A playing frequency below the acoustic resonance frequency is the behaviour expected from an inward-striking reed (see Sect. 3.2.4). In this case the phase relationship between reed and air column resonance makes a feedback loop with positive regeneration impossible above f_{ac}, and there was no evidence of a playable note at 58 Hz. The results of the Bouasse experiment have been confirmed by physical modelling simulation of a trombone with a saxophone mouthpiece (Velut et al. 2017a).

The fact that an essentially linear description can be given of the threshold sounding of the pedal note does not mean that nonlinear coupling of harmonic upper modes is irrelevant to pedal note behaviour. The player's experience is that the pitch of a pedal B$^{\flat}$1 is relatively uncentred when played very quietly, but gains stability at higher dynamic levels. The increasing participation of the upper modes during a crescendo such as that illustrated in Fig. 7.67 helps the player to keep the pitch constant while also contributing to the spectral enrichment which is such a spectacular characteristic of a fortissimo pedal note (Sect. 7.8.5).

In Sect. 5.4.5 bifurcation diagrams will be used as maps to explore the nonlinear landscape which unfolds once we leave behind the foothills of threshold behaviour.

5.4.5 Bifurcation Diagrams of Reed and Brass Instruments

In the quest for a theory of wind instrument behaviour which is both musically and scientifically convincing, the holy grail is the understanding of the principles which determine the intonation and ease of playing of the instrument. We have seen in Sect. 5.4.3 that linear stability analysis of the physical model of a brass instrument can supply information about the oscillation threshold frequency f_{thr} which is relevant to intonation, while data on the corresponding threshold pressure p_{thr} can be related to the ease of playing of the instrument. However this information is only directly applicable to the periodic oscillations obtained in musical performance when p_m is near p_{thr}, and cannot be used to explain how intonation and playability evolve as the amplitude of oscillation increases. Instruments which display an inverse Hopf bifurcation behaviour at the initiation of a note present a particular problem for LSA, since the oscillation begins immediately at a finite amplitude (Sect. 5.3.2).

In Sect. 5.3.1 examples were given of time domain solutions of the physical model equations using numerical methods. Such simulations can be carried out many times with different sets of control parameters, allowing the investigation of periodic solutions over a wide range of amplitudes. Since a fresh simulation has to be carried out for each change in a control parameter, this method is relatively laborious and computationally expensive. In Sect. 5.4.1 continuation methods for calculating bifurcation diagrams were introduced. A bifurcation diagram gives an overview of the evolution of periodic regimes as a function of one of the control parameters: a typical wind instrument bifurcation diagram plots the amplitude and frequency of the oscillation as a function of the mouth pressure. In the examples which follow, we will see that these simple curves distil much useful information about intonation and ease of playing in wind instruments.

In Chap. 4 it was noted that the bores of brass instruments are usually designed in such a way that the impedance peak frequencies are as close as possible to a harmonic series. This harmonic alignment of resonances is clearly important for intonation, since many musically significant intervals are found within the harmonic series (see Sect. 1.2.2). A more subtle advantage of designing an instrument whose resonance frequencies are harmonically related is that notes played on it are generally found to be more stable and easier to sound than on an instrument with inharmonic resonances. The desirability of harmonically aligned resonances was accepted and developed by Arthur Benade in his groundbreaking textbook *Fundamentals of Musical Acoustics* (Benade 1976), but he noted generously that the idea originated elsewhere:

> The usefulness of the harmonically related air column resonances in fostering stable oscillations sustained by a reed-valve was first pointed out by the french physicist Henri Bouasse in his book *Instruments à Vent*. (Benade 1976)

For this reason we call the recommendation of harmonically related resonances as a recipe for good playability the 'Bouasse-Benade prescription'.

To provide a practical demonstration of the validity of the Bouasse-Benade prescription, Benade designed an instrument which spectacularly failed to follow it: a short horn with a clarinet mouthpiece (Fig. 5.20) whose resonance frequencies were chosen to avoid as much as possible integer relations between them. Since the instrument was designed to be unplayable, it was dubbed the 'tacet horn' by members of Benade's research team (Benade and Gans 1968). It was indeed found impossible to sound a note at the frequency of the lowest resonance of the horn. A note was however playable at a frequency close to the third resonance peak, because for that note, the Bouasse-Benade prescription had in fact been fulfilled: the sixth resonance frequency was almost exactly twice the third.

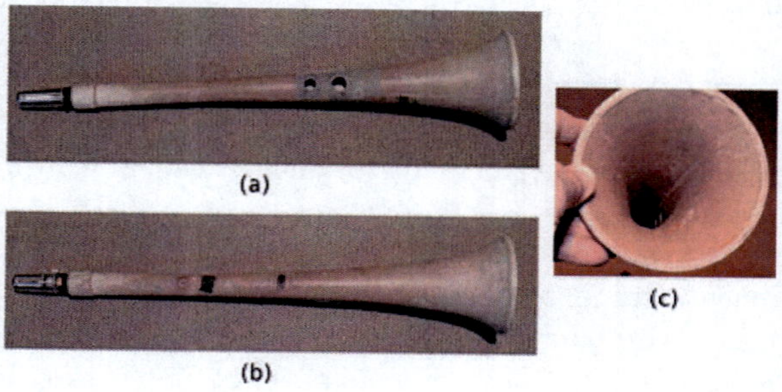

Fig. 5.20 Photographs of the tacet horn described in Benade and Gans (1968): (**a**) front; (**b**) back; (**c**) bell. Courtesy of Steven Thompson

The theoretical foundation for the Bouasse-Benade prescription can be explored by calculating bifurcation diagrams for models describing instruments with different degrees of resonance inharmonicity. To clarify the basic underlying principles, it is useful to begin with the simplest possible case: a multimode resonator having only two quasi-harmonic resonances with frequencies f_{res1} and $f_{res2} \simeq 2f_{res1}$. The deviation from perfect harmonicity is described by the equation

$$f_{res2} = 2f_{res1}(1 + \epsilon), \tag{5.43}$$

where ϵ is the inharmonicity parameter. To simplify the theoretical investigations, the sounding mechanism is chosen to be an inward-striking reed operating in the stiffness-dominated low-frequency regime.

If the resonances are exactly harmonic ($\epsilon = 0$), this theoretical problem can be analysed analytically (Dalmont et al. 2000). A bifurcation diagram obtained from such an analysis for the case in which the impedance amplitude of the first resonance is slightly higher than the amplitude of the second resonance is shown in Fig. 5.21a. The bifurcation branch marked by circles and labelled 'standard' shows the playing regime with the lowest threshold pressure. This is an inverse Hopf bifurcation of the type illustrated in Fig. 5.13, which means that the start of a note with infinitesimal amplitude cannot be achieved with a rising mouth pressure. At the threshold p_{th1}, the oscillation begins with an RMS amplitude more than 30% of the minimum reed closing pressure p_M. Once this sound is initiated, the oscillation amplitude can be reduced by lowering the pressure below p_{th1}, until the sound cuts out again at p_{sc}. The curves marked 'octave' and 'inverted' are bifurcation branches describing other possible oscillation states of the model, which are discussed in Dalmont et al. (2000).

If the inharmonicity is not equal to zero, the problem can no longer be resolved analytically, and numerical methods must be used. Bifurcation diagrams have been calculated for the two-mode system with different degrees of inharmonicity using a continuation method (Sect. 5.4.1). For this purpose the elementary model for the two-mode system was reformulated using the 'real mode' approach (Gilbert et al. 2020). The bifurcation diagram calculated numerically for the case $\epsilon = 0$, shown on the upper graph in Fig. 5.21b, is qualitatively consistent with the analytically derived diagram in Fig. 5.21a. The continuation method gives information on the stability nature of the periodic oscillations which is not available from the analytical solution, showing that only the standard branch from its point of inflexion upwards represents a stable oscillatory state. On this branch the oscillation frequency f_{osc} is locked at the value f_{res1} for all values of p_m, as shown on the lower graph in Fig. 5.21b.

Figure 5.22a and b illustrate how the bifurcation diagram changes as the inharmonicity of the two resonances increases. Figure 5.22a shows the case for $\epsilon = 0.02$, corresponding to a widening of the octave relationship between the two resonances by 34 cents. The bifurcation diagram is quite similar to the diagram for $\epsilon = 0$ shown in Fig. 5.21b, but there are some changes which have implications for ease of playing. At the threshold $p_m = p_{th1}$, the green curve representing the fundamental regime has the appearance of a direct Hopf bifurcation, but just above

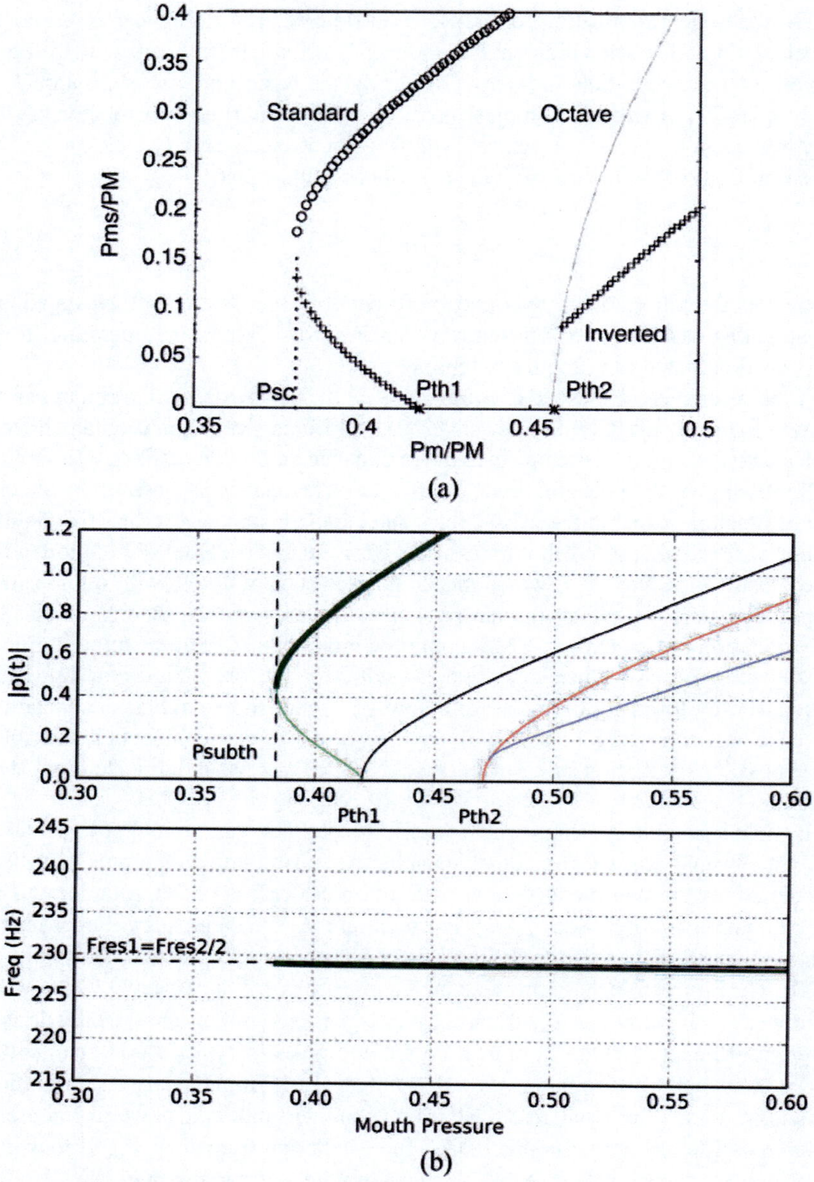

Fig. 5.21 Bifurcation diagrams corresponding to a model instrument with two exactly harmonic impedance peaks at f_{res1} and $f_{\mathrm{res2}} = 2f_{\mathrm{res1}}$, and with $Z_1 > Z_2$. All pressures are divided by the minimum reed closing pressure p_{M}. (**a**) Analytical calculation: acoustic pressure amplitude p_{rms} versus mouth pressure p_{m} (Dalmont et al. 2000). (**b**) Numerical calculation using continuation method. Green, standard regime; red, octave regime; blue, inverted Helmholtz regime (thick lines, stable states; thin lines, unstable states). Upper graph, amplitude $|p(t)|$ of oscillating pressure versus mouth pressure p_{m}. Lower graph: frequency of oscillation f_{osc} ($f_{\mathrm{osc}}/2$ for octave branch). Dashed horizontal line: reference frequency f_{res1}. For comparison, the black curve illustrates a direct Hopf bifurcation branched at $p_{\mathrm{m}} = p_{\mathrm{th1}}$, corresponding to an air column having only one resonance at the frequency f_{res1}. Adapted from Gilbert et al. (2020) (Color figure online)

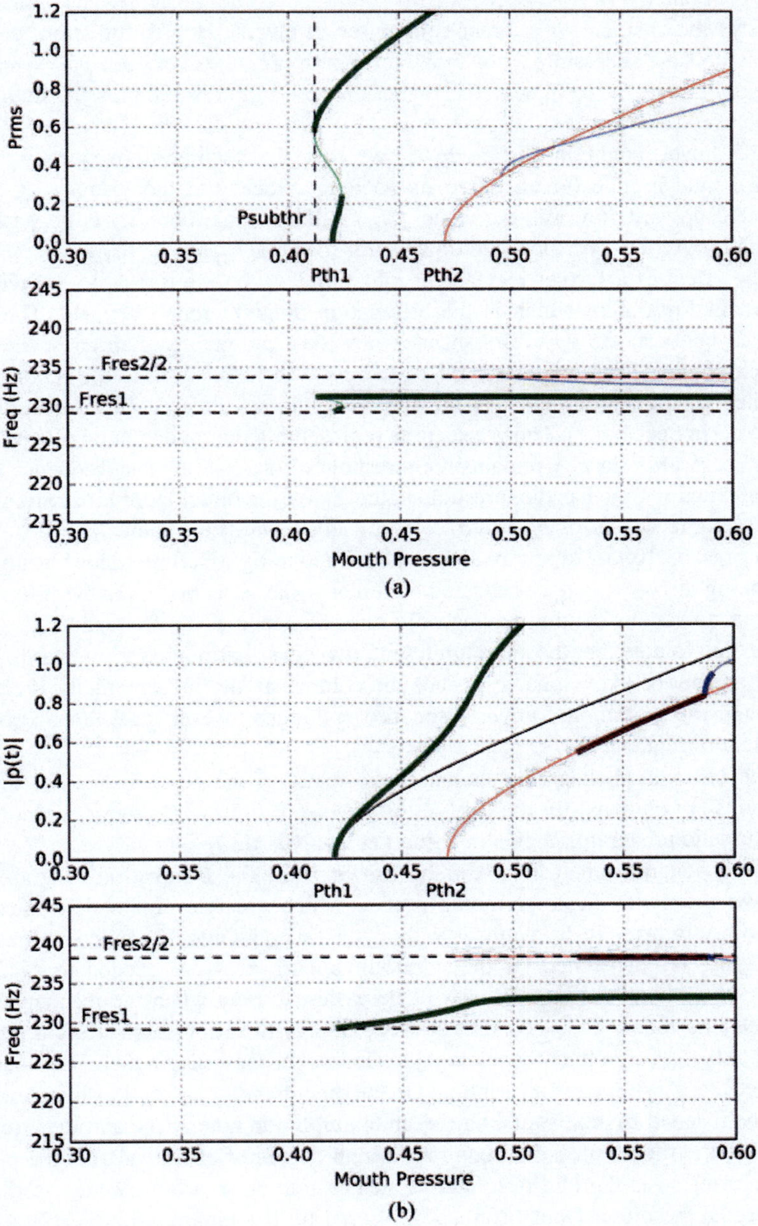

Fig. 5.22 Bifurcation diagrams for an air column with two quasi-harmonic resonances as a function of the control parameter p_{m}. All pressures are divided by the minimum reed closing pressure p_{M}. Green, standard regime; red, octave regime; blue, inverted Helmholtz regime (thick lines, stable states; thin lines, unstable states). Upper graphs: amplitude $|p(t)|$ of oscillating pressure. Lower graphs: frequency of oscillation f_{osc} ($f_{\mathrm{osc}}/2$ for octave branch). Dashed horizontal lines: reference frequencies f_{res1} and $f_{\mathrm{res2}}/2$. **(a)** $\epsilon = 0.02$. **(b)** $\epsilon = 0.04$. For comparison, the black curve illustrates a direct Hopf bifurcation corresponding to an air column having only one resonance at the frequency f_{res1}. Adapted from Gilbert et al. (2020) (Color figure online)

p_{th1}, the branch turns to the left, and the solution becomes unstable. The branch then displays the character of a branch just after an inverse Hopf bifurcation: periodic oscillations are accessible from above for mouth pressures between $p_{\text{m}} = p_{\text{th1}}$ and a lower limit $p_{\text{m}} = p_{\text{subth}}$ which is higher than the subthreshold pressure in the case $\epsilon = 0$.

The lower graph in Fig. 5.22a shows that the oscillation frequency of the fundamental regime (green curve) is no longer locked at the value f_{res1}, but is attracted upwards towards the value $f_{\text{res2}}/2$. If the inharmonicity is negative, the oscillation frequency is deviated downwards towards $f_{\text{res2}}/2 < f_{\text{res1}}$.

The effect of a further increase in inharmonicity on the threshold behaviour is shown in Fig. 5.22b, which is the bifurcation diagram for $\epsilon = 0.04$. The pitch interval between the two resonances is now 68 cents greater than an octave. The branch coming from this threshold $p_{\text{m}} = p_{\text{th1}}$, corresponding to the fundamental regime, has the appearance near threshold of a classic direct Hopf bifurcation (see Fig. 5.12 in Sect. 5.4.1). The green curve representing the fundamental regime never turns backwards, so p_{th1} remains the threshold of oscillation. The frequency of the fundamental regime has the threshold value f_{th1} at the direct Hopf bifurcation point and for higher values of p_{m} is attracted upwards towards the value $f_{\text{res2}}/2$.

In Figs. 5.21b, 5.22a and b a branch corresponding to a direct Hopf bifurcation beginning at $p_{\text{m}} = p_{\text{th1}}$ is indicated by a black line. This branch is the bifurcation diagram for an air column having only one resonance at the frequency f_{res1}. It is interesting to note that the air column with two quasi-harmonic resonances behaves more and more like a one resonance air column as the inharmonicity increases, although the inharmonic upper resonance continues to exert an influence on the oscillation frequency.

The bifurcation diagrams shown above relate to the case $Z_2 < Z_1$. Similar diagrams are obtained for the case $Z_2 > Z_1$ (see Fig. 10 of Dalmont et al. (2000), or bifurcation diagrams shown in Gilbert et al. (2020)).

Some general conclusions can be drawn from the bifurcation diagrams for the two resonance quasi-harmonic model, which are relevant to discussions of playability in brass instruments as well as reed woodwinds. If we accept that ease of playing is determined by the minimum mouth pressure needed to sustain a pianissimo sound, the instrument will be easiest to play when the resonances are perfectly harmonic ($\epsilon = 0$). The inverse Hopf bifurcation then offers a playing threshold p_{subth} significantly below p_{th1}. This can be viewed as a theoretical support for the Bouasse-Benade prescription. On the other hand, this very low threshold can only be reached by first sounding the finite amplitude note at p_{th} and then quickly reducing the mouth pressure, requiring a high degree of expertise from the player. If the criterion is the ability to start a steady note at an arbitrary low oscillation amplitude, the direct Hopf bifurcation offered by the inharmonic $\epsilon = 0.04$ case might be preferred.

Questions of intonation must also be considered. For $\epsilon = 0$ the oscillation frequency is independent of the mouth pressure and therefore the oscillation amplitude. The oscillation frequency is also largely independent of amplitude for $\epsilon = 0.02$, although this small degree of inharmonicity does result in an upward pressure jump just above the threshold. When the inharmonicity is further increased to $\epsilon = 0.04$, the advantage of a direct bifurcation comes at the cost of pitch instability, since the oscillation frequency increases with amplitude over a considerable range of mouth pressures. The balance of low minimum sounding mouth pressure and constancy of pitch with changing amplitude thus favours the perfectly harmonic case recommended by the Bouasse-Benade prescription.

The application of continuation methods to the case of a brass instrument, with a lip valve and many significant resonances, poses additional theoretical and technical challenges. A promising preliminary study by Fréour et al. (2020) calculated bifurcation diagrams for a number of different trumpets using the asymptotic numerical method (ANM) and the harmonic balance technique (Karkar et al. 2013). The calculation was based on the elementary model of brass playing, with the resonator represented as a sum of complex modes whose parameters were derived from analysis of the measured input impedance. Bifurcation diagrams for seven trumpets were calculated, and all show the classic behaviour of an inverse Hopf bifurcation (Fig. 5.13). The example in Fig. 5.23a illustrates a method for describing the shape of the bifurcation diagram using two parameters H and D. The hysteresis parameter H measures the range of mouth pressures below the onset threshold accessible to the player because of the inverse nature of the bifurcation. The dynamic range parameter D measures the increase in the amplitude of the mouthpiece acoustic pressure as the mouth pressure rises from its subthreshold value to a fixed value of 5 kPa. The diagram in Fig. 5.23b shows that the bifurcation behaviours of different trumpets can be differentiated by their positions in the (H, D) space.

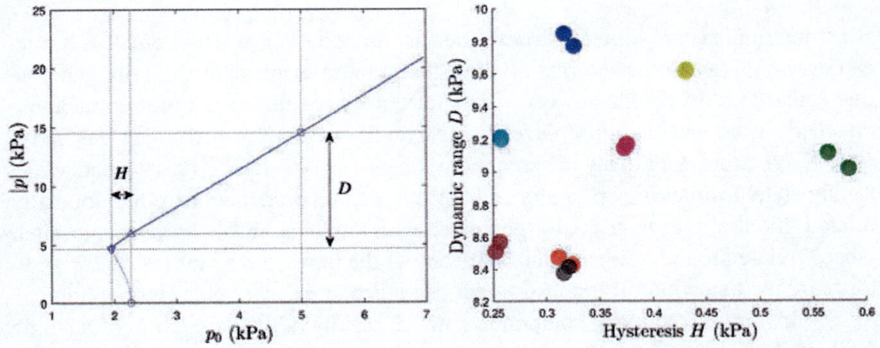

(a) (b)

Fig. 5.23 **(a)** Bifurcation diagram for the note B$^\flat$4 on a B$^\flat$ trumpet. Amplitude of the mouthpiece pressure p is plotted as a function of mouth pressure p_0. The dotted blue line indicates unstable portions of the branch, while the solid blue line indicates the stable branch. **(b)**: Categorisation of trumpets in the (H, D) space. The different colours correspond to the different trumpets. To each trumpet, two points are associated, corresponding to two impedance measurements of the instrument. From Fréour et al. (2020), reproduced with the permission of the Acoustical Society of America (Color figure online)

The bifurcation diagrams in Figs. 5.21 and 5.22 show that continuation methods are able to find branches of periodic solutions and also to detect bifurcation points from them. For example, the red line in Fig. 5.22a represents an octave regime with oscillation frequency equal to that of the higher of the two quasi-harmonic acoustic resonances. The blue line shows a period doubling bifurcation to an inverted Helmholtz motion state. When the inharmonicity between the two resonances is sufficiently large, quasi-periodic regimes can be observed (see a study of saxophones in Dalmont et al. (1995) and Doc et al. (2014)). Continuation methods are in principle able to predict the occurrence of quasi-periodic regimes by detecting a Neimark-Sacker bifurcation point localised on the branch of a periodic solution.

Nonlinear dynamical systems typically display a large variety of oscillation behaviours (see, e.g. Nayfeh (1995) and Manneville (2010)). Several different types of oscillation state which can occur in brass instruments are illustrated by simulations in Vergez and Rodet (2001b)). The 'route to chaos' through a succession of period doubling bifurcations (known as the Feigenbaum process) has been simulated in reed instruments using elementary models (Maganza et al. 1986). Measured non-periodic regimes in wind instruments have been extensively analysed using the phase space representation (Gibiat 1988; Gibiat and Castellengo 2000). Quasi-periodic regimes are of great interest to musicians, since they correspond to the special sounds known as multiphonics. In Sect. 5.4.6 some of the multiphonics available on the trombone are described and simulated.

5.4.6 Multiphonics

Brass instruments are sometimes described as 'monodic' or 'monophonic'. Is it true, as the etymology of these terms implies, that a brass instrument can only generate one sound at a time? The answer to this question depends on the precise meaning attached to the word 'sound'. Every note played on a wind instrument has a frequency spectrum with many different components, but in standard performance, the steady-state oscillation is periodic and the frequency components are harmonically related. Such a note is normally perceived as a single sound, with a unique pitch closely related to the fundamental frequency of the harmonic series (Sect. 2.2.3). In this case the behaviour of the instrument can indeed be described as monophonic.

The fusion of frequency components which results in the perception of a single pitch is ineffective if the components are not perceived as members of the same harmonic series. Two or more pitches may then be heard. Although the result is still strictly speaking one sound, the impression is of several separate sound sources acting simultaneously, and such sounds are described as multiphonics.

Many wind instruments are capable of generating multiphonics (Castellengo 1981). Two different categories of multiphonic can be distinguished, based on the acoustic phenomena involved in their production. One category involves the generation of multiple simultaneous pitches through an extension of conventional playing techniques. Woodwind multiphonics come into this category, since they rely on quasi-periodic regimes corresponding to specific choices of embouchure and/or fingering (Gibiat and Castellengo 2000). Sounds of this type on brass instruments are described as ‘lip multiphonics’. Brass instrument players can also produce multiphonic sounds by singing while simultaneously buzzing the lips (Campbell and Greated 1987; Sluchin 1995). Sounds in this second category are described as ‘sung multiphonics’. Didgeridoo players make extensive use of sung multiphonics in their distinctive playing style (Wolfe and Smith 2008).

An example of a lip multiphonic played on a tenor trombone is illustrated by the spectrogram in Fig. 5.24. The trombonist first sounded the fourth natural note of the instrument, $B^{\flat}3$, in the normal way: the spectrogram of this note shows the expected set of harmonics of 233 Hz, the nominal fundamental frequency of $B^{\flat}3$. The embouchure was then subtly adjusted and the note resounded. A pitch close to $B^{\flat}3$ is still evident, and the spectrogram still shows strong bands at the harmonic frequencies of that note. However, many additional frequency components which are not integer multiples of 233 Hz are now also present. Sidebands accompanying the original spectral components at multiples of $\pm$48 Hz are evidence of a modulation of the signal at 48 Hz. The pitch corresponding to this frequency, which is close to $G^{\flat}1$, was discernible in the sound.

A quasi-periodic oscillation is characterised by the simultaneous existence of two or more periodicities whose ratios are irrational. In the example shown in Fig. 5.24, the ratio of the two periodicities is close to 5. If an exact integer relationship is achieved by the player, the oscillation becomes strictly periodic, and the two-component harmonic series coalesce into a single set of harmonics. The strength and stability of the oscillation regime is then enhanced, in accord with the Bouasse-Benade prescription. The multiphonic character of the sound is preserved if the

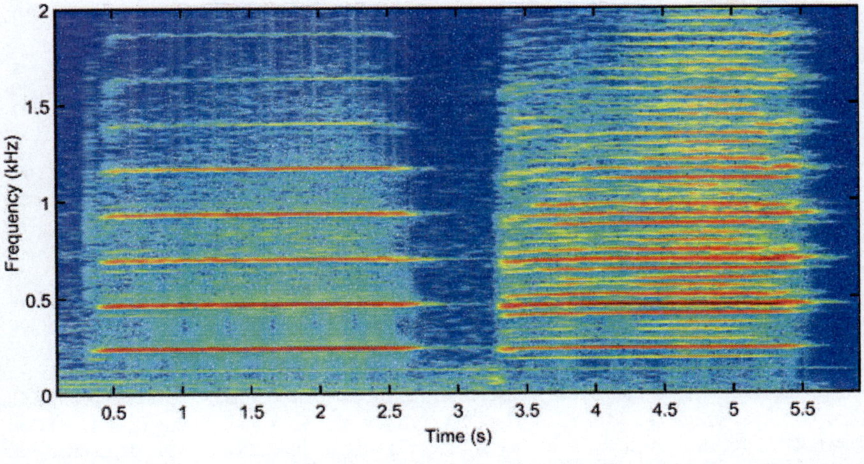

Fig. 5.24 Spectrogram of a conventionally played trombone $B^{\flat}3$, followed by a multiphonic which includes this pitch. From Campbell (2014a)

harmonics corresponding to the higher pitch are sufficiently strong to maintain the perception of a distinct sound at that pitch.

The choice of pitches in lip multiphonics is restricted by the necessity of identifying suitable quasi-periodic states. This restriction does not apply to sung multiphonics, since in principle any sung pitch can be combined with any played note. Successful execution of a sung multiphonic clearly requires good control of simultaneous vocal fold and lip vibrations. Depending on the voice range of the performer, the pitch of the sung note can be chosen to be either above or below that of the lip vibration.

It is not obvious that our elementary model of brass playing is capable of describing the production of a sung multiphonic. A full scientific treatment would involve discussion of the mutual interactions of two pressure controlled valves (the vocal folds and the lips) and two multimode acoustical resonators (the player's windway and the instrument air column). To explore the applicability of the elementary model to sung multiphonics, a study including both experimental measurements on trombone players and numerical simulations was undertaken by Velut et al. (2016). One of the multiphonics examined, which requires the player to sound the pitch F3 using the lips while singing the pitch C4, is described as the 'F3-C4 multiphonic'. This is one of the most common multiphonics played by trombonists and is the first exercise presented in the manual on sung multiphonics written by Sluchin (1995). In physical terms, the player buzzes the lips at the frequency f_{buzz} of the third natural note of the trombone while simultaneously causing the vocal folds to vibrate at a frequency $f_{\mathrm{sing}} = 1.5 f_{\mathrm{buzz}}$.

In the experimental phase of the study, the musician sang the pitch C4, then played the note F3 on the trombone and finally performed an F3-C4 multiphonic. Acoustic pressures were measured in the player's mouth (p_{m}), the mouthpiece (p) and the external sound field (p_{ext}). Figure 5.25 shows the spectrograms

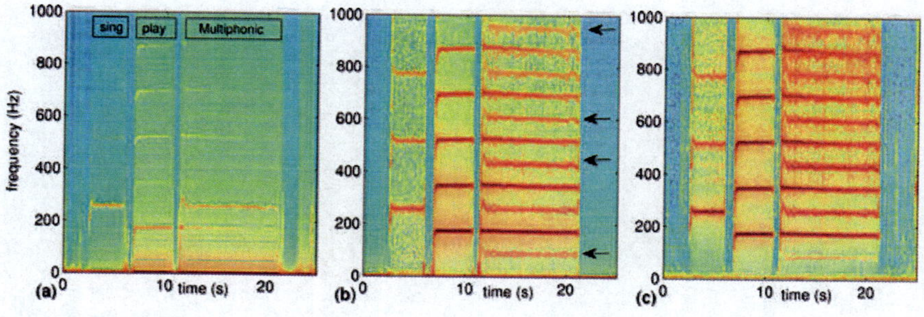

Fig. 5.25 Experiment: spectrograms of the pressures in (**a**) the mouth p_{m}, (**b**) the mouthpiece p and (**c**) the external sound field p_{ext}, measured in a trombone performance. The musician sings C4 ($T = 2.5 - 6.5$ s), plays F3 ($t = 7 - 11$ s) and performs an F3-C4 multiphonic ($t = 12 - 21$ s). Spectral components which do not belong to either the sung or the played note are designated with arrows in (**b**). Adapted from Velut et al. (2016) with the permission of the Acoustical Society of America

corresponding to these measurements, with the successive tasks – singing, playing, multiphonic – successively appearing on each spectrogram.

During the singing of C4, its fundamental frequency component at $f_{\text{sing}} = 259.8\,\text{Hz}$ appears in all three microphones, with its upper harmonics also evident in the mouthpiece and the external field. Similarly, $f_{\text{buzz}} = 173.6\,\text{Hz}$ and its upper harmonics appear in the mouthpiece and the external field while the musician is playing F3; a component at f_{buzz} can also be observed in the mouth spectrogram, because of the coupling with the vocal tract of the musician (see Sect. 6.3). When the multiphonic is played, p and p_{ext} contain two sets of harmonics, one with fundamental f_{sing} and the other with fundamental f_{buzz}. In addition, other frequency components appear which are not members of either harmonic series. These components are shown by arrows in Fig. 5.25b. Figure 5.26 superimposes the spectra of $p(t)$ during the three phases of the performance. This highlights that some peaks of the multiphonic spectrum clearly do not belong to the played signal or to the sung signal, with the lowest multiphonic component below both f_{sing} and f_{buzz}.

The study of the F3-C4 trombone multiphonic by Velut et al. (2016) compared the experimental results with simulations using physical modelling. For this purpose the measured input impedance of the trombone was represented as a sum of 13

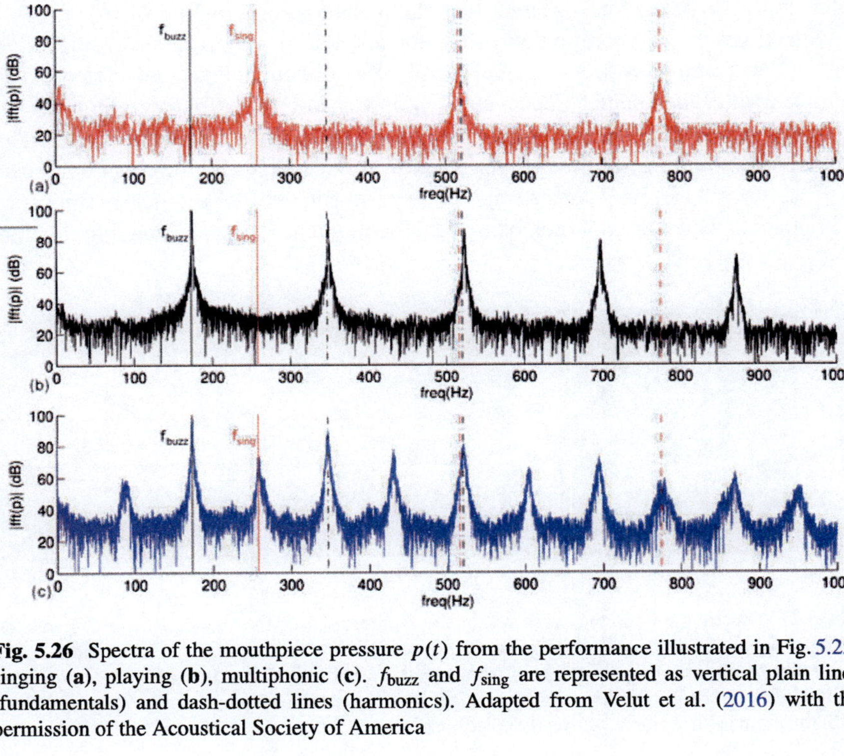

Fig. 5.26 Spectra of the mouthpiece pressure $p(t)$ from the performance illustrated in Fig. 5.25: singing (**a**), playing (**b**), multiphonic (**c**). f_{buzz} and f_{sing} are represented as vertical plain lines (fundamentals) and dash-dotted lines (harmonics). Adapted from Velut et al. (2016) with the permission of the Acoustical Society of America

acoustic modes, and the elementary model equations were solved in the time domain using the MoReeSC software package (Silva et al. 2014). The lip resonance frequency $f_l = 140\,\text{Hz}$ and other parameters of the model were chosen so that oscillation occurred at a frequency f_{buzz} corresponding to the third natural note of the trombone. In the simulation $f_{\text{buzz}} = 189\,\text{Hz}$, which is more than a semitone higher than the third natural note played in the experimental phase of the study; this discrepancy is a consequence of the simplified treatment of the lips as a 1DOF outward-striking valve (Campbell 2004; Silva et al. 2007).

To avoid the necessity of modelling the behaviour of the larynx and the player's windway, the effect of the modulation of the air flow by the vibration of the vocal folds was represented by adding an oscillating component to the mouth pressure:

$$p_{\text{m}}(t) = p_{\text{m}}^0 + p_{\text{m}}^1 \sin(2\pi f_{\text{sing}}\, t), \tag{5.44}$$

where p_{m}^0 is the constant mouth pressure assumed in the elementary model and p_{m}^1 is the amplitude of the pressure signal in the mouth during singing.

The simulated performance began with the lip oscillation, which was initiated by raising the mouth pressure p_{m}^0 to 4500 Pa, slightly above the threshold calculated by linear stability analysis (Velut et al. 2017a). After an interval of 3 s to allow the oscillating state to stabilise, the multiphonic was generated by adding the forcing sinusoidal component to the static mouth pressure, at a frequency $f_{\text{sing}} = 1.5 f_{\text{buzz}}$. The amplitude of the forcing sinusoidal component was set to 30% of p_{m}^0.

Spectrograms of the simulation results for p_{m}, p and p_{ext} are displayed in Fig. 5.27a, b and c, respectively. Before $t = 3$ s only components generated by the lip buzz with fundamental frequency f_{buzz} appear in Fig. 5.27b and c. Figure 5.27a does not display any spectral component at f_{buzz} because the elementary model does not include coupling with the player's windway. At $t = 3$ s the forcing component at $f_{\text{sing}} = 283\,\text{Hz}$ appears. As in the experiment, p and p_{ext} show frequency components which are not members of the harmonic series corresponding to either

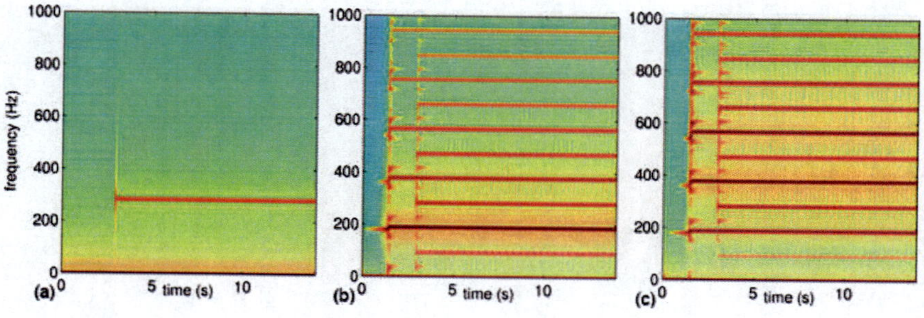

Fig. 5.27 Spectrograms of the pressures in the mouth p_{m} **(a)**, the mouthpiece p **(b)** and the external sound field p_{ext} **(c)**, during simulation of the multiphonic F3-C4. The mouth has a stationary component ($p_{\text{m}}^0 = 4500\,\text{Pa}$) from $t = 1.3$ s. After $t = 3$ s an oscillating component (amplitude $p_{\text{m}}^1 = 1350\,\text{Pa}$, frequency $f_{\text{sing}} = 283\,\text{Hz}$) is added. Adapted from Velut et al. (2016) with the permission of the Acoustical Society of America

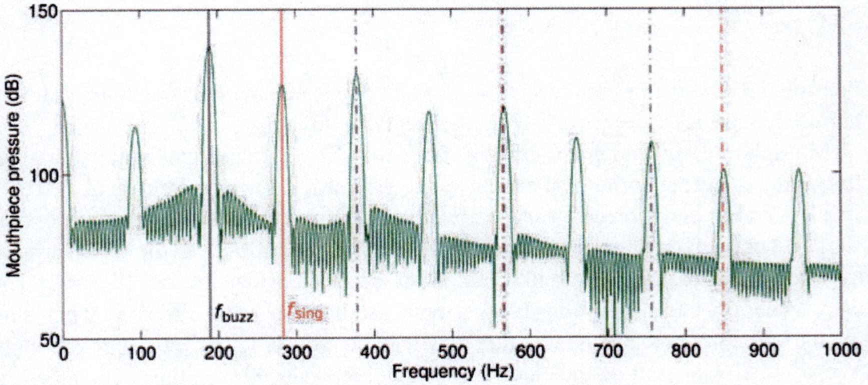

Fig. 5.28 Spectrum of the mouthpiece pressure p from the simulated multiphonic. $f_{\text{buzz}} = 189\,\text{Hz}$, $f_{\text{sing}} = 283\,\text{Hz}$ and their harmonics are shown. Adapted from Velut et al. (2016) with the permission of the Acoustical Society of America

the lip buzzed note nor the sung note. These additional frequency components can also be seen in the p spectra displayed in Fig. 5.28: some peaks of the multiphonic signal do not match the fundamentals f_{sing}, f_{buzz} (solid lines) or any of their upper harmonics (dashed lines).

The frequencies of the peaks appearing in the spectra of the multiphonics, both simulated (Figs. 5.27 and 5.28) or measured (Figs. 5.25 and 5.26), can be identified as either members of the two harmonic series

$$f_{\text{buzz}}^m = m f_{\text{buzz}}^1, \qquad f_{\text{sing}}^n = n f_{\text{sing}}^1, \qquad (m, n \text{ integers } > 0), \tag{5.45}$$

or sum and difference frequencies of members of the two series:

$$f_{\text{comb}}^{m,n} = f_{\text{buzz}}^m \pm f_{\text{sing}}^n \qquad (f_{\text{comb}}^{m,n} > 0). \tag{5.46}$$

Such linear combinations of harmonic components are the classic products of the nonlinear mixing of two periodic signals. In his discussion of the consequences of nonlinearity in the human hearing process, Helmholtz employed the useful term ‘combination tone’ to describe all such additional components (Campbell and Greated 1987).

A sung multiphonic corresponds to a quasi-periodic oscillation state if the ratio $f_{\text{sing}}/f_{\text{buzz}}$ is irrational. The F3-C4 multiphonic is one of the exceptional cases, described as an internal resonance, in which the ratio of sung frequency to buzzed frequency can be expressed as a rational fraction: the oscillation state is then strictly periodic. In general, if $f_{\text{sing}}/f_{\text{buzz}} = n/m$, with n and m integers and $n > m$, all of the components are members of one harmonic series with fundamental frequency

$$f_{\mathrm{p}} = \frac{f_{\mathrm{sing}} - f_{\mathrm{buzz}}}{n - m}. \tag{5.47}$$

For the F3-C4 multiphonic, $n/m = 3/2$. The fundamental frequency of the harmonic series is $F_{\mathrm{p}} = f_{\mathrm{sing}} - f_{\mathrm{buzz}}$, an octave below f_{buzz}.

Multiphonics in which the buzzed frequency is held constant while the sung frequency is swept continuously downwards were also studied by Velut et al. (2016). In all cases the mouthpiece pressure spectra were found to contain harmonics of the sung and played frequencies, together with combination tones from the set described by Eq. 5.46. Measurements and simulations confirmed that the oscillation states were generally quasi-periodic, but periodic oscillations were observed when the pitch of the sung note reached a value at which the ratio $f_{\mathrm{sing}}/f_{\mathrm{buzz}}$ became rational. The simulations based on the elementary model reproduced well the frequencies of the spectral components in the multiphonics, but were less successful in predicting the relative amplitudes of the components. The amplitude of the sinusoidal forcing pressure p_{m}^{1} used in the simulation was much larger than the mouth pressure signal measured in the experiments, reflecting the limitation of the simplified model used to replicate the effect of the sung note. Nevertheless, the results of the study by Velut et al. (2016) demonstrate that the basic physics of sung multiphonics can be captured using the elementary model of brass playing described in this chapter.

Chapter 6
Shocks and Surprises: Refining the Elementary Model

In Chap. 5, an elementary model of a brass instrument coupled with a player was developed and analysed. This elementary model does not attempt to include all the complexities of human brass performance, and the three equations on which the model is based rely on several simplifying approximations. It is assumed that the propagation of sound waves in the air column of the instrument can be described using linear acoustics, that the resonances of the player's windway can be neglected, that the player's lips can be modelled as a one degree of freedom oscillator, and that the walls of the instrument are completely rigid. These rather drastic approximations are in accord with the philosophy of modelling explained in Sect. 2.2.1, which seeks as a first step to find the simplest possible physical model which captures at least some of the behaviour of the system under study. We saw in Chap. 5 that the elementary model is indeed capable of reproducing and explaining many of the important features of brass instrument performance. The next stage in the modelling approach is to reconsider the assumptions on which the elementary model is based, with the aim of developing a refined model which incorporates some of the previously neglected phenomena. In this chapter each of the four assumptions listed above is revisited and discussed.

6.1 Why Brass Instruments Sound Brassy

In Sect. 6.1 it is explained that the linear approximation to the theory of sound propagation is no longer adequate when describing a brass instrument played at a level above mezzoforte. The distortion of the pressure waveform due to nonlinear propagation is shown to be the main cause of the 'brassy' sound characteristic of loudly played trumpets and trombones. An important result is the possibility of deducing a parameter (the brassiness potential parameter B) which can be used

M. Campbell et al., *The Science of Brass Instruments*, Modern Acoustics and Signal Processing, https://doi.org/10.1007/978-3-030-55686-0_6

to classify different brass instruments (see Chap. 7). A more detailed explanation of nonlinear propagation theory is presented in the Going Further Sect. 6.2.

6.1.1 Brassy Sounds in Music

The tone colour of a brass instrument note is strongly related to the steady state frequency spectrum, although the starting transient and other time-dependent factors are also significant timbral features. One of the most characteristic aspects of brass instrument timbre is the way in which the spectrum is increasingly enriched by the growth in amplitude of higher harmonics as the loudness increases. Figure 6.1 shows the spectrogram of the note F4 played on a tenor trombone. The note starts at a piano dynamic level, with only two or three of the lowest harmonics appearing as horizontal lines in the spectrogram. As the dynamic level rises, higher harmonics become increasingly evident, until at the fortissimo level 28 harmonics are shown. The upper limit of the spectrogram is 10 kHz: other measurements of fortissimo trombone playing have demonstrated the presence of harmonics up to 40 kHz, well beyond the upper limit of human hearing.

The dramatic growth in spectral enrichment as the loudness approaches the fortissimo level is perceived musically as a categorical change of tone colour, from the fairly smooth and rounded sound of normal playing to a sound which is described as 'hard', 'blaring' or 'brassy' (French 'cuivré', German 'schmetternd'). The rate of spectral enrichment depends on the type of instrument being played:

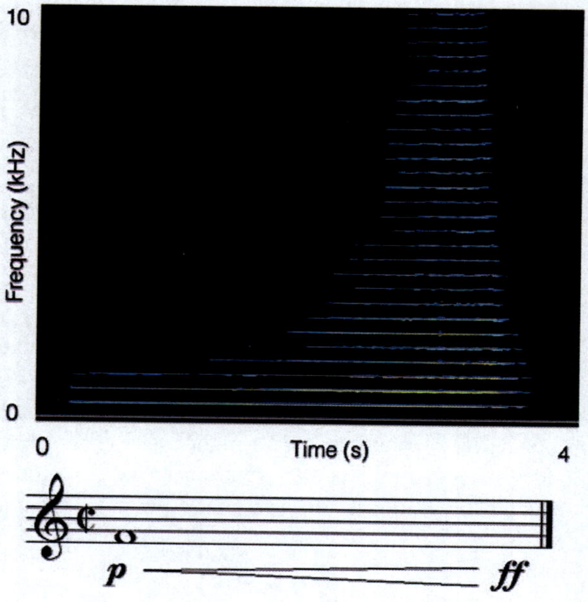

Fig. 6.1 Spectrogram of a crescendo played on a trombone

trumpets, trombones and french horns develop a brassy edge at much lower dynamic levels than do bugles, euphoniums and tubas.

The dramatic possibilities of brassy timbre have been imaginatively exploited by many composers, notably Gustav Mahler. Sometimes the player is specifically instructed to ensure that the sound is brassy: the muted trumpet call which interrupts the famous flugelhorn solo in the third movement of Mahler's Third Symphony is marked *schnell und schmetterd wie ein Fanfare* ('fast and brassy like a fanfare'). It is not however necessary to specifically request brassiness when a trumpet is asked to play fortissimo, as in the example from Mahler's Fifth Symphony shown in Fig. 6.2a. Spectral enrichment at high dynamic level is unavoidable in an instrument with a conventional trumpet bore.

Some degree of brassiness can also make a significant contribution to timbre at more moderate dynamic levels. The solo for the second trombone in the second

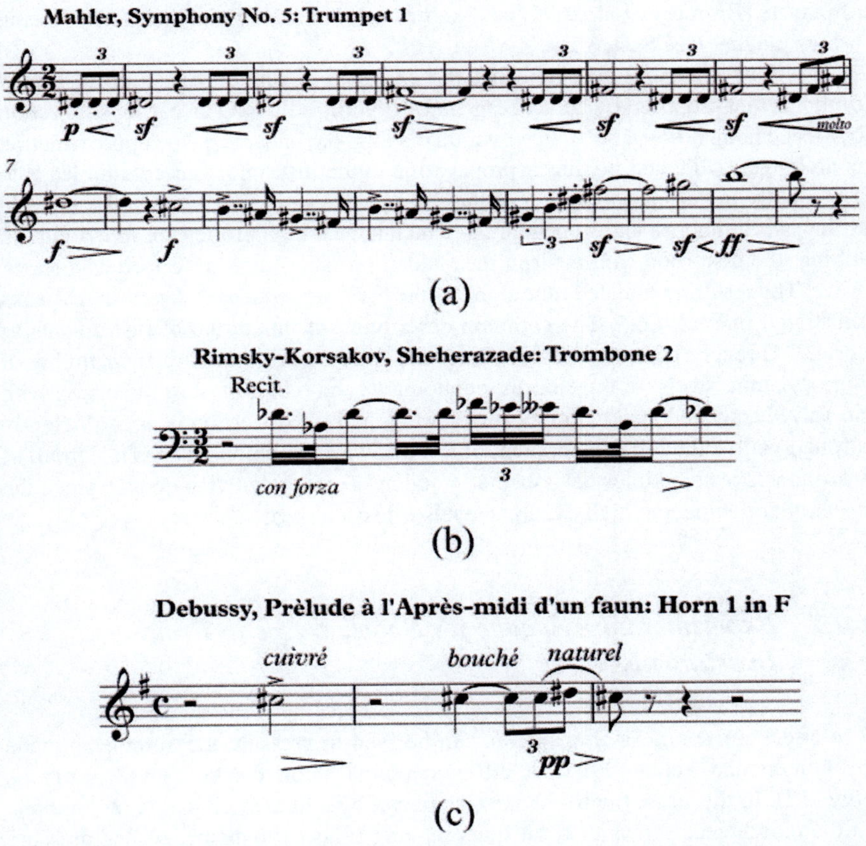

Fig. 6.2 (**a**) Excerpt from the first trumpet solo in Mahler's *Fifth Symphony*. (**b**) Excerpt from the second trombone solo in Rimsky-Korsakov's *Shéhérazade*. (**c**) Excerpt from the first horn part in Debussy's *Prélude à l'après-midi d'un faun*

movement of *Shéhérazade* by Rimsky-Korsakov, reproduced in Fig. 6.2b, does not have a specific dynamic marking, but the instruction *con forza* is a strong hint that a brassy sound is appropriate. A touch of brassiness can add salience to an otherwise fairly quiet accented note; in the excerpt from Claude Debussy's *Prélude à l'après-midi d'un faun* shown in Fig. 6.2c, the first horn is specifically asked to play such a note *cuivré*, before hand-stopping the instrument to play the following passage pianissimo (Del Mar 1983).

The French term 'cuivré', like the English 'brassy', suggests that the effect is due to forced vibrations in the metal bell. It is clear, however, from a variety of experimental and theoretical studies, that the amplitudes of bell vibrations are much too small to explain the phenomenon (see Sect. 6.6). It has also been suggested that the open area of the aperture between the player's lips, which has an approximately sinusoidal variation at low dynamic levels and medium to high pitches, might saturate for very loud playing, resulting in a clipped waveform and the generation of additional upper harmonics. Experiments involving high-speed filming of the lips of players (Bromage et al. 2006) have shown no evidence of such saturation during brassy playing (see Sect. 3.1).

It is now generally accepted that the explanation for the transition to brassy timbre is that proposed by Hirschberg et al. (1996b), following an earlier suggestion by Beauchamp (1969, 1980). In very loud playing, the linear acoustic approximation is no longer valid, and nonlinear propagation in the instrument air column leads to a gradually increasing distortion of the sound wave as it travels down the tube. In trumpets, trombones and french horns, which have a high proportion of cylindrical tubing, the distortion can result in the sudden pressure jump described as a shock wave. The resulting radiated time signal is the periodic sequence of very sharp peaks illustrated in Sect. 4.6.5. The common description of this group of instruments as 'bright' (French 'cuivres clairs') reflects the extreme brightness of their timbre at high dynamic levels. In flugelhorns, euphoniums and tubas, most of the tubing with no valves activated is either conical or flaring, and the distortion is not sufficiently strong to allow the full development of a shock wave within the tube. This group of instruments is sometimes described as 'mellow' (French 'cuivres doux'), since the spectral enrichment at high dynamic levels is less marked.

6.1.2 *Experimental Evidence for Shock Waves in Brass Instruments*

The effects of nonlinear propagation can be seen in pressure measurements made by Gilbert and Petiot (1997) at different points along the bore of a trombone (Fig. 6.3). In this case, the trombone was blown by a human player. A microphone in the mouthpiece shows that, for quiet playing (musical dynamic *p*), the pressure variation is almost sinusoidal. In contrast, for very loud playing (musical dynamic *ff*), the point in the vibration cycle at which the lips open is marked by a rapid

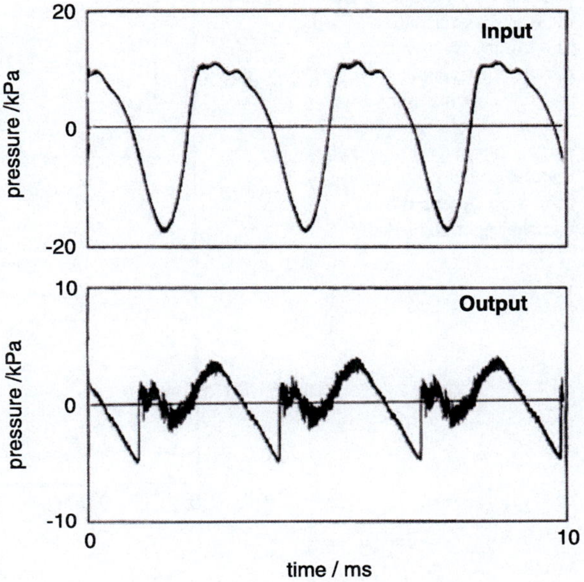

Fig. 6.3 Internal acoustic pressure in a trombone played *ff*. Upper curve, mouthpiece pressure; lower curve, pressure at the output of the slide section. Adapted from Gilbert and Petiot (1997)

rise in the pressure in the mouthpiece, which is a consequence of the nonlinear dynamics of the lip valve. At the downstream end of the cylindrical slide section, a second microphone shows that this rise has developed into the sudden pressure jump characteristic of nonlinear propagation in the air column (see Sect. 6.2).

Even more dramatic is the almost instantaneous rise and fall of the pressure signal measured by a microphone just outside the bell of the trombone (see Fig. 6.4). The resulting change in the density of the air as the wavefront passes is so large that it can be made visible using the schlieren optical method, as explained in Sect. 4.6.5. Figure 6.5 is a photograph of the spherical impulsive wavefront radiated from a trumpet bell. This is an acoustic wave, travelling at around 350 m/s; the turbulent mean air flow emerging from the bell with a speed of a few metres per second is also visible in this impressive photograph.

The physical processes which give rise to nonlinear distortion and brassiness are not related to the material of the instrument, but they are sensitive to the length and shape of the bore, the geometry of the mouthpiece and the embouchure of the player. A brass player can readily experiment with the effects of some of these parameters on brassiness by making a plastic hosepipe trumpet using a 3- m-long tube, a conical funnel and a trumpet or trombone mouthpiece (Fig. 6.6, left-hand side). Buzzing the lips loudly into the mouthpiece can generate an impressively brassy sound, since the tube is long enough to allow the shock waves to form. Playing the same loud note using a shorter 0.3 m tube (Fig. 6.6, right-hand side), it is impossible to get the same level of brassiness, since the effect is only spectacular when the tube is long (Gilbert et al. 2010). The increase of brassiness with length can also be demonstrated by playing the note F4 at the same dynamic level on a trombone with the slide in

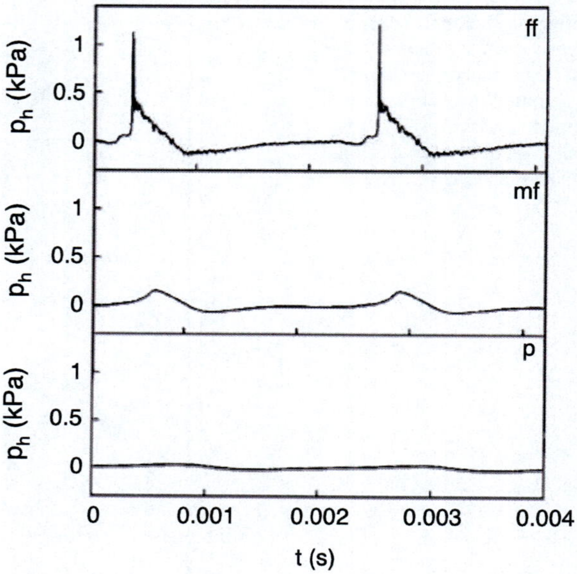

Fig. 6.4 Acoustic pressure measured just outside the bell of a trombone. Lower and middle curves, note played *pp* and *mf*, showing smoothly varying waveform; upper curve, note played *ff*, showing impulses due to shock wave formation. Adapted from Hirschberg et al. (1996b)

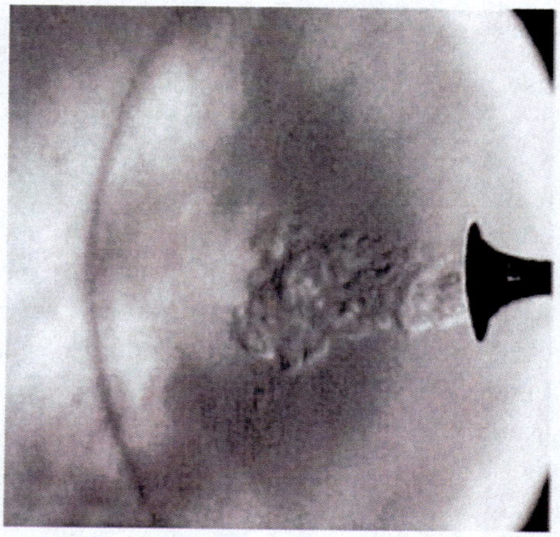

Fig. 6.5 Schlieren image of the region in front of a trumpet during playing of the note G5 fortissimo, showing impulsive wave and turbulent mean flow. Adapted from Pandya et al. (2003) with the permission of the Acoustical Society of America

first position and then in sixth position. The sound is more brassy in sixth position, with the slide almost fully extended, than in first position, when the instrument has its minimum length.

6.1.3 *To Infinity and Beyond: Nonlinear Propagation in Tubes*

In this section the theory of nonlinear sound propagation is summarised; Sect. 6.2 gives a fuller mathematical description of the theoretical background to brassiness.

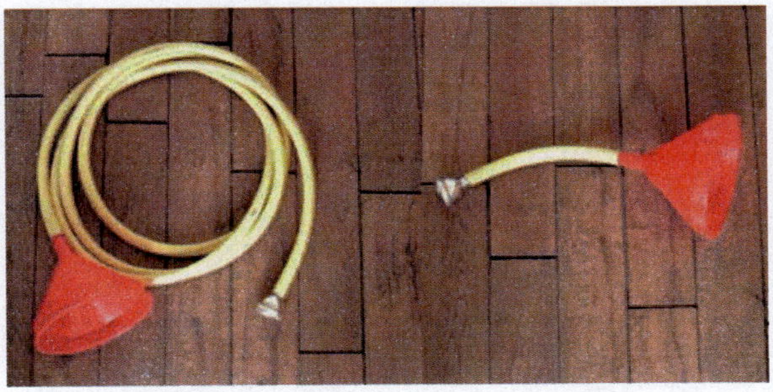

Fig. 6.6 Two plastic hosepipe trumpets. Left: tube length 3 m. Right: tube length 0.3 m

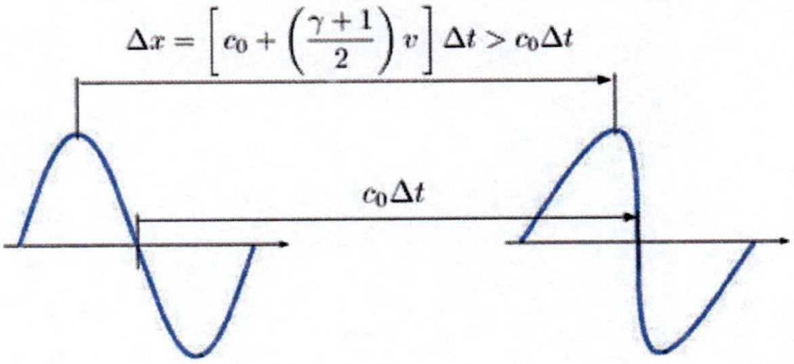

Fig. 6.7 Illustrating the effect of nonlinear propagation on an initially sinusoidal wave travelling from left to right. Each point on the wave moves with a characteristic speed dependent on the amplitude. Since its maximum moves faster than its minimum, the wave is distorted during propagation

The high acoustic pressure generated inside a loudly played brass instrument induces a local unsteady increase in speed of sound c in the compression part of the propagating pressure wave, corresponding to the adiabatic increase in temperature during compression. The speed of sound in air is proportional to the square root of the absolute temperature. Small perturbations therefore propagate with increased velocity c compared to the average speed of sound c_0. For outgoing acoustic waves, the increase in pressure also induces a particle velocity v in the direction of the pressure wave propagation. The expansion part of the wave will be slowed down by the same physical effects. As the expansion part of the wave is slow, it will tend to be overtaken by the fast compression part of the wave. This results in a gradual steepening of the wavefront (Fig. 6.7).

The equation for the speed of travel of a given point on the wave is

$$\frac{dx}{dt} = c_0 + \left(\frac{\gamma+1}{2}\right)v. \tag{6.1}$$

The derivation of this equation is fully explained in Sect. 6.2. The steepness of the wave is measured by the time rate of change of the pressure rise, which tends to infinity after a critical distance called L_s (shock length formation distance). At this point the disturbance is classed as a shock wave.

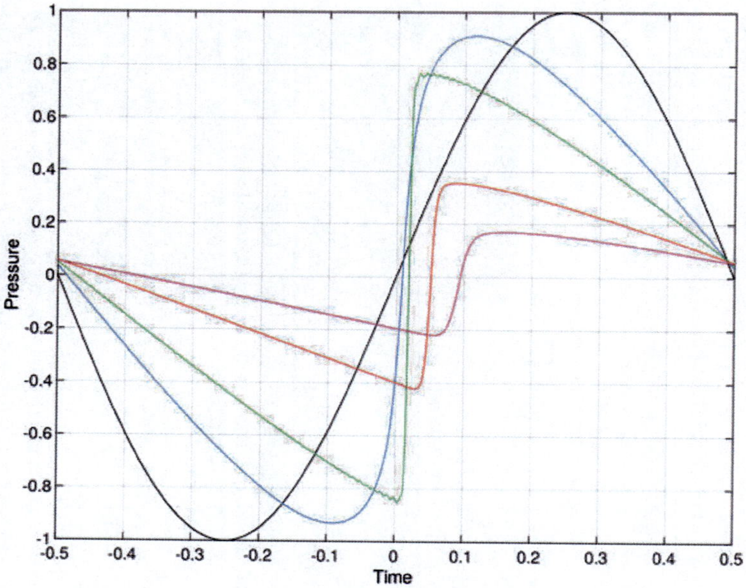

Fig. 6.8 Illustration of the distortion of an initial sinus as the function of the propagation distance

The gradual development of nonlinear distortion as a wave propagates is charted in Fig. 6.8, which is based on a calculation taking viscothermal losses into account. The black line represents the original undistorted sine wave. After travelling some distance, the partially distorted wave shape is illustrated by the blue line. In contrast to water waves at the seashore, where the peak of the surface water wave overtakes the trough and the wave breaks, the compression wave cannot actually overtake the expansion wave. The green line represents the wave after it has travelled the distance L_S: the time rate of change of pressure is theoretically infinite, marking the formation of a shock wave. Beyond 'infinity' the wave retains the basic 'N-wave' shape characteristic of a shock wave (Fig. 4.79), but the red and purple curves show that both the amplitude and steepness gradually decrease.

To avoid possible confusion, it should be noted that the shape of the partially distorted wave illustrated in Fig. 6.7 resembles an inverted capital N because the horizontal axis represents distance rather than time. As the wave travels past a fixed point in space, the almost vertical wavefront corresponds to a rapid rise in the pressure at that point, and a graph of pressure against time has the form of an upright capital N seen in Fig. 6.8.

The distortion in the time domain illustrated in Fig. 6.8 has a consequence in the frequency domain: energy is transferred from the fundamental component to upper harmonics. This process, sometimes called the harmonic cascade phenomenon, corresponds to the increase in perceived brightness of sound which we have called brassiness. When the shock wave reaches the pipe exit, the high frequencies in the wave are radiated, while the low frequencies are reflected back into the instrument. The waveform of the radiated pressure corresponds roughly to the time derivative of the pressure signal arriving at the pipe exit (see Fig. 6.4). The result is a periodic emission of sharp pressure peaks, corresponding to a spectrum with uniform harmonic amplitudes up to very high frequencies. This is the signature of brassy sounds.

The necessity for nonlinear treatment of propagation is related to the value of the mouthpiece pressure amplitude. For a sine wave (amplitude $\hat{p}$, frequency f), a shock wave is formed at a distance x from the mouthpiece which is larger than a critical distance L_s given for a cylindrical bore by

$$L_s \simeq \frac{2\gamma p_{atmos} c}{[(\gamma + 1)(2\pi f \hat{p}]}, \tag{6.2}$$

where γ is the ratio of specific heats of air, p_{atmos} the atmospheric pressure and c the speed of sound (Pierce 1989). The shock length is inversely proportional to both the mouthpiece pressure amplitude and the frequency. The larger the input pressure amplitude, the shorter the L_s is and the brassier the sound is. For a given amplitude, increasing the frequency reduces the shock length and makes the sound brassier.

The expression for the shock length given in Eq. 6.2 is valid only if the input wave is sinusoidal. Hirschberg et al. (1996b) pointed out that this is not the relevant parameter for judging the severity of the nonlinear wave steepening of an arbitrary input waveform. A more general expression for L_s is

$$L_s \simeq \frac{2\gamma p_0 c}{[(\gamma + 1)(\partial p_m / \partial t)_{max}]}. \tag{6.3}$$

The distance to shock formation is thus inversely proportional to the maximum rate of change of pressure in the mouthpiece, rather than to the pressure amplitude. For the pressure gradients measured in very loud trombone playing, L_s is of the same order as the length of the slide section of the tenor trombone, so shock wave formation is indeed to be expected.

Brassiness does not suddenly appear at a particular dynamic level but develops gradually once L_s is of the same order as the length of the bore. Some brass players have found that by employing slight changes in embouchure, they can exert a degree of control over the level of brassiness at a constant dynamic level. Experimental data from playing tests on a french horn, reported in Norman et al. (2010), suggest that this technique is based on the player's ability to modify the rate of change of the input pressure wavefront as it is formed in the mouthpiece, without significantly changing the amplitude of the mouthpiece pressure. This permits a modification of L_s and therefore the amount of distortion obtained during nonlinear propagation.

The player's control over the mouthpiece waveform is limited and results in a subtle change in the shock length. Its use as a musical technique is therefore restricted to moderate dynamic levels. At low dynamic level (*pp*), L_s is much larger than the sounding length L, and the sound is smooth and cannot be brassy. Between *pp* and *mf*, the brightness increases with dynamic level, but this is mainly due to the nonlinear behaviour of the lip valve (see Chap. 3). At high dynamic level (*ff*), L_s is shorter than L and the sound cannot be other than brassy. At intermediate dynamic level (*mf* or *f*), the brightness of the sound is mainly due to the effect of nonlinear propagation: even if the shock waves are not fully developed in the tube, the sound is already significantly affected by wave steepening. At these intermediate dynamics, the player can slightly control the sound brightness by controlling the shape of the input pressure waveform (Norman et al. 2010).

The shock length formation distance L_s is theoretically derived from nonlinear propagation in cylindrical pipes. If the guide is nonuniform, as it is in all realistic brass instruments, nonlinear propagation is present, but its effect is reduced as the cross-sectional area of the tube increases. By quantifying the relationship between the bore profile of an instrument and its relative susceptibility to nonlinear propagation, it is possible to use this property in order to classify brass instruments (Myers et al. 2012). An approach along these lines is described in Sect. 6.1.4.

6.1.4 The Brassiness Potential Parameter

The relative importance of spectral enrichment due to nonlinear propagation in brass instruments with different bore profiles can be estimated using a brassiness potential parameter B which can be calculated from the bore dimensions (Gilbert et al. 2007). We consider an instrument with a typical axially symmetric bore profile similar to that illustrated in Fig. 6.9. The tube has a sounding length L with the mouthpiece removed and a diameter $D(x)$ which increases from a relatively narrow mouthpiece receiver at $x = 0$ to a much wider bell at $x = L$. A mouthpiece is not included in the calculation of B because different players often choose mouthpieces with different internal profiles for use on the same instrument and mouthpieces are much shorter than sounding lengths.

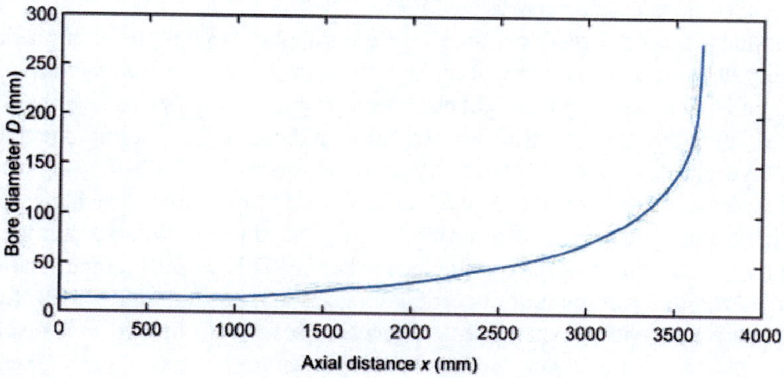

Fig. 6.9 The bore profile of a typical brass instrument (an Alexander contrabass tuba)

As the sound energy spreads across the increasing cross-section of this outwardly flaring horn, the particle velocity decreases. If we neglect losses within the horn and assume that the sound energy is spread uniformly across the tube, conservation of energy leads to

$$\frac{v(x)}{v_{\min}} = \frac{D_{\min}}{D(x)}, \tag{6.4}$$

where $D_{\min}$ and $v_{\min}$ are $D(x)$ and $v(x)$ at a reference point. For an entire brass instrument, the reference point is defined to be the point near the input end where the diameter of the instrument is minimum. This is not *at* the input end, which forms the mouthpiece receiver, but (usually) just beyond the point where the mouthpiece stem ends when the mouthpiece is inserted into the instrument.

According to Eq. 6.1, the rate of nonlinear distortion of the wave depends on the value of the particle velocity v. The maximum rate of distortion will occur at the point of minimum diameter, where the particle velocity has its largest value $v_{\min}$. Substituting the value of $v(x)$ derived from Eq. 6.4 in 6.1 gives an expression for the speed of travel of a given point on the wave at a distance x along the bore:

$$\frac{dx}{dt} = c_0 + \left(\frac{\gamma + 1}{2}\right)\left(\frac{D_{\min}}{D(x)}\right) v_{\min}. \tag{6.5}$$

To achieve the same amount of nonlinear distortion, the sound must travel farther in an outward-flaring horn than in a cylindrical tube, since the rate of nonlinear distortion diminishes as the diameter increases.

Figure 6.10 compares the development of nonlinear distortion in a cylinder, a relatively narrow trumpet bell and a more rapidly expanding flugelhorn bell. If a sine wave with the same particle velocity is injected at the left end of each duct, it

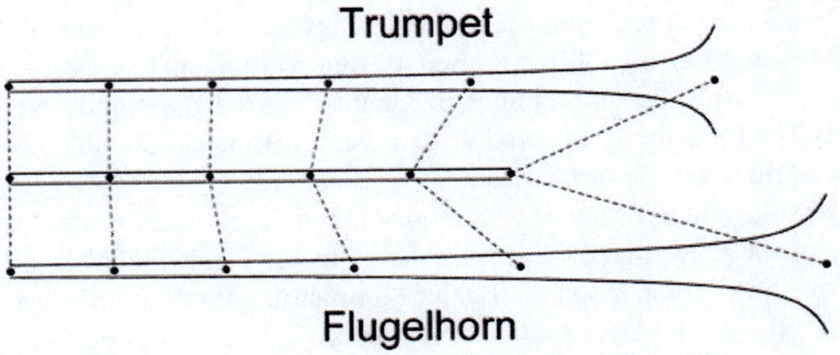

Fig. 6.10 Three ducts with the same total distortion. At the top is a Vincent Bach model 37 B♭ trumpet bell, in the centre a cylindrical tube and at the bottom the final portion of a Salvation Army St. Albans flugelhorn bell. The dashed lines connect intermediate points at which the distortion is the same in all three ducts. Reproduced from Myers et al. (2012) with the permission of the Acoustical Society of America

will have undergone the same degree of nonlinear distortion when it arrives at the right end.

The length of a cylindrical tube of diameter $D_{\min}$ which gives the same degree of nonlinear distortion as a flaring tube with the same minimum diameter, sounding length L and bore profile $D(x)$ is

$$L_{\mathrm{NDeq}} = \int_0^L \frac{D_{\min}}{D(x)} dx. \tag{6.6}$$

For brass instruments that flare outwards (as virtually all do), it is clear that the length of the equivalent cylinder will be less than the geometrical length of the instrument.

The brassiness potential parameter B is defined as the ratio L_{NDeq} to L_{ec}, the equivalent cone length of the instrument (Sect. 4.3.4). That is,

$$B = \frac{1}{L_{\mathrm{ec}}} \int_0^L \frac{D_{\min}}{D(x)} dx. \tag{6.7}$$

The brassiness potential parameter B is thus a dimensionless number that for normal brass instrument contours always lies between zero and unity. It is higher for 'cylindrical' instruments like the trumpet and trombone and lower for 'expanding' instruments like the flugelhorn and euphonium. For the same input signal, larger values of B mean greater nonlinear distortion and therefore a greater tendency for the timbre to acquire a brassy edge at louder dynamics.

For a perfect cylinder, $D(x) = D_{\min}$ and $B = L/L_{\mathrm{ec}}$. Because of the open end correction, the value of B for a cylinder is slightly less than 1. Brassiness potentials for most common brass instruments fall within the range 0.25–0.85.

The brassiness potential parameter defined by Eq. 6.7 can be used as a taxonomic tool for the classification of brass instruments. Scatter plots of brassiness parameter against minimum bore diameter, such as those shown in Fig. 6.11a, b, clearly separate brass instruments into identifiable subgroups. Once the areas which are occupied by instruments of recognised design within such a scatter plot are established, instruments of problematic identity can be measured and B values computed. Their positions in relation to other instruments should then give an indication of their timbral properties.

Equations 6.2 and 6.3 show that the shock length is inversely proportional to the frequency of the played note. The normalising factor L_{ec} is included in Eq. 6.7 to facilitate the direct comparison of B values for instruments with different sounding lengths (and therefore playing ranges). A modern B$^\flat$ trumpet is half the sounding length of a B$^\flat$ trombone, but the frequency of the fourth natural note of the trumpet is twice that of the trombone; if the two instruments have comparable bore shapes, they will have similar B values, and their equivalent natural notes will have comparable brassiness.

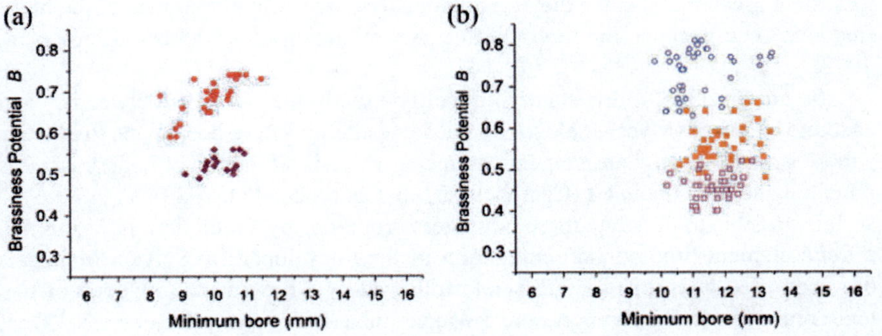

Fig. 6.11 Scatter plots of the brassiness potential parameter B computed from physical measurements, plotted against the minimum diameter $D_{\min}$. **(a)** Typical C and B$^\flat$ trumpets (red circles) and flugelhorns (magenta circles). **(b)** Typical tenor trombones (blue squares), baritone saxhorns and Tenorhorns (orange squares) and euphoniums (violet squares)

In practice, a continuous measurement of the bore is not necessary to calculate B. Good results can be obtained by measuring the bore at some 20 to 30 places over the length of the instrument and performing the sum

$$B \approx \sum_{1}^{N} \frac{l_n}{L_{ec}} \left(\frac{2D_{\min}}{D_n + D_{n+1}} \right), \tag{6.8}$$

where the sounding length is divided into N sections with arbitrary lengths l_n $(1 \leq n \leq N)$ and D_n is the bore diameter at the start of the nth section, $D_{\min}$ is the minimum bore close to the mouthpiece receiver, and D_{N+1} is the exit diameter at the bell.

It is evident from the plots shown in Fig. 6.11, and from similar scatter plots presented in Chap. 7, that 'bright' instruments are generally associated with high values of B, while 'mellow' instruments have lower B values. There is however a strong effect of absolute bore size on spectral enrichment in the far field for a given dynamic output of instruments with identical values of B. It is well-known by brass players that, for comparable bores, the narrowest tube will be the brassiest. Because the transfer function, which determines the radiated pressure amplitude for a given mouthpiece pressure, increases with increasing bore diameter, a radiated fortissimo on a narrow bore instrument will require a higher mouthpiece pressure than the same radiated dynamic on a wide bore instrument. The rate of nonlinear distortion will therefore be greater in the narrower tube, making the sound brassier at a prescribed high dynamic level (see Sect. 7.2.6).

There is, however, a competing process which tends to counteract the increase in brightness caused by nonlinear distortion. Viscothermal losses drain acoustic energy from a sound wave propagating in a tube; the decay constant which determines the magnitude of these losses is proportional to the square root of the frequency and inversely proportional to the tube radius (Eq. 4.110). In a narrower bore, the losses are greater, and since they predominantly reduce the amplitudes of the high-frequency components, the timbral effect is a reduction in brightness (Chick et al. 2012).

It is not possible at this stage to quantify exactly the trade-off between B and absolute bore size (Myers et al. 2012), but some attempts have been made to identify a more general spectral enrichment parameter which is a function of both brassiness potential and bore diameter (Campbell 2014b; Campbell et al. 2020)

It is possible to illustrate these brassiness properties by simulation. For example, a finite element time domain simulation method developed by Sylvain Maugeais has been used to compare a crescendo followed by a diminuendo on each of two tenor brass cousins, a trombone and a Wagner tuba (Maugeais and Gilbert 2017). It is evident from the spectrograms displayed in Fig. 6.12 that the spectral enrichment during the crescendo is much more important in the trombone simulation than in the Wagner tuba. Characteristic brassy effects are typical of bright brass instruments like trombones or trumpets but are less noticeable in the mellow brass instruments like tubas or flugelhorns.

In each of the cases shown in Fig. 6.12, the maximum pressure amplitude simulated in the mouthpiece was 8.5 kPa, corresponding to an rms value of 6 kPa. This is a typical level for very loud playing. The maximum pressures in the radiated sounds are however very different: 1993 Pa for the trombone but only 243 Pa for the Wagner tuba. The reason for this gross disparity is evident from Fig. 6.13, which shows the detail of the waveforms of the radiated sounds at the peak of the crescendo. The Wagner tuba waveform, which started as a sine wave at the mouthpiece, has clearly been significantly distorted, but the distortion in the trombone is much more severe, resulting in extremely short pulses with an amplitude of almost 2 kPa in the radiated sound.

In time domain simulations of the type described in the previous paragraph, the input pressure signal is prescribed, and the instrument is driven in forced oscillation. It is assumed that the nonlinear distortion which occurs in the air column of the instrument does not influence the input signal. Msallam et al. (2002) simulated self-sustained oscillations in a trombone, but the nonlinear propagation was taken into account only in the cylindrical section of the instrument. A simulation of vibrating lips coupled to a cylindrical tube by Berjamin et al. (2017) confirmed the importance of nonlinear propagation in modifying the timbre of the output signal but also noted significant influence on mouthpiece pressure, playing frequency and time envelope. These results point to the importance of developing simulations of the playing of realistic brass instruments which incorporate nonlinear propagation throughout the bore, including the flaring bell section.

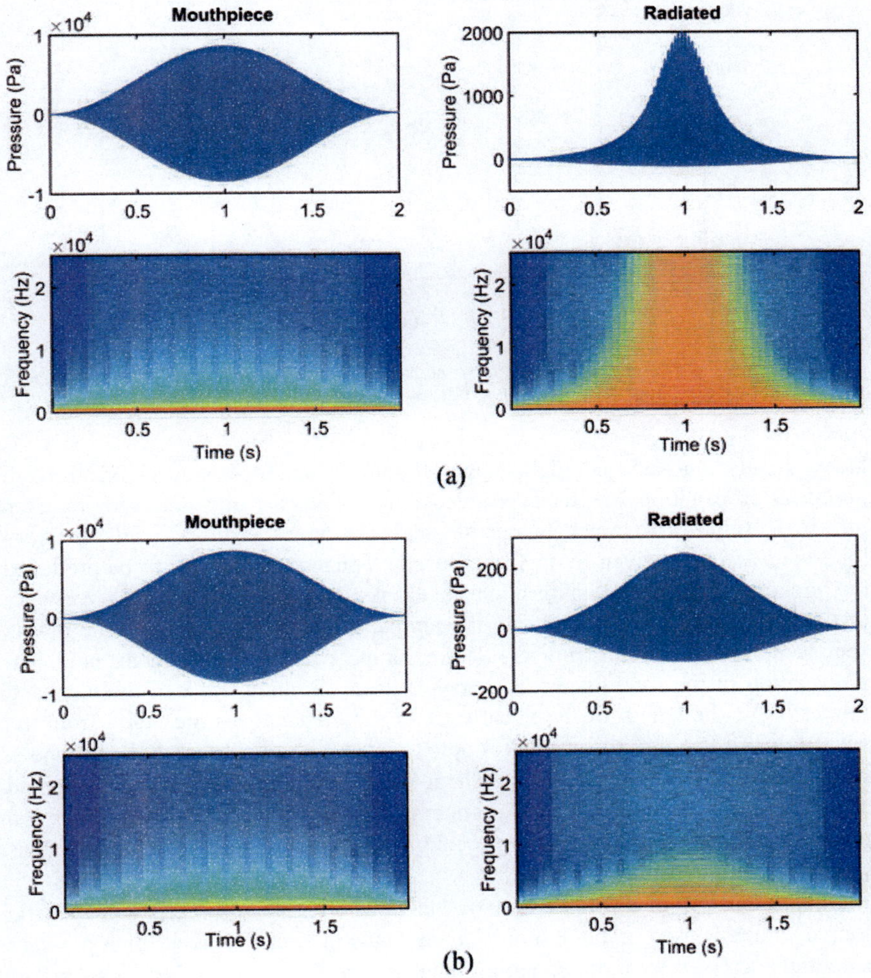

Fig. 6.12 Simulation of a crescendo and diminuendo for (**a**) a Hawkes trombone and (**b**) an Alexander Wagner tuba playing the note F4. For each instrument the mouthpiece (left) and the radiated (right) pressure signals are displayed as amplitude envelopes (above) and spectrograms (below). For simplicity reasons, the mouthpiece pressure has been assumed to be a sinus. The simulations include the nonlinear propagation effect. Courtesy Sylvain Maugeais

6.1.5 Elephants, Exhausts and Angels: Some Surprising Sources of Brassy Sounds

Voice production is very similar to brasswind sound production: the vibrating vocal folds are modulating a volume flow coming from the lungs into the vocal tract. Unfortunately, because of the short length of the vocal tract, it is not possible to get

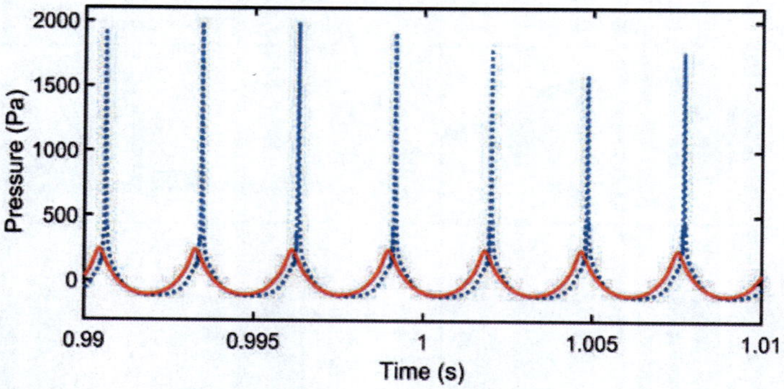

Fig. 6.13 Zoom to a 20 ms time segment of the radiated sound waveforms shown in Fig. 6.12. Blue dotted line: trombone. Red dotted line: Wagner tuba (Color figure online)

brassy shouts. Another mammal is however able to generate brassy sounds through nonlinear propagation. Elephants produce a broad diversity of sounds ranging from infrasonic rumbles to much higher-frequency trumpets. Trumpet calls are very loud voiced signals given by highly aroused elephants and appear to be produced by forceful expulsion of air through the trunk. Some trumpet calls have a very distinctive quality that is unique in the animal kingdom but resemble the brassy sounds of brass musical instruments. These trumpet calls are caused by the nonlinear propagation of the sound wave as it travels down the long trunk of the elephant. Interestingly, the length of the vocal tract of the elephant (as measured from the vocal folds to the end of the trunk) approximates the critical length L_s for shock wave formation (Gilbert et al. 2014). The relevance of nonlinear propagation should also be considered in other species producing high-intensity, high-pitched calls and possessing relatively unusually long vocal tracts, including birds with long tracheas such as geese and trumpeter swans.

The modulation of a volume flow by the valve effect is a very efficient acoustic source, which is used in the design of many kinds of siren. The very high pressure amplitudes created by these sounding devices can also generate brassy sounds if the source is connected to a long horn. Automotive exhaust pipes can also exhibit brassy sounds, and some motor bike exhausts use this effect to create a specific sound giving the impression of powerfulness.

Finally we must mention the instrument known as the tromba marina, or trumpet marine, which is able to produce strikingly brassy sounds. Despite its name, this is not a brass instrument but a member of the bowed string family. In its simplest form, a single string is mounted on a narrow trapezoidal soundboard; a fifteenth-century painting of an angel playing an instrument of this type is illustrated in Fig. 6.14a. The string is bowed close to the end nearest the player, and harmonics can be sounded by touching the string at different nodal points with the fingers of the left hand. In the seventeenth century, the construction of the tromba marina was improved to allow a wide range of harmonics to be sounded; an instrument of this type is shown in Fig. 6.14b.

(a) (b)

Fig. 6.14 (**a**) Angel playing a tromba marina. Detail from Hans Memling, *Christ Surrounded by Angel Musicians* (left panel of triptych), c.1480, Koninklijk Museum voor Schone Kunsten, Antwerp, photograph: Rik Klein Gotink. (**b**) Tromba marina played by Canon Francis Galpin

Many writers from the sixteenth century onwards have commented on the remarkable similarity between the sound of the natural trumpet and that of the tromba marina (Praetorius 1619). An important aspect of that similarity is the 'brassy' timbre which each instrument can generate. In the case of the tromba marina, the brassiness is not a consequence of nonlinear sound propagation, but the result of a deliberately induced nonlinearity in the mechanism which transmits the vibration from the string to the soundboard. One of the two feet of the asymmetric bridge is not pressed against the soundboard, but rests in its equilibrium state just above it. When the string is bowed, this foot raps once every cycle against the soundboard, adding high-amplitude impulsive components to the vibration transmitted by the other foot. This results in the dramatic increase in the spectral centroid of the radiated sound which we associate with the brassy timbre of a loudly blown trumpet (Gibiat and Padilla 2005).

6.2 Going Further: Nonlinear Propagation

In Sect. 6.1, it was noted that the acoustic pressure amplitude in the pipe of a loudly blown trombone can reach 10% of the ambient atmospheric pressure. In this case the small-amplitude approximation used to derive the linear acoustic wave equation is not valid, and it is necessary to return to the fundamental fluid dynamic equations.

6.2.1 From the Fundamental Fluid Dynamic Equations to the Nonlinear Wave Propagation Equation

A criterion for judging the relevance of the linear approximation is that the dimensionless acoustic Mach number $M = v'/c_0$ is much smaller than unity. Here v' is the acoustic velocity amplitude (the medium being assumed to be quiescent), and c_0 is the speed of propagation in the linear approximation. For $M \simeq 1$, we are in the strongly nonlinear regime, and the approach through linearisation of the fundamental equations fails (Hamilton and Blackstock 1998). Even for $M \ll 1$, however, it is possible to observe very pronounced nonlinear distortion effects. These effects are locally negligible: over distances small compared to a wavelength, it is still possible, for example, to make the assumption that in the case of a travelling plane wave, $p' = \rho c v'$. However, the nonlinear effects are cumulative over distances of the order of several wavelengths and can generate strongly distorted waveforms and even the pressure discontinuities called shock waves. This is the weakly nonlinear regime which we consider here.

We start our analysis by considering plane wave propagation in a tube with uniform diameter. The mass conservation and force equations in one dimension, assuming an ideal fluid without friction or heat conduction, have quadratic nonlinear terms:

$$\frac{\partial \rho}{\partial t} + \frac{\partial(\rho v)}{\partial x} = \frac{\partial \rho}{\partial t} + v\frac{\partial \rho}{\partial x} + \rho\frac{\partial v}{\partial x} = 0, \tag{6.9}$$

$$\rho\left(\frac{\partial v}{\partial t} + v\frac{\partial v}{\partial x}\right) = -\frac{\partial p}{\partial x}. \tag{6.10}$$

In the above equations and the derivations which follow, $p = p_0 + p'$, $\rho = \rho_0 + \rho'$, and $v = v_0 + v'$, the variables including both mean values (subscript 0) and fluctuating values (primed).

The two equations above yield the system of equations:

$$\frac{d\rho}{dp}\frac{\partial p}{\partial t} + \frac{d(\rho v)}{dp}\frac{\partial p}{\partial x} = 0, \tag{6.11}$$

$$\rho\frac{dv}{dp}\frac{\partial p}{\partial t} + \left(\rho v\frac{dv}{dp} + 1\right)\frac{\partial p}{\partial x} = 0. \tag{6.12}$$

These equations have non-trivial solutions for $\partial p/\partial t$ and $\partial p/\partial x$ if the determinant of the system of equations is zero:

$$\frac{d\rho}{dp}\left(\rho v\frac{dv}{dp}+1\right)-\frac{d(\rho v)}{dp}\rho\frac{dv}{dp}=0\longrightarrow\frac{d\rho}{dp}-\rho^2\left(\frac{dv}{dp}\right)^2=0. \tag{6.13}$$

Two solutions emerge, for waves travelling in the $\pm x$ direction, with

$$\frac{dv}{dp}=\pm\frac{1}{\rho}\left(\frac{d\rho}{dp}\right)^{1/2}. \tag{6.14}$$

We can use the conservation of entropy, since in the present discussion, friction and heat transfer have been neglected. At constant entropy,

$$\frac{dp}{d\rho}=c^2, \tag{6.15}$$

(see, e.g. (Pierce 1989, Sect. 1–5). Equation 6.14 can thus be rewritten as

$$\frac{dv}{dp}=\pm\frac{1}{\rho c}. \tag{6.16}$$

For the case of a forward travelling wave, described in the nonlinear acoustics literature as a simple wave, some further algebraic operations lead to the nonlinear partial differential equation (Earnshaw 1860):

$$\frac{\partial p}{\partial t}+(c+v)\frac{\partial p}{\partial x}=0. \tag{6.17}$$

This equation demonstrates the effect of convection (fluid entrainment). To a fixed observer, a given point on the wave (such as a crest) travels not at the local speed of sound c, but at a speed

$$\frac{dx}{dt}=c+v'; \tag{6.18}$$

the wave crests are therefore speeded up and the troughs slowed down.

The rate of nonlinear distortion is increased by variations in the local speed of sound c due to the variations in acoustic pressure. In a region of compression, the temperature rises, and c increases, while in a region of expansion, the temperature falls, and c decreases. An expression for c can be derived from the equation relating pressure and density in a perfect adiabatic fluid (Kinsler et al. 1999), which is itself nonlinear:

$$c^2=\frac{dp}{d\rho}=\frac{\gamma p}{\rho}=\gamma\frac{p_0}{\rho_0^\gamma}\rho^{\gamma-1}, \tag{6.19}$$

where γ is the ratio of heat capacities. For a calorically perfect gas, γ is constant (equal to 1.4 for air at room conditions), and

$$\frac{p}{p_0} = \left(\frac{\rho}{\rho_0}\right)^{\gamma}, \tag{6.20}$$

where (p_0, ρ_0) is a reference state. Expanding Eq. 6.19 as a power series in $\dfrac{\rho'}{\rho_0}$

$$c^2 = \gamma \frac{p_0}{\rho_0^{\gamma}} \rho^{\gamma-1} = \gamma \frac{p_o}{\rho_o}\left(1 + \frac{\rho'}{\rho_0}\right)^{\gamma-1} = \gamma \frac{p_o}{\rho_o}\left[1 + (\gamma - 1)\frac{\rho'}{\rho_0} + \cdots\right], \tag{6.21}$$

from which it follows that

$$c \simeq c_0 \left[1 + \left(\frac{\gamma - 1}{2}\right)\left(\frac{\rho'}{\rho_0}\right)\right], \tag{6.22}$$

with $c_0^2 = \gamma p_0/\rho_0$.

Recalling that $p'/p_0 = \gamma\rho'/\rho_0$, and that for a forward travelling wave $p' = +\rho_0 c_0 v'$, the local speed of sound c becomes

$$c \simeq c_0 \left[1 + \left(\frac{\gamma - 1)}{2}\right)\left(\frac{p'}{\gamma p_0}\right)\right] = c_0 \left[1 + \left(\frac{\gamma - 1}{2}\right)\left(\frac{\rho_0 c_0 v'}{\gamma p_0}\right)\right]$$
$$= c_0 + \left(\frac{\gamma - 1}{2}\right) v'. \tag{6.23}$$

Thus we arrive at the weakly nonlinear plane wave equation:

$$\frac{\partial p'}{\partial t} + [c + v']\frac{\partial p'}{\partial x} = \frac{\partial p'}{\partial t} + \left[c_0 + \left(\frac{\gamma - 1}{2}\right) v' + v'\right]\frac{\partial p'}{\partial x}$$
$$= \frac{\partial p'}{\partial t} + \left[c_0 + \left(\frac{\gamma + 1}{2}\right) v'\right]\frac{\partial p'}{\partial x} = 0. \tag{6.24}$$

The same equation applies to the acoustic velocity v'.

Figure 6.7 illustrates the distortion of the wave along the propagation axis in space, which occurs because of the dependence of the speed of propagation on the local acoustic velocity v'. Each point on the waveform travels with a given characteristic velocity

$$\frac{dx}{dt} = c_0 + \left(\frac{\gamma + 1}{2}\right) v' \tag{6.25}$$

which depends on the amplitude. A point of maximum pressure, corresponding to a crest on the wave, moves faster than a point of minimum pressure, corresponding to a trough on the wave. Portions of the waveform for which $dp/dx < 0$ (where pressure is increasing with time) become steeper with increasing time and propagation distance. At a particular point, called the shock formation point (or after a distance called the shock formation distance) , the derivative of the pressure becomes infinite, and the wave has a sawtooth profile.

The nonlinear travelling wave equation obtained can be solved accurately by the method of Riemann invariants, also called the method of characteristics. In Sect. 6.1.3 it was noted that for a periodic but not necessarily sinusoidal input pressure p_m, the shock formation distance is

$$L_s \simeq \frac{2\gamma p_0 c}{[(\gamma + 1)(\partial p_m/\partial t)_{max}]}. \tag{6.3}$$

For times later than the onset time of the shock, the plot of p versus x derived from the method of Riemann invariants becomes multivalued, which cannot happen in reality. In this model without friction or heat conduction, the dilemma is resolved by using the so-called equal-area rule (see, e.g. Pierce (1989)). The wave profile retains a sawtooth profile as the propagation increases, its amplitude decreasing towards 0.

6.2.2 The Burgers Equations

The model outlined in Sect. 6.2.1, which ignores friction and heat transfer, predicts the gradual distortion of the wave and the formation of the sawtooth profile described as an N-wave. After the shock formation point, the model predicts a continuous decrease in the amplitude of the propagating shock, despite the fact that no dissipative mechanism is taken into account (Pierce 1989). Including viscothermal losses helps in obtaining a model which is closer to reality. The goal here is to estimate a priori the order of magnitude of competing phenomena: nonlinear effects and viscothermal losses (Menguy and Gilbert 2000). If losses dominate they may damp the signal before it has time to distort significantly. In this case, the context of linear acoustics is sufficient to model the phenomena. Strong nonlinear phenomena can lead to the formation of a shock wave, but beyond the shock point, the amplitude of the N-wave may decrease until dissipative effects dominate over nonlinear distortion. Since the viscothermal losses at walls increase with frequency, the sawtooth profile is damped and deformed over time, tending ultimately to a sinusoidal signal of very low amplitude. During the formation of the shock wave, both the volume viscothermal losses (usually neglected in the pipe) and viscothermal losses in the boundary layer describe correctly the shock wave shape. In practice, the angles of the N-wave are rounded by the dissipation. It is the aim of the following paragraphs to explore this using the Burgers equations (Hamilton and Blackstock 1998).

The exact analytical solution of the equations of nonlinear acoustics including losses is not possible. An approximate way to tackle the problem is to use a perturbation method: the 'method of multiple scales' (see, e.g. Menguy and Gilbert (2000)). This method is based on the presence in the equation of a parameter small compared to unity. A suitable parameter in the present case is a Mach number defined in the vicinity of the acoustic source of the system. This number serves as a basis for the definition of 'fast' and 'slow' scales. The fast scale, which is a short spatial scale, is a dimensionless delayed time θ that describes the wave propagation locally. The slow scale is a long spatial scale, represented by the variable $\sigma = x/L_s$, that represents the cumulative effects of nonlinear distortion and losses.

After some mathematical operations which will not be detailed here, the weakly nonlinear lossless wave equation, applied to a function q which is one of the dimensionless acoustic parameters, takes the form

$$\frac{\partial q}{\partial \sigma} - q\frac{\partial q}{\partial \theta} = 0. \tag{6.26}$$

For a sine wave input, the above equation has an exact solution for $\sigma < 1$ given by Fubini (see, e.g. Pierce (1989)):

$$q(\sigma, \theta) = \sum_{n=1}^{\infty} Q_n(\sigma)\sin(n\theta), \tag{6.27}$$

with

$$Q_n(\sigma) = 2\frac{J_n(n\sigma)}{n\sigma}, \tag{6.28}$$

where J_n is the Bessel function of order n. The Fubini approximation is able to describe the harmonic cascade phenomenon which occurs before shock wave formation (see Sect. 6.1.3). After the shock formation, for a distance more than three times the shock length distance, the behaviour of the harmonic components corresponding to a N-wave is well represented by the Fay-Blackstock approximation (Hamilton and Blackstock 1998). The analytic Fourier coefficients Q_n are written as follows:

$$Q_n(\sigma) = 2A\frac{1}{\sinh(n(1+\sigma)A)}. \tag{6.29}$$

These two asymptotic behaviours are recovered by the solutions of the generalised Burgers equations discussed below (see Fig. 6.15).

The original 'Burgers equation' was developed by the Dutch physicist J.M. Burgers in his work on turbulence. Burgers included the effect of volume viscothermal losses on plane wave propagation by adding a term to the right-hand side of Eq. 6.26. Generalised Burgers equations are extensions of the basic Burgers equation which take additional physical phenomena into account. As explained in Sect. 4.1, for air in tubes, wall losses are more important than volumic losses. Wall losses display a square root frequency dependence, which is equivalent to a fractional derivative

$\partial^{1/2}$ in the time domain. A generalised Burgers equation including wall losses takes the form

$$\frac{\partial q}{\partial \sigma} - q\frac{\partial q}{\partial \theta} = C_1\frac{\partial^2 q}{\partial \theta^2} + C_2\frac{\partial^{1/2} q}{\partial \theta^{1/2}}, \tag{6.30}$$

where C_1 and C_2 are constants depending on the thermodynamic constants of the gas and the geometric characteristics of the system under study. The first term on the right-hand side of Eq. 6.30 corresponds to volume viscothermal phenomena and the second to viscothermal losses near the walls.

In practice, the volume losses term can be ignored everywhere except in the shock region when discussing wind instrument tubes. Volume losses are however essential when discussing nonlinear propagation in free space. It is possible to separate and measure the relative influence of the different effects using dimensional analysis (see, e.g. Menguy and Gilbert (2000)).

The most general plane wave solution involves two waves propagating in opposite directions. A major hypothesis of the weakly nonlinear approximation is that the two travelling waves behave independently, so that the wave in one direction does not influence the nonlinear distortion of the wave in the opposite direction. Even with this simplification, the problem is nonintegrable, and there is no general analytical solution (Blackstock 1985; Hamilton and Blackstock 1998). The generalised Burgers equation can however be solved numerically in the frequency domain as described in Menguy and Gilbert (2000) (see also Thompson and Strong (2001)). A spatial finite difference method is used, with the boundary condition at the source ($\sigma = 0$) being a time periodic function. The evolution in time is explored in the frequency domain by using a harmonic balance approach.

The results of the numerical method for the case in which wall losses are ignored ($C_2 = 0$) are shown by the curves labelled (a) in Fig. 6.15. The harmonic cascade is well described, and the behaviour after the shock formation distance is in agreement with the Fay-Blackstock approximation. The curves labelled (b) in Fig. 6.15 show the additional effects due to the inclusion of wall losses ($C_2 \neq 0$): harmonic cascade at the beginning, shock formation around $x/L_s = 1$ and decrease of the harmonic amplitudes illustrating the decrease in the amplitude of the sawtooth profile at large distances.

The frequency domain numerical model is a very good approximation in the case of a cylindrical bore, although subtle effects arising from the mutual interactions between forward and backward waves require further study (Harrison 2018). The generalised Burgers equation dedicated to weakly nonlinear wave propagation in uniform ducts is well-established (see, e.g. the theoretical demonstration and the comparison between experimental and numerical results in Menguy and Gilbert (2000)). However if we want to deal with brass instruments, we have to find an extension of the present work which deals with nonuniform ducts. This is the aim of the following section.

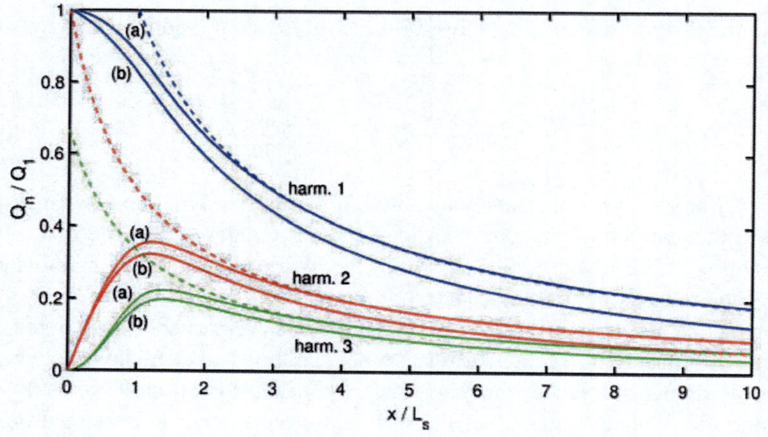

Fig. 6.15 Fourier coefficients Q_n for harmonics 1 (blue), 2 (red) and 3 (green) vs the dimensionless propagation distance $\sigma = x/L_s$ for a weakly dissipative fluid with $C_1 = 1/100$ in uniform duct. Two cases are illustrated: **(a)** $C_2 = 0$, and **(b)** $C_2 = 10$. The fluid is excited at $\sigma = 0$ by a sinusoidal source. There are two calculations: results simulated from the generalised Burgers equation (solid lines) and Fay-Blackstock approximation (dashed lines) (Color figure online). Adapted from Gilbert et al. (2008) with the permission of the Acoustical Society of America

6.2.3 *Brassiness of Flaring Bells*

Equation 6.24 describes a simple travelling wave in a uniform duct, assuming the lossless approximation. It is possible to extend the scope of this nonlinear equation to include the case of a duct with a varying bore diameter $D(x)$ by including an extra linear term which depends on $D(x)$. To do this, it is useful to rewrite Eq. 6.24, replacing the time variable t by the retarded time $\tau = t - \dfrac{x}{c_0}$:

$$\left[1+\left(\frac{\gamma+1}{2\rho_0\, c_0^2}\right) p'\right]\left(\frac{\partial p'}{\partial x}\right) - \left(\frac{\gamma+1}{2\rho_0\, c_0^3}\right) p' \left(\frac{\partial p'}{\partial \tau}\right) = 0. \tag{6.31}$$

A useful simplification can be made by noting that

$$1+\left(\frac{\gamma+1}{2\rho_0\, c_0^2}\right) p' = 1+\left(\frac{\gamma+1}{2c_0}\right)\left(\frac{p'}{\rho_0\, c_0}\right) = 1+\left(\frac{\gamma+1}{2}\right)\left(\frac{v'}{c_0}\right) \simeq 1,$$

since in the weakly nonlinear approximation adopted here the Mach number

$$M = \frac{v'}{c_0} \ll 1. \tag{6.32}$$

Equation 6.31 then becomes

$$\frac{\partial p'}{\partial x} - \frac{\gamma + 1}{2\rho_o c_o^3} p' \frac{\partial p'}{\partial \tau} = 0. \tag{6.33}$$

The flaring expansion of the tube is taken into account by adding to the left-hand side of Eq. 6.33 the additional term $p' \frac{1}{D}\frac{dD}{dx}$:

$$\frac{\partial p'}{\partial x} - \frac{\gamma + 1}{2\rho_o c_o^3} p' \frac{\partial p'}{\partial \tau} + p' \frac{1}{D}\frac{dD}{dx} = 0. \tag{6.34}$$

Generalised Burgers equations including this extra term have been used (Gilbert et al. 2008) to simulate the behaviour of weakly nonlinear wave propagation in nonuniform ducts in periodic regimes. The linear limit of these coupled equations does not lead exactly to the Webster-Lagrange equation, since the dispersive effect of the flare is not included in the approximation (Gilbert et al. 2007). This is equivalent to the assumption that the forward travelling wave is reflected only at the output plane, with the consequence that the model does not correctly reproduce the resonance frequencies of a flaring brass instrument (Harrison 2018). It does however capture the main effects of nonlinear propagation, since the major contribution to the cumulative distortion of the waveform takes place in the earlier portions of the bore where the rate of flare is very small. Comparisons between simulations and measurements are presented in Myers et al. (2012), Campbell et al. (2014b) and Maugeais and Gilbert (2017).

Using the following double change of variables:

$$z(x) = \int_0^x \frac{D_{\text{min}}}{D(x)} dx, \tag{6.35}$$

$$w(x, \tau) = D(x) p'(x, \tau), \tag{6.36}$$

Equation 6.34 can be rewritten:

$$\frac{\partial w}{\partial z} - \frac{\gamma + 1}{2\rho_o c_o^3} w \frac{w}{\partial \tau} = 0, \tag{6.37}$$

which has the same form as Eq. 6.33.

The foregoing discussion has reviewed the theoretical background to the definition of the brassiness potential parameter presented in Sect. 6.1.4:

$$B = \frac{1}{L_{\text{ec}}} \int_0^L \frac{D_{\text{min}}}{D(x)} dx, \tag{6.7}$$

where L is the sounding length of the instrument air column and L_{ec} the equivalent cone length. As explained in Part III, the utility of the brassiness potential parameter in characterising brass instruments is established, and the graphical presentation of results in a 2D space defined by bore size and brassiness potential is very fruitful to compare brass instruments, and for historical evolution, taxonomy and quality evaluation of brass instruments.

6.3 The Player's Windway

The act of producing a sound from a brass instrument is complex, requiring control of respiratory muscles and adjustment of the lip embouchure to get the lips to vibrate. The coupling between the buzzing lips (giving the valve effect by periodic modulation of the volume flow) and the downstream acoustic resonator (the instrument itself) is the main factor determining the playable notes. There is, however, another resonator in the story: the player's windway, between the lungs and the lips. The assumption that there are no significant acoustic resonances in the player's windway leads to the prediction that the acoustic pressure in the mouth is zero. The experimental measurements described in Sects. 2.1.2 and 2.1.3 show that in a realistic playing situation, the acoustic mouth pressure amplitude can be more than 10% of the acoustic pressure amplitude in the mouthpiece. In this section we explain how the elementary model can be extended to include the coupling of the lips to both upstream and downstream resonances and review the role played by modifications of windway resonances and respiratory activity in brass playing.

6.3.1 *Coupling of Upstream and Downstream Resonances*

The windway of a brass instrument player can be seen as an auxiliary resonator, upstream of the lips, in series with the downstream resonator formed by the tubing of the instrument (Elliott and Bowsher 1982; Benade 1983). The influence of this additional resonance can be included in the elementary model by adding a second feedback loop, as shown in Fig. 6.16.

Following Elliott and Bowsher (1982) and Hoekje (1986) in assuming the continuity of acoustic flow u from upstream to downstream leads to the following impedance relationship:

$$u(\omega) = \frac{p_d(\omega)}{Z_d(\omega)} = -\frac{p_u(\omega)}{Z_u(\omega)}, \tag{6.38}$$

where p_d and p_u are the acoustic pressures in the mouthpiece (downstream) and in the mouth (upstream), respectively, and Z_d and Z_u the input impedances of the instrument and the windway, respectively.

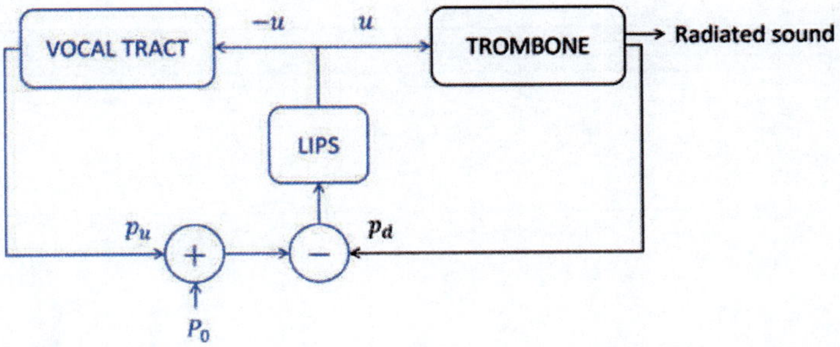

Fig. 6.16 Feedback loop representing the brass playing model with a secondary loop including the vocal tract interaction. Adapted from Fréour (2013)

This model is plausible and has been used in simulations of single reed woodwind instruments (Clinch et al. 1982; Sommerfeldt and Strong 1988). However, a drawback of the model is that it requires estimates of parameters which are not easy to measure (e.g. Z_u). Modifications of the vocal tract during trumpet playing have been studied by magnetic resonance imaging, and a model of the vocal tract derived from this information has been used in a time domain simulation (Kaburagi et al. 2011). Procedures have also been developed to estimate directly the upstream input impedance Z_u of a wind instrument (Fritz and Wolfe 2005) or to estimate the value of Z_u relative to the downstream input impedance Z_d at frequencies where acoustic energy is produced (Scavone et al. 2008; Guillemain et al. 2010). In both situations, the operation involves measurements of physical quantities in the mouth of the player and/or at the interface with the instrument.

The significance of player windway resonances in trombone playing was investigated by Fréour (2013). In part of this study, nine experienced trombonists were asked to play the ascending arpeggio shown in Fig. 6.17a. The acoustic pressures in the mouth (p_u) and in the mouthpiece (p_d) were measured during the performance, and in Fig. 6.17b the ratio p_u/p_d expressed in decibels is illustrated by a vertical bar for each player and each note of the arpeggio. These measurements show that the ratio of mouth to mouthpiece pressure is player dependent, but the systematic increase of the ratio with the pitch of the played note is clearly visible. For F3, the lowest note studied, the average value of the ratio p_u/p_d is approximately 0.1 (−20 dB). The fact that in the low range of the tessitura there is an order of magnitude difference between p_u and p_d provides some support for the approximation made in the elementary model that p_u is negligible compared with p_d. This assumption is clearly no longer valid for very high notes: the average value of p_u/p_d for the five players who were able to sound the note F5 is 1.6 (+4 dB). This result implies that in the extreme upper range of the trombone, the input impedance of the player's windway can equal or even exceed the input impedance of the instrument.

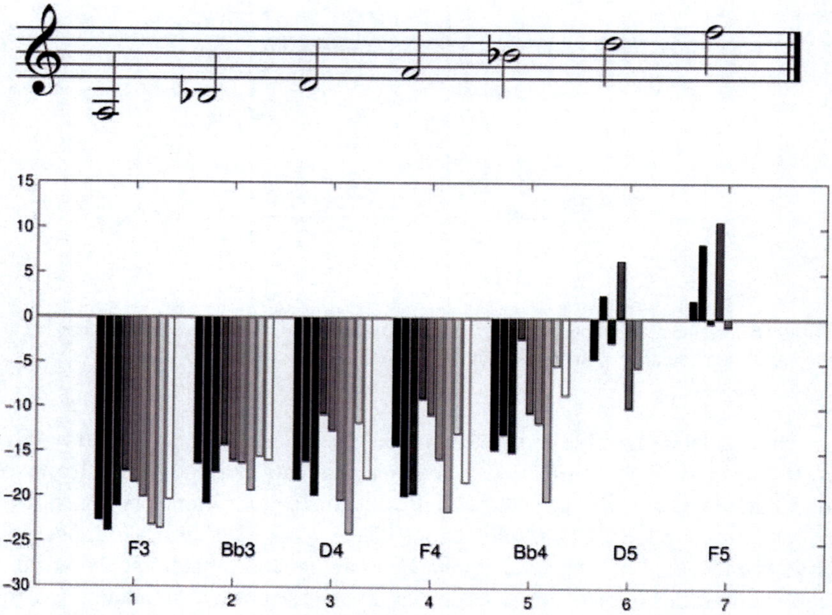

Fig. 6.17 (**a**) The trombone arpeggio F3-B^{b}3-D4-F4-B^{b}4-D5-F5 used as a test piece. (**b**) The ratio of mouth to mouthpiece pressure (expressed in dB as $20\log_{10}(p_u/p_d)$) at the fundamental frequency against the arpeggio index for nine trombone players. Adapted from Fréour and Scavone (2013) with the permission of the Acoustical Society of America

6.3.2 *Tuning of Windway Impedance Peaks*

The parallel between brass instruments and the human voice was introduced in Sect. 2.2.4 and further discussed in Sect. 3.4. An important difference between these two sound-generating systems is that in brass playing the lips are strongly coupled to the downstream acoustic resonator, while in singing the vocal folds are only weakly coupled to the downstream vocal tract. In voice production the resonator is mainly excited in forced oscillation, and its resonances exert a filtering effect on the radiated sound spectrum without significantly influencing the pitch. Nevertheless professional singers can tune their vocal tract resonances with precision, and there are considerable differences in the resonance strategies used by individual singers (Henrich et al. 2011).

The mechanisms of sound generation in voice production and brass playing can be observed and analysed using similar experimental setups and procedures (Hézard et al. 2014). The extent to which reed instrumentalists tune windway resonances to the frequencies of played notes has been extensively studied (see, e.g. Chen et al. 2008; Scavone et al. 2008; Guillemain et al. 2010; Wolfe et al. 2015). It has been concluded that windway resonances have only modest effects on the sounding pitches over much of the instrument's range. However, to play very high-pitched

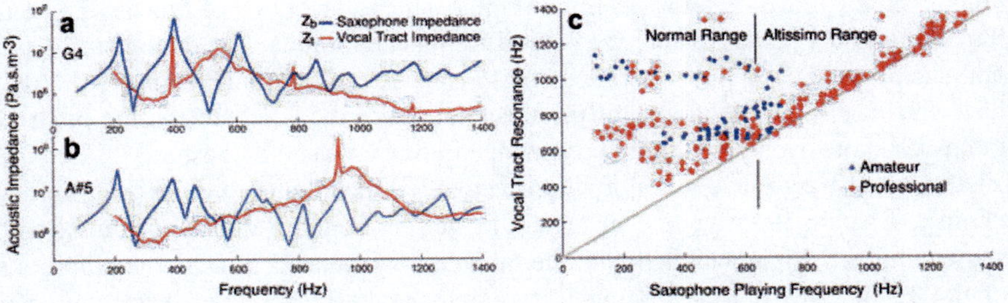

Fig. 6.18 Red curves: representative examples of acoustic impedance Z_t of the windway of a professional saxophone player (**a**) for the note G4 (near 400 Hz, normal range) and (**b**) for the note A$^\sharp$5 (near 940 Hz, altissimo range). The narrow peaks are harmonics of the notes played; the broad peaks are windway resonances. Blue curves: acoustic impedance Z_b of the saxophone with fingering for (**a**) G4 and (**b**) A$^\sharp$5. (**c**) Frequencies of the relevant resonance in the windway plotted against the frequency of the note played. The diagonal line is where the windway resonance frequency equals the sounding frequency (Color figure online). Adapted from Chen et al. (2008) with the permission of the Acoustical Society of America

notes in the altissimo register of the saxophone, experienced players learn to tune a resonance of the windway near to the frequency of note to be played (see Fig. 6.18). Similar effects are likely to be important in other single and double reed instruments, whose players also report the importance of windway modifications for special effects including high register playing.

Is it possible to see the same kind of coincidence between resonances of the windway and the frequency of the note played on a brass instrument? Measurements of the impedance of the windways of trumpeters during playing were reported by Chen et al. (2012); strong impedance peaks were found, but their frequencies were not tuned to match the frequency of the note being played. The authors comment, however, that the measurement technique used inhibited the normal mouth configuration used for very high note playing. A study of high register trombone playing by Fréour and Scavone (2013) provided some evidence suggesting that for the highest-pitched notes, the players tune the vocal tract resonance close to the played frequency. This conclusion was supported by experimental studies using an artificial mouth and active control methods and by simulations in which the magnitude and phase of the upstream pressure were varied to model changes in the windway impedance (Fréour et al. 2015).

6.3.3 Other Effects of the Player's Windway

Preliminary studies (Fréour and Scavone 2013) have suggested that skilled trombonists may employ continuous adjustment of the windway resonance as an aid to the clean execution of slurred note transitions. Similar windway resonance changes may also play a part in the technique known as 'lipping', in which the pitch of a note is pulled above or below its normal pitch centre. The basis of this technique is the ability of the player to alter the natural resonance frequency of the lips using the embouchure muscles, but experiments with an artificial mouth (Fréour et al. 2015) have shown that modifications of the upstream resonance can change the pitch of the played note even when the lip resonance frequency is held constant.

The vocal tract plays a major role in extended techniques such as sung multiphonics (Sluchin 1995; Velut et al. 2016). In the spectrograms of a sung multiphonic shown in Fig. 5.25, it is evident that the modulation of the air flow by the vibrations of the vocal folds during singing is not sinusoidal, since several harmonics are visible in the frequency spectrum of the mouthpiece pressure when the lips are not vibrating. The relative strengths of these components, and therefore the timbre of the multiphonic created when the lips also buzz, are influenced by the resonances of the vocal tract.

An even more spectacular illustration of the effect of windway changes on timbre is provided by the didgeridoo (Sect. 1.1.1). A single drone note, usually the lowest natural note of the tube, is sounded continuously; the musical interest comes from rhythmic variations in timbre produced by changes in the vocal tract configuration, including those required for the cyclic breathing technique that allows continuous playing. Similar techniques are sometimes used in avant garde brass performances.

The influence of the player's windway is much more effective in the didgeridoo than in most conventional brass instruments because the didgeridoo does not have a mouthpiece with a constricted throat: the lips of the player vibrate directly against the open end of the tube, typically with diameter 30–50 mm. The impedance peaks are not harmonically related because of the irregular bore of the tube, and their magnitudes are typically of the same order of magnitude as the input impedance peaks in the player's windway (Tarnopolsky et al. 2005, 2006). The upper resonances of the instrument are therefore very weakly coupled to the lips, and their effect is to filter the sound in a way analogous to the creation of formants in singing. Maxima in the input impedance of the player's windway corresponds to minima in the acoustic volume flow into the instrument and therefore results in antiformants (regions of low amplitude) in the frequency spectrum of the radiated sound.

In Chap. 3 we described how the ability of a player to buzz the lips without a brass instrument or mouthpiece could be explained as a 'flutter effect', occurring due to the coupling of two mechanical modes of the lips (Cullen et al. 2000). A similar mechanism is responsible for driving vocal fold vibrations in voiced sound production. An alternative interpretation of the buzzing lips phenomenon (Fletcher and Rossing 1998) considers that the player's mouth impedance is coupled to the lips by the air flow. From this point of view, the destabilisation occurs because of coupling between a mechanical mode of the lips and an acoustic mode of the vocal tract. It seems likely that both mechanisms play some role in the buzzing lips phenomenon.

Even where there is no deliberate use of the windway resonances in brass playing, we have seen that a coupling between upstream and downstream exists, particularly for high notes. It is possible that the brass player adapts the mouth geometry in order to control the embouchure by modifying the equivalent mechanical parameters of the vibrating lips. The vocal tract is at the end of the respiratory system, which is the supply of air of the brass player, and modifications of the vocal tract can help the player to control the high static mouth overpressure required when playing very high notes.

New investigation techniques such as real-time magnetic resonance imaging are very promising. Pioneering studies by Iltis et al. (2015) investigated arpeggio playing on several brass instruments, charting the evolution of the player's vocal tract internal shape (oropharyngeal cavity size and tongue conformation) when moving from lower to higher notes. Tuba and trombone playing showed a progressive decrease in oropharyngeal area, with an upward and forward displacement of the tongue. Trumpet playing showed progressive increases in oropharyngeal area, with the posterior compartment showing the largest change. There was essentially no change on oropharyngeal area in horn playing.

6.3.4 Respiratory Control

The discussion of the player's windway so far has concentrated on its role as an acoustic resonator upstream of the lips. We should keep in mind however that the vocal tract is part of the pulmonary airways leading to the lungs (through the trachea), all these subsystems constituting the pulmonary apparatus. The lungs, a cluster of small elastic bags (alveoli) surrounded by the pleura, constitute an air reservoir that a brass player fills before playing (inhalation) and empties progressively during sound production (exhalation). The action of respiratory muscles allows air to circulate in the pulmonary apparatus by modulating the air pressure in the lungs. It is therefore clear that the respiratory muscles are intimately involved in the regulation of the quasi-static mouth pressure that the player controls to generate oscillations of the lips.

Following the pioneering work of Bouhuys (1964), and taking advantage of technological advances such as optoelectronic plethysmography (OEP) (Aliverti 1996; Cala et al. 1996), a systematic study of the control of the respiratory system during brass instrument performance has been undertaken by Vincent Fréour (2013). Pressures at different points within the respiratory system were monitored using transnasal balloon-catheter systems, and variations of chest wall volumes were recorded using OEP. These measurements allowed for the quantification of the activity of different groups of respiratory muscles during performance. Although the results obtained come from only one player, self-consistent patterns were observed during basic playing tasks of varying pitch and loudness.

During ascending and descending arpeggios with no dynamic constraints, Fréour observed a sequential decay of abdominal and rib-cage volumes resulting in a parallel activation of rib-cage and abdominal muscles during the playing task. Abdominal muscles appear to be the primary generators of pressure, while rib-cage muscle action is overall less, though certainly associated with fine control of the pressure. The diaphragm, very active during inspiration, appears to be relaxed during tone production; this mechanically guarantees an optimal coupling between the abdominal and rib-cage compartments and an optimal transmission of the abdominal pressure generated by abdominal muscles to the lungs. Interestingly, an almost linear relationship was observed between the net pressure developed by expiratory muscles and the playing frequency, as if a linear mapping of the respiratory muscular activity on playing frequency was adopted by the performer (when no constraints on the sound dynamic were applied).

The respiratory pressure measurements also allowed an estimate of the glottal resistance, which is associated with the opening area of the glottis. An increase of glottal resistance for high notes suggests that control of the glottis could possibly be correlated with the acoustical control of the player's windway.

6.4 Improving the Lip Model

The mechanical behaviour of the brass player's lips, which was described in detail in Chap. 3, plays a crucial role in the generation of sound by the instrument. It is therefore important that a model of brass playing includes an adequate treatment of lip dynamics. In the elementary model developed in Chap. 5, the vibrating lips are represented by a single mass-spring system with only one degree of freedom. This simple 1DOF model has been widely used in simulations, and realistic sounds have been generated using physical modelling synthesis based on a 1DOF outward-striking lip valve. It was noted however in Sect. 3.4 that in describing the quite similar physiological systems involved in singing and snoring, it has been found necessary to employ models with two masses and two or more degrees of freedom (Ishizaka and Flanagan 1972; Lous et al. 1999; Auregan and Depollier 1995). In this section we review the limitations of the 1DOF model and outline some approaches which extend the model to describe more accurately the observed behaviour of brass playing lips.

6.4.1 Evidence from Mechanical Response Measurements

The mechanical response of the lips at an angular frequency ω was defined in Sect. 3.3 as

$$H_{\mathrm{mr}}(\omega) = \frac{h(\omega)}{\Delta p(\omega)}, \tag{3.16}$$

where $h(\omega)$ is the lip opening height resulting from forcing by a sinusoidal pressure difference $\Delta p(\omega)$ across the lip opening. An ideal 1DOF oscillator has only one mode of vibration, and its mechanical response curve has a single peak at the modal

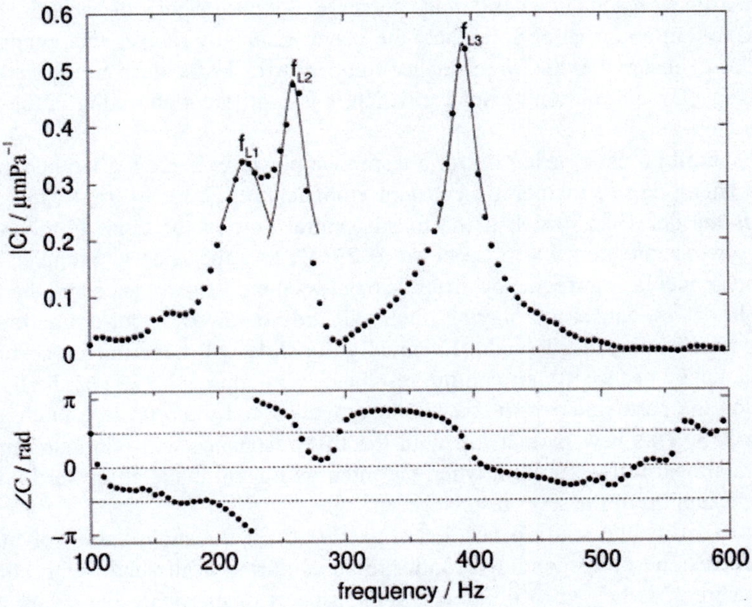

Fig. 6.19 Measured mechanical response $C = H_{\mathrm{mr}}(\omega)$ of artificial lips forming an embouchure on a tenor trombone mouthpiece. Top: magnitude. Bottom: phase. A static mouth pressure of 1.83 kPa was applied. Adapted from Cullen et al. (2000)

resonance frequency. The measured mechanical response of a pair of artificial lips shown in Fig. 6.19 exhibits a pair of mechanical resonances at frequencies 223 Hz and 260 Hz. The existence of two peaks close to each other suggests that a model with two degrees of freedom is necessary to represent the playing behaviour in this frequency range. The phase of the lower-frequency resonance is characteristic of an outward-striking valve, while the higher-frequency resonance shows inward-striking behaviour.

Resonance pairs with this outward/inward phase relationship are frequently found in response measurements of artificial lips (Lopez et al. 2006). Mechanical response measurements on the lips of human brass players (Newton et al. 2008) have also revealed a couple of outward-inward resonances, although they are much more heavily damped than the resonances of artificial lips (see Sect. 3.3.2).

6.4.2 *Evidence from Measurements of Threshold Playing Parameters*

In playing a descending glissando, a trombone player moves the slide steadily outwards while the lips buzz continuously. The increase in the total tube length of the instrument results in a continuous decrease in the frequency of each air column resonance. In a normal performance, the player gradually relaxes the embouchure muscles to ensure that the lip resonance frequency tracks the descending frequency of the specific air column resonance which is primarily collaborating in the sound production.

It is an illuminating exercise for a trombone player to make a glissando gesture while attempting to maintain a constant embouchure. If, for example, the player sounds the note $B^{\flat}3$, which is the fourth natural note of the tenor trombone, the lip resonance frequency will be close to 233 Hz. As the slide is extended, there will be an increasing frequency difference between the lip resonance and the fourth air column resonance: the playing pitch will drop below $B^{\flat}3$, but not as much as in a normally played glissando. Eventually the slide will have moved so far that the frequency of the fifth air column resonance will approach 233 Hz; the lips will abandon the partnership with the fourth resonance and jump up to a pitch a little above $B^{\flat}3$. This new partnership with the fifth resonance will continue until the slide is almost fully extended, when the pitch will again make an upward jump to gain the support of the sixth resonance.

An experimental study by Cullen et al. (2000) of threshold values of playing frequencies and corresponding mouth pressures during a glissando with constant embouchure has helped to illuminate some aspects of the phase behaviour of the lips. Since it is difficult for a human player to avoid involuntary embouchure changes which are the fruit of extensive practice in normal glissando playing, the experiment was carried out using an artificial mouth.

The first step in the procedure was to choose a set of lip parameters such that the mouth was capable of sounding the instrument over the complete range of tube lengths available. The mechanical response for this embouchure was then measured with no mouth overpressure, the forcing sinusoidal acoustic pressure being generated by the loudspeaker mounted on the mouth cavity. A second mechanical response measurement was made with a mouth overpressure just below the threshold for self-sustained oscillation. Finally the loudspeaker was switched off, and the mouth overpressure increased until a clear note was sounded; the overpressure was then reduced until the oscillation died out. This procedure was repeated several times to establish the experimental threshold pressure, defined as the lowest overpressure at which a stable self-sustained oscillation could be obtained. The frequency of this oscillation, taken to be the threshold playing frequency, was also recorded.

Results of this investigation, displayed in Fig. 6.20a, demonstrate how the threshold values of mouth pressure and playing frequency vary as the trombone slide is extended. Comparable results of simulations using inward- and outward-striking 1DOF models are shown in Fig. 6.20b.

The measured range of threshold pressure is similar to the range predicted by the inward-striking model. The outward-striking model predicts much higher threshold pressure values, but they are still within a range which a human player could easily produce. Both measurements and simulation demonstrate distinct regimes corresponding to the fourth, fifth and sixth modes of the acoustic resonator. The

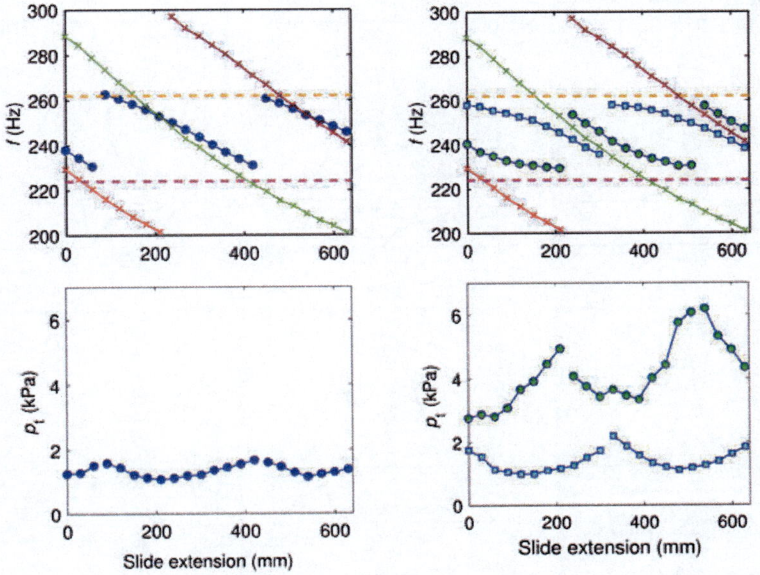

(a) Measured threshold behaviour (b) Modelled threshold behaviour

Fig. 6.20 Variation of threshold values of playing frequency (upper plots) and mouth pressure (lower plots) with slide position for a fixed artificial mouth embouchure. (**a**) Measured values (blue-filled circles). (**b**) Predicted values using an outward-striking (green-filled circles) and inward-striking (cyan-filled squares) one-mass lip model. Frequencies of acoustic modes 4 (orange solid line), 5 (green solid line) and 6 (brown solid line) and lip resonances 1 (magenta dashed line) and 2 (yellow dashed line) are also shown in the upper plots (Color figure online). Adapted from Cullen et al. (2000)

regimes are apparent in the variation of threshold frequency and, to a lesser extent, in the variation of threshold pressure. The experimental measurements of slide position at the transitions between regimes do not coincide with the regime transition slide positions predicted by either the inward-striking or outward-striking models.

With a fixed embouchure, it is found that for some slide positions, the threshold frequency is above the measured acoustic resonance frequency; for others the situation is reversed. This agrees with earlier playing frequency measurements on cup mouthpiece instruments, which indicate that the playing frequency may be either above or below the instrument resonance frequency, depending on playing conditions (Chen and Weinreich 1996; Gilbert et al. 1998; Cullen et al. 2000). Simple 1DOF models cannot explain this behaviour: with such models the threshold frequency is always below the acoustic resonance frequency (inward-striking case) or always above the acoustic resonance frequency (outward-striking case).

A complementary study by Neal et al. (2001) replaced the trombone by a telescopic pipe resonator attached to the mouthpiece. A systematic study of the behaviour of the lips when driving the telescopic pipe in the vicinity of the self-

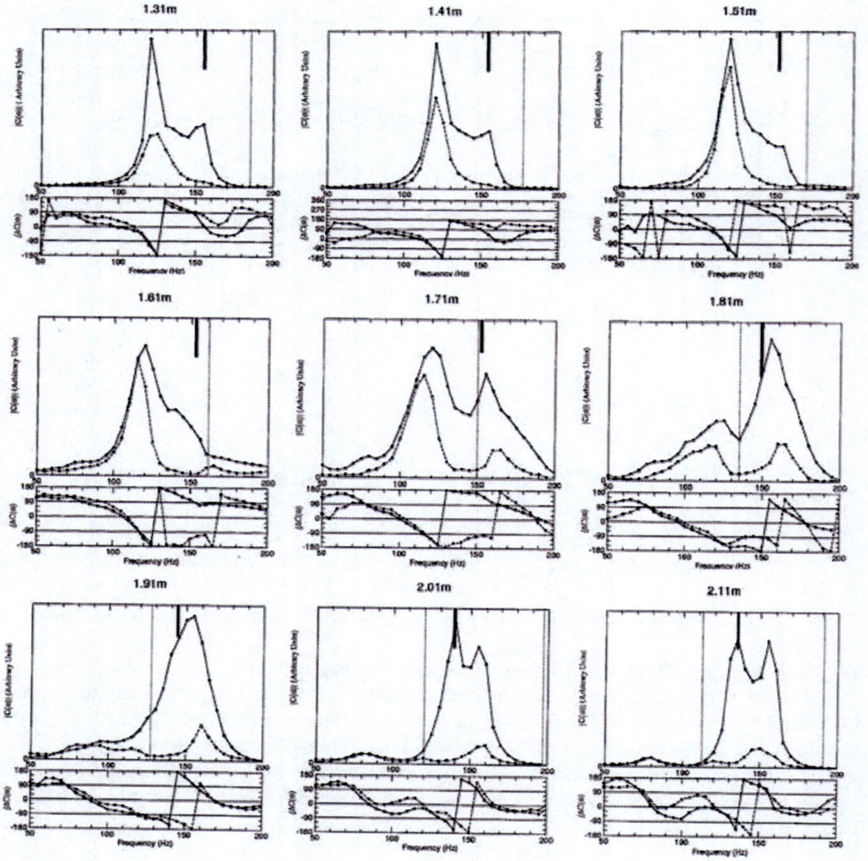

Fig. 6.21 Measured mechanical responses of artificial lips driving a telescopic cylindrical tube. Adapted from Neal et al. (2001)

oscillation threshold was carried out for a number of different pipe lengths. In a separate experiment, the input impedance of the mouthpiece and telescopic pipe was determined for each pipe length employed. Figure 6.21 shows mechanical response measurements taken with the telescopic pipe resonator attached to the mouthpiece and extended to lengths between 1.3 m and 2.1 m. For each pipe length, a response measurement is shown for both the case of no mouth overpressure (dotted line) and a mouth overpressure just below the threshold playing frequency (solid line). The complete vertical lines indicate the decreasing acoustic resonance frequency as the tube length is increased, and the short vertical lines mark the threshold playing frequencies.

Looking at the first graph of Fig. 6.21 for the 1.31 m pipe length, it can be seen that in the case of no mouth overpressure, there is only one visible resonance of the lips at 120 Hz, with an outward-striking behaviour. When the mouth overpressure is increased, another resonant peak appears at 160 Hz. This resonance is also close to the playing frequency of the lips and has an inward-striking behaviour. Although this peak cannot be seen in the amplitude response curve for the case of no mouth overpressure, its presence can be predicted by the phase response at this frequency which has a value of $-\pi/2$ at 152 Hz. The lack of an amplitude peak at this frequency can be attributed to the fact that it corresponds to an impedance maximum of the pipe; this additional load on the lips reduces the amplitude response. Both peaks are seen in the amplitude response curve when the measurement is carried out with the pipe removed.

As the mouth pressure is increased, the 160 Hz resonance approaches destabilisation, the height of the amplitude response peak growing much more dramatically than that of the 120 Hz resonance. This gives the expected behaviour that the lips must behave as an inward-striking reed if the playing frequency is below the acoustic resonance frequency of the instrument. A contrasting case can be seen in the graph for the 2.11 m pipe length. Here the resonance responsible for the playing of the instrument (135 Hz) has an outward-striking character, and the playing frequency is above the acoustic resonance of the instrument.

In the intervening lengths between these two cases, there is a much more complicated interaction between the different lip resonances, and there is clearly some coupling between the resonant modes of the lips. Despite this change in the dominant lip resonance, there is a smooth change in playing frequency as the pipe is extended. Although the change in frequency is continuous, some interesting features of the coupling between lip resonances can be found in the region in which neither lip resonance is clearly dominant. As the pipe is extended, the playing frequency does not always show a monotonic drop in frequency; for some embouchures there is a small region in which the playing frequency is found to rise as the telescopic pipe is extended. This anomalous behaviour occurs when the playing frequency is very close to the acoustic resonance frequency.

The work of Cullen (2000) and Neal (2002) shows that the near-threshold behaviour of the lips cannot be completely explained using a lip model with only a single degree of freedom. In a 1DOF model, the threshold playing frequency for a fixed embouchure should be always above the coupled acoustic resonance frequency in the outward-striking case and always below it in the inward-striking case; however Fig. 6.21 shows a continuous transition from one case to the other. In general, mechanical response measurements of the lips show several distinct mechanical resonances. However, the experimentally observed behaviour described above suggests that important features of the lip motion may be reproduced by a model involving only two mechanical modes, the lower-frequency mode having outward character and the higher-frequency mode having inward character.

6.4.3 Models with More Than One Degree of Freedom

A year after Adachi and Sato published their seminal paper describing a 1DOF lip model (Adachi and Sato 1995), the authors extended the model by allowing the pendulum arm on which the single mass swung to have finite stiffness (Adachi and Sato 1996). The result is a one-mass model with two degrees of freedom, one corresponding to the rotational mode and one corresponding to the vibrational mode. Although this model, and a one mass 2DOF model including translation and rotation (Strong and Dudley 1993), have been successfully used in brass instrument simulations, they are capable of reproducing the alternation between converging and diverging lip channels described in Sect. 3.4.

Numerical simulations by Neal et al. (2001) and Richards et al. (2003) have demonstrated that a 2DOF model with two masses, in which a mode with outward-striking characteristics is coupled to a mode with inward-striking characteristics, allows the threshold frequency to lie above or below the acoustic resonance frequency depending on the acoustic resonator conditions. Using linear stability analysis of a simple two-mass 2DOF model, a similar effect to the experiments has been obtained showing the continuous transition from playing above the acoustic resonance frequency to playing below.

The 2DOF two-mass model is able to explain how a brass player can buzz the lips in the absence of a mouthpiece or instrument (see, e.g. Boutin et al. (2015b)) without invoking the intervention of an upstream resonance (see Sect. 3.4). This type of behaviour, in which sound is produced by self-sustained oscillations of a mechanical system destabilised and then driven by air flow, is an example of the phenomenon of 'flow-induced vibration', which occurs in many different contexts: for an overview see Blevins (1986).

6.5 Playing Frequencies of Brass Instruments

Increasing the number of degrees of freedom in the lip model can improve its ability to predict realistic playing behaviour, as explained in Sect. 6.4.3. Several additional refinements, such as the inclusion of nonstationary flow terms (Elliott and Bowsher 1982; Saneyoshi et al. 1987), can also be readily introduced. Increasing the complexity of the lip model does however exacerbate the problem of selecting appropriate values for the additional parameters (Velut et al. 2017a). A further complication arises when we include the effects of coupling to resonances of the player's windway, described in Sect. 6.3, since these depend strongly on the physiology and technique of the individual performer. To simulate the playing of a note by a particular musician on a given instrument, it would be necessary to provide the model with detailed information about the mechanical resonances of the lips, the acoustic resonances of the player's windway, the quasi-static pressure in the mouth and input impedance of the instrument.

For a musician, the intonation of the instrument is a critical quality factor. The manufacturer can determine the frequencies of the input impedance peaks by a suitable choice of bore profile (Chap. 4), but when the instrument is played, the sounding frequencies will also depend on the performer's choice of lung pressure, windway shape and embouchure. We have seen that notes can be 'lipped' both above and below a 'pitch centre' (Sect. 1.2.2), which is usually close to one of the impedance peak frequencies (see, e.g. Boutin et al. (2015b)). Good intonation thus implies that the frequencies of the pitch centres (rather than the impedance peaks) should correspond to the musically desirable notes, usually taken to be frequencies on the equally tempered (ET) scale with A4 = 440 Hz.

To establish the relationship between impedance peak frequencies and playing frequencies in musical practice, Eveno et al. (2014) carried out a systematic study in which four performers were recorded playing three trumpets with different bore profiles. On each trumpet four different valve fingerings were used (000, 100, 110 and 111). For each fingering, the natural notes from the second to the sixth were sounded at three dynamic levels (*mf*, *p* and *f*). The performers were asked to play the notes with the easiest emission, without trying to correct the intonation: in other words, to find the pitch centre for each note. To establish consistency each test was repeated three times, giving a total of 2160 recorded notes. The input impedances of the 12 different bore profiles (4 fingerings on each of 3 trumpets) were also measured.

An analysis of this large corpus of data was used to obtain Fig. 6.22. Each graph in this figure corresponds to one of the played regimes (natural notes) and includes the results from all three trumpets and four fingerings. The horizontal axis, labelled $\mathrm{F_{res}}$ (cents), is the pitch difference between the impedance peak nearest to the played note for a given fingering and the nominal ET pitch of the note. The vertical axis, labelled $\mathrm{F_{play}}$ (cents), is the pitch difference between the played pitch and the nominal ET pitch.

A striking feature of these results is the wide range of the playing pitches for a given fingering and playing regime. Taking the example of regime 2, the pitches for the fingering 000 are represented by black crosses. All of the crosses are slightly to the left of the $F_{res} = 0$ vertical line, indicating that the second impedance peak is approximately 10 cents below the nominal pitch of $\mathrm{B^b3}$ for all three trumpets. The playing pitches range from 25 cents below $\mathrm{B^b3}$ to 25 cents above. For the other three fingerings, the vertical spread of the data exceeds 50 cents. The dependence of playing pitch on dynamic level was found to be weak. It thus appears that two players asked to play the same note on the same instrument may differ in assessing the pitch centre by up to half a semitone.

It is also noteworthy that most (though not all) of the data points are above the black line representing $F_{play} = F_{res}$. This is true for all the regimes, although the playing frequencies of the second regime are shifted up to the greatest extent with respect to the impedance peak frequencies. This observation can be related to the inharmonicity of the resonances corresponding to the second regime. For regimes 3 to 6, playing frequencies are, on average, around 15 cents above the impedance peak frequencies.

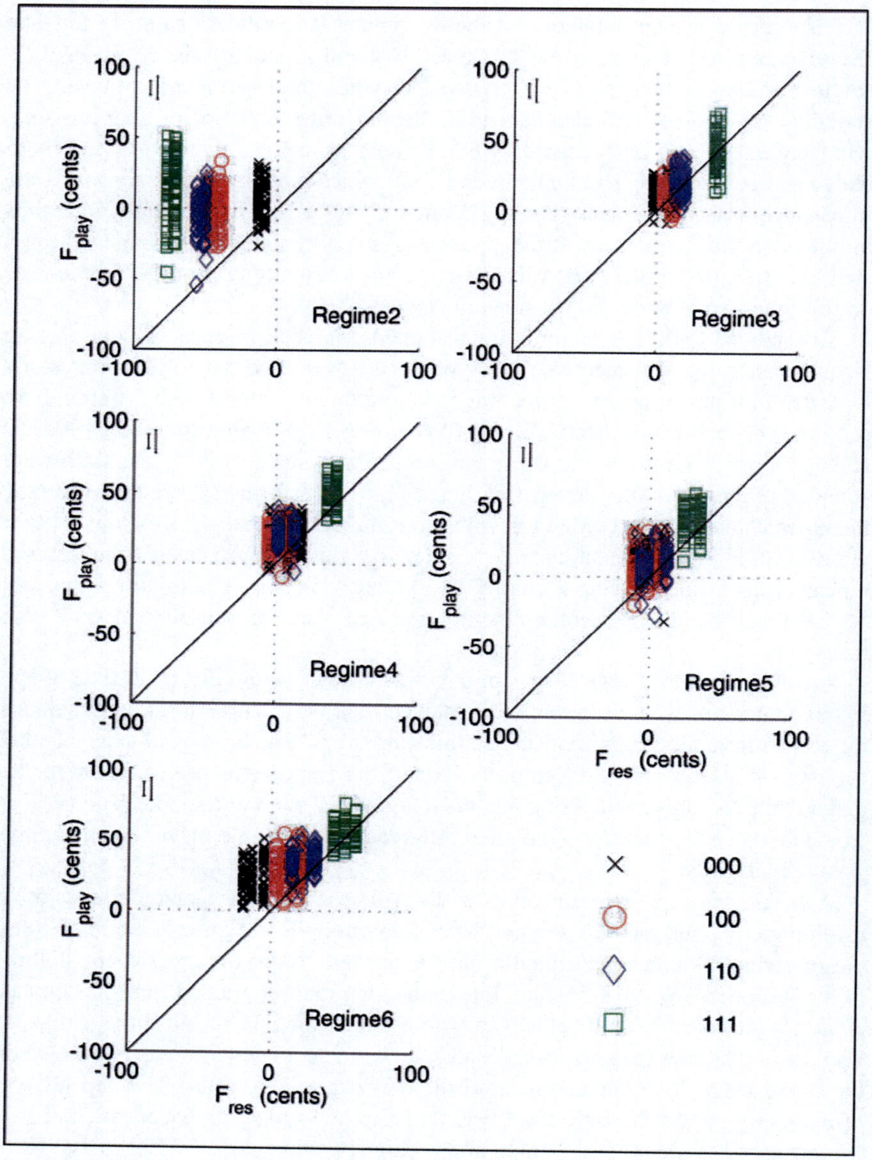

Fig. 6.22 Playing frequencies F_{play} as functions of resonance frequencies F_{res}, given in cents (taking the equally tempered scale as a reference) for 2160 recorded notes. There are one figure for each regime (natural note), from regime 2 to regime 6, and a different marker for each fingering: 000 in crosses, 100 in circles, 110 in diamonds and 111 in squares. The error bar on the left represents twice the average standard deviation of a note. The error bar on the right represents twice the average reproducibility of the trumpet players. Adapted from Eveno et al. (2014)

We can conclude from the global view presented by Fig. 6.22 that most of the playing frequencies are higher than the bore resonance frequencies over the playing regimes studied here. This is consistent with the predominantly outward-striking lip behaviour which has been observed in previous studies (see Chap. 5). Nevertheless, Fig. 6.22 also shows that musicians can sometimes play below the resonance frequencies, strengthening the view that a single mechanical oscillator cannot model the complete behaviour of the lip reed.

6.6 The Influence of Wall Material on Brass Instrument Performance

Relaxation of the assumption of wall rigidity acknowledges the fact that the walls of brass instruments vibrate when the instruments are played. The controversial issue of the musical significance of these wall vibrations is the principal topic of Sect. 6.6.

When sounding a fortissimo note on a trumpet or trombone, it is easy for the player to feel the vibration of the metallic structure, and tempting to identify the characteristic brassy blare with this metal vibration. As explained in Sect. 6.1, this intuitively attractive idea is incorrect: the brassy timbre is a consequence of nonlinear acoustic propagation in the air column of the instrument. A brassy note can be easily sounded using a trumpet mouthpiece on a suitable length of plastic hose whose wall has a negligible vibration amplitude.

The only critical requirement for the wall material of a brass instrument is that it must be capable of providing a smooth and rigid boundary to define the shape of the internal air column. In Sect. 6.6.1 we review briefly some other considerations which have influenced the choice of labrosone wall materials. Section 6.6.2 presents experimental evidence showing that wall vibrations can exert a subtle influence (of the order of 1 or 2 decibels) in the amplitudes of particular harmonics in the spectra of brass instruments. Section 6.6.3 describes some studies of thin-walled metal organ pipes which demonstrate that an injudicious choice of wall material properties can result in much more dramatic effects, including the disruption of the normal playing behaviour. The coupling between structural and acoustic modes in brass instruments is discussed in Sect. 6.6.4, and the possible influence of the direct communication of mechanical vibrations between the mouthpiece and the lips of the player is considered in Sect. 6.6.5. The complex bore geometry of brass instruments makes it difficult to develop an accurate theoretical model of structural vibrations, but vibroacoustic theoretical and experimental results from a study of cylindrical tubes, presented in Sect. 6.7, provide some insights.

6.6.1 Factors Affecting the Choice of Wall Material

Although many different materials have been used in the manufacture of members of the labrosone family (see Sect. 1.1), the most common choice by far has been the alloy of copper and zinc known as brass. One obvious practical reason for this preference is that the metal allows the construction of a rigid, strong but relatively lightweight instrument. The durability of metal instruments is also impressive: the silver and bronze trumpets from the tomb of Tutankhamun were still playable more than 3000 years after they were made. Whether plastic trumpets will achieve such longevity remains an open question.

Another important consideration is the workability of the material (see Sect. 8.1). The properties of the alloy affect the ease with which internal corners can be rounded and also influence the extent to which a bell formed on a mandrel retains its shape when removed. Residual internal roughness can influence the damping coefficient of the air modes, leading to a divergence between the playing experience and the predictions of theory for a perfectly reflecting and rigid wall (Watkinson et al. 1982).

Other characteristics of the material may affect the sound of the instrument in an indirect way. Silver, for example, has a very high thermal conductivity and warms up more quickly than a low thermal conductivity material such as wood or plastic (see 2.1.7). The breath from the mouth of the player enters the instrument at a temperature of around 35°C and includes significant proportions of water vapour and carbon dioxide as well as oxygen and nitrogen. During the warm-up exercise usually performed by the player, a temperature gradient is established along the bore of the instrument (van Walstijn et al. 1997), and changes in the proportions of the different gases in the tube influence the frequencies of the acoustic modes and hence the played pitches (Boutin et al. 2013; Campbell 2014a). These effects could in principle depend on the thermal conductivity of the wall material, but this topic requires further study.

6.6.2 Experimental Studies of Brass Instrument Wall Vibrations

The dominant source of sound radiation from a brass instrument is undoubtedly the transmission of sound energy from the internal air column through the instrument bell and any other open apertures. The vibrating wall of the instrument also radiates sound, but it was pointed out in Sect. 2.1.6 that the direct contribution of wall vibrations to the acoustic output has been estimated to be around 40 dB lower than the contribution due to the air column. The relative insignificance of wall vibrations is a property of wind instruments which differentiates them strongly from string and percussion instruments, in which the dominant sound source is the vibration of the structure. On the other hand, when the walls vibrate, they may also influence the sound output by coupling to and modifying the resonances of the internal air column. The musically important question is whether the wall vibrations exert an influence on the overall sound which can be perceived by the player or listener, and if so under what circumstances.

In a groundbreaking series of experiments in the 1970s, Richard Smith investigated the influence of wall thickness on the structural modes and radiated sound spectra of tenor trombones (Smith 1986). Six bells were formed on the same mandrel, with wall thickness varying from 0.3 mm to 0.5 mm. Using these bells, six otherwise identical trombones were manufactured. In a first experiment, the instruments were excited using a sinusoidal pressure signal generated in the mouthpiece by an acoustic driver. Using the recently developed technique of interference holography, Smith obtained images showing the wall vibration patterns corresponding to different structural resonances.

Figure 6.23 reproduces interference holography images of three bells with different wall thicknesses, excited into vibration by the same acoustic input amplitude. Each image represents the same structural vibration mode, with a frequency close to 240 Hz. Since the separation of the white fringes represents equal steps of wall displacement, it is evident that the amplitude of the wall vibration increases dramatically when the thickness is reduced below 0.4 mm.

Having identified the structural resonant frequencies at which significant wall vibrations could be measured, Smith carried out a second experiment to explore the possible influence of these vibrations on the radiated sound of the trombone. To achieve stable and reproducible results, the six instruments were driven in forced oscillation by a siren acoustic driver (Wogram 1972). The input signal to

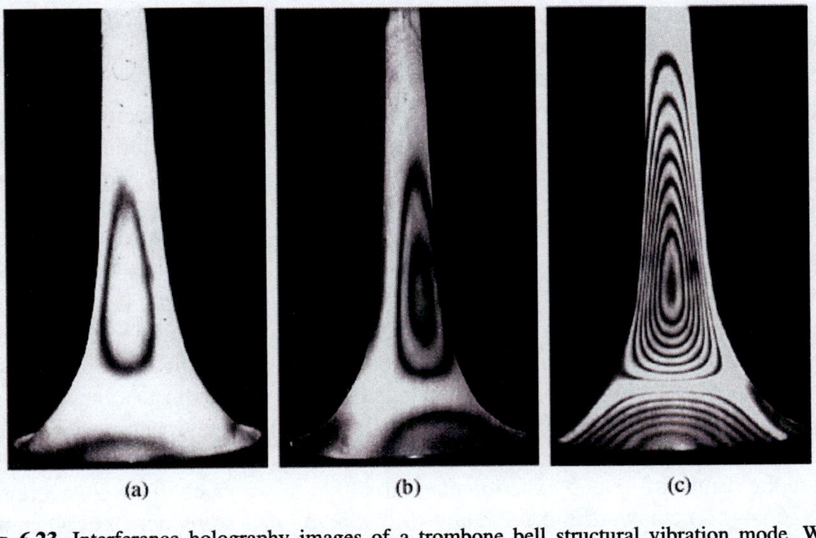

(a) (b) (c)

Fig. 6.23 Interference holography images of a trombone bell structural vibration mode. Wall thickness w: (**a**) 0.5 mm; (**b**) 0.4 mm; (**c**) 0.3 mm. Courtesy Richard Smith the instrument mouthpiece had an amplitude similar to that generated by a human player and a frequency spectrum rich in upper harmonics.

The frequency of the structural resonance illustrated in Fig. 6.23 was close to the fourth harmonic of the note $B^{\flat}1$. When this pedal note was played by the acoustic driver, a microphone measuring the sound output on axis in front of the bell did not register a systematic change in the amplitude of the fourth harmonic for different bell thicknesses, suggesting that the wall vibration was having a negligible effect on the sound. However a microphone in the left ear of an artificial head in the normal human playing position showed that the amplitude of the fourth harmonic was several dB higher for the 0.3 mm bell than for the 0.5 mm bell. A similar result was found when the note $B^{\flat}2$ was played: the structural resonance frequency was close to the second harmonic of this note, and at the position of a human player's left ear, the amplitude of the second harmonic was several dB higher for the thin bell than for the thick bell.

The radiated sound measurements reported by Smith suggest that the timbral modification caused by a coincidence between a structural mode and an acoustic mode might well be perceived by a trombone player, although it would probably be inaudible to a distant listener. To test this hypothesis, ten professional trombonists were invited to play and evaluate the six instruments. The players were blindfolded, and precautions were taken to equalise the weight and balance of the bells (see Sect. 1.3.3). Under these circumstances none of the players were able to distinguish between thick and thin bells (Smith 1986). Thus although there were clearly measurable spectral differences at specific frequencies between the sounds generated using different bell thicknesses, these differences could not be considered musically significant in a realistic performance situation.

In the first decade of the twenty-first century, Thomas Moore and Wilfried Kausel revisited the question of the musical effects of brass instrument wall vibrations in a series of experimental studies (Moore et al. 2005; Kausel et al. 2008, 2010). Structural modes of trumpets and horns were observed using electronic speckle pattern interferometry, and investigations were undertaken on the influence of wall vibrations on the sound radiated from instruments excited both by acoustic drivers and artificial mouths (see Sect. 3.1.6). The experimental arrangement used by Moore at Rollins College to study an artificially blown King 'Silver Flair' trumpet is illustrated in Fig. 6.24a. The artificial lips were in this case made from solid rubber and with a mouth pressure of around 20 kPa played the note $B^{\flat}4$ (frequency 466 Hz). The bell of the trumpet was placed directly in front of a circular aperture in the wall of a small anechoic chamber which contained the measuring microphone. The vibrations of the bell could be heavily damped by packing sandbags around it.

The relative strengths of the harmonics in the measured spectra were found to depend strongly on the embouchure setting, but a significant trend emerged when the results of several experiments with different embouchures were averaged (Fig. 6.24b). The powers in the first and second harmonics were increased when the bell was damped, while the powers in the harmonics from the third to the eighth were diminished by damping.

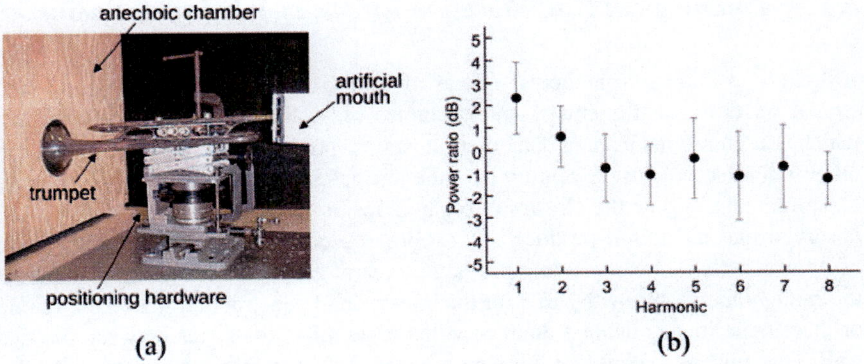

Fig. 6.24 (**a**) Experimental setup for studying the effect of wall vibrations on a trumpet played by an artificial mouth. (**b**) Power in each harmonic of the played note $B^\flat 2$ with the bell damped, divided by the corresponding power with the bell free. From Moore et al. (2005)

Fig. 6.25 Experimental apparatus for measuring the effect of damping wall vibrations on a french horn sounded by an artificial mouth. Courtesy Wilfried Kausel

Similar results were obtained in measurements on a french horn carried out at the Institute of Musical Acoustics in Vienna (Kausel et al. 2010). This experiment was performed in a large anechoic chamber (Fig. 6.25a), and the artificially blown horn was encased by a wooden box which could be filled by sand to damp the wall vibrations (Fig. 6.25b). A dependence of the relative sound powers of harmonics on damping over a wide frequency band was again observed, with the powers of the lowest harmonics augmented by damping and the powers of higher harmonics diminished. This small but measurable influence of wall damping on the spectrum of the radiated sound was also found when the artificial lips were replaced by an acoustic driver.

6.6.3 *Pathological Wall Vibration Effects in Wind Instruments*

Although no dramatic perceptual effects of wall vibrations have yet been demonstrated on brass instruments, some examples of 'pathological' circumstances in which structural resonances can disrupt sound production have been found in other wind instruments. A century ago, the journal *Science* published a spectacular experimental study on the physics of organ pipes in which Dayton C. Miller (1909) observed that the sound produced by an organ pipe can depend on the vibration of its walls. He compared flue pipes having rectangular cross-section, of identical internal geometry, but with different thicknesses and construction materials. In one of these experiments using a double-walled organ pipe, the space between the two walls (see Fig. 6.26) could be filled with water while the pipe was sounded. Miller observed that the filling led to unusual and clearly audible behaviour of the pipe, presumably related to modification of the wall vibrations. Some heights of the water jacket produced pitch changes or inharmonic and unstable tones (Hoekje 2003). Recently Gautier et al. (2012) repeated the experiment and obtained the same kind

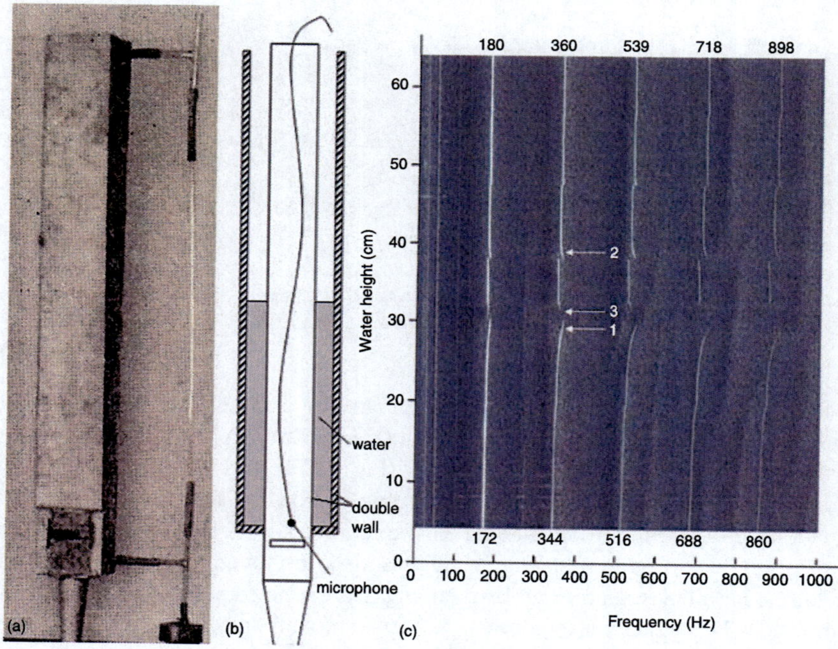

Fig. 6.26 (**a**) Photograph of Miller's historic experiment (from Miller (1909)). (**b**) Sketch of experimental setup using an organ pipe, as similar as possible to the one used by Miller. (**c**) Time-frequency analysis of the sound. In this spectrogram, time has been converted to water height (vertical axis) using the constant filling rate. Arrows at critical water heights indicate a strong pitch change (1), an unstable tone (2) and silence (3). Adapted from Gautier et al. (2012) with the permission of the Acoustical Society of America

of results as Miller. The spectrogram in Fig. 6.26 illustrates several situations in which the sound production of the pipe is severely modified and even extinguished.

Similar disruptive effects are sometimes observed during the manufacture of organ pipes. An example of an organ pipe giving a quasi-periodic regime because of a coincidence between an acoustic mode and a structural mode is described by Nederveen and Dalmont (2004). The resulting jarring sound resulting from the coupling of the modes can be described as a 'wolf', by analogy with the similar perturbation which is caused on a bowed string instrument by a coincidence between a powerful body resonance and the fundamental string mode. To get such spectacular effects in wind instruments, one needs a resonator having a large internal diameter and a thin and elastic tube with low damping. Through trial and error, organ builders have learned to make the walls of axisymmetric metal organ pipes sufficiently thick to avoid wolfing. Coupling between acoustic and structural modes is more of a problem in tubes which do not have axial symmetry (see Sect. 6.6.4); probably for this reason, square cross-section pipes of the type investigated by Miller usually have thick wooden walls with high internal damping.

6.6.4 Frequency-Localised and Broadband Effects of Structural Resonances in Brass Instruments

We have seen that when a brass instrument is played, it vibrates tangibly in the player's hands. Like any metallic structure, a brass bell has a set of lightly damped structural modes, each mode corresponding to a specific vibrational pattern. Taking advantage of technological advances over the last half century, many authors have measured, visualised and calculated structural modes of brass instruments. Experimental techniques used include holographic interferometry (Smith 1986) and electronic speckle pattern interferometry (Moore et al. 2002), while fruitful calculations have been performed using finite element techniques (Watkinson and Bowsher 1982; Balasubramanian et al. 2019). Calculations and measurements (see Fig. 6.27) show many structural modes; these usually come in pairs, with associated eigenfrequencies and normal modal shapes which are slightly different because of asymmetries in the system.

One axisymmetric mode is shown in more detail in Fig. 6.28. This mode, sometimes called the piston mode, is particularly important in the context of axisymmetric brass instrument bells, since it is the most appropriate to be coupled with the internal acoustic field. In this mode the bell vibration consists of a periodic expansion and contraction in the axial direction, with a single nodal plane cutting the axis at the centre of mass. For the dimensions typical of a brass instrument bell, the vibration amplitude is significant only very close to the bell rim, as shown by the colour scale in Fig. 6.28.

The mechanisms of the vibroacoustic couplings involved in brass instruments are difficult to investigate, as fluid-structure interactions are weak. To help clarify the

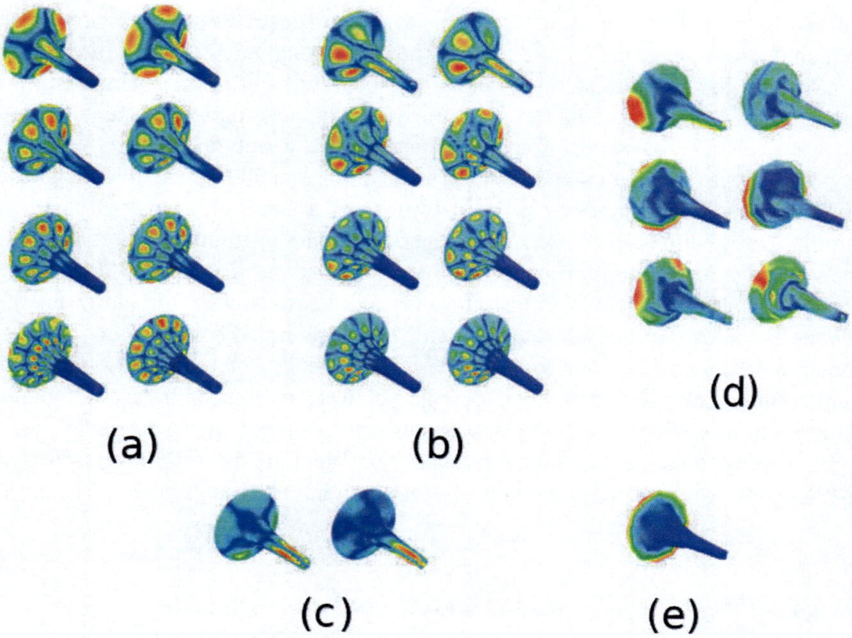

Fig. 6.27 Mode patterns of a trombone bell obtained by roving hammer technique and experimental modal analysis (Sécail-Géraud et al. 2018). Localised shell modes (**a**) with maxima at the bell rim, (**b**) with nodes at the bell rim, (**c**) with significant amplitudes only in the narrow part of the bell. (**d**) Bending beam-like modes of the entire bell. (**e**) Axisymmetric piston mode

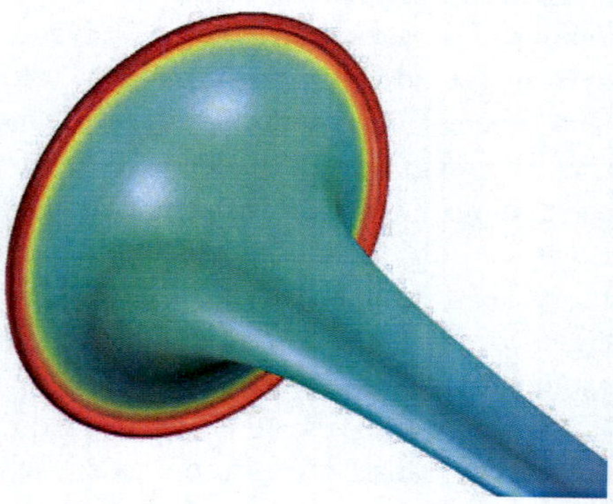

Fig. 6.28 The lowest frequency axisymmetric mode ('piston mode') of a trumpet bell, modelled by a finite element COMSOL computation (Balasubramanian et al. 2019). Courtesy of Wilfried Kausel

nature of these interactions, some studies have focused on the relatively simple case of the coupling between the internal acoustic field and the mechanical behaviour of a cylindrical duct (Gautier and Tahani 1998; Pico Vila and Gautier 2007). Figure 6.29 illustrates the case of a brass tube, 240 mm long, 7.5 mm radius and 0.2 mm wall thickness, with an 8% ellipticity. The input impedance of this tube was calculated using a vibroacoustic model described in detail in Sect. 6.7.1.

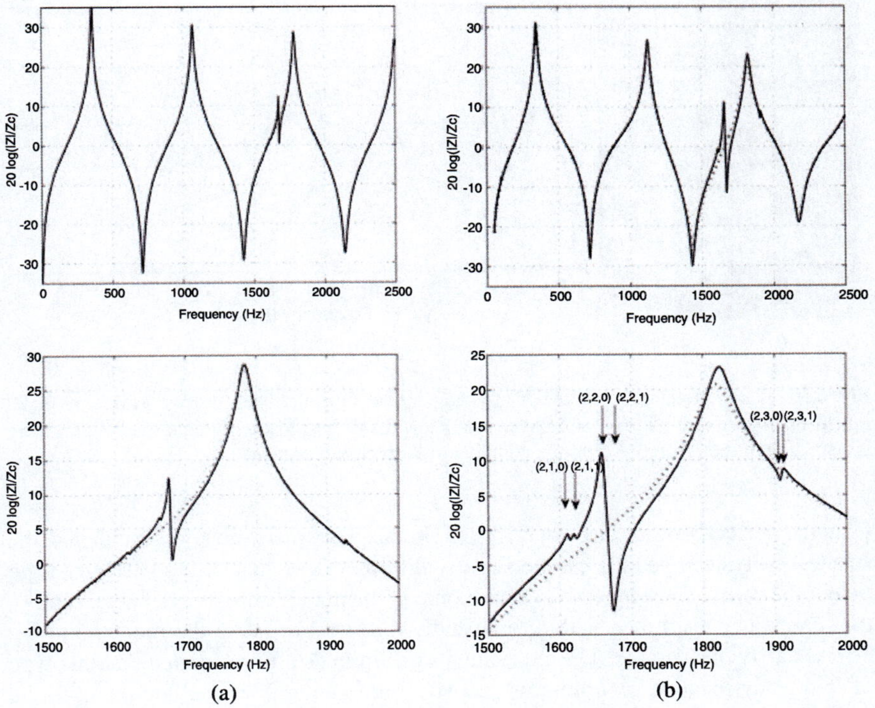

Fig. 6.29 (**a**) Calculated acoustic input impedance magnitude on a dB scale for a brass tube, 24 cm long, 0.2 mm thick, 7.5 mm radius, with an 8% ellipticity. Above: broad frequency band. Below: zoom on the frequency range containing mechanical modal frequencies implying perturbations on the input impedance. Perfectly rigid tube (dotted line). Vibrating tube (solid line): additional peaks correspond to mechanical ovalling mode resonances. (**b**) Measured input impedance of the thin brass tube. Above: broad frequency band. Below: zoom on the frequency band containing mechanical modal frequencies. Quasi-cylindrical tube (dotted line): no coupling with mechanical modes. Slightly flattened tube (solid line): coupling with mechanical modes. Adapted from Nief et al. (2008) with the permission of the Acoustical Society of America

Figure 6.29 shows that the theoretical prediction of a coupling between the mechanical ovalling mode of a duct with an oval cross-section and the plane propagating acoustic mode is confirmed experimentally. Due to this coupling, the acoustic input impedance is perturbed; the perturbation is significant close to the eigenfrequencies of the ovalling modes of the duct structure. The importance of asymmetry already noticed in calculations by Watkinson and Bowsher (1982) has also been shown experimentally and theoretically; it has been demonstrated that for a perfectly symmetrical tube such a coupling cannot occur. In practice, for real instruments, perfect symmetry is not achievable; the greater the asymmetry, the more the input impedance will be perturbed.

These results confirm that when the eigenfrequency of a mechanical mode matches an acoustic resonance or antiresonance, particular behaviours different

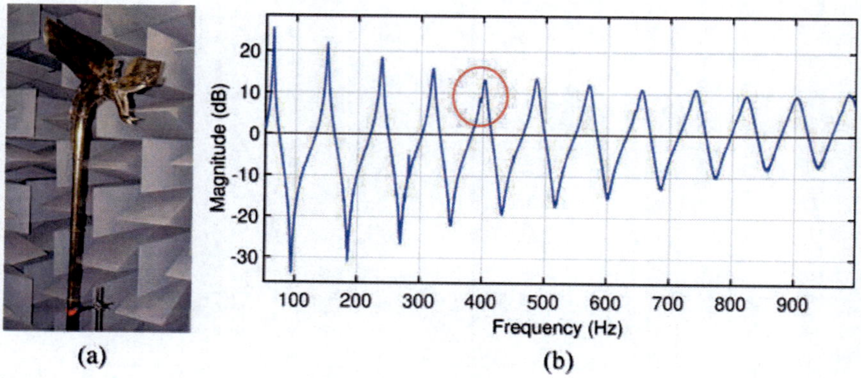

Fig. 6.30 (**a**) Reproduction of the Tintignac carnyx (see Sect. 9.1.2). (**b**) Measured input impedance of the reproduction of carnyx: just below the fifth acoustic resonance at 402 Hz, there is a small perturbation at 396 Hz corresponding to a structural resonance

from the perfectly rigid case can be found. In this case of frequency coincidence, the spectral content of sounds can be slightly different from that of the perfectly rigid case if the coincidence is close to a harmonic of the played note. In Nief et al. (2008), the tube was constructed with an unusually thin wall to make the mode coupling effects easily measurable. The experiments of Smith (1986) suggested that this type of frequency-localised coupling of acoustic and structural modes is unlikely to be of musical importance in realistic brass instruments.

Nevertheless, sometimes it is possible to see small perturbations on impedance measurements of real brass instruments (see, e.g. Fig. 6.30, around the fifth resonance of a carnyx input impedance). Such small-magnitude secondary peaks, which arise from coupling of structural resonances with the internal acoustic field, have been measured on a trumpet by Macaluso and Dalmont (2011). It was found that when the structural resonances of the bell were damped by wrapping it in a light plastic film, the main acoustic resonances were not significantly altered but the small secondary peaks were suppressed.

Such coincidence effects are influenced by details of the mechanical boundary conditions. The effect of clamping the brace of a trombone is illustrated in Fig. 6.31. The additional damping caused by the clamp suppresses a mechanical mode which coincided with an acoustic mode, and as a consequence, a small perturbation near 780 Hz disappears from the measured input impedance.

Experiments involving the heavy damping of brass instrument bells to investigate the influence of wall vibrations on the radiated sound were described in Sect. 6.6.2. The effects of structural resonances on the transfer function and input impedance of a B^b trumpet have also been studied using this method by Kausel et al. (2010). In the transfer function experiments, the instrument was excited sinusoidally by an acoustic driver; the input pressure was measured by a microphone in the delivery tube between the driver and the mouthpiece and the transmitted pressure by a second

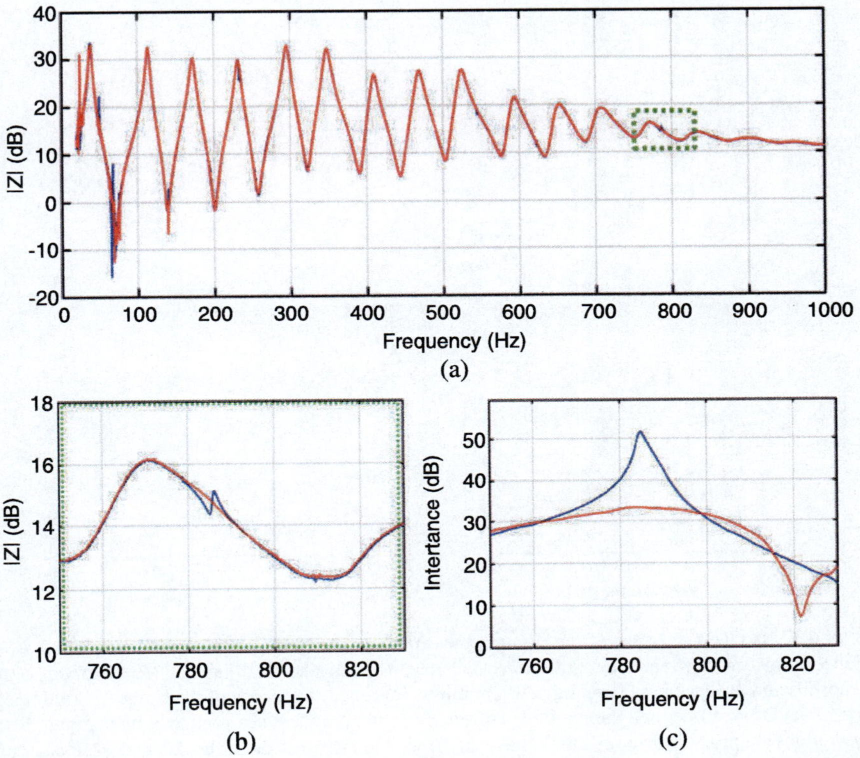

Fig. 6.31 A modification of a boundary limit condition (adding damping by clamping the brace of a trombone) suppresses a mechanical piston mode responsible for a coincidence effect with an acoustic mode. (**a**) Acoustic input impedance. (**b**) Expansion of dotted area in (**a**). (**c**) Axial mechanical inertance measured at bell rim. Blue curves: brace free. Red curves: brace clamped. Courtesy of Mathieu Sécail-Géraud and François Gautier

microphone on axis in the plane of the bell. The input impedance was measured using the BIAS apparatus (see Sect. 4.2.1).

The transfer function curves illustrated in Fig. 6.32a, b do not display any features which can be identified as effects of coupling with lightly damped structural modes. A broadband frequency-dependent effect of the damping is however evident: the transfer function magnitude at frequencies below 500 Hz is increased when the damping is applied but reduced at frequencies above 500 Hz. This behaviour is consistent with the measurements of the effects of damping on the spectral content of the sound radiated from instruments played by an artificial mouth (Fig. 6.24b). A broadband effect can also be seen in the input impedance curves shown in Fig. 6.32c, d: impedance peak magnitudes are reduced by damping below 900 Hz but increased above this frequency.

An explanation of these broadband effects based on coupling between the acoustic field and relatively highly damped axial structural modes, including the

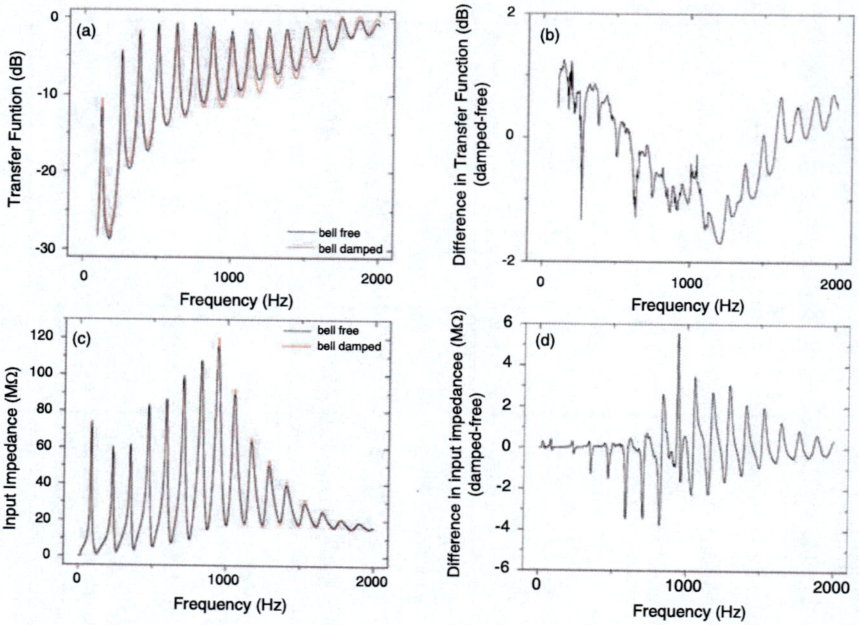

Fig. 6.32 (**a**) Transmission of the trumpet measured at the output plane of the bell plotted as a function of driving frequency. Solid line: bell free. Dashed line: bell damped. (**b**) Difference in transmission produced by damping bell vibrations. (**c**) Input impedance of the trumpet. Solid line: bell free. Dashed line: bell damped. (**d**) Difference in input impedance produced by damping bell vibrations. Reproduced from Kausel et al. (2010) with the permission of the Acoustical Society of America

piston mode, has been proposed (Kausel et al. 2015). Finite element and finite difference calculations have been shown to be capable of reproducing the main features of the experimental measurements (Moore et al. 2015), but the material damping coefficient required is a factor of 50 greater than that expected for brass. There is thus as yet no totally satisfying model to interpret the experimentally observed broadband frequency dependence of wall vibration effects, and further research in this area is required.

A study of the effect of wall damping on a brass instrument bell has also been undertaken by Gautier et al. (2013). Inspired by the Miller (1909) experiment described in Sect. 6.6.3, François Gautier and colleagues at the University of Le Mans used a water tank surrounding a trombone bell which could be filled progressively in order to modify the mechanical modes of the bell in a continuous manner (see Fig. 6.33). The bell was excited in turn by a loudspeaker and a mechanical shaker. The acoustical and mechanical responses were measured for different levels of water, allowing an analysis of the vibroacoustic couplings. The measurements demonstrated that the vibrational modes of the bell were excited by the internal acoustic field, but did not lead to significant changes in the radiated

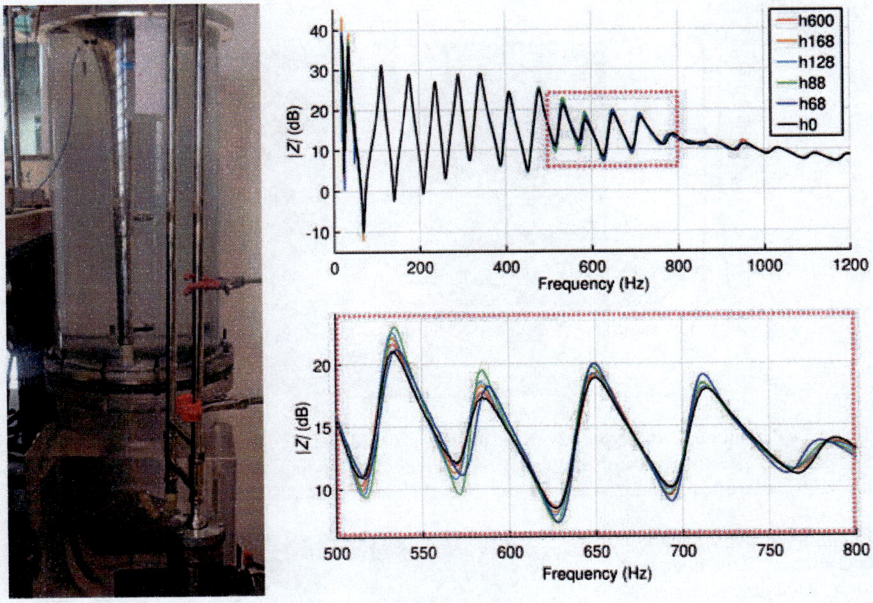

(a) (b)

Fig. 6.33 (**a**) Photograph of the Miller experiment replicated on a trombone bell. (**b**) Input impedance magnitude of the trombone measured for different heights of water. Legend: water height in mm (h0 empty, h600 full). The area marked by the dotted line in the upper graph is expanded in the lower graph. Adapted from Gautier et al. (2013)

acoustic field. Measurements of the acoustic input impedance for different water levels showed that the impedance was significantly modified by the wall vibration, with shifts of acoustic resonance frequencies up to 14 cents, and 1 dB change in magnitude, in the high pitch range of the trombone between resonance peaks 9 and 12 (see Fig. 6.33). While these effects are measurable over a large frequency band of the input impedance, they are not strong enough to influence the radiated sound in low- and medium-frequency playing.

Another experiment which showed a spectacular effect of a coincidence between structural and acoustic modes without any significant modification of the radiated sound has been described by Sécail-Géraud et al. (2018). This study, based again on the experiment of Miller (1909), used a trombone which had a double-walled prototype bell (Fig. 6.34). With the slide in first position, the note B$^{\flat}$3 (fundamental frequency 235 Hz) was played by an artificial mouth while the space between the two bells was gradually filled with water. The increase in the effective mass of the wall due to the added water continuously modified the set of mechanical resonances of the bell. The resulting descending glissando in the mechanical resonance frequencies led to the possibility of coincidences with harmonics of the played note.

Fig. 6.34 Trombone prototype having a double skinned bell (courtesy of Stéphane Gaudet), allowing the effective mass of the wall to be increased by adding water while the instrument is played by an artificial mouth. Adapted from Sécail-Géraud et al. (2018)

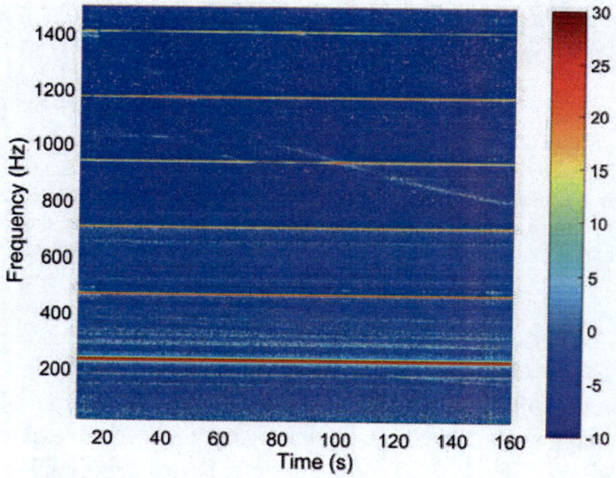

Fig. 6.35 Spectrogram of the acceleration of the inner bell in the axial direction, while the space between the bells is gradually filled with water. As a consequence one descending line, corresponding to the piston mode, is clearly visible in the spectrogram, crossing the fifth and then the fourth harmonic of the played note $B^{\flat}3$ (fundamental frequency 235 Hz). Adapted from Sécail-Géraud et al. (2018)

A spectrogram of the acceleration of the bell measured in the axial direction is shown in Fig. 6.35. Forcing of the bell by the internal acoustic field gives rise to acceleration components at multiples of 235 Hz, which appear as horizontal lines in the spectrogram. As a consequence of the gradually added water, a descending line, corresponding to the piston mode, is clearly visible in the spectrogram, crossing the fifth and then the fourth harmonic of the played note.

The acceleration signal undergoes a modification of around 15 dB at the coincidence between the mechanical piston mode and the fourth harmonic, as shown in Fig. 6.36. At the same time, the acoustic signal remains almost perfectly stable: Fig. 6.37 displays only a very small variation of 0.2 dB in the fourth harmonic of the acoustic pressure in the bell exit plane of the trombone.

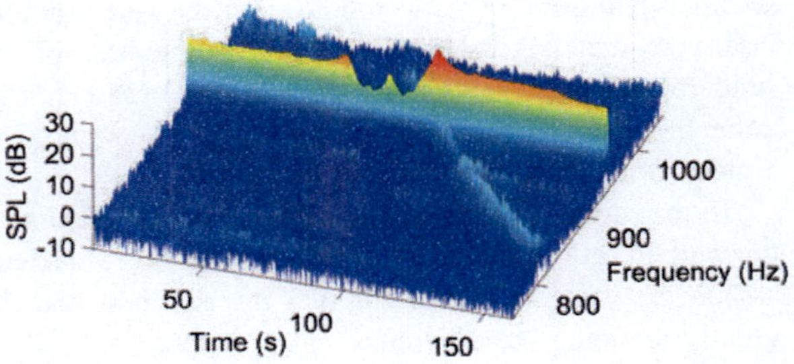

Fig. 6.36 Detail of the spectrogram of trombone bell axial acceleration shown in Fig. 6.35, illustrating modification of the fourth harmonic due to coincidence with the mechanical piston mode. Adapted from Sécail-Géraud et al. (2018)

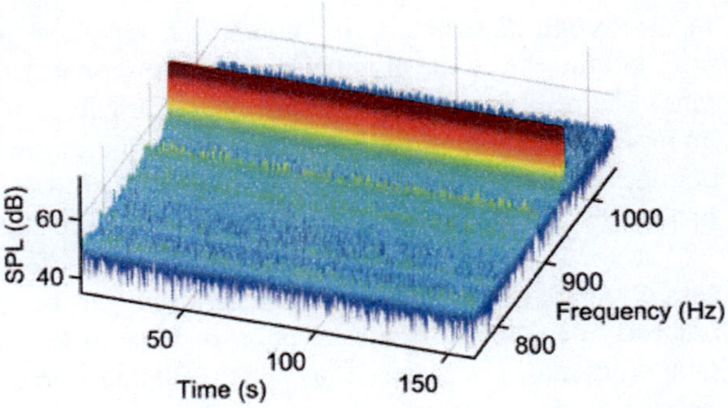

Fig. 6.37 Spectrogram of bell pressure (focused on the fourth harmonic), corresponding to the experiment described in Fig. 6.36. Adapted from Sécail-Géraud et al. (2018)

It is evident from the examples discussed above that wall vibrations can in some cases have undesirable effects on the radiated sound. Are there other circumstances in which wall vibrations can enhance the playing properties of the instrument? There is disagreement on this issue among players and makers of brass instruments. In a survey of manufacturers' claims about the effects of different wall materials, Bowsher and Watkinson (1982) reported that some makers boasted of 'live' bells, implying that wall vibration was desirable; one well-known manufacturer declared that 'as far as the overall instrument is concerned, the more inert it is to vibration the better it is'. From the scientific viewpoint, this question is open, but the principle of energy conservation requires that energy given to the wall vibration is lost from the internal air column, which is the primary source of radiated sound.

6.6.5 *Mechanical Vibration at the Lip-Mouthpiece Interface*

Section 6.6.4 considered wall vibrations arising from forcing by the air column acoustical oscillation in the internal bore. It is also possible that the vibrating lips of the player could directly excite wall vibrations. The interaction of a human player's lips with the mouthpiece is difficult to study directly, but some insight into the relative significance of these two possible excitation mechanisms has been provided by experiments carried out by James Whitehouse et al. (2003) using artificial lips. A simple instrument (a trombone mouthpiece coupled to a cylindrical metal pipe) was sounded using an artificial mouth, and its wall vibrations were measured using a laser vibrometer. A velocity amplitude of the order of 1 mm s^{-1} was observed at the fundamental frequency of the sounded note (333 Hz); the corresponding operating deflection shape resembled the shape of the second bending mode of the pipe, with a mode frequency of 230 Hz.

To investigate the relative importance of mechanical and acoustical excitation of the wall vibrations, two further experiments were carried out. In the first experiment, a short piece of flexible tubing was inserted between the mouthpiece and the pipe, greatly reducing the efficiency of the mechanical coupling of the lips to the pipe. It was found that this reduced the amplitude of the pipe vibration by a factor of 5. In the second experiment, the mouthpiece remained mechanically coupled to the pipe, but the end of the mouthpiece stem was connected to a second tube inside the pipe. The resulting removal of the acoustical pumping of the pipe had little effect on its vibration amplitude. The main conclusion of the study was therefore that the dominant mechanism in exciting this type of wall resonance was the motion of the lips against the mouthpiece, rather than air pressure changes within the pipe.

The focus of Sect. 6.6.4 was on the possible effects of wall vibrations on the sound radiated by a brass instrument. From the perspective of the listener, this is clearly paramount. From the point of view of the performer, the story may be rather different. The player is in contact with the vibrating instrument, most directly through the lips pressed against the mouthpiece rim. Newton's second law tells us that if the lips exert an oscillating force on the mouthpiece, the mouthpiece exerts an equal and opposite force on the lips. The sensitivity of the player to subtle vibrations communicated through the lip-mouthpiece interface is a topic which needs much further study. The difference between the listener and the player can be expressed in the language of measurement technology: the listener is like a microphone, sensing only the sound, but the player is like a microphone sensitive to the sound and also an accelerometer sensitive to the vibrations. This could be one reason why, although the listener may not hear any difference between two brass instruments whose mechanical behaviour is slightly different, the player could detect these differences by sensing the vibrations.

6.7 Going Further: Analytical Modelling of Vibroacoustic Coupling in Ducts

Several different approaches to the modelling of the interaction between the internal acoustic field and the wall vibrations in a brass instrument were outlined in Sect. 6.6.4. Section 6.7 introduces an analytical approach to the theory of vibroacoustic coupling in ducts. To illustrate the basic principles and some general conclusions, only the simple case of a tube of uniform cross-section is considered. It is assumed that the tube is rigidly supported and otherwise isolated from external mechanical forces, so that wall vibration can arise only from coupling to the acoustic field. The basic theory is explained in Sect. 6.7.1 and used in Sect. 6.7.2 to develop an analytical expression for the input impedance of a duct whose cross-section is slightly elliptical. Some experimental results obtained using clarinet-like instruments are compared with predictions of the vibroacoustic model in Sect. 6.7.3. The theoretical and experimental results presented in this section come principally from Backus and Hundley (1966), Gautier et al. (2007), Pico Vila and Gautier (2007) and Nief et al. (2008).

6.7.1 Basic Vibroacoustic Theory

The easiest way to take into account the influence of wall vibrations on the internal acoustic field is to model the pumping effect due to changes in the cross-sectional area of the duct (Backus and Hundley 1966). We consider a tube with yielding walls such that under internal pressure p, the fractional change in area S from the undistorted area S_0 is given by

$$(S - S_0)/S_0 = \beta p. \tag{6.39}$$

The 'yield' parameter β, which is dimensionally an inverse pressure, is assumed to have a value such that $\beta p \ll 1$. As a consequence of Eq. 6.39, the conservation of mass is modified, and after some mathematics the classical wave equation is obtained with a speed of sound c slightly different from its value c_0 in a rigid-walled tube:

$$c = c_0/\sqrt{1 + \beta\rho_0 c_0^2} \approx c_0(1 - \beta\rho_0 c_0^2/2). \tag{6.40}$$

This correction implies that the resonance frequencies of a cylindrical tube are shifted from those of a rigid tube of the same length by a fractional amount

$$\Delta f/f = \Delta c/c_0 = -\beta\rho_0 c_0^2/2. \tag{6.41}$$

For a copper tube, Backus found that $\beta = 1.210^{-10}$, which implies a frequency change of 0.0084%: this corresponds to a pitch difference of 0.14 cents, which is too small to be significant.

Are there situations where strong effects of coupling between vibrating walls and the internal pressure field can be visible, for example, on the acoustic input impedance? A thin stretched membrane in the form of a cylindrical tube can exhibit strong effects on sound waves propagating through the tube. In some frequency ranges, the waves can become evanescent, leading to the presence of stop bands (see the experimental and theoretical work based on Korteweg's equation in Gautier

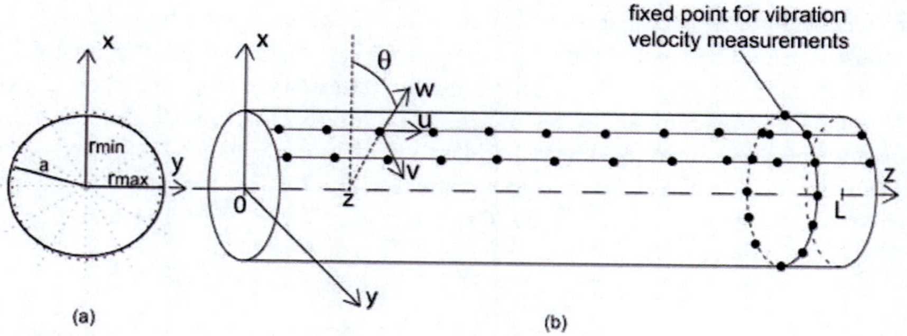

Fig. 6.38 (**a**) Polar graph of the elliptical cross-section of the tube. (**b**) Notations and coordinate systems used for the quasi-cylindrical shell. The positions of the measured points on the tested tube for modal analysis are added. From Nief et al. (2008), with permission of the Acoustical Society of America

et al. (2007)). This is an extreme case which is not realistic from the musical acoustic point of view, but which is helpful in clarifying vibroacoustic issues in tubes.

We now consider the more realistic case of a homogeneous, isotropic, thin-walled quasi-cylindrical shell of length L, mean radius a and wall thickness h. The importance of allowing for a small deviation from perfect cylindrical symmetry will be explained in Sect. 6.7.2. The shell material has a Young's modulus E, a Poisson's ratio ν and a density ρ_S. The internal cavity is filled with a fluid characterised by its density ρ_0 and its sound speed c_0. Surfaces S_0, S and S_L described in Fig. 6.38 correspond to the quasi-circular end surface at coordinate $z = 0$, the lateral surface of the shell (mean radius $r = a$) and the quasi-circular end surface at coordinate $z = L$, respectively.

The external fluid/structure coupling leads to added mass and added damping effects for the shell and also to intermodal coupling effects (see Gautier and Tahani (1998)). In the case of a low-density external fluid, these effects can be neglected for the calculation of the shell response, and their influence on the acoustic input impedance can be considered as negligible compared to the influence of inner fluid/shell coupling. Thus, we consider only the inner fluid/shell interaction.

In an oscillatory regime, two equations govern the situation. The first is the acoustic Helmholtz equation, and the second is the shell motion equation. The motion of the middle surface of the quasi-cylindrical shell is described by displacement vector **X**, whose components u, v and w denote the longitudinal, circumferential and radial displacements, respectively (see Fig. 6.38). Using the usual cylindrical coordinates, the dynamic behaviour of the shell is described by the Donnell operator denoted by $\mathbb{L}$; see Leissa (1973):

$$\mathbb{L} = \begin{pmatrix} a^2\frac{\partial^2}{\partial z^2} + \frac{1-\nu}{2}\frac{\partial^2}{\partial \theta^2} & \frac{1+\nu}{2}a\frac{\partial^2}{\partial z \partial \theta} & \nu a \frac{\partial}{\partial z} \\ \frac{1+\nu}{2}a\frac{\partial^2}{\partial z \partial \theta} & \frac{\partial^2}{\partial \theta^2} + \frac{1-\nu}{2}a^2\frac{\partial^2}{\partial x^2} & \frac{\partial}{\partial \theta} \\ -\nu a \frac{\partial}{\partial z} & -\frac{\partial}{\partial \theta} & -1 - \eta\left(a^2\frac{\partial^2}{\partial z^2} + \frac{\partial^2}{\partial \theta^2}\right) \end{pmatrix}, \tag{6.42}$$

where $\eta = \dfrac{h^2}{12a^2}$ is a non-dimensional thickness parameter. Using this shell operator, the equation of motion of the shell loaded by the air can be written in the following form (Leissa 1973):

$$\rho_s h(\omega_a^2 \mathbb{L} + \omega^2)\mathbf{X} = -p\mathbf{n}, \tag{6.43}$$

where $\omega_a = 2\pi f_a$ is the shell ring angular frequency. Equation 6.43 is associated with mechanical boundary conditions: the shell is supposed to be simply supported at both ends ($z = 0, L$). There are acoustical boundary conditions as well: an acoustic velocity distribution at $z = 0$ is considered as the excitation source of the system, while the surface at $z = L$ is assumed to be open. As a first approximation, we do not take into account radiation from the open end of the tube, setting $p = 0$ at $z = L$. On the lateral surface S, continuity of the normal velocities of the fluid and of the shell is imposed. This set of boundary conditions leads to an analytical tractable solution for the normal in vacuo modes of the shell (Leissa 1973).

6.7.2 *Effect of Vibroacoustic Coupling on Input Impedance*

This work aims to evaluate the wall vibration effect of a cylinder on its acoustical behaviour. To achieve this objective, the input acoustic impedance of the wall-vibrating cylinder is calculated by solving the coupled Helmholtz equation and equation of motion presented in Eq. 6.43, associated with boundary conditions for the acoustic cavity and the shell. The acoustic impedance is deduced from the calculation of the coupled system response to the known excitation source; for details see Pico Vila and Gautier (2007).

For a cylindrical and perfectly rigid tube open at the end, considering only the plane acoustic wave, the acoustic input impedance can be written as

$$Z_r(\omega) = \frac{\rho_0 c_0}{S} j \tan[k(\omega)L], \tag{6.44}$$

where c_0 is the speed of sound, ρ_0 the air density, L the equivalent length of the tube taking into account the end correction due to radiation and S its cross-sectional area. $k(\omega)$ is the complex wave number, taking into account viscous and thermal losses and dispersion effect at the walls (see Sect. 4.7.3). The subscript in $Z_r(\omega)$ indicates that it describes the input impedance of a rigid tube. The modulus of this input impedance shows peaks corresponding to the acoustic resonances of the closed-open air column.

The previous description does not take into account possible wall vibrations. In order to describe those vibrations, it is necessary to consider that the tube behaves as an elastic shell. Following the theoretical approach described in Sect. 6.7.1, the instrument is modelled as a quasi-cylindrical shell, in this case clamped at one end and free at the other. Structural vibrations and inner acoustic pressure are coupled: the shell is excited by the internal acoustic pressure field, and the resulting vibrations induce a disturbance of the initial pressure field. As a consequence of this modification of the air column oscillation, the input impedance of the vibrating tube can differ significantly from that of a perfectly rigid cylinder (Pico Vila and Gautier 2007).

The central point for modelling the acoustic input impedance of the vibrating tube is to take into account a slight ellipticity of the tube, which models the asymmetry which is in practice unavoidable. In a first approximation, its radius r can be written using the polar equation, which is plotted in Fig. 6.38,

$$r(\theta) = a[1 + \epsilon \cos(2\theta)], \tag{6.45}$$

where a denotes the mean radius and ϵ an ellipticity parameter, which is small compared to unity. With the notation of Fig. 6.38, ϵ is equal to $(r_{\max} - r_{\min})/2a$.

The consequence of this asymmetry is the existence of a vibroacoustic coupling between the acoustic plane wave and the ovalling modes of the structure. Ovalling modes, characterised by a circumferential modal shape in a $\sin(2\theta)$ form, are often the modes of lowest frequency for geometries similar to those of wind instruments. Physically, a section of the tube wall is subjected to a uniform pressure distribution due to the plane wave. If the tube is perfectly circular, this tends to dilate and contract it only in a cylindrically symmetric manner, and the section of the tube is only subjected to tension forces. If the tube is oval shaped, then the isotropic pressure distribution implies also bending forces that tend to round the tube by enlarging the small diameter and shortening the bigger one. This movement is directly linked to the ovalling deformation of the pipe. A detailed description of the vibroacoustic coupling is given in Pico Vila and Gautier (2007). The conclusion of this is an analytical expression for the acoustic input impedance of the vibrating tube, which can be written as

$$Z(\omega) = Z_r(\omega)[1 + C(\omega)], \tag{6.46}$$

where C is a correction factor describing the wall vibration effect. Considering only the interaction between the internal acoustic pressure and a single ovalling mode, it is shown that

$$C(\omega) \propto \epsilon^2 / \{(1 - e^{-2jkL}) \cos(kL) m_\mu [\omega_\mu^2 (1 - j\eta_\mu) - \omega^2]\}, \tag{6.47}$$

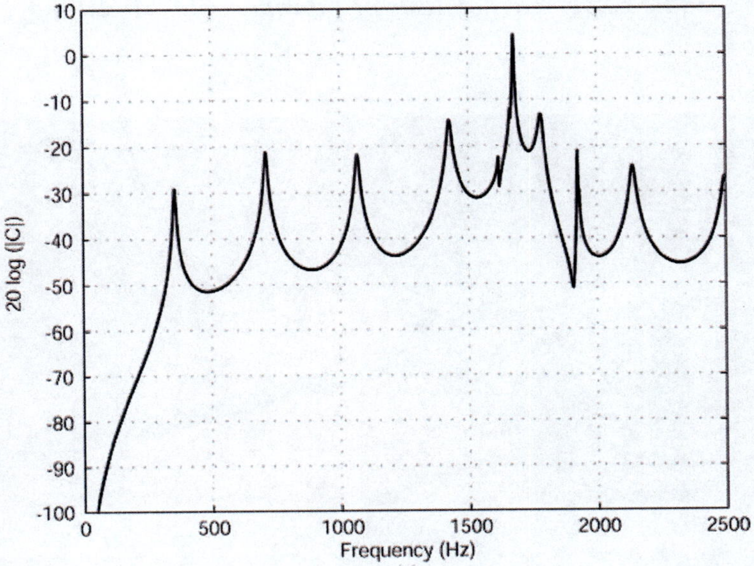

Fig. 6.39 Calculated correction factor C of a brass tube, length 240 mm, wall thickness 0.2 mm, inner radius 7.5 mm, ellipticity 8%. Reproduced from Nief et al. (2008) with the permission of the Acoustical Society of America

where m_μ is the modal mass, ω_μ is the modal natural angular frequency, and η_μ is the modal damping, which is the natural decay rate of the mode. These parameters are obtained experimentally from the modal testing of the tube used in experiments.

Equation 6.47 allows a direct interpretation of the correction factor. Firstly, the bigger the ellipticity ϵ, the more the input impedance is affected. This is due to the increase in the coupling between the ovalling mode and the inner pressure field. Secondly, C is inversely proportional to the modal mass m_μ, which means that disturbance will be more important when the cylinder is thin and light. Thirdly, when the driving angular frequency ω approaches the mechanical angular eigenfrequency, the disturbance increases (coincidence effect). Finally, at the acoustic resonances of the tube, when $\cos(kL)$ is minimum, the disturbance is maximum. The correction factor C may then take significant values when a frequency coincidence is realised between an acoustical frequency and a mechanical ovalling eigenfrequency.

In Fig. 6.39, the modulus of the correction factor C for the brass tube discussed in Sect. 6.6.4 is plotted versus frequency. The regularly spaced peaks correspond to the effect of acoustic resonance, and the other peaks correspond to the effects of coincidences with the resonances of the ovalling modes. The modulus of the computed input impedance of the vibrating tube, illustrated in Fig. 6.29a, also shows additional peaks due to the mechanical resonances of ovalling modes.

6.7.3 Some Experimental Tests of Vibroacoustic Modelling

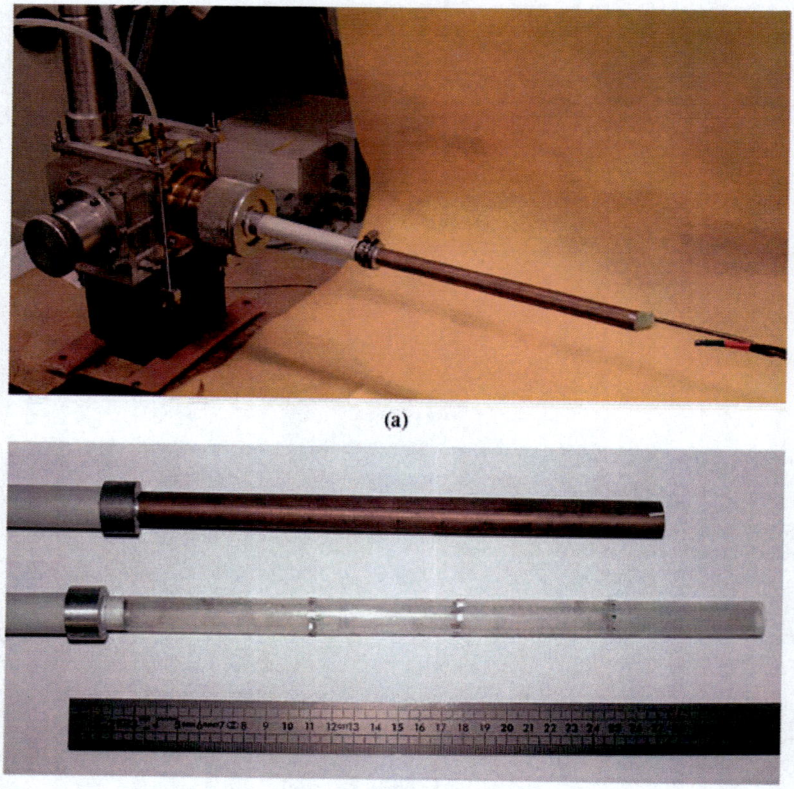

(a)

(b)

Fig. 6.40 (**a**) Clarinet artificial mouth sounding a sliding tube with a thin-walled brass extension. (**b**) Brass and plastic extension tubes used in experiment. Courtesy of Guillaume Nief

A study reported by Nief et al. (2008) was designed to provide experimental confirmation of the predictions of the vibroacoustic model described in Sect. 6.7.2. The object of study was a clarinet-like instrument (a pure cylindrical tube adapted with a clarinet mouthpiece) designed to show obvious wall vibration effects on radiated sounds. A clarinet-like instrument was chosen because it was easy to model and readily playable using an artificial mouth (Fig. 6.40).

The influence of wall vibrations on the input impedance Z is described by the correction factor C defined in Eq. 6.47. Since this correction factor is very small for normal wind instrument tubes, no significant effect of wall vibration is generally observed. To determine the geometrical characteristics and material of a tube whose acoustic input impedance could be significantly modified by wall vibrations, a design study was undertaken.

For slender tubes ($L/a \gg 1$), the first eigenfrequencies of shells of finite length are close to the cut on frequency f_m^c of the flexural wave associated with the circumferential index $m \geq 2$ ($m = 2$ for ovalling modes) of infinite cylinders. This frequency is given by

$$f_m^c = \frac{m(m^2-1)}{4\pi\sqrt{3}\sqrt{m^2+1}} \frac{h}{a^2} \sqrt{\frac{E}{\rho_s(1-\nu^2)}}, \tag{6.48}$$

which can be used with $m = 2$ to estimate the eigenfrequency of the first ovalling mode.

As experiments were to be carried out on a clarinet-like instrument, the sounding length of the system was chosen to be about 50 cm and the internal radius $a = 7.5$ mm. This gave the following series of acoustical eigenfrequencies for the first five acoustic modes: 170, 510, 850, 1190 and 1530 Hz. Geometrical and material parameters of the tube were determined so that the eigenfrequency of the first ovalling mode was close to one of these values.

It was found to be possible to obtain the coincidence between the second acoustic mode and the first ovalling mechanical mode, as illustrated in Fig. 6.41, by using a thin plastic tube having the following characteristics: $h = 0.2$ mm, $a = 7.5$ mm, $E = 1.8$ GPa, $\rho_s = 1350$ kg/m^3 and $\nu = 0.3$. The dimensioning is indicative: the cut on frequency f_m^c gives an order of magnitude of the ovalling mechanical mode frequency, and the mechanical parameter values are not precisely known. The plastic tube was connected to a rigid slide, making it possible to vary continuously the acoustic resonance frequencies, thanks to a variable sounding length of the tube, without changing the fixed mechanical resonance frequencies of the vibrating tube.

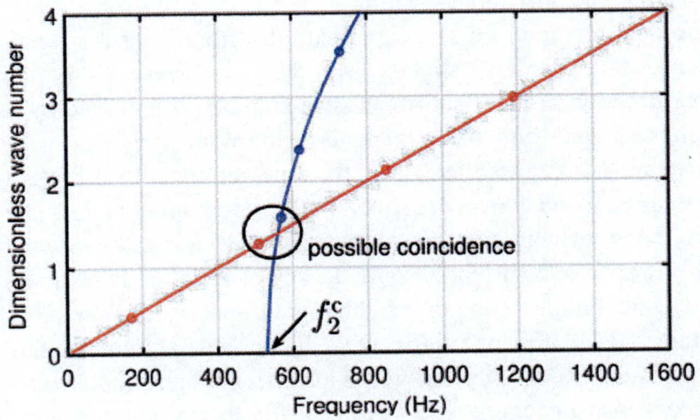

Fig. 6.41 Dispersion diagram representing the variations of the acoustic wave number (in red) and of the flexural wave number associated with the circumferential number $m = 2$ (in blue) (Color figure online). Adapted from Nief et al. (2006)

This device, making it possible to satisfy exactly the frequency coincidence or, on the contrary, to avoid it, was connected to a clarinet mouthpiece in an artificial mouth. The length of the thin-walled plastic section was 24 cm, and the additional lengths of the rigid slide in open position and the mouthpiece inside the artificial mouth gave a sounding length of about 50 cm. The experimental setup thus had a set of fixed mechanical modes and a set of variable acoustic modes. The second acoustic resonance frequency could be adjusted between 400 Hz and 550 Hz by moving the slide, allowing upward and downward glissandi to be played.

With the slide at its minimum length, a periodic regime corresponding to the normal clarinet functioning on its second natural note was obtained. As the slide was pulled slowly outwards, the fundamental frequency of the periodic regime decreased until a crucial point (around the middle of the slide) was reached, at which point a bifurcation towards a quasi-periodic regime was obtained. The resulting sound resembled the 'wolf note' described in 6.6.3. When the slide was further extended, the periodic regime was re-established. Similar behaviour was observed in upward glissandi. Since the sounding mechanism was an artificial mouth, the phenomena were very reproducible.

It was demonstrated by Nief et al. (2008) that the bifurcation was obtained when the coincidence occurred between the first ovalling mechanical mode and the second acoustic mode. As a consequence of the coincidence, the input impedance Z was drastically perturbed, the second peak being split into two peaks.

To bring the experimental conditions a little closer to those in a real brass instrument, the same kind of experiment was carried out with a sliding brass cylindrical tube in place of the plastic one (Fig. 6.40). The material parameters of brass are $E = 110\,\text{GPa}$, $\rho_s = 8700\,\text{kg/m}^3$ and $\nu = 0.3$. According to Eq. 6.48, the only free parameter in determining the frequency of the first ovalling mode is the thickness h. A thickness of $h = 0.2\,\text{mm}$ gives a first ovalling frequency of 1630 Hz, which is in the vicinity of the fifth acoustic resonance. The vibrating tube was therefore made from brass tubing used in musical instrument making, carefully machined to a thickness of about 0.2 mm.

The theoretical prediction that a significant disturbance of the input impedance would occur if the tube was slightly oval was confirmed by experiment. The measured perturbation of the input impedance could be suppressed by pinching the tube between the fingers, and the corresponding harmonic (the fifth) was sufficiently altered to give an audible modification of the tone colour.

The work summarised above, described in detail in Nief et al. (2008), has clearly established a theory of vibroacoustic coupling of a vibrating cylindrical tube and its inner air. It has been shown that a coincidence between an acoustic mode and an ovalling mechanical modes can perturb the input impedance of the tube, modifying the self-sustained oscillation of a clarinet-like instrument, although audible effects in cylindrical tubes are found only for wall thicknesses much smaller than those used in normal brass instruments. A full vibroacoustic theory dealing with the flaring bells found on many brass instruments is not yet available.

Part III
Historical Evolution and Taxonomy of Brass Instruments

In Part III we draw on some of the techniques discussed in the previous chapters to explore the diversity of brass instruments from the smallest to the largest, giving reasons why they sound different and illustrating members of the families which have made important contributions to music making. In Chap. 7 we look at instruments from the last 500 years, giving examples of typical bore profiles, and also at typical mouthpieces. Chapter 8 outlines some of the technical processes employed in making brass instruments. In conclusion, Chap. 9 presents brief glimpses of two very different fields of brass instrument research in which scientists and musicians are collaborating. In Sect. 9.1 we look at studies aimed at reconstructing playable examples of instruments from the ancient world. In Sect. 9.2 we hint at possible avenues for exploitation of digital technology in improving existing instruments and inventing new ones.

Chapter 7
The Amazing Diversity of Brass Instruments

In this chapter we discuss the factors which distinguish the commonly recognised species of brass instrument. It is asserted that bore profile is the principal determinant of the character of an instrument. Parameters which can be derived from physical measurement of instruments (or from instrument makers' designs) are identified. The main kinds of brass instrument are discussed, going from the shortest to the longest, with illustrations of typical examples and their bore profiles, and plots of the brassiness potential and minimum bore of larger populations of the common species. The phenomenon of instruments made in families (such as saxhorns) is examined. Parameters which can be used to characterise the mouthpieces appropriate for the different kinds of instrument are presented. The chapter concludes with a case study of the bass brass instruments composed for (and commented on) by Hector Berlioz.

7.1 What Are Important Features of Brass Instruments?

We saw in Chap. 1 how musicians—not only players but also composers, arrangers and conductors—consciously or unconsciously assess brass instruments by certain criteria. The most important of these are compass, dynamic range, timbre, responsiveness, intonation control, directivity, wrap, weight, ergonomics, and appearance. Any visitor to a major collection of historic brass instruments must surely be struck by the remarkable variety of the shapes and sizes of the instruments which have evolved through this interaction between makers and performers. Before commencing a systematic discussion of the relationship between design parameters and musical properties, we consider the most important features which can be used to identify different subgroups of the labrosone family.

M. Campbell et al., *The Science of Brass Instruments*, Modern Acoustics and Signal Processing, https://doi.org/10.1007/978-3-030-55686-0_7

7.1.1 Taxonomic Labels Based on Tube Length

In the later sections of this chapter instruments are discussed in subgroups based on their nominal pitch. We adopt the widely used convention, explained in Sect. 1.2.3, in which the nominal pitch is specified by a number of feet (ft) coupled with a musical pitch name. Table 7.1 shows the nominal fundamentals and equivalent cone lengths for the more common nominal pitches, and gives some examples of instruments commonly made at these pitches. Various pitch standards have been adopted at different times and in different places, some more than a semitone above A4 = 440 Hz and some significantly lower, resulting in a corresponding variation in the equivalent cone lengths corresponding to a nominal pitch. The values in Table 7.1 are calculated for A4 = 440 Hz.

The nominal equivalent cone length L_{ecN} of an instrument is related to the frequency F_{N} of its nominal fundamental by the formula

$$L_{\mathrm{ecN}} = \frac{c}{2F_{\mathrm{N}}}, \tag{7.1}$$

Table 7.1 Nominal fundamental frequency F_{N}, nominal fundamental pitch P_{N}, and nominal equivalent cone length L_{ecN} for common brass instruments with different nominal pitches (at A4 = 440 Hz)

Nominal pitch	F_{N} (Hz)	P_{N}	L_{ecN} (m)	Examples
$2\frac{1}{4}$-ft B$^\flat$	233.1	B$^\flat$3	0.74	Piccolo trumpet
$2\frac{3}{4}$-ft G	196.0	G3	0.83	Piccolo trumpet
3-ft F	174.6	F3	0.99	High F trumpet
$3\frac{1}{4}$-ft E$^\flat$	155.6	E$^\flat$3	1.11	Soprano cornet, soprano flugelhorn
$3\frac{1}{2}$-ft D	146.8	D3	1.18	Bugle horn, 'Bach' trumpet
4-ft C	130.8	C3	1.32	Orchestral trumpet
$4\frac{1}{2}$-ft B$^\flat$	116.5	B$^\flat$2	1.48	Bugle, cornet, flugelhorn, B$^\flat$ valve trumpet
5-ft A	110.0	A2	1.57	Post-horn
$5\frac{1}{2}$-ft G	98.0	G2	1.77	Keyed and early valve trumpets; American bugle
6-ft F	87.3	F2	1.98	Low F trumpet, English slide trumpet, mellophone
$6\frac{1}{2}$-ft E$^\flat$	77.8	E$^\flat$2	2.22	Alto trombone, tenor horn
7-ft D	73.4	D2	2.36	Natural trumpet
8-ft C	65.4	C2	2.65	Serpent, ophicleide, high C tuba
9-ft B$^\flat$	58.3	B$^\flat$1	2.97	Tenor trombone, B$^\flat$ french horn, baritone, euphonium
11-ft G	49.0	G1	3.53	G bass trombone, alphorn
12-ft F	43.7	F1	3.96	F french horn, F bass tuba
13-ft E$^\flat$	38.9	E$^\flat$1	4.45	E$^\flat$ bass tuba
14-ft D	36.7	D1	4.71	Trompe de chasse
16-ft C	32.7	C1	5.29	C contrabass tuba
18-ft B$^\flat$	29.1	B$^\flat$0	5.94	french horn crooked in B$^\flat$ basso, B$^\flat$ contrabass tuba

where c is the speed of sound (see Sect. 4.3.4). Values of L_{ecN} are calculated from Eq. 7.1 with $c = 346\,\mathrm{m\,s^{-1}}$, a typical figure for air inside wind instruments. The instruments are taken to have toneholes all covered, slides in closed position, and no keys or valves operated.

The nominal equivalent cone length can differ significantly from the physical length of the instrument tubing, depending on the bore profile and the assumed pitch standard. It should also be noted that the value of L_{ecN} does not necessarily agree exactly with the distance in feet specified in the pitch label. For example, the nominal equivalent cone length for a trumpet in '4-ft C' is given in Table 7.1 as $L_{ecN} = 1.32$ m: this is 10 cm longer than four feet (1.22 m). The nominal pitch label must therefore be considered as defining a category rather than a specific length.

7.1.2 *Bore Profile and Brassiness*

Once the basic necessities of sufficiently inflexible tubing, a stable structure and manageable wrap and ergonomics have been provided, the most important design features which instrument designers control are those of bore profile. The bore profile is the principal determinant of an instrument's compass, dynamic range, timbre, responsiveness, and to some extent intonation control. It is the bore profile which makes a B$^\flat$ trumpet different from a B$^\flat$ flugelhorn and a tenor valve trombone different from a euphonium. It is also the finer details of bore profile which largely distinguish the different models of instrument a maker will offer, the instruments of one maker from those of another, and good instruments from bad. This was realised in the middle of the nineteenth century; for example, the inventor and maker Adolphe Sax wrote (Sax 1850):

> Proportions are the governing laws and constitute the nature of the instrument; indeed it is not the form that gives them their voice, their sound quality: it is only the proportions. These proportions are, therefore, different for each species of instrument.

The design of the mouthpiece is an integral part of the design of a brass instrument; however the mouthpiece has specific acoustical functions, and as the interface between instrument and player is of critical importance on several counts. Although instrument makers often include a mouthpiece when they supply an instrument, players usually make a separate choice and will often use a single mouthpiece to play different instruments. For these reasons it is helpful to consider instrument taxonomy and mouthpiece taxonomy separately (see Sect. 7.4).

The bore profile of an instrument can be infinitely varied by the designer, and a complete description using a small number of parameters is impossible. In Sects. 4.3 and 6.1 we discussed the effects of bore size (tube diameter) and nonlinear propagation (brassiness) on timbre. The values of the two parameters D_{min} and B effectively determine much of the nature of an instrument of a given tube length. The timbre of an instrument depends of course on playing technique and mouthpiece choice, but together D_{min} and B largely characterise the contribution of the instrument itself to the final sound. Traditionally instrument makers will have made a conscious decision in specifying D_{min}, but will have followed convention, experience and creativity in designing the bore shape which of which B is one measure. Indeed, small adjustments to bore shape may have negligible effect on the value of B but important effects on intonation.

The mechanics of the instrument—whether natural or with toneholes, slide or valves, or played with a hand in the bell—limit the possibilities for bore profile. The application of toneholes has been most successful with instruments which have a bore which expands throughout, often loosely referred to as 'conical' instruments. Opening a tonehole effectively removes a section of the tube at the end further from the mouthpiece from the sounding length, and a cone when truncated remains a cone. Conversely, the application of slides has been most successful with instruments which have a bore much of which is close to cylindrical. In order to obtain a substantial chromatic compass from the 2nd natural note upwards, the sliding parts have to add up to one-third of the minimum sounding length which requires a substantial proportion of near-cylindrical sliding tube. The invention of the valve opened up a wide range of possibilities for effective instruments, and many types of instrument were created in the nineteenth century as makers exploited these possibilities and introduced types such as the cornet and the tuba which had no pre-valve equivalents. Not all points in the three-dimensional space defined by sounding length (L), bore size (D_{min}) and brassiness potential (B) are occupied by recognised instruments; indeed, some points may represent instruments which could be useful but have not yet been invented.

In this chapter the values of L and B are given for the instruments with all toneholes covered (in the case of finger-hole and keyed instruments), or the shortest length (for slide and valved instruments). In most cases the latter is the length giving the nominal pitch of the instrument. Obviously opening toneholes, moving a trombone slide, or operating valves will give a different bore shape and thus change the value of B to some extent. Strictly speaking, a chromatic instrument should be represented by a series of points on a plane defined by constant D_{min} in this three-dimensional space, and it is apparent to players if not to audiences that the timbre changes as hole coverings, slides and valves are changed. For an instrument to have a clear identity, the variations in timbre between adjacent notes cannot be too great. The increases in B when slides are extended or valves operated are small, however, especially for the more 'cylindrical' instruments with high values of B, but are more significant for 'conical' instruments where slides and valves introduce sections of approximately cylindrical tubing.

7.2 The Different Kinds of Brass Instrument

It is a valid question whether the hundreds of nominally different instruments really all respond to the player and sound differently. After discussing valved brass instruments in some detail, Carse (1939) states provocatively:

> No doubt other instruments, all in varying degree akin to the cornet, bugle or ophicleide, could be found if the records, catalogues and patent files of last century were searched and the museums ransacked. The collection of varied types could be enlarged by including some solitary instruments made only in one size and named according to register, kinship to type, or perhaps with the idea of perpetuating the name of some optimistic inventor. Most of these would probably fail to establish a claim to any individual existence, for the field is limited, and there is not room for any great variety between the tone quality of the cornet and that of the bugle, whether large or small; nor does the admixture of trumpet, horn, or trombone bore, and their characteristic mouthpieces, supply sufficient variety to provide very many new and clearly different tone qualities. Many claimants to a separate existence within this restricted field have had to give up their pretended individuality and throw in their lot with the common types that are in use today. The flügelhorns and contralto saxhorns, the tenorhorns and baritones, the tubas and bombardons may be differently named in each country, or may even be differently named in the same country, but their nomenclature is always more varied than their tone qualities. Different widths of bore and diversity of mouthpiece cup will give variety of tone quality within a certain radius, but that radius is limited in extent. In the highest register, the field of brass instruments in high E flat, it matters little to the hearer whether the instrument be a trumpet, cornet, saxhorn or flügelhorn. In the contralto or B flat register, there is room enough for the cornet and the flügelhorn, but hardly for anything in between the two. So it is in the tenor or E flat register, the baritone and the bass registers; we can admit instruments which are large sized cornets or large sized bugles, but anything between these two makes the distinction too fine for ordinary ears, and therefore too fine for practical use ...

Carse may be correct in suggesting that 'ordinary ears' can distinguish no more than two differing types of brasswind in each register. There is no doubt that trained musicians can recognise more than two types, if not as auditors then certainly as performers. The continued production by individual manufacturers of a wide range of instruments as well as differing models of the most popular instruments, nominally the same type, is commercially justified only by purchasers perceiving differences.

In the following subsections we examine some of the characteristics of the more widely used modern and historical brasswind species. Rather than a typological or chronological arrangement our approach is organised by sounding length (L), bore size (D_{min}) and brassiness potential (B). The photographs are of examples in the Edinburgh University Collection of Historic Musical Instruments (EU) and the collection of the Royal Conservatoire of Scotland (RCS). In most cases the instruments are photographed without mouthpieces. The scatter diagrams are populated by data from measurements of instruments typical of their species in the collections of some seventy musicians and institutions.

The bore profiles presented in the following subsections give for typical instruments the bore diameter plotted against the distance from the mouthpiece receiver, the instruments being treated as if straightened out. The axial lengths are measured along the centre line of the bore, which is generally approximately circular in cross-section. Figure 7.1 shows the complete bore profiles of the sixteenth-century trombone (sackbut) and the twentieth-century trombone pictured in Figs. 7.29 and 7.30. The dramatic difference in the diameters of the bell ends is an important distinction between early and modern trombones: the influence of bell size on sound radiation is discussed in Sect. 4.6. The sounding lengths also differ: both instruments

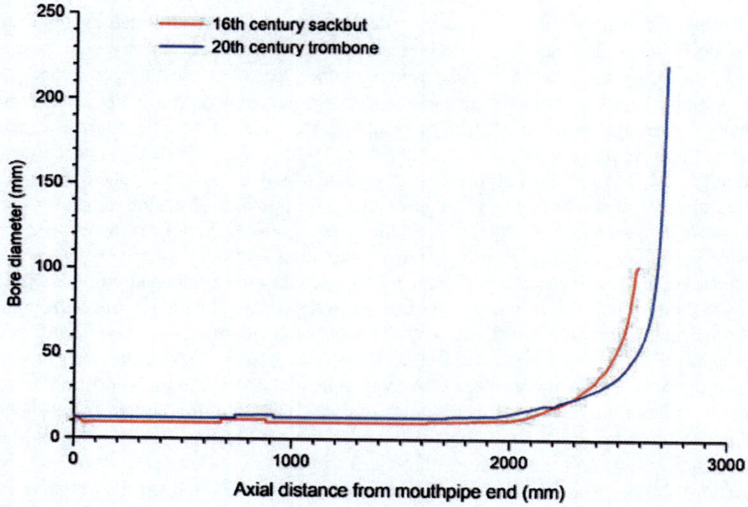

Fig. 7.1 Bore profiles of tenor sackbut in 9-ft B$^\flat$ (Anton Schnitzer II, Nürnberg, 1594) EU (2695) and tenorbass trombone in 9-ft B$^\flat$ (Herbert Gronitz, Hamburg, mid-late 1950s) EU (4115)

have a nominal pitch of B$^\flat$, but the early trombone was built to the higher pitch (A4 = 465 Hz) prevalent in the sixteenth century and is shorter than that of a modern trombone (A4 = 440 Hz). The difference in bore profile also accounts for some difference in length. The abrupt step up in bore diameter at around 700 mm and back down again around 900 mm are the discontinuities in the bore where the inner slides of the trombone end and the necessarily wider outer slides continue. These steps can be seen in all the slide trombone bore profiles in the following subsections.

Although the profile of the bell flare has importance for radiation of sound, associated parameters such as the bell diameter and the cut-off frequency do not provide a reliable means of distinguishing consistently between all species of brass instrument. The brassiness potential B is a more useful parameter, and has been used systematically in the following subsections. The brassiness potential is derived from the bore profile over the entire length of the instrument, but wide sections of the tube make relatively little contribution (see Sect. 6.1). In order to show the narrower sections of tubing more clearly, the diameters in the bore profile plots are shown on a larger scale than the lengths and the widest part of the bell is generally omitted.

7.2.1 *Instruments with the Shortest Tube Lengths*

Here we consider instruments with basic tube lengths less than approximately 500 mm.

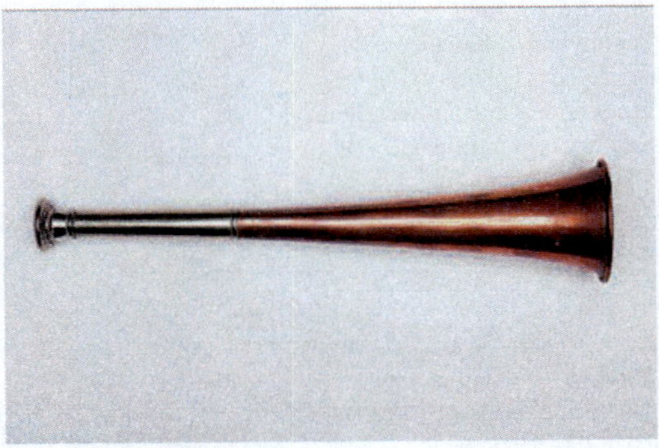

Fig. 7.2 English fox-hunting horn (Keat, London, c. 1960). EU (2346)

Fig. 7.3 Cornettino (anonymous, date uncertain). RCS (263)

Brass instruments with basic tube lengths shorter than 2-ft C are generally either signalling instruments on which one note is sounded (such as the English fox-hunting horn shown in Fig. 7.2) or tonehole instruments, most prominently the cornettino. The cornettino shown in Fig. 7.3 has a tube length of 424 mm without mouthpiece. When the thumb-hole and all six finger-holes are covered the lowest natural note D4 can be sounded, and by appropriate fingering a complete chromatic scale upwards for over two octaves obtained. The lowest note can be 'lipped' down by two semitones (Sect. 1.2.2), so the cornettino is sometimes referred to as an instrument 'in C'. However with all toneholes closed the natural notes from the second upwards are close to harmonics of $E^{\flat}4$, which is therefore the nominal fundamental pitch of the instrument. The upper end of the range is limited by the difficulty of sustaining a high-frequency vibration of human lips: even with the tiny mouthpiece used for the cornettino the vibrating portion of the lip is too massive and insufficiently rigid to oscillate easily much above 1500 Hz. The upper limit of the compass is given in Walther's *Musicalisches Lexicon* (Walther 1732) as D6, with optional extension to G6 or even A6.

7.2.2 Instruments with Very Short Tube Lengths in C and B$^{\flat}$

Here we consider instruments in 2-ft C and $2\frac{1}{4}$-ft B$^{\flat}$, basic tube lengths between approximately 500 and 700 mm.

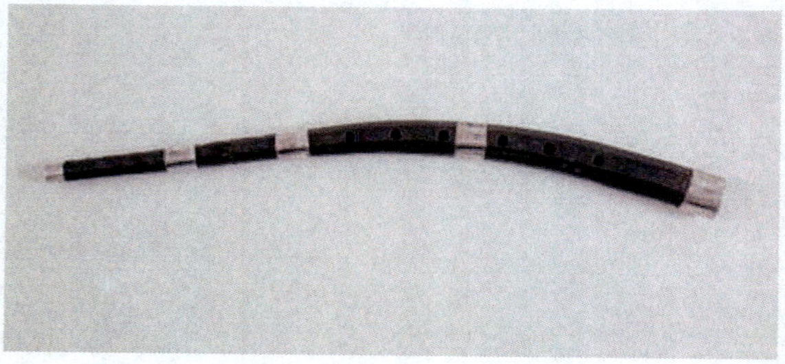

Fig. 7.4 Cornett (anonymous, Italian, early seventeenth century); silver rings added during twentieth century restoration. EU (3189)

Fig. 7.5 Piccolo trumpet (Mahillon, Brussels, 1929). EU (3900)

The pitch standard for wind instruments in northern Italy in the first half of the seventeenth century was A4 = 465 Hz (Haynes 2002); at this standard the first natural note of the treble cornett illustrated in Fig. 7.4 with all toneholes closed is A3. This note can be lipped down two semitones to G3. The natural notes from the second to the sixth are close to harmonics of $B^\flat 3$, so the appropriate nominal pitch label is $2\frac{1}{4}$-ft $B^\flat$. As with other finger-hole labrosones, the fingering for notes in the lowest register is not exactly replicated in the first overblown register. In the sixteenth century the compass of the treble cornett was G3–A5, but in the following century the upper limit was extended to D6 and beyond. A late seventeenth-century opera score by Bononcini contains soloistic cornett passages up to $F^\sharp 6$, but it is possible that this and similar extremely high parts were played on cornettini (Collver and Dickey 1996).

The timbre of the cornett has often been compared to that of a treble voice, and the blend of cornetts and sackbuts with voices was exploited successfully by Giovanni Gabrieli and Claudio Monteverdi. The flexibility and expressive potential of the treble cornett made it the supreme virtuoso wind instrument in the late sixteenth and early seventeenth centuries, and most of the cornetts surviving from this period are trebles with nominal fundamental pitch $B^\flat 3$. Like the violin, concert flute and tenor trombone, the treble cornett can be seen as occupying a 'sweet spot' where making the instrument any smaller detracts from the tone quality while scaling the instrument up in size makes the instrument difficult to manipulate.

The piccolo trumpet is, like the treble cornett, an instrument in $2\frac{1}{4}$-ft $B^\flat$, with a nominal fundamental pitch of $B^\flat 3$. Unlike the treble cornett, the first natural note (described as the 'pedal') is not part of the normal compass; with no valves operated the lowest pitch commonly used is the second natural note $B^\flat 4$. The lowest note of a three-valve piccolo trumpet, played with all three valves operated, is E4. Many piccolos have four valves, allowing notes such as D4 required by some of the repertoire of the instrument to be played (Fig. 7.5).

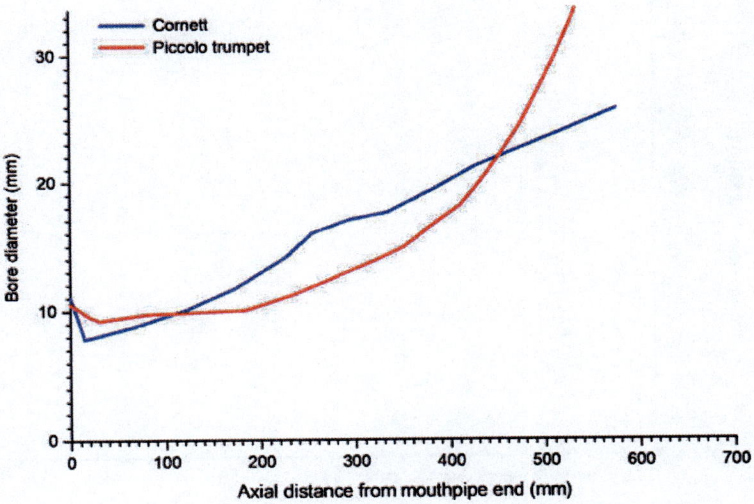

Fig. 7.6 Bore profiles for the cornett and piccolo trumpet shown in Figs. 7.4 and 7.5. The vertical (bore diameter) scale is exaggerated to show the differences in the significant narrower parts of the instruments; the wider part of the trumpet bell flare is omitted

Because the upper end of the range is dictated by the physical properties of the human lip rather than by the limitations of the instrument, the piccolo trumpet does not have a compass extending significantly higher than trumpets with longer sounding lengths: it is employed primarily to give the player the greater security resulting from a higher cutoff frequency (see Sect. 2.2.5) and from using lower, wider spaced, members of the series of natural notes. Spectral enrichment in piccolo trumpets is considerably less than in longer trumpets playing the same pitches, other things being equal (see Sect. 6.1.4). Since nonlinear propagation and brassiness are of relatively minor importance in these short instruments, there is no real distinction in terms of timbre between piccolo trumpets and differently-named instruments which have been built at this tube length, the piccolo cornet and the 'petit saxhorn suraigu' (the highest member of the saxhorn family). The piccolo bore profile shown in Fig. 7.6 has minimal cylindrical bore; the scatter diagram Fig. 7.7 shows piccolo trumpets to differ widely in bore profile.

7.2.3 *Instruments with Short Tube Lengths in G, F, E$^\flat$ and D*

Here we consider instruments in $2\frac{3}{4}$-ft G, 3-ft F, $3\frac{1}{4}$-ft E$^\flat$ and $3\frac{1}{2}$-ft D, basic tube lengths between approximately 700 and 1100 mm (Figs. 7.8–7.12); Fig. 7.13 shows the bore profiles for some of these.

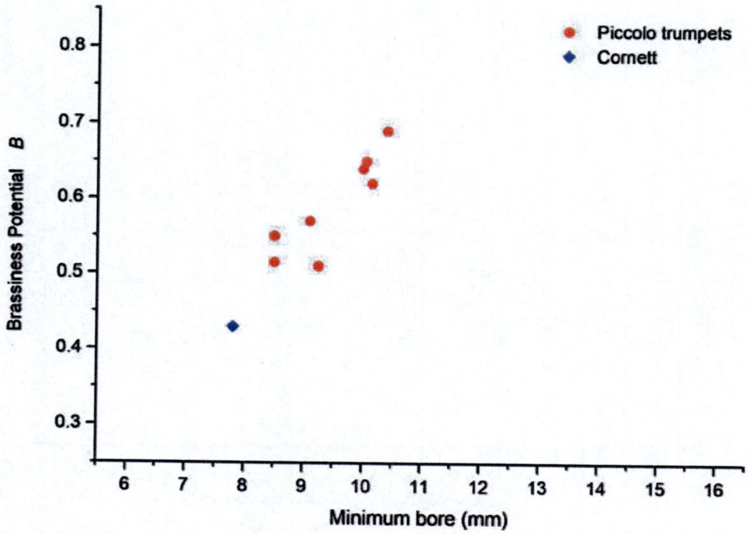

Fig. 7.7 B / $D_{\min}$ scatter plot for a variety of piccolo trumpets and the cornett shown in Fig. 7.4

Fig. 7.8 Keyed bugle in $3\frac{1}{4}$-ft E^{b} (Wigglesworth, Otley, mid nineteenth century). RCS (241)

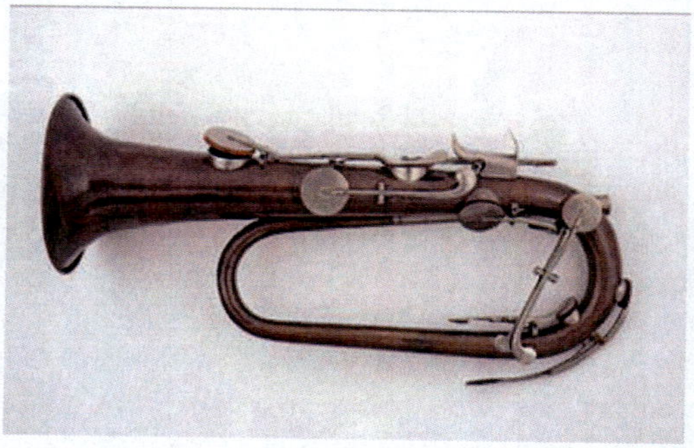

Finger-hole instruments longer than the standard treble cornett in $2\frac{1}{4}$-ft B^{b} have been made, but have been relatively rare, perhaps due to compromises necessary to accommodate the limited stretch of the hand. Keyed brass instruments in $3\frac{1}{4}$-ft E^{b}, however, have been widely used and at one time provided the main melody voice of the American brass band. The key nearest to the bell of the E^{b} keyed bugle stands open at rest, and when operated gives the lowest note D4 (the first natural note E^{b}3 not normally being used). Six further keys provide a complete chromatic compass upwards from D4, though not all bugles have seven keys (when fewer, the less frequently required notes are obtained by fork fingering).

Valve instruments at this pitch show some effect of bore profile on timbre, though the character of different kinds of instrument are not always very obvious. This is particularly apparent when instruments sold as trumpets have been made with mouthpiece receivers as narrow as those of cornets. The scatter diagram Fig. 7.14

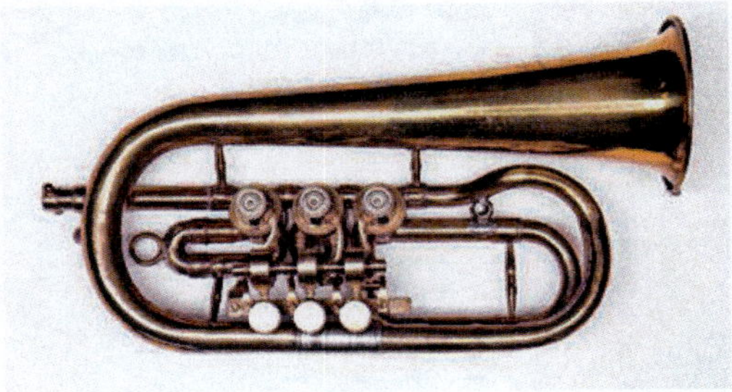

Fig. 7.9 Swedish kornett in $3\frac{1}{4}$-ft $E^{\flat}$ (Ahlberg and Ohlsson, Stockholm, late nineteenth century). EU (3499)

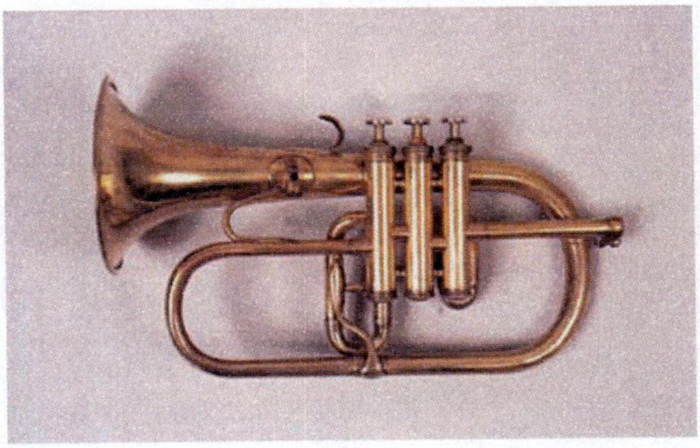

Fig. 7.10 Petit bugle (soprano flugelhorn) in $3\frac{1}{4}$-ft $E^{\flat}$ (Ouvriers Réunis, Paris, late nineteenth century). EU (3856)

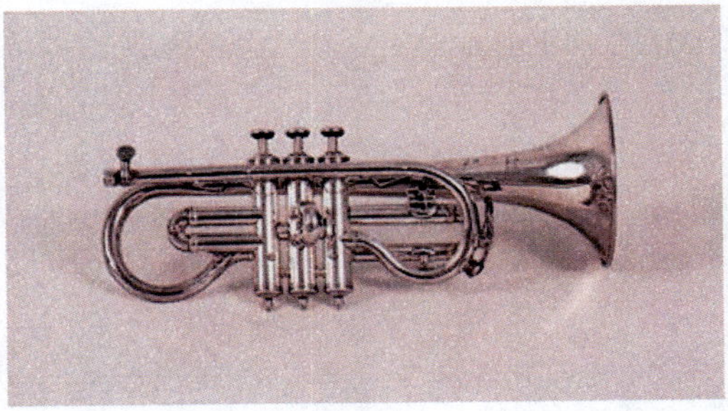

Fig. 7.11 Soprano cornet in $3\frac{1}{4}$-ft $E^{\flat}$ (Antoine Courtois, Paris, 1872–1878). EU (2553)

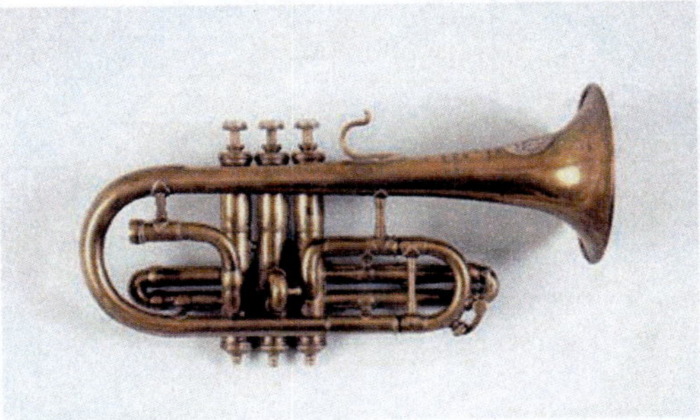

Fig. 7.12 Valve trumpet in $3\frac{1}{4}$-ft $E^{\flat}$ (Mahillon, Brussels, early twentieth century). EU (4259)

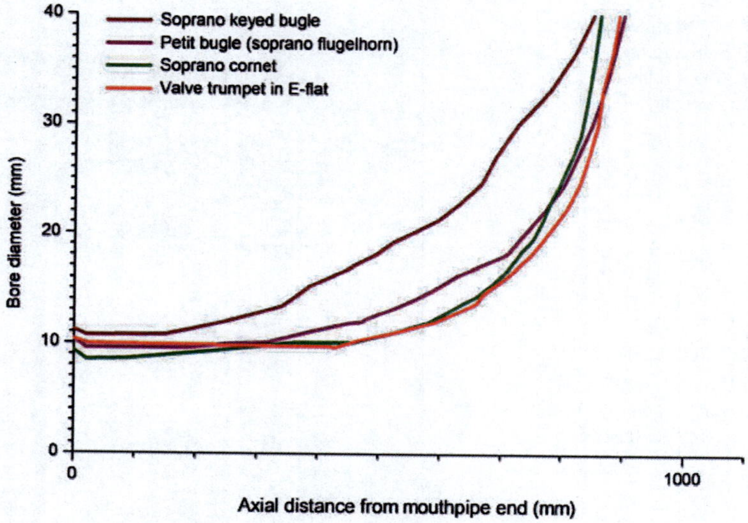

Fig. 7.13 Bore profiles for four instruments shown above: small keyed bugle, petit bugle (soprano flugelhorn), soprano cornet, and small valve trumpet. The vertical (bore diameter) scale is exaggerated to show the differences in the significant narrower parts of the instruments; the wider parts of the bell flares are omitted

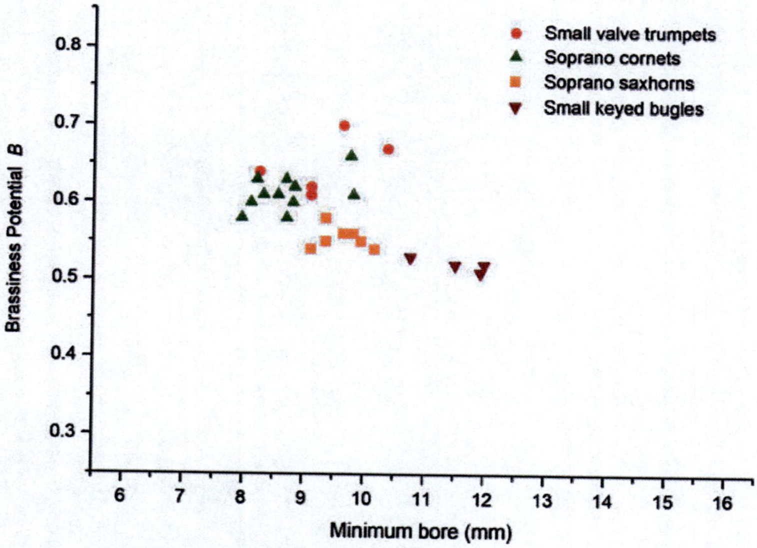

Fig. 7.14 $B / D_{\min}$ scatter plot for some typical small valve trumpets, soprano cornets, soprano saxhorns and soprano keyed bugles

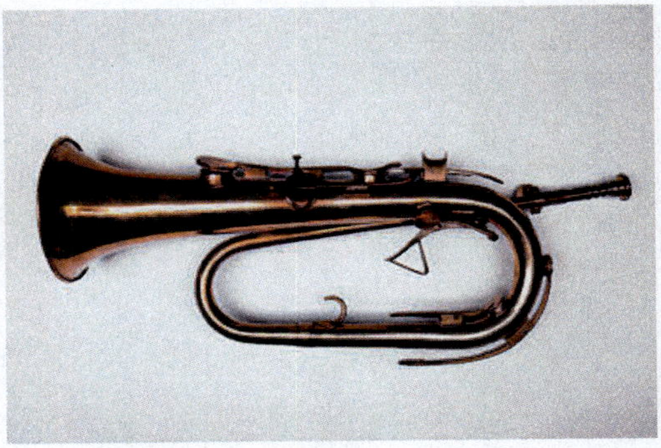

Fig. 7.15 Keyed bugle in 4-ft C (Halari, Paris, c. 1830). EU (907)

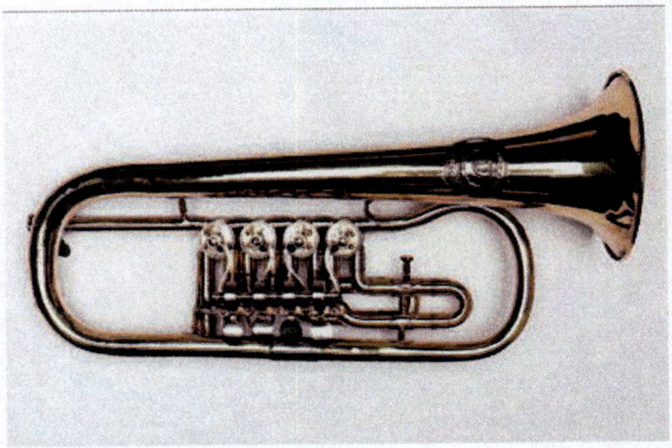

Fig. 7.16 Flugelhorn in $4\frac{1}{2}$-ft B$^\flat$ (Červený, Königgrätz, c. 1900). EU (3664)

shows some overlap between the family members, and musicians have often used instruments interchangeably with impunity, for example playing soprano cornet parts on E$^\flat$ trumpets.

7.2.4 *Instruments with Short Tube Lengths in C and B$^\flat$*

Here we consider instruments in 4-ft C and $4\frac{1}{2}$-ft B$^\flat$ with basic tube lengths between approximately 1.1 and 1.5m (Figs. 7.15–7.19); Fig. 7.20 shows their bore profiles. Orchestral trumpets are most commonly in B$^\flat$ or C; B$^\flat$ trumpets take the lead in jazz, swing, and many wind bands; the characteristic voice of the brass band is provided by the B$^\flat$ cornet. The B$^\flat$ flugelhorn is also much used .

Fingerholes are rather widely spaced for this and longer tube lengths: the tenor cornett in 4-ft C (using its fundamentals) has seen very limited use. Keyed brass instruments in 4-ft C were for a short time from the 1810s widely used, though soon giving way to valved instruments especially the cornet which was developed in the 1830s. Slide instruments in this size range have been used to a small extent: the soprano trombone has slide positions so close together that placing the slide accurately enough for good intonation and fast enough for fluent delivery is

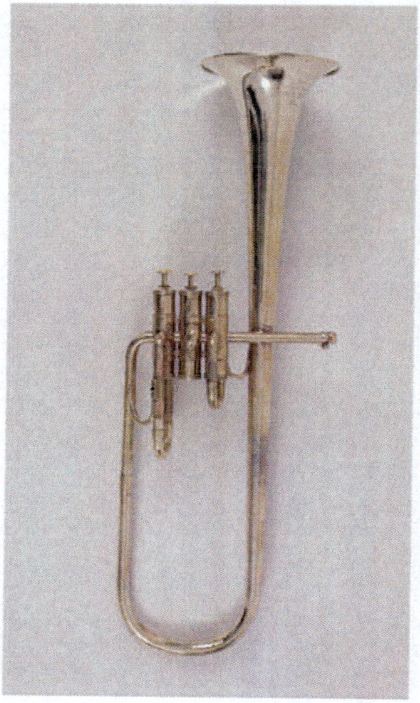

Fig. 7.17 Contralto saxhorn in $4\frac{1}{2}$-ft $B^{\flat}$ (Adolphe Sax, Paris, 1867). EU (4253)

Fig. 7.18 Cornet in $4\frac{1}{2}$-ft $B^{\flat}$ (Antoine Courtois, Paris, 1856-1858). EU (3475). Shown with shank for $B^{\flat}$ (inserted), shank for A, and crooks for $A^{\flat}$ and G

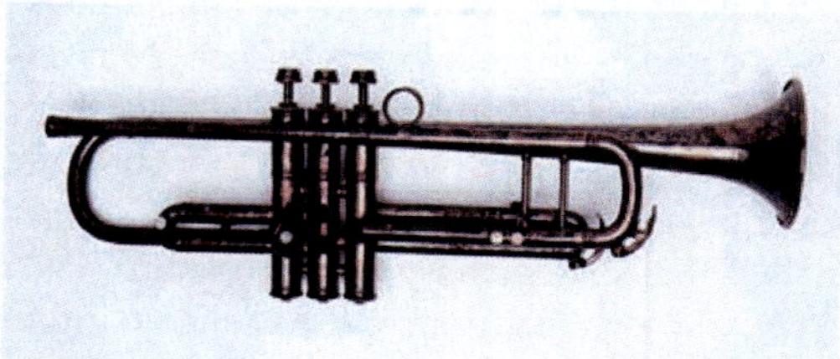

Fig. 7.19 Valve trumpet in $4\frac{1}{2}$-ft $B^{\flat}$ (York, Grand Rapids, 1925-27). EU (5772)

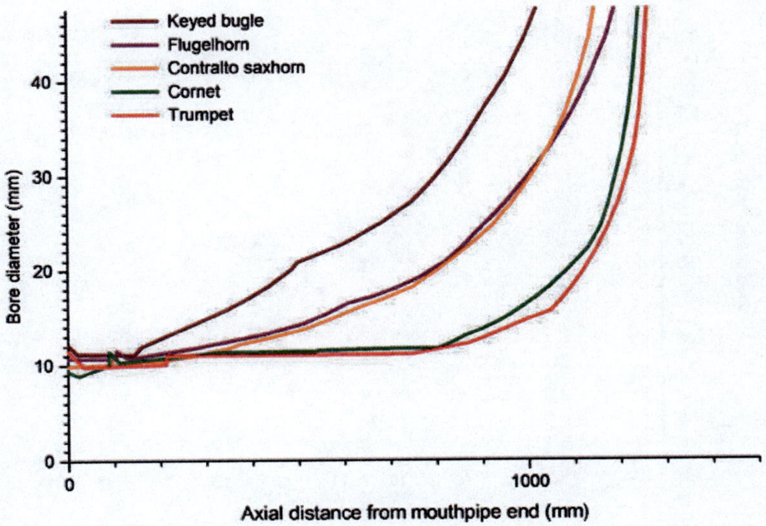

Fig. 7.20 Bore profiles for the five instruments shown above: keyed bugle, flugelhorn, contralto saxhorn, cornet, and valve trumpet. The vertical (bore diameter) scale is exaggerated to show the differences in the significant narrower parts of the instruments; the wider parts of the bell flares are omitted

impracticable. The most widely used valved melody instruments have been pitched in 4-ft C or $4\frac{1}{2}$-ft $B^{\flat}$. The flugelhorn, cornet and trumpet display some overlap in the scatter diagram Fig. 7.21, but each has its own home territory. The evolutionary convergence of the trumpet and cornet is discussed in Sect. 7.5. The contralto saxhorns, on the other hand, did not establish an individual acoustical identity, there being no space between cornets and flugelhorns for a distinct sonority.

7.2.5 *Instruments with Medium Tube Lengths in G, F, $E^{\flat}$ and D*

Here we consider instruments in $5\frac{1}{2}$-ft G, 6-ft F, $6\frac{1}{2}$-ft $E^{\flat}$ and 7-ft D, basic tube lengths between approximately 1.5–2.25 m (Figs. 7.22–7.25); Fig. 7.26 shows their bore profiles.

The instruments in this range of tube lengths with lesser expansion of bore have been natural trumpets and alto sackbuts and trombones. With a tube length of around 2 m, a bore of 10–11 mm, and a high bell cutoff frequency of around 1500 Hz, the natural trumpet is able use the upper natural notes. With an appropriate mouthpiece (discussed in Sect. 7.4) and highly developed skill, notes from the 8th to the 16th natural notes are possible, and even higher notes were written for virtuoso players. A skilled natural trumpet player can not only play reliably in this high

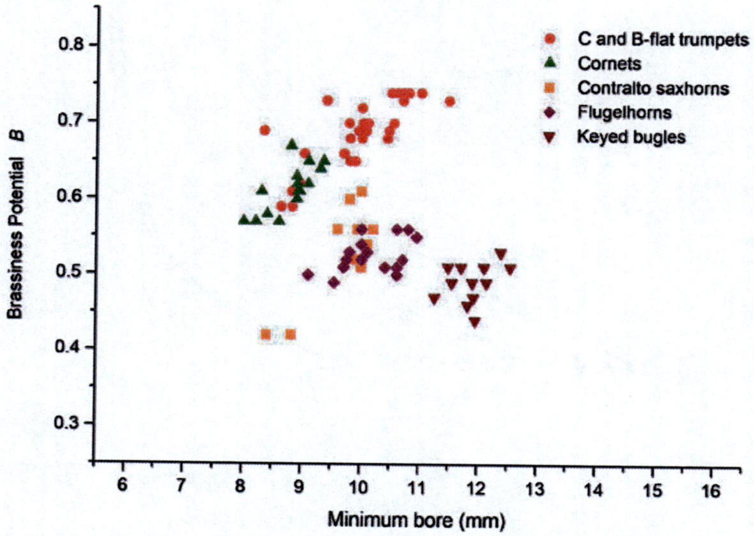

Fig. 7.21 $B / D_{\min}$ scatter plot for some typical valve trumpets, cornets, contralto saxhorns, flugelhorns, and keyed bugles

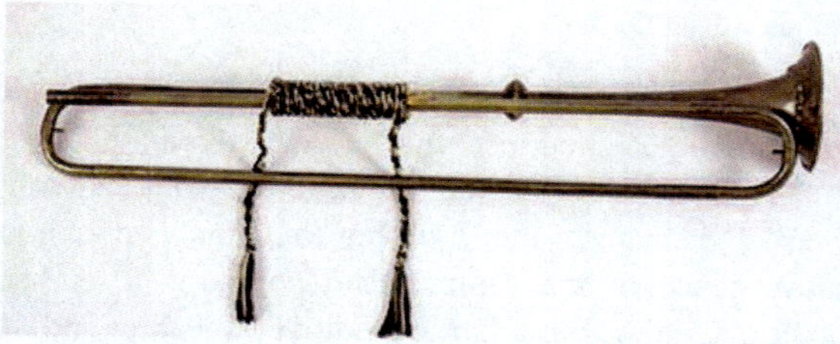

Fig. 7.22 Trumpet in 7-ft D (Joseph Huschauer, Vienna, 1794). EU (3247)

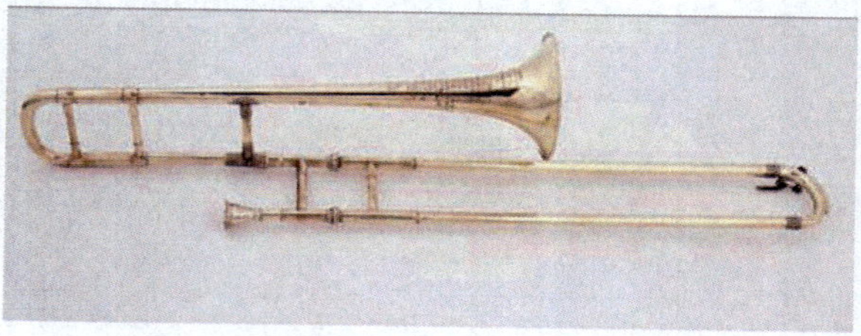

Fig. 7.23 Alto trombone in $6\frac{1}{2}$-ft E$^\flat$ (Courtois and Mille, Paris, 1878-1900). RCS (508)

(clarino) register, where the difference in lip muscle setting between the desired note and those above and below is very small, but also 'lip' the 7th, 11th, 13th and 14th and other notes into agreement with the temperament adopted by an ensemble, and play 'out-of-series' notes such as the written B$^\natural$ between the 7th natural note (written B$^\flat$) and the 8th (written C). This clarino technique is only feasible at low and moderate dynamics; the natural trumpet used as a fanfare instrument at high dynamics, giving a brassy sound, has generally exploited the lower tessitura from the 3rd to the 6th natural notes. The security of attack and the bright timbre are

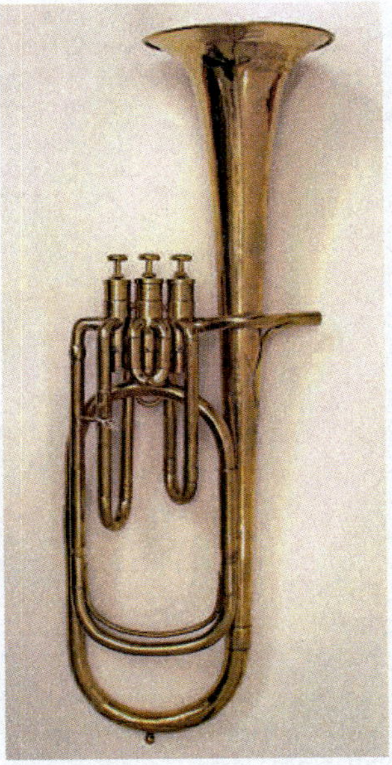

Fig. 7.24 Tenor saxhorn in $6\frac{1}{2}$-ft E$^{\flat}$ (Adolphe Sax, Paris, 1855). EU (4543)

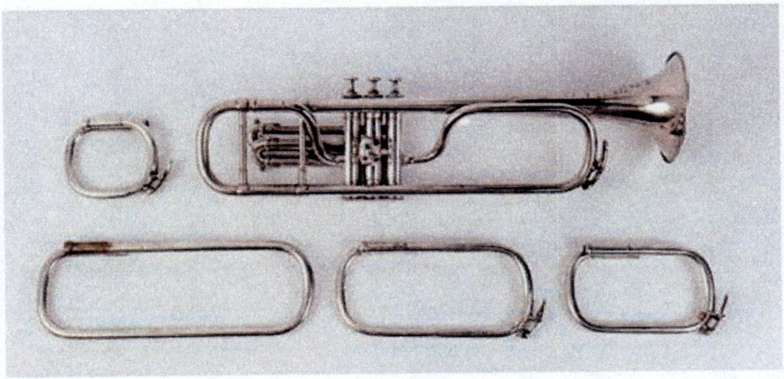

Fig. 7.25 Valve trumpet in $5\frac{1}{2}$-ft G (Antoine Courtois, Paris, c. 1885). EU (6078). Shown with crooks for F, E, E$^{\flat}$ and D

supported by the close to harmonic higher resonances (see Sect. 2.2.3) which form a more extensive series in the natural trumpet than in the bugle (Myers and Campbell 2006).

The areas of the $B\,/\,D_{\text{min}}$ scatter diagram (Fig. 7.27) populated by natural trumpets and alto sackbuts largely overlap. The alto sackbut has a fully chromatic compass of over two octaves without going higher than 8th natural note, best served by a different mouthpiece and playing technique from those of the natural trumpet.

The instruments with greater expansion of bore correspond in tessitura to the alto voice and have been instruments with a primarily harmonic function such as tenor saxhorns, and mellophones. This supporting role is probably the reason for the rather

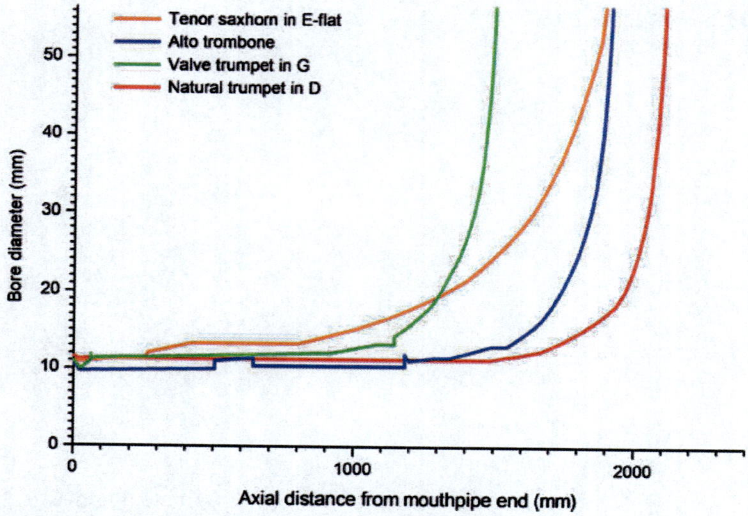

Fig. 7.26 Bore profiles for the four instruments shown above: natural trumpet, alto trombone, tenor saxhorn, and valve trumpet. The vertical (bore diameter) scale is exaggerated to show the differences in the significant narrower parts of the instruments; the wider parts of the bell flares are omitted

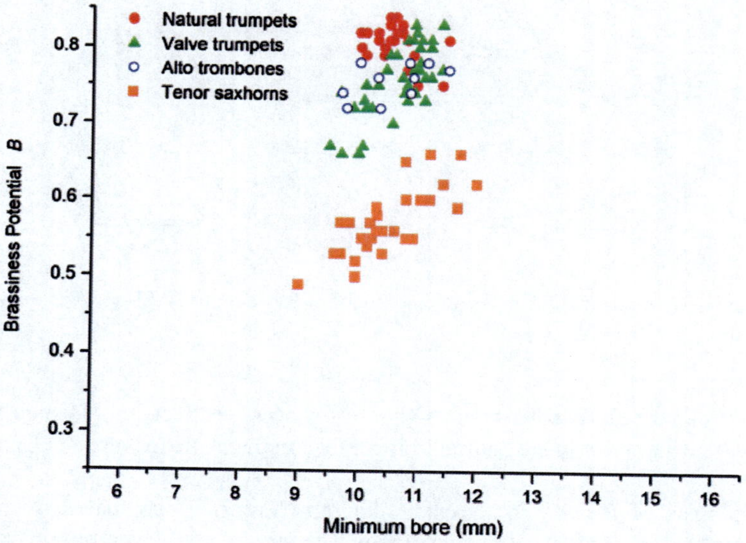

Fig. 7.27 B / $D_{\min}$ scatter plot for some typical natural trumpets, valve trumpets, alto trombones, and tenor saxhorns

weak distinction between the kinds of instrument, which include not only tenor saxhorns and mellophones but also alto cornophones, néocors, clavicors, Koenig horns, and other instruments differentiated more by their name than any acoustical properties (Figs. 7.29 and 7.30).

7.2.6 Instruments with Long Tube Lengths in C and B$^{\flat}$

The principal tenor and harmony instruments are today in 9-ft B$^{\flat}$. Tenor and bass trombones (both with nominal fundamental pitch B$^{\flat}$1) form the trombone sections in orchestras and bands of many kinds; the B$^{\flat}$ french horn and the B$^{\flat}$ side of the double horn are the most regularly used horns; the tenor solo instrument in the brass band and many military bands is the euphonium, and the B$^{\flat}$ baritone is a traditional member of the brass band.

In this Section we consider instruments in 8-ft C and 9-ft B$^{\flat}$, with basic tube lengths between approximately 2.25 and 3 m (Figs. 7.28–7.33); Fig. 7.34 shows the bore profiles for some of these. Historical bass trombones in F and G and the F side of the double horn are discussed in Sect. 7.2.7.

The basic tube lengths of 8-ft C and 9-ft B$^{\flat}$ have displayed the greatest variety of instrument type. The tonehole instruments (serpent and ophicleide) have bores which, unconstricted by valves, expand rapidly from the mouthpiece giving very low values of B. The lowest members of the series of natural notes are used, so the compass descends to a little below C2. Since there is minimal spectral enrichment due to nonlinear propagation the low-frequency components of the radiated sound are strong and the instruments are effective in delivering a bass in an ensemble;

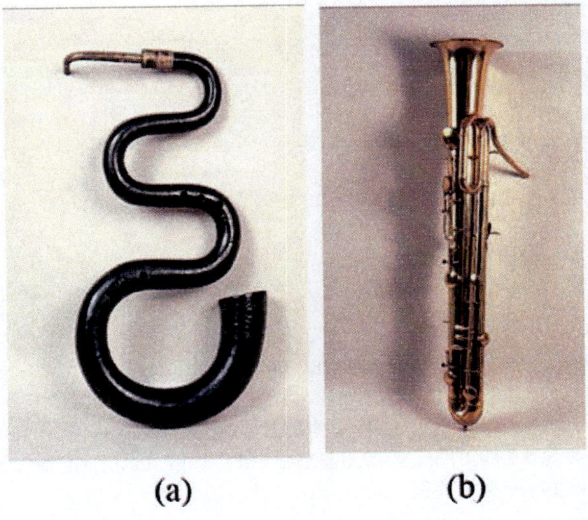

Fig. 7.28 (**a**) Serpent d'eglise in 8-ft C (C. Baudouin, Paris, c. 1820). EU (3606). (**b**) Ophicleide in 9-ft B$^{\flat}$ (Gautrot, Paris, c. 1860). EU (3590)

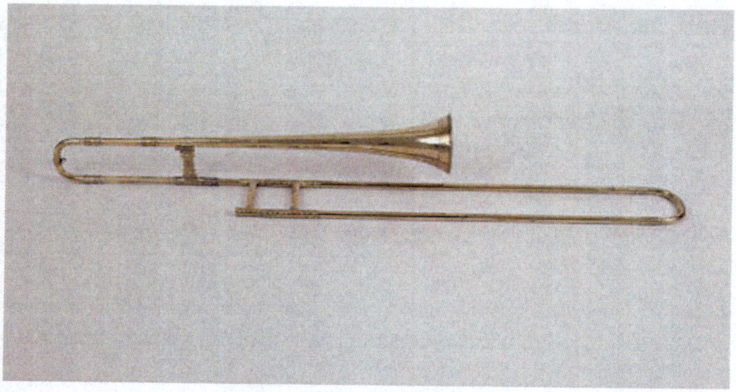

Fig. 7.29 Tenor sackbut (Anton Schnitzer II, Nürnberg, 1594). EU (2695)

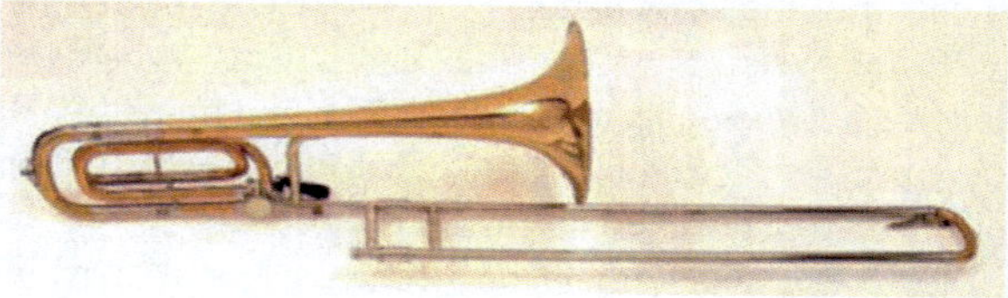

Fig. 7.30 Tenorbass trombone in 9-ft B$^\flat$ with thumb valve for F (Herbert Gronitz, Hamburg, mid-late 1950s). EU (4115)

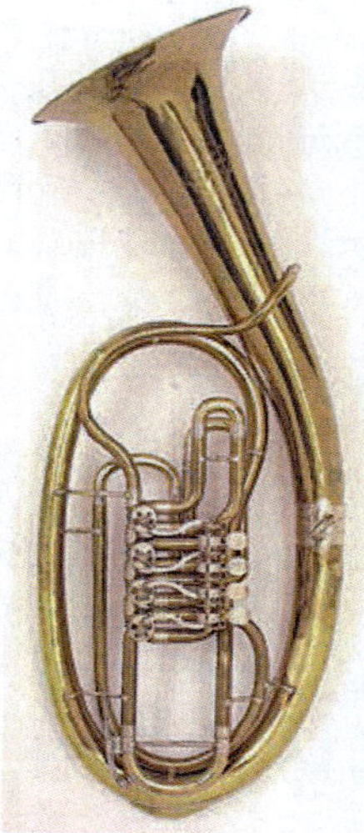

Fig. 7.31 Wagner tuba in 9-ft B$^\flat$ (Alexander, Mainz, early 1930s). EU (2515)

they give an effect of depth comparable to that given by the contrabassoon which is pitched an octave lower but has weak low-frequency spectral components.

The instruments with least expansion of bore are the slide trombones. The sackbut with its very narrow bore, high value of B, and small bell flare has a low transfer function: it requires a significantly higher acoustic pressure amplitude in the mouthpiece than the wide-bore modern trombone to produce a given performance

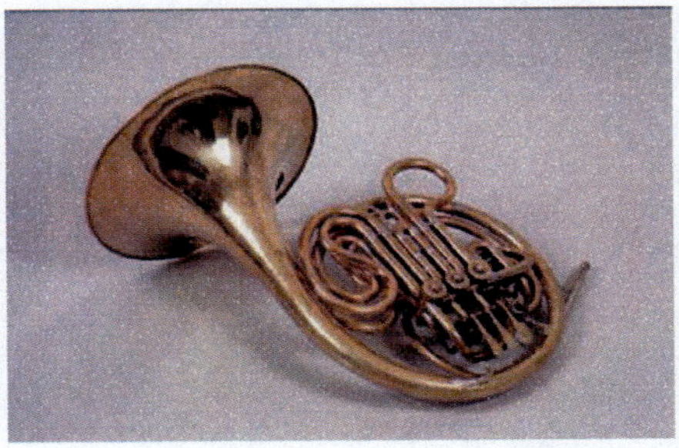

Fig. 7.32 Double horn (Alexander, Mainz c. 1950). EU (1804)

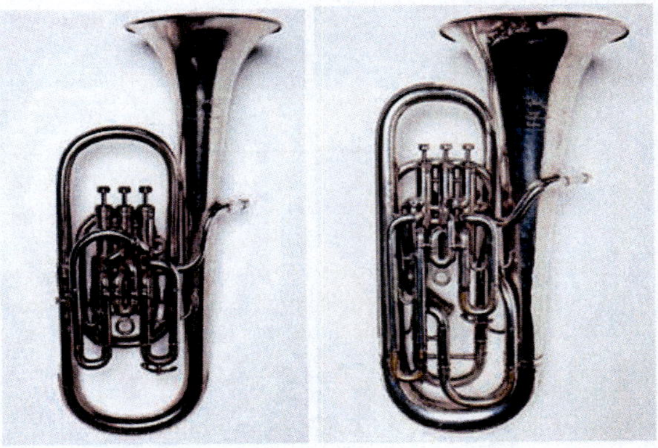

(a) (b)

Fig. 7.33 (**a**) Baritone saxhorn in 9-ft B^b (Boosey and Hawkes, London 1962). EU (3887). (**b**) Euphonium in 9-ft B^b (Boosey and Hawkes, London, 1961-62). EU (4667)

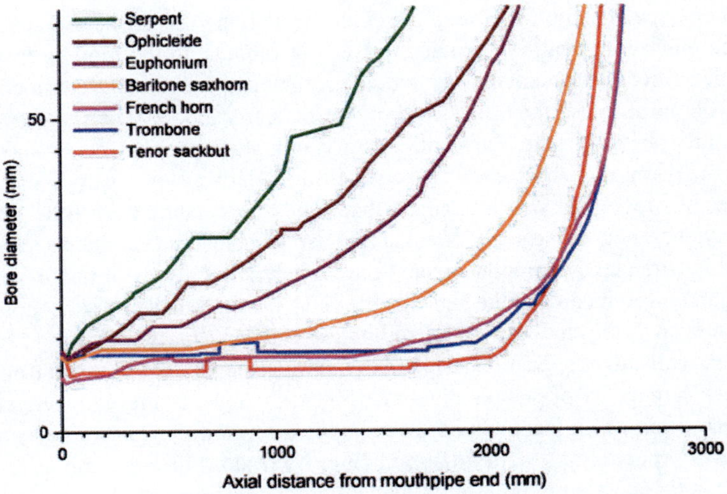

Fig. 7.34 Bore profiles for seven of the instruments shown above: serpent, ophicleide, euphonium, baritone, french horn (the B^b side of the double horn), modern trombone, and tenor sackbut. The vertical (bore diameter) scale is exaggerated to show the differences in the significant narrower parts of the instruments; the wider parts of the bell flares are omitted

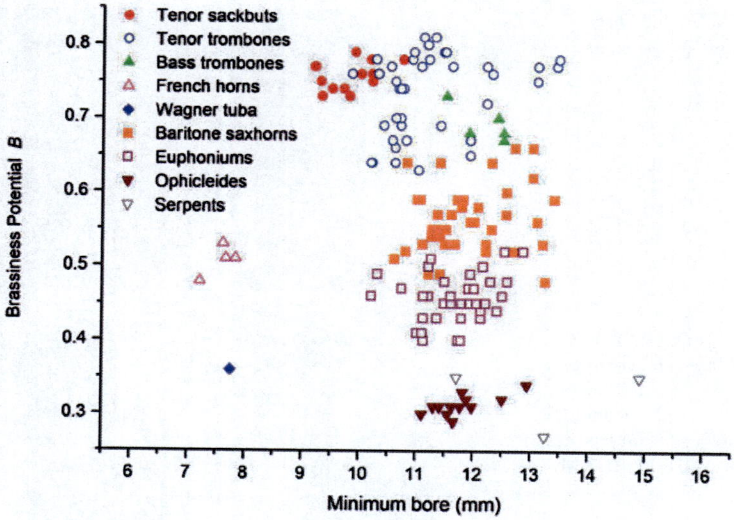

Fig. 7.35 B / D_{min} scatter plot for some typical tenor sackbuts, tenor trombones, modern bass trombones, french horns (double horns, B$^{\flat}$ side); the B$^{\flat}$ Wagner tuba shown above; typical baritone saxhorns, euphoniums, ophicleides, and serpents

dynamic (Myers et al. 2012; Campbell et al. 2014a). As a result, the sackbut has an enriched timbre at low and moderate dynamic levels making it a good match for other Renaissance period instruments and for the voice. Although there is a continuous line of evolution from the sackbut through the classical period trombone, then the nineteenth-century French and German models, to the modern trombone, the Renaissance instrument and the modern instrument are so different in character that it is helpful to use the term 'sackbut' for the early model. The valve trombone was highly popular in the nineteenth century, though rather neglected today. It has been considered that the inevitable bends in the windway of a valved instrument impaired the tone. We saw in Chap. 4 that bends in the tube have little effect on low frequencies but some small effect at high frequencies: the enriched spectra with strong high-frequency components and the high dynamic levels of much trombone playing may well mean that the bends in the valve passages have an effect noticeable to the player. Many modern slide trombones with one or two thumb valves are equipped with valves specially designed to minimise bends. The fact that trombones occupy a large area of scatter diagram (Fig. 7.35) reflects the wide variation in trombone design as it has evolved from classical to modern. The valve trombone, not shown separately in scatter diagram (Fig. 7.35), does have certain advantages: valves can be operated more rapidly than slides can be moved. Also, since without the constraint of cylindrical or near-cylindrical tube in the slide section, valve trombones can be designed with more bore expansion than slide trombones, and some valve trombones (such as those shown in Fig. 1.17e) have a bore profile approaching that of the baritone saxhorn (Fig. 7.33a).

The valve instruments with greater expansion of bore correspond to the tenor and baritone voices. The tenor Wagner tuba in 9-ft B$^\flat$ has a narrow minimum bore, being designed to be used with a french horn mouthpiece, but the bore expands rapidly and terminates in a bell similar to that of a baritone saxhorn. There is room for two distinct members of the saxhorn family (the baritone and the wider-bored bass or euphonium); this distinction is also present in the German and east European equivalents, the Tenorhorn and the Bariton. The relatively rare but distinguishable species of vocal (or ballad) horn in 8-ft C has a bore profile similar to that of the cornet but an octave lower, occupying a position intermediate between the french horn and the baritone saxhorn (Myers 2016).

7.2.7 *Instruments with Long Tube Lengths in G, F, E$^\flat$ and D*

Here we consider instruments in 11-ft G, 12-ft F, 13-ft E$^\flat$ and 14-ft D, basic tube lengths between approximately 3 and 4.5 m (Figs. 7.36–7.39).

The alphorn has the bore profile nearest to a cone of all brass instruments, giving a series of natural notes very close to a harmonic series without the need for any

Fig. 7.36 Alphorn in 11-ft G (Adolf Oberli, Zwischenflüh, Canton Bern, c. 1935). EU (3097)

Fig. 7.37 Cor solo (Marcel Auguste Raoux, Paris, 1823). EU (6144). Shown with tuning slide for G (inserted), and tuning slides for F, E, E$^\flat$, and D

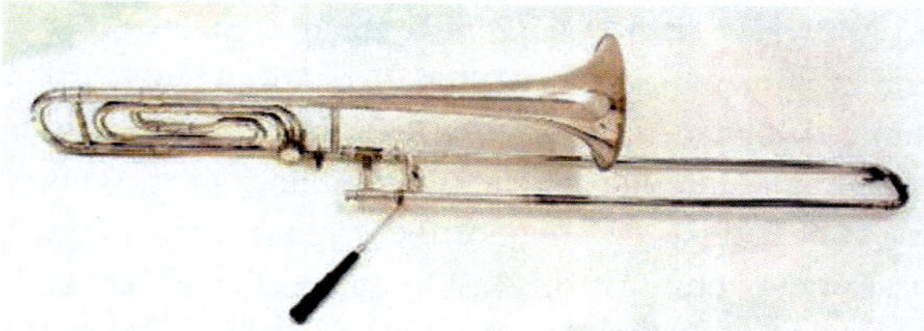

Fig. 7.38 Trombone in 11-ft G with thumb valve for D (Besson and Co, London, 1960). EU (1866)

Fig. 7.39 Tuba in 12-ft F (Besson & Co, London, c. 1931). EU (5848)

pronounced bell flare. The angle of the cone is wide enough to give a mellow timbre with little enrichment by nonlinear propagation, delivering strong lower spectral components, but the cone angle is narrow enough to allow a wide compass to be produced easily. Traditional repertoire uses the 2nd to the 16th natural notes, and the notes which diverge from Western tempered scales (corresponding to the 7th, 11th, 13th and 14th members of a harmonic series) are played without attempting to bend them to align with conventional scales: their intonation is accepted as characteristic alphorn idiom.

The french horn, in contrast, is played with the hand in the bell to give fine control of intonation to play in tune with other instruments. The natural horn played with hand-in-the-bell technique is often now referred to as a 'hand horn' and the effect of the hand (described in Sect. 4.5.6) is capable of giving not only control of intonation but also notes filling some of the gaps in the series of natural notes. A hand horn has an almost complete chromatic compass from the 2nd natural note upwards, but even with skilled hand technique there is a difference in timbre between notes in the natural series (unstopped) and notes lowered in pitch by the hand (stopped). The hand horn is equipped with a series of crooks to give a choice of basic sounding lengths for the instrument so that the natural notes are aligned with the tonality of the music. Composers (such as Beethoven) wrote idiomatically for the hand horn, exploiting the stronger and weaker notes.

There are various systems of crooks. Where a crook contributes to the sounding length by an appropriately tapered tube which is inserted between the mouthpiece and the body of the instrument, it is called a terminal crook (the crooks shown in Figs. 7.18 and 7.25 are terminal crooks). An orchestral hand horn has a set of terminal crooks giving most tonalities from 9-ft B♭ (occasionally 8-ft C) down to 18-ft B♭. Lower pitches, even down to 22-ft G, are called for in some Italian operas of the 1830s and 1840s. Where the different sounding lengths are obtained by using tuning-slides of different tube lengths the crooks are called 'inventionshorn crooks' or simply 'tuning-slide crooks'. These crooks are essentially cylindrical, and the bore taper from the mouthpiece to the tuning-slide is built into a fixed mouthpipe. The horn shown in Fig. 7.37 has tuning-slide crooks for 11-ft G, 12-ft F, 13-ft E and E♭, and 14-ft D; this model is called a 'cor solo' because these tonalities were those favoured for solo (as opposed to orchestral) horn music (Scott et al. 2019). The crookings higher or lower than this central group are less well served by a single mouthpipe taper.

With valve horns such as that shown in Fig. 7.32 the hand technique is not needed to supply the 'missing' notes, but hand-in-bell technique is used by valve horn players for fine control of intonation and timbre. For many decades after the invention of the valve and its application to horns, instruments continued to be supplied with sets of crooks for facility in playing in different tonalities, though fewer crooks than the ten or so of a fully-equipped orchestral hand horn. As late as the mid-twentieth century, french horns for military band used crooks for 12-ft F and 13-ft E♭.

The horn shown in Fig. 7.32 is a 'double horn'; the most commonly used model today is a double horn in both 9-ft B♭ and 12-ft F. In this model the mouthpipe leads from the mouthpiece receiver to the fourth valve (operated by the left thumb) which in one position directs the windway through the three valves for the left-hand fingers, back through the fourth valve and on to the bell. In the other position the fourth valve directs the windway through separate, longer tubing through the three valves (with longer valve loops), back through the fourth valve and on to the bell. Players of a double horn in 9-ft B♭ and 12-ft F typically use the B♭ 'side' most of the time, and always for high passages; they use the F side for some low passages and when the tone quality of longer tubing (with higher brassiness potential) is preferred. Some double horns are in 6-ft F and 9-ft B♭, others are in 6-ft F and 12-ft F. A double horn in 9-ft B♭ and 12-ft F incorporates tubing with sounding lengths from 9-ft B♭ down to 18-ft B, equivalent to most of the crooks of the orchestral hand horn—though players do not generally think in terms of the valves offering an instantaneous change of crook.

The instruments with greater expansion of bore correspond to the bass voice and lower; they include members of the tuba family. The bass Wagner tuba in 12-ft F has a narrow minimum bore, like that of the tenor Wagner tuba, to be used with a french horn mouthpiece; the bore expands rapidly and terminates in a euphonium-like bell.

Figure 7.39 shows an orchestral bass tuba in 12-ft F with five valves, allowing the lowest notes of the repertoire to be played. Many tubas, both for orchestral and band use, are pitched in 13-ft E♭. Although there is a solo repertoire for the tuba, the

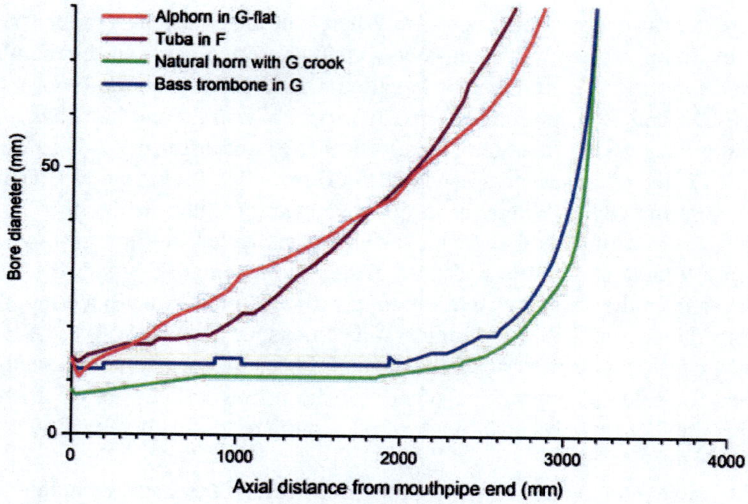

Fig. 7.40 Bore profiles for the four instruments shown above: alphorn, tuba, natural horn (cor solo crooked in G), and bass trombone. The vertical (bore diameter) scale is exaggerated to show the differences in the significant narrower parts of the instruments; the wider parts of the bell flares are omitted

design of the instruments optimises the strong low-frequency sound output required in ensemble playing. The spectral enrichment which might be expected from such a long tube length is minimised by the marked expansion in the bore and a wide bore diameter (Figs. 7.40 and 7.41).

The brassiness potential of the tuba is raised when valves are operated, since they introduce additional cylindrical tubing into the windway. Nevertheless, because of their wide bore tubas can produce a loud sound output with a relatively small acoustic pressure in the mouthpiece, so that nonlinear distortion remains modest even when the length is considerably increased by the operation of valves.

Historically, bass sackbuts and bass trombones were built in 11-ft G or 12-ft F, occasionally in 13-ft E$^\flat$. The modern B$^\flat$/F bass trombone is a wide-bore instrument in 9-ft B$^\flat$, with a rotary thumb valve which inserts sufficient additional tubing to put the instrument into 12-ft F. The slide has the same length as a standard B$^\flat$ trombone, which means that only six of the normal seven positions are obtainable when the F valve is operated; many bass trombones have a second valve to extend the compass downwards giving typically an instrument in B$^\flat$/F/G$^\flat$/D.

Parts written with older forms of bass trombone in mind can surprise present day players. Playing the glissando from B1 to F2 in the Trombone III part of Béla Bartók's *Concerto for Orchestra* (Fig. 7.42) is not straightforward on a modern bass trombone, since the slide is too short to cover the interval of a diminished fifth once a valve is operated. Bartók may have been recalling orchestration manuals referring to the 12-ft F bass trombone with its longer slide, although this instrument was no longer in use when the *Concerto for Orchestra* was written in 1943.

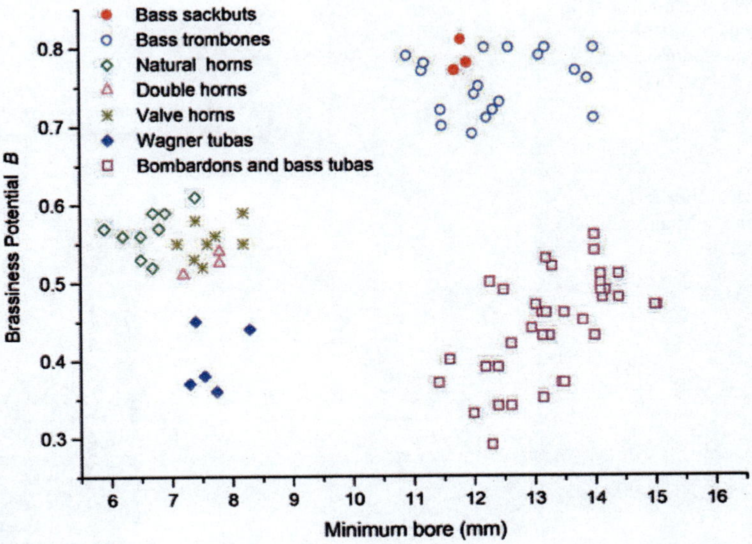

Fig. 7.41 B / D_{min} scatter plot for some typical bass sackbuts, bass trombones, natural horns, double horns (F side), valve horns, F Wagner tubas; bombardons and bass tubas

Fig. 7.42 Trombone parts from the fourth Movement of Bartók's *Concerto for Orchestra*. Reproduced by permission of Boosey and Hawkes Music Publishers Ltd

Fig. 7.43 Trombone parts from the fourth Movement of Elgar's *Symphony No. 1*

In the fourth movement of Elgar's *Symphony No. 1*, written in 1908, the first and second (tenor) trombones play a fortissimo A$^{\flat}$1, while the third (bass) trombone and tuba are silent (Fig. 7.43). Elgar clearly wanted the brassy sound of the trombone

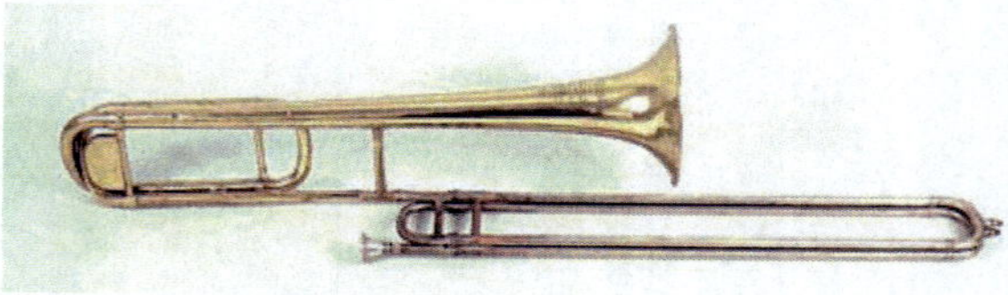

Fig. 7.44 Contrabass slide trombone in 18-ft B$^\flat$ (Antoine Courtois, Paris, c. 1895). EU (4215)

Fig. 7.45 Tuba in 16-ft C (Alexander, Mainz, c. 1980). EU (4283)

pedal note at this point rather than the mellower timbre of the tuba, but why was the bass trombone not involved? The reason is probably that Elgar expected this note to be unplayable on the bass trombone: throughout his lifetime the British bass trombone was in 11-ft G, and its compass had a gap from C2 down to A$^\flat$1.

7.2.8 *Instruments with Very Long Tube Lengths in C and B$^\flat$*

Here we consider instruments in 16-ft C and 18-ft B$^\flat$, basic tube lengths between approximately 4.5 and 6 m (Figs. 7.44 and 7.45).

With basic tube lengths longer than 14-ft D, nonlinear propagation and 'brassiness' can be very important since the tube is so long that spectral enrichment can fully develop as sound travels over the length of the tube. Apart from orchestral hand horns in their longest 'basso' crookings, only two distinct instrument types have been built with in this size: the contrabass trombone (slide or valve) and the contrabass tuba.

With their considerable length of near-cylindrical tube, contrabass trombones in 16-ft C and 18-ft B$^\flat$ have high values of brassiness potential and in forte and

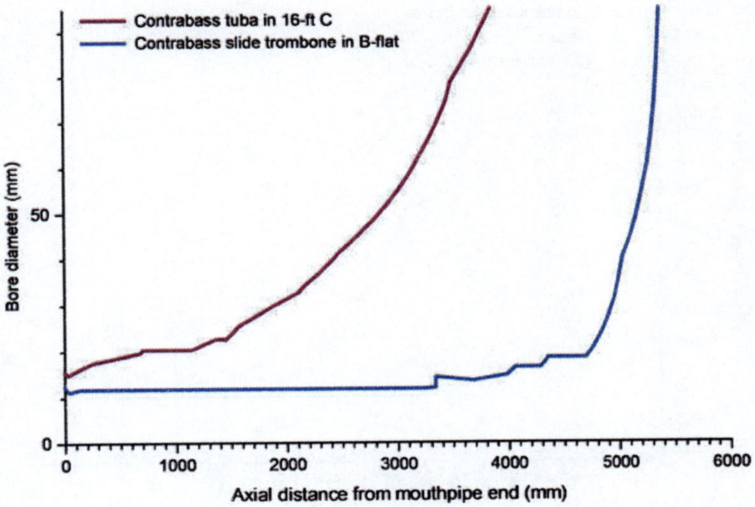

Fig. 7.46 Bore profiles of the two instruments shown above: contrabass tuba and contrabass slide trombone. The vertical (bore diameter) scale is exaggerated to show the differences in the significant narrower parts of the instruments; the wider parts of the bell flares are omitted

fortissimo playing produce a rich harmonic spectrum. This gives the listener a strong sense of pitch but is not very effective in blending and providing the bass for an ensemble. For this reason, and because it is tiring to play, the contrabass trombone has tended to be used for special effect rather than a standard orchestral or band instrument. Many modern contrabass slide trombones do not have such long basic tube lengths but are built in 12-ft F and equipped with two valves to give a complete compass.

The contrabass tuba in 16-ft C or 18-ft B$^{\flat}$ can blend and provide the bass for an ensemble; it is widely used as an alternative to smaller tubas in the orchestra and in addition to them in bands. Its effectiveness in delivering sound with significant energy at low frequencies is increased with a greater expansion of bore and with a greater overall bore size, but these can achieved only with an increase in size and weight. Tuba design is a trade-off between acoustical optimisation and ease of holding and portability (Myers 2019).

7.2.9 *Instruments with Very Long Tube Lengths in G, F, E$^{\flat}$ and D*

Here we consider instruments in 22-ft G, 24-ft F, 26-ft E$^{\flat}$ and 28-ft D, basic tube lengths greater than approximately 6 m.

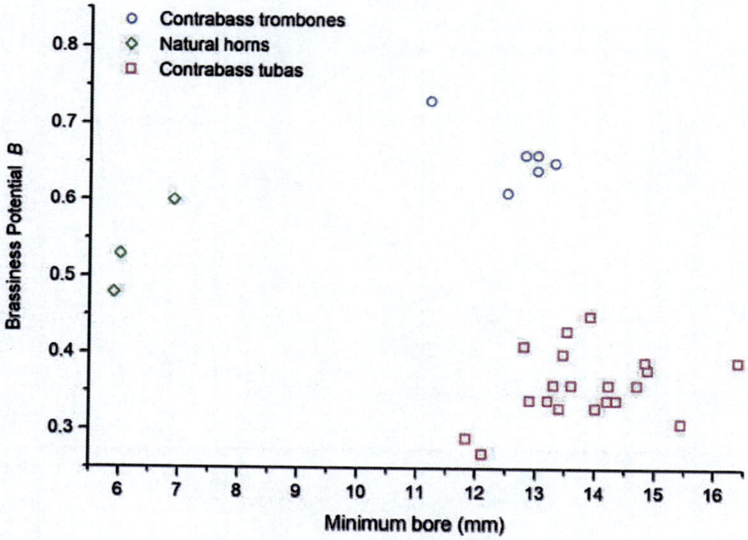

Fig. 7.47 $B\,/\,D_{\mathrm{min}}$ scatter plot for some typical contrabass trombones, natural french horns with low (basso) crooking, and contrabass tubas

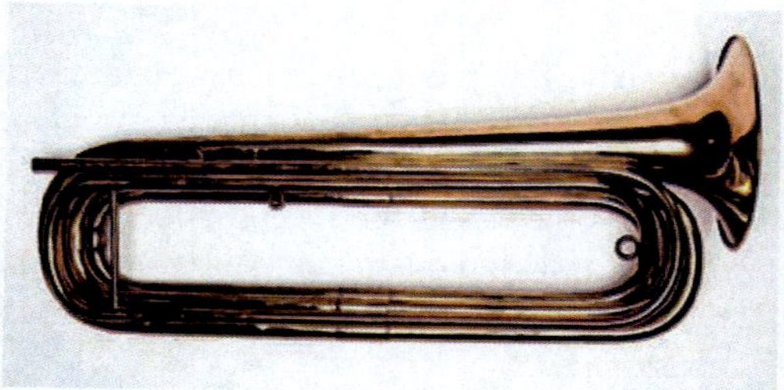

Fig. 7.48 Contrabass trumpet in 26-ft E$^\flat$ (Buffet Crampon, Paris, c. 1920). EU (4546)

At this size instruments are expensive, heavy and cumbersome, and stretch the capabilities of human lips and lungs. A small number of giant tubas have been built (Bevan 2000) but they have been more effective in providing publicity than in their contribution to music. The natural instrument shown in Fig. 7.48 was built to provide a 'bass' for trumpet fanfare ensembles in which the higher parts were played on natural trumpets in $6\frac{1}{2}$-ft E$^\flat$; normally only the tonic (E$^\flat$1) and dominant (B$^\flat$1) would be used.

7.3 Families

Families of instruments, in which family members differ in size and compass but are similar otherwise, are well defined and easily recognised for many instrument kinds including recorders, clarinets and saxophones. Because the boundaries between different kinds of brass instrument are not clear-cut, establishing brasswind families can be more a matter of tradition rather than science. Some things cannot be scaled, including the arms, fingers, lips, and vocal tract of the player and the highest and lowest perceptible frequencies, and the dominant region for pitch perception (approximately 500–2000 Hz) of the listener. There is little evidence that formants contribute significantly to brasswind character. Factors which can be scaled in instrument families include the size and proportions of pitch-determining elements and the size and proportions of timbre-determining elements. The principal scaling variables in brasswind families are the air column length (determining pitch), absolute bore size (affecting timbre and dynamic response), and the bore profile geometry (affecting timbre), and where applicable, tonehole size and placement (affecting pitch and timbre). Also important are mouthpiece cup volume (mainly affecting timbre), and mouthpiece cup shape (mainly affecting transients) discussed in Sect. 6.4. The elements that we expect a brasswind family to have in common are the mechanics of pitch control (whether by finger-holes, keys, slide, or valves) and some aspects of the geometry of the bore profile which determine timbre.

Some brasswind families have resulted from a process of evolution, including cornetts, sackbuts, slide trombones, valve trombones and tubas. Other families have been created more or less simultaneously, including saxhorns, saxotrombas, and cornophones; this was a nineteenth-century phenomenon, exploiting the freedom given by the valve to create novel bore profiles.

The best-known and longest-lasting family has been the saxhorns. As devised by Adolphe Sax, these were actually two families: from the 9-ft B$^\flat$ baritone upwards (narrower bore), and from the 9-ft B$^\flat$ bass downwards (wider bore) (Fig. 7.49). The need for two distinct families derives from different musical functions: the narrower bore instruments (tenor horns and baritones) typically fill out the harmonies, with the wider bore instruments (essentially euphoniums and tubas) providing a firm bass and needing to be effective in delivering low-frequency sound. This distinction is also present in band instrument designs not influenced by Sax: German and eastern European instrument makers have produced both a 'Tenorhorn' and a 'Bariton' in 9-ft B$^\flat$. Within families a longer air column length is accompanied by slightly wider absolute bore size, while retaining a similar bore profile geometry and brassiness potential. This is true for created families as well as for evolved families such as slide trombones (where the mechanics of the slide dictate the similar bore profile geometry and brassiness potential).

The ophicleide, illustrated in Figs. 7.28b and 7.57, was originally conceived as a member of a family, the smallest members of which were the keyed bugles (Figs. 7.8 and 7.15). An intermediate size, the quinticlave or alto ophicleide, never achieved the popularity of the keyed bugle or the ordinary (bass) ophicleide. A few contrabass ophicleides were made, scaled up from the ordinary C and B$^\flat$, the largest (at least of surviving examples) being pitched in 13-ft E$^\flat$ (lowest note D1). The unique contrabass instrument shown in Fig. 7.50 is a taxonomic challenge. It was made circa 1840, nicknamed in modern times 'The Anaconda', and is now in the possession of one of the present authors. Although at first glance a double-size serpent, it has the essential acoustical features of an ophicleide: eleven toneholes

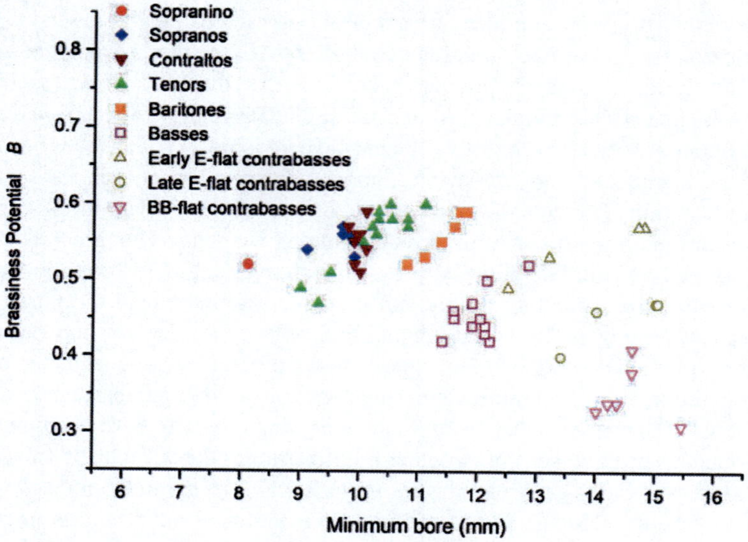

Fig. 7.49 $B / D_{\min}$ scatter plot for instruments (all made by Adolphe Sax) in the two saxhorn families: sopranino saxhorns, soprano saxhorns, contralto saxhorns, tenor saxhorns, and baritone saxhorns (narrower bore and higher brassiness potential); bass saxhorns, early 13-ft E^b contrabass saxhorns, later 13-ft E^b contrabass saxhorns, and 18-ft B^b contrabass saxhorns (wider bore and lower brassiness potential)

Fig. 7.50 Andrew van der Beek playing 'The Anaconda'. Courtesy of the University of Edinburgh

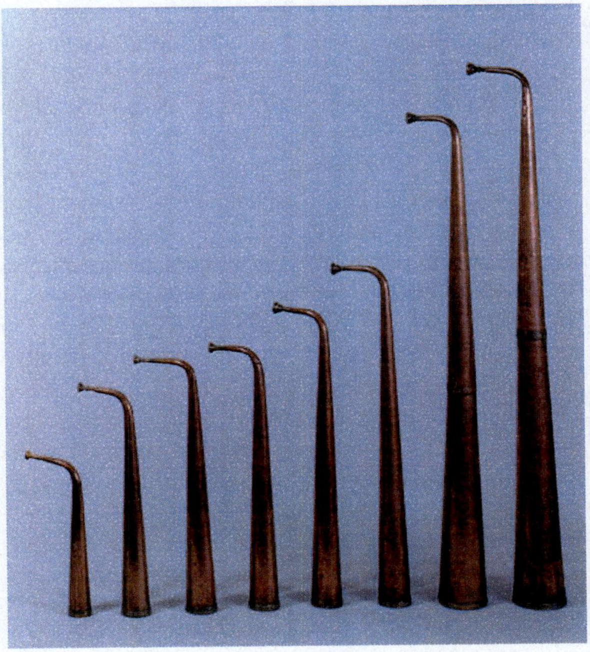

Fig. 7.51 Russian horn band instruments: eight instruments from a set (Russia, c 1810-20). Musical Instrument Museum, University of Leipzig (1630, 1634, 1636-41), photograph: Janos Stekovics

spaced over the length of the tube, each tonehole large compared with the local bore diameter and covered by a key. On the other hand, the body construction is that of a serpent and its lowest note is C1, an octave below the lowest regular note of the ordinary serpent. Despite being an effective contrabass instrument, no further contrabass side-hole brass instrument appears to have been made until the late twentieth century.

A special case of a family is the Russian horn band (see Fig. 7.51), in which each instrument is designed to sound just one note, the music being arranged to be played by 24 or more musicians who played their one note as required. In each case the instrument sounds the fundamental, so as to maximise efficiency the horns are very wide-bore conical natural instruments (with integral mouthpieces).

7.4 Mouthpieces

On one view, the mouthpiece is an integral part of a brass instrument. It has specific acoustical functions, however, and as the interface between instrument and player it is of critical importance on several counts and needs to be given specific consideration. The behaviour of the player's lips and the flow of air in the mouthpiece are complex and are less well understood than the behaviour of the air column in the body of a brass instrument.

Although instrument makers often provide a mouthpiece when they supply an instrument, players most usually make a separate choice. Interchangeability of mouthpieces is generally catered for by makers: the tapers of instrument mouthpiece receivers and of mouthpiece stems are not completely standardised but do, for most instruments, allow players to choose from a considerable range of mouthpieces. The bore in the instrument just beyond the end of the mouthpiece stem is itself acoustically significant, so each kind of instrument tends to have mouthpieces with a narrow range of stem diameters. There are exceptions, such as the rare models of trumpet which are designed to take cornet mouthpieces.

Mouthpieces cannot be specified adequately by a small number of parameters. The parameter most often cited is the cup diameter, because of its great influence on the embouchure and the feel to the player. A relatively wide cup is generally considered to facilitate 'lipping' wayward notes into tune on the natural trumpet. Cup diameters vary from 12 mm or less in cornettino mouthpieces to 30 mm or more in the largest tuba mouthpieces, the limits dictated by the size of the human lip rather than by the nature of the instrument. The edge between the rim (placed against the lips) and the cup is known as the 'bite' of the mouthpiece, and its curvature is also important for a comfortable and flexible embouchure.

The aerodynamic and acoustical functions of a mouthpiece may be usefully characterised by cup volume and cup shape, parameters which might have a greater effect on timbre and response and are not constrained in any way by the size of the human lip. Working definitions of cup volume and cup shape are given in Sect. 7.7. Similar mouthpieces are in some cases frequently used for quite different instruments. For example, mouthpieces for baritone saxhorn and for tenor trombone are almost completely overlapping categories. Similarly, no distinction can consistently be made between mouthpieces for bass trombone and for euphonium.

In Sect. 4.3.6 we discussed the effect of the mouthpiece cup on the resonances of a system consisting of a tube with inserted mouthpiece. The amplitudes of the input impedance peaks with frequencies in the region of the resonance frequency of the detached mouthpiece are boosted. The mouthpiece cup volume has a strong effect on the response of a brass instrument as a whole, as players will confirm. The boost given to the resonances of the instrument plus mouthpiece system needs to be in the region of the impedance peaks employed in the normal playing tessitura of the instrument. A simple experiment is to attempt to play a tuba with a trumpet mouthpiece or vice-versa: in either case notes can be sounded, but it is impossible to produce a musically acceptable sound in the normal playing range of the instrument.

To characterise a mouthpiece we need to establish one or more parameters which reflect the contribution of the cup volume. In Sect. 4.3.5 we discussed the 'popping frequency' of a detached mouthpiece. This is clearly related to the cup volume, and also to the throat cross-sectional area, but popping frequency also depends on the length and shape of the backbore of the mouthpiece. When the mouthpiece is inserted into an instrument the backbore becomes part of the overall bore profile of the instrument, and the point where the mouthpiece ends and the inside the instrument is of minor importance. A simple experiment is to extend the stem of a mouthpiece with a roll of paper: a small increase in backbore length audibly lowers the popping frequency. The popping frequency of a detached mouthpiece can give an approximate indication of the region of resonance boost, but although it is readily measurable, it is not a reliable parameter for comparative taxonomy. The cup volume, although more difficult to measure, provides a more robust parameter. In defining the cup volume as a taxonomic parameter we sometimes disregard the protrusion of a player's lips into the cup when playing, as it is highly variable (and under the control of the player to some extent). We also need to establish where the cup ends and the backbore begins. Although this is necessarily somewhat artificial, it is not of critical importance as the throat is much narrower than the cup diameter, and any error determining the position of the bottom of the cup will have a relatively small effect. Cup volume can be measured by blocking the throat and filling the cup with water, or it can be computed from physical measurements. In Sect. 7.7 we employ the latter method, since physical measurements also allow us to characterise the shape of the cup.

The listener's perception of timbre depends not only on the steady state spectrum of the sustained sound, but also on the starting transient. Cup shape, especially the shape of the throat where the bottom of the cup leads into the minimum bore, seems likely to affect transients even if steady state sound is largely independent of cup shape detail. Some players consider that throat curvature affects the response of the system to the 'attack' of a note—the ease for the player of commencing a note and the heard starting transient. When air flows from the cup though the throat into the minimum bore ('grain') past the sharp edge as in hemispherical cup mouthpieces, flow separation occurs, with toroidal vortices inside the throat. These absorb some of the energy, and reduce the effective cross-sectional area of the throat (Vereecke and Carter 2014). Mouthpiece aerodynamics is a topic very ripe for further research.

Players are extremely sensitive to small changes in mouthpiece shape, whether changes to the rim and bite affecting the placement of the mouthpiece on the lips or changes to the cup, throat and backbore affecting air flow. The material of construction is possibly also a factor: most mouthpieces are metal, but ivory, wood and plastic have also been used. Changes detectable by a skilled player can be too small and too diverse to be easily explained in scientific terms (see Sect. 6.6.5).

7.5 Going Further: Trumpets and Cornets—Are They Different?

The difference between trumpets and cornets is often a source of confusion. Part of the problem is that what has been meant by the terms 'Cornet' and 'Trumpet' has differed from decade to decade and from country to country. Even at one time and in one place, the names used by instrument makers, players, composers and arrangers have not always been the same. It is generally accepted that the designs of both instruments have converged since the introduction of the cornet in the 1830s, with the differences between trumpets and cornets becoming less obvious (Campbell et al. 2020).

In the mid–late nineteenth century a cornet was pitched in $4\frac{1}{2}$-ft B$^\flat$ or higher, possibly with crooks for lower nominal pitches, had an approximately conical bore profile between mouthpiece receiver and valves, and had a short, horn-influenced wrap; at this time a valve trumpet was commonly pitched in 5-ft G, 6-ft F or lower, had a predominantly cylindrical bore profile from mouthpiece receiver for at least half the sounding length, and had a long, natural trumpet-influenced wrap. The mouthpieces, too, were not interchangeable and differed in cup shape. By the early twentieth century both were commonly pitched in $4\frac{1}{2}$-ft B$^\flat$ and had an expanding bore profile between mouthpiece receiver and valves. Cornets dispensed with the detachable shanks and crooks (which accommodated much of the bore expansion between mouthpiece receiver and valves), but the fixed-mouthpipe models have retained the narrower mouthpiece receiver. The trumpet has always been used with a mouthpiece with a wider tapered stem than a cornet and thus had a wider mouthpiece receiver. Modern B-flat trumpets have a constriction in the mouthpipe which is followed by an approximately conical bore expansion to the valves. Mouthpieces will often be provided by makers with exactly the same cup and rim shape for cornets and trumpets (to ease transition for musicians who play both).

The cornet in $4\frac{1}{2}$-ft B$^\flat$ shown in Fig. 7.18 is typical of mid–late nineteenth century models; the valve trumpet in $4\frac{1}{2}$-ft B$^\flat$ shown in Fig. 7.19 is a typical of early twentieth century models. The complete bore profiles of these two instruments, illustrated in Fig. 7.52, are very similar, and it would be hard to claim that the cornet is significantly more conical than the trumpet. Trumpet bell end diameters are on average slightly smaller than those of cornets, but both range significantly and the ranges overlap, so bell size cannot be used as a distinguishing feature. The most reliable distinction between trumpets and cornets is in the width of the mouthpiece receiver.

Figure 7.53 shows a plot of brassiness potential and minimum bore for typical cornets and trumpets; a group of flugelhorns is included for comparison. There is a clear distinction between the group of cornets (B in the range 0.57 to 0.61) and the group of relatively early valve trumpets in B$^\flat$ (B in the range 0.70 to 0.76). While cornet bore profiles have changed relatively little, the later valve trumpets show a progression through the twentieth century to more constricted mouthpipes and lower values of B, probably to allow playing at higher dynamic levels without excessive cuivré effect. The remaining differences between cornets and these later trumpets are limited to wrap (and thus general appearance) and mouthpiece receiver taper. The timbral difference between a cornet and a late valve trumpet with the same minimum bore and brassiness potential depends not so much on properties of the instrument but on the player's mouthpiece choice and technique.

In the late nineteenth century, orchestral trumpet parts were frequently played on cornets: even with careful mouthpiece selection and playing techniques, the timbre

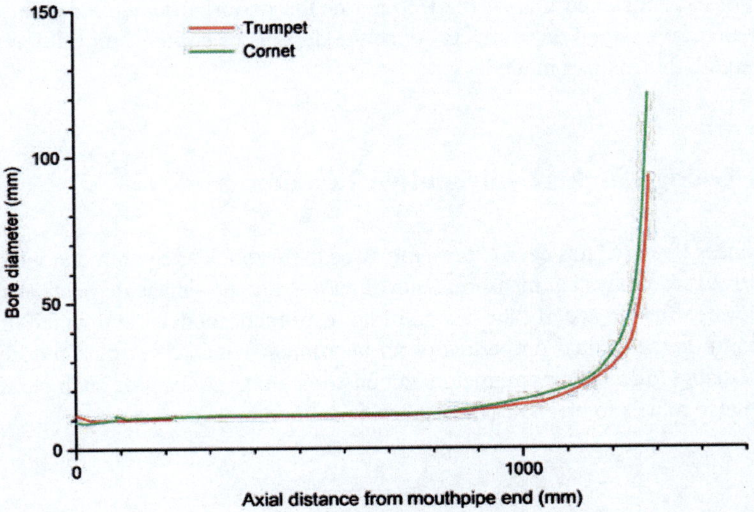

Fig. 7.52 Bore profiles of the trumpet in $4\frac{1}{2}$-ft B^b (York, Grand Rapids, 1925-27) EU (5772) shown in Fig. 7.19 and the cornet in $4\frac{1}{2}$-ft B^b (Antoine Courtois, Paris, 1856-1858) EU (3475) shown in Fig. 7.18

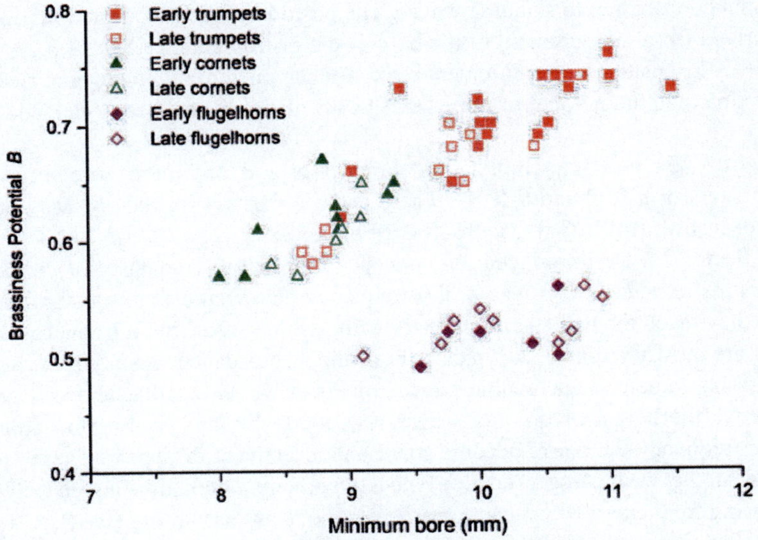

Fig. 7.53 B / D_{min} scatter plot for typical nineteenth-century B^b trumpets, twentieth-century B^b trumpets, nineteenth-century cornets, twentieth-century cornets, nineteenth-century flugelhorns, and twentieth-century flugelhorns

will not have matched that of the trumpet of the period. Today, orchestral cornet parts are often played on trumpets: since the designs of cornets and trumpets have converged, this matters much less.

7.6 Going Further: Alternative Taxonomies

Brassiness Potential has been extensively used in this book as a parameter which can be derived from physical measurements of an instrument—measurements such as a maker might use to specify the design of an instrument model—and which directly relates to the acoustical properties of an instrument which determine timbre. It is also possible to use other parameters to construct a taxonomy. One such parameter, which also relates to brightness of timbre, is defined by

$$N = \frac{2L_{\text{peak}}}{\pi D_{\text{peak}}}, \tag{7.2}$$

where L_{peak} is the distance from the mouthpiece receiver to the point on the bore axis at which the horn function peaks, and D_{peak} is the bell diameter at the same point. The values of length and diameter at the horn function peak are used rather than the values at the bell exit because the widest parts of a flaring bell make a minimal contribution to standing waves. The position of the horn function peak can be derived from measurement of the bore of the instrument (see Sect. 4.3.7), and is relatively insensitive to measurement error. The parameter N is found empirically to approximate to the number of impedance peaks of significant size (Pyle and Myers 2006).

Figure 7.54 is a scatter plot of parameter N and minimum bore for typical instruments in 8-ft C and 9-ft B$^\flat$. There is a similarity with the $B\,/\,D_{\text{min}}$ scatter plot for instruments in 8-ft C and 9-ft B$^\flat$, Fig. 7.35.

In Sect. 2.2.5 we introduced the concept of the cutoff frequency of the bell of a brass instrument. Sound waves of a frequency below cutoff are mostly reflected when they reach the bell end of an instrument, while waves with a frequency above cutoff are mostly radiated. A typical plot of input impedance against frequency for a brass instrument, such as that shown in Fig. 4.12, shows that above a certain frequency the height of the peaks decreases markedly, and in a region around a higher frequency the peaks become small and effectively disappear. These are the effects of the weakening reflection and strengthening radiation at the bell with increasing frequency. The player experiences the peaks as slotting (see Sect. 1.2.2); at the high pitches corresponding to frequencies where the peaks are insignificant or absent, the notes are difficult to sound and to play in tune.

Assigning a single cutoff frequency to take account of these effects is not always possible. Several methods giving figures for cutoff frequency have been proposed. For bells with pronounced flares, such as those of trumpets and trombones, a cutoff frequency can be derived from the peak value of the horn function, which in turn

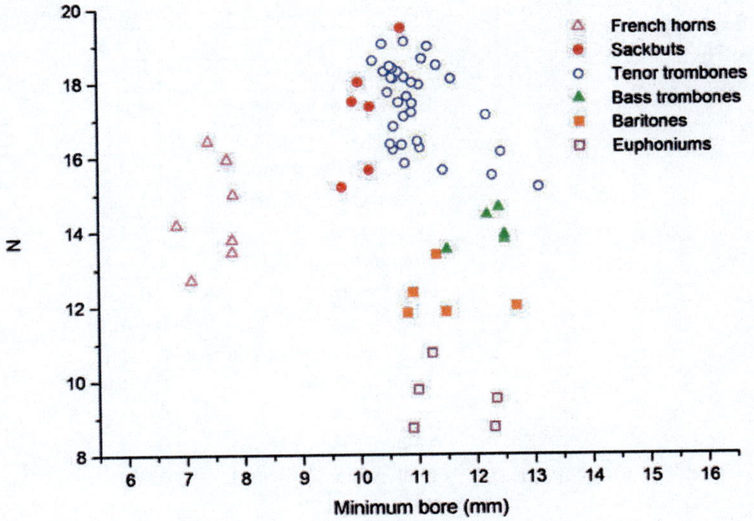

Fig. 7.54 N / D_{min} scatter plot for typical french horns, sackbuts, tenor trombones, bass trombones, baritones, and euphoniums

can be computed by fitting a Bessel horn to the profile of the bell near the end (see Sect. 4.3.7). For a bell with an abrupt termination, such as a Swedish kornett or an alphorn, a cutoff frequency can be defined as the frequency at which the reflection coefficient drops below a specified value. For moderately flaring bells an approximation to the cutoff frequency has been suggested by Hélie and Rodet (2003). Although cutoff frequency is an important determinant of the character of a brass instrument, reflecting both the response of an instrument to the player and the timbre of the radiated sound, the fact that the cutoff effect is imprecise and the lack of a single derivation applicable to all bells limit the usefulness of cutoff frequency as a taxonomic parameter to differentiate different species of instrument (Myers, 1997a). Table 7.2 gives typical values of cutoff frequency for common species of instrument derived from computed horn functions, following an approximation given by Benade and Jansson (1974).

7.7 Going Further: Mouthpiece Parameters

Although mouthpiece cup volume and cup shape are considered to be important for response and timbre, defining parameters to represent these is not straightforward. There is often no clear point at which the cup ends and the grain or backbore begins, and the degree to which a player's lips occupy part of the volume of the mouthpiece is variable. The following method allows a statement of cup volume which is consistent and provides a parameter relating to cup shape.

Table 7.2 Typical values of cutoff frequency for common species of instrument

Instrument species	Cutoff frequency (kHz)
Piccolo trumpets	2.0–2.5
Natural trumpets	1.0 –1.9
Low F valve trumpets	1.0 – 1.7
English slide trumpets	1.3 – 1.6
High E^b valve trumpets	1.2–1.6
B^b and C valve trumpets	1.2 – 1.6
Soprano cornets	1.2–1.5
Cornets	1.1–1.5
Alto trombones	1.0–1.4
Flugelhorns	0.9–1.2
Tenor saxhorns (alto horns)	0.7–1.1
Tenor trombones	0.8–1.0
Baritone saxhorns	0.7-0.9
Modern bass trombones	0.7–0.8
Euphoniums	0.6–0.8
Natural horns	0.5–0.8
Valve horns	0.5–0.7

Physical measurements of mouthpieces can be manipulated using a computer-aided drafting software. Nearly all brass instrument mouthpieces have circular symmetry; after plotting a longitudinal cross-section containing the centre line (axis of symmetry), a cone can be constructed which touches the mouthpiece rim and throat. The point where this reference cone touches the rim is a small distance into the mouthpiece cup, a little less than a depth of one bite radius. In mouthpieces with a cup shape approaching a hemisphere, such as the serpent mouthpiece shown in Fig. 7.55a, there is an annular part of the cup volume which lies outside this cone. For a conical cup shape, such as the nineteenth-century French trombone mouthpiece shown in Fig. 7.55b, there is no annulus. For a cuspoidal (inward-curving) cup shape, such as that of the french horn mouthpiece shown in Fig. 7.55c, there is no throat. In this case the cone apex is placed at the point of minimum bore; the cup walls lie inside the reference cone and the annulus volume can be regarded as taking a negative value.

The volume V_{cone} of the cone plus the annulus volume V_{ann} is adopted as a measure of the acoustically effective volume of the mouthpiece. The fact that the cone terminates at a point a small distance into the mouthpiece cup rather than on the face of the rim can be regarded as an allowance made for the fact that a player's lips occupy part of the volume of the mouthpiece. The protrusion of the lips into the mouthpiece varies considerably from one player to another and from embouchures for high notes to embouchures for low notes; inspection of embouchure formation with transparent mouthpieces confirms that this allowance is of the right order of magnitude. That the apex of the cone is not necessarily exactly at the point of minimum bore reflects the fact that the position of minimum bore is often not well defined and can be at some distance from the throat.

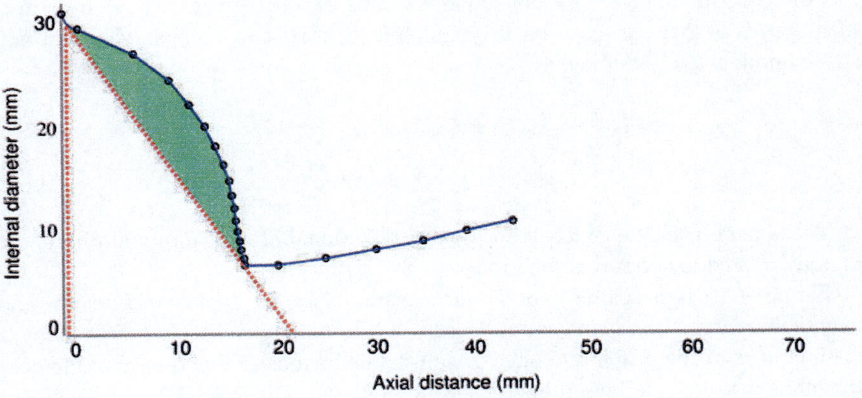

(a)

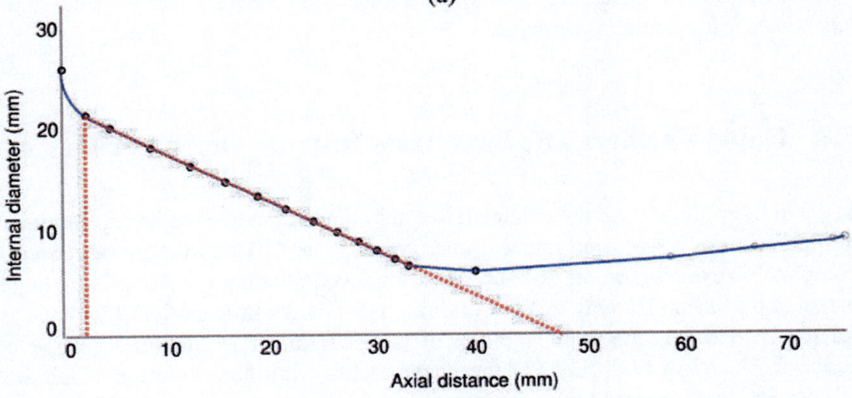

(b)

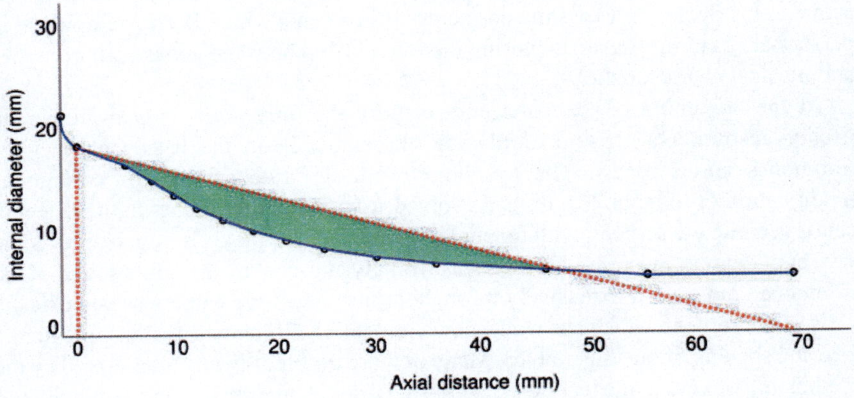

(c)

Fig. 7.55 Three typical mouthpiece cup shapes: (**a**) outward-curving, (**b**) conical, (**c**) inward-curving. Solid blue line: measured internal profile. Dashed red line: fitted cone. Green filled area: cross-section of annular volume (see text)

The construction of this cone and annulus readily gives two well-defined measurements that can be used in taxonomic studies: cup volume V_{cup} and cup shape quotient Q_{cup}, defined as

$$V_{cup} = V_{cone} + V_{ann} , \tag{7.3}$$

$$Q_{cup} = 1 + V_{ann}/V_{cone} . \tag{7.4}$$

These parameters, together with the cross-sectional area of minimum bore, are directly related to acoustical function.

Figure 7.56 is a scatter plot of these parameters for typical nineteenth and twentieth-century brasswind mouthpieces. While mouthpieces for tuba occupy a distinct area of the graph and have a well-defined acoustic function, mouthpieces for other instruments have properties which overlap. Distinguishing mouthpieces can depend on other characteristics such as the external diameter of the stem at the end where it fits into an instrument.

7.8 Going Further: The Bass Brass Instruments of Berlioz

In the first few decades of the nineteenth century there were two major developments in the design of brass wind musical instruments. The first was the introduction of completely keyed wide-bore side-hole instruments, including the Royal Kent bugle patented by Joseph Haliday in 1810 (although invented a little earlier). The catalyst for the second was the demonstration of a successful brass instrument valve by Heinrich Stölzel in 1814. Each of these approaches led to the evolution of families of completely chromatic brass instruments, and composers such as Hector Berlioz were quick to take advantage of the potential of the new instruments in orchestral scores. This Section reviews the bass brass instruments which Berlioz employed in performances of his music, exploring the relationship between acoustical properties and musical characteristics.

At the beginning of the nineteenth century the only commonly available lip excited instruments capable of playing chromatically in the bass register were trombones and serpents. The trombone, with its slide system for continuous modification of the tube length, had evolved from the slide trumpet in the fifteenth century, while the serpent, with toneholes covered by the player's fingers, was a late sixteenth-century invention. Berlioz was strongly attracted by the noble sound of the trombone, and used it extensively throughout his career. He had a less enthusiastic view of the serpent, which has well-known intonation difficulties associated with the size and spacing of the finger holes. Many of these problems were ameliorated by the application of keys to the toneholes, which led through intermediate stages including the bass horn to the completely keyed ophicleide patented by Halary in 1821. Berlioz scored for ophicleides in many of his major works, including the *Symphonie Fantastique*. By the 1840s he came to recognise the potential of valved saxhorns and

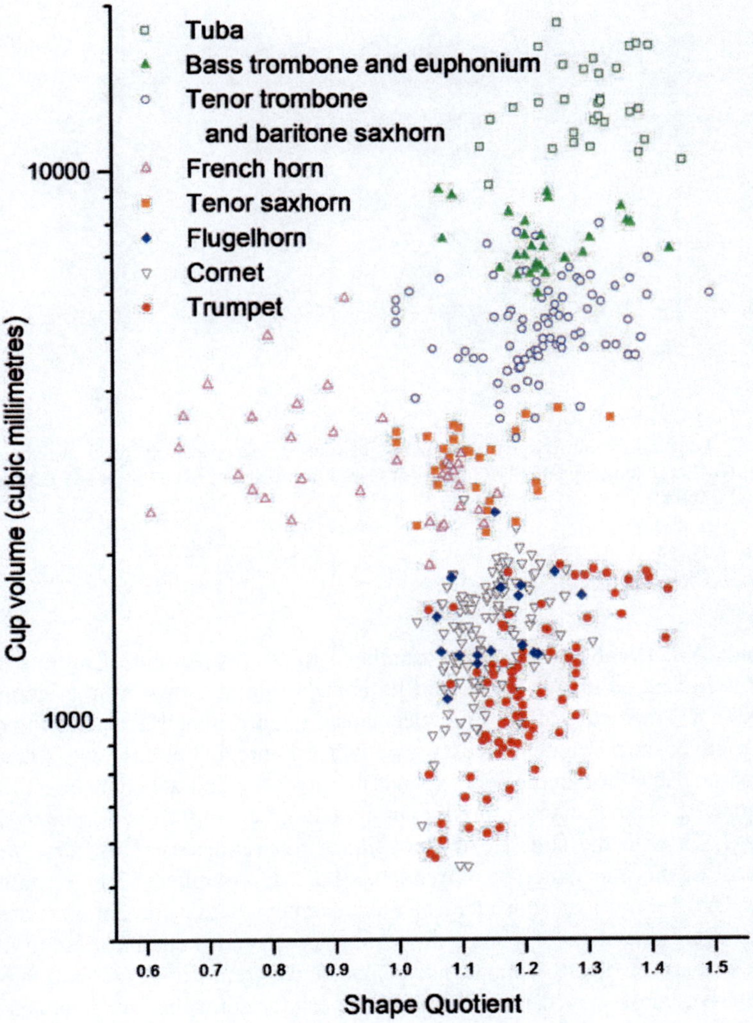

Fig. 7.56 Scatter plot of cup volume V_{cup} and shape quotient Q_{cup} for typical nineenth and twentieth-century brasswind mouthpieces: tuba, bass trombone and euphonium, tenor trombone and baritone saxhorn, french horns, tenor saxhorn, flugelhorn, cornet, and trumpet. To the right hand side are the more hemispherical cup shapes, to the left the more inward-curving (cuspoidal); those with a shape quotient of 1.0 have conical cup shapes

tubas. The musical possibilities and limitations of the trombone, serpent, ophicleide and tuba are closely related to characteristic acoustical features of the instruments, which are discussed in the following sections.

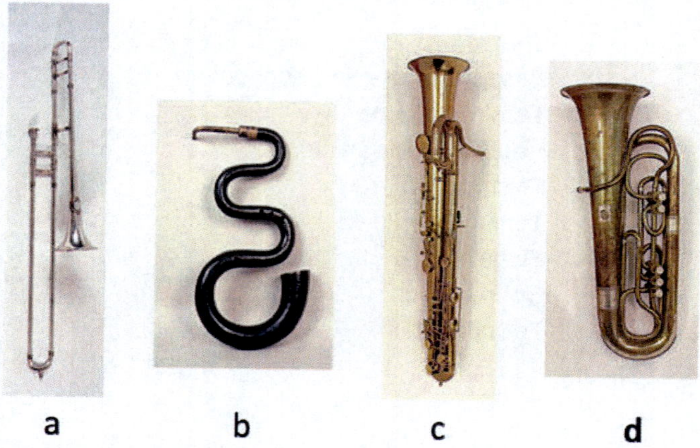

Fig. 7.57 Four nineteenth-century bass brass instruments: (**a**) Courtois tenor trombone, EU (3747); (**b**) Baudouin serpent, EU (3606); (**c**) Gautrot ophicleide, EU (4287); (**d**) Zetsche bass tuba, EU (4091)

7.8.1 The Trombone

A typical nineteenth-century tenor trombone in $B^\flat$ (by Antoine Courtois, Paris, 1865) is illustrated in Fig. 7.57a, and its bore profile is shown by the blue curve in Fig. 7.58. The nature of the slide mechanism requires that a large fraction of the total length of the instrument tube has a cylindrical bore, in this case with a diameter of 11.5 mm. Near the exit the bore expands rapidly into a bell with diameter 148 mm.

Figure 7.59 shows the measured input impedance curve for the trombone shown in Fig. 7.57a with the slide in first position (fully retracted). This curve reveals that most of the impedance peak frequencies are approximately integer multiples of 59.5 Hz; the musical equivalent of this statement is that the natural notes are approximately members of a harmonic series whose fundamental is close to $B^\flat 1$.

The extent to which the impedance peaks are different from exact harmonics of $B^\flat 1$ can be quantified by calculating for each peak the equivalent fundamental pitch (see Sect. 4.3.4). The EFP plot for the trombone, shown in Fig. 7.60, confirms that for $n \geq 4$ the peaks are close to exact harmonics of a note around 40 cents above $B^\flat 1$ (at the modern pitch standard of A4 = 440 Hz). The $n = 2$ peak is 143 cents flatter than the second harmonic of the $B^\flat 1$ series, while the $n = 1$ peak is over 8 semitones flatter than the nominal fundamental.

The inharmonicity of the lower resonances of the trombone, which is a consequence of its largely cylindrical bore profile, helps to explain one of the characteristic musical properties of the instrument. It is possible for a skilled player to sound the note $B^\flat 1$ on a tenor trombone, but since there is no resonance to support the fundamental of the spectrum, this 'pedal note' has a unique timbre dominated by upper harmonics (see Sect. 5.4.4).

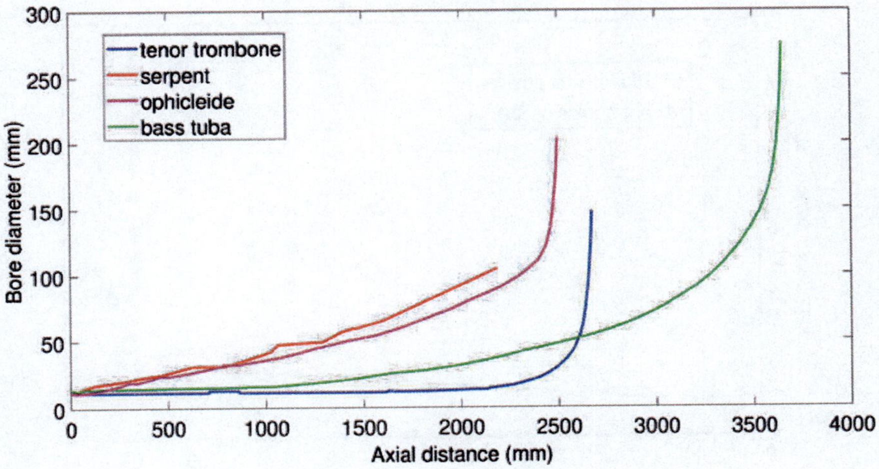

Fig. 7.58 Bore profiles of the four instruments illustrated in Fig. 7.57

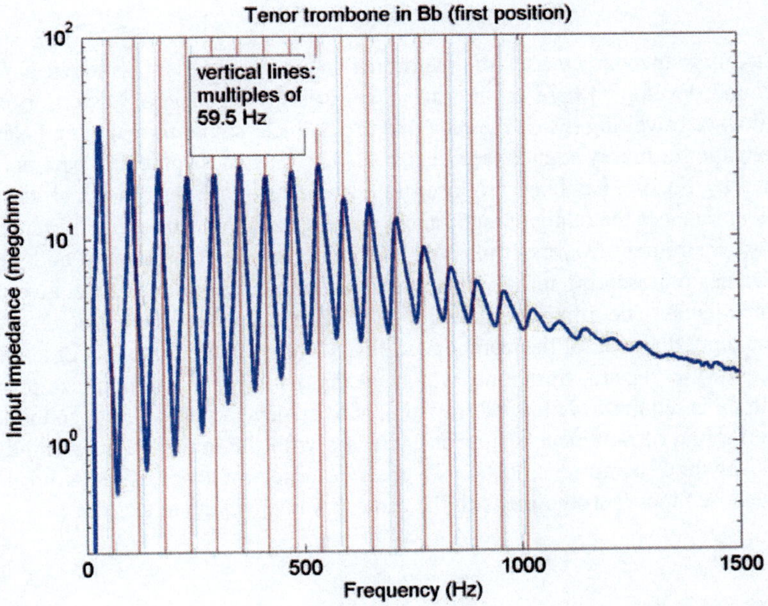

Fig. 7.59 Input impedance curve for the tenor trombone shown in Fig. 7.57a, with the slide fully retracted

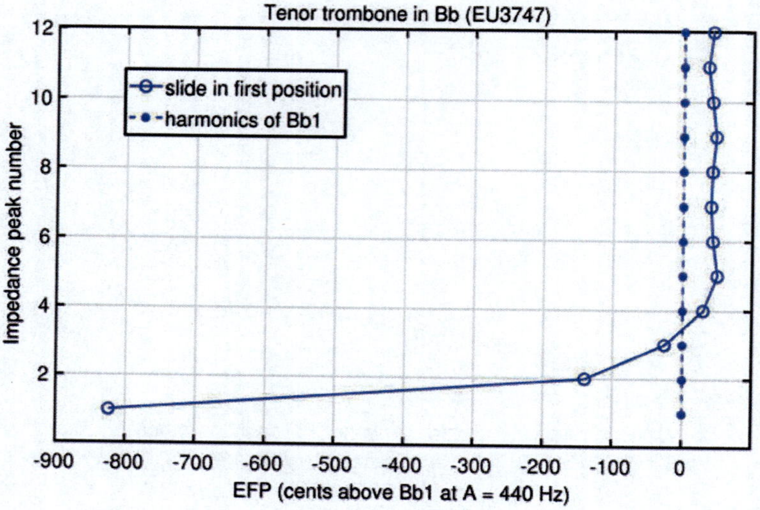

Fig. 7.60 Open circles: EFP plots for the tenor trombone shown in Fig. 7.57. Filled circles: harmonics of B^b1

An important characteristic of the timbre of a brass instrument is the rate at which the sound develops a hard or 'brassy' edge during a crescendo. Distortion of the sound wave travelling inside the tube due to nonlinear sound propagation is largely responsible for brassy sounds (see Sect. 6.1). The brassiness potential parameter B, defined by Eq. 6.7, has been proposed as a measure of the way in which the bore profile influences the relative importance of nonlinear propagation in different types of brass instrument (Myers et al. 2012). B can take values between 0 and 1, with high values representing instruments which develop brassy sounds at relatively low dynamic levels. The trombone shown in Fig. 7.57a has a value $B = 0.8$.

The input diameter of the trombone in Fig. 7.57a is smaller than that of a twenty-first-century orchestral trombone, which means that a higher mouthpiece pressure amplitude is required in the older instrument to achieve a given output sound level. Since the rate of nonlinear distortion increases with the input pressure amplitude, the narrow bore instrument develops a brassy timbre at a lower dynamic level than a large bore modern trombone with the same B value (Myers et al. 2012).

7.8.2 The Serpent

Figure 7.57b is an illustration of a nineteenth-century French serpent (by C. Baudouin, Paris, c. 1820). Its bore profile is shown by the red curve in Fig. 7.58. Unlike the trombone, the serpent does not have a lengthy cylindrical section. Figure 7.61 demonstrates that the profile of the instrument shown in Fig. 7.57b can

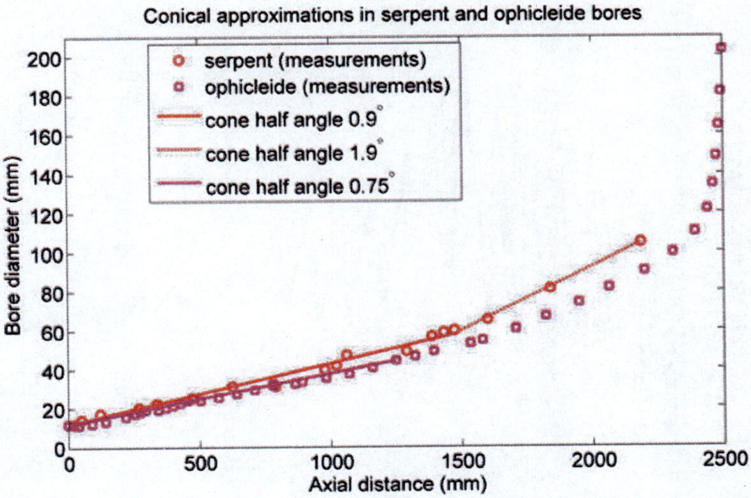

Fig. 7.61 Red circles and lines: approximation of the bore of the Baudouin serpent by two conical sections. Magenta squares and line: approximation of the first half of the bore of the Gautrot ophicleide by a cone

be approximately modelled as a cone with a half angle of 0.9° for three quarters of its length, terminated by a second cone of around twice this angle.

The instrument has six toneholes which can be closed by the fingertips of the player. Figure 7.62 illustrates the input impedance of the serpent with all six holes closed (blue curve) and with the lowest two holes open (red curve). D was the common nominal pitch of early nineteenth-century French serpents. While some of the surviving instruments appear to have been designed to play at the late eighteenth century 'ton de l'opéra' (A4 = 392 Hz) (Eveno and Le Conte 2013), Fig. 7.63 shows that the impedance peak frequencies of the EU (3606) serpent with all holes closed are fairly close to integer multiples of 69.2 Hz, the frequency of D2 at A4 = 415 Hz.

In contrast with the trombone, the first resonance of the serpent is well placed to reinforce the fundamental of the series. However it is evident from the results shown in red in Figs. 7.62 and 7.63 that when the lowest two toneholes are opened the regular spacing of the impedance peaks is disrupted. This fingering is prescribed for playing F♯2 on a serpent in D, so ideally opening the two holes would raise all the impedance peak frequencies by 400 cents. The red EFP plot in Fig. 7.63 shows that this is approximately true for the first two peaks, but the higher peaks have been raised by a much smaller interval. The reason is that the diameter of a serpent finger hole is much smaller than the bore diameter; in consequence the tonehole cutoff frequency, above which open holes no longer vent effectively (Nederveen 1998a), is only around 200 Hz.

A keyless serpent has to have an irregular spacing of toneholes to make it possible for the fingers of a human player to close them. This further increases the inharmonicity of the resonances. In consequence, the spectra of serpent notes are

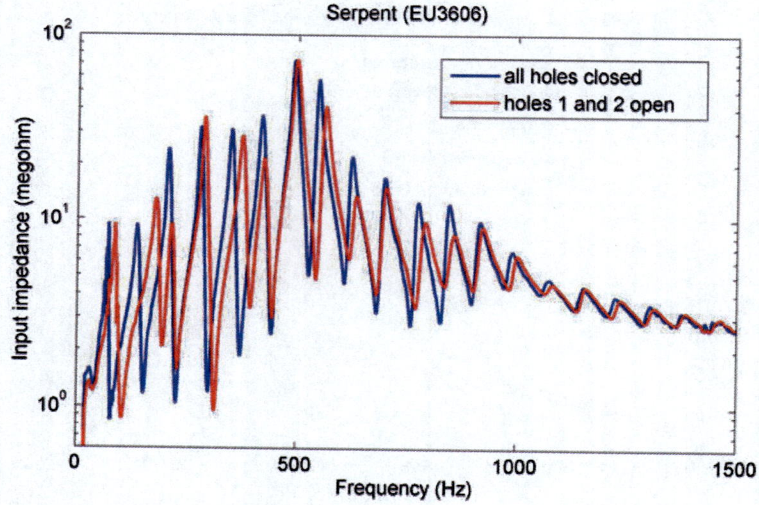

Fig. 7.62 Input impedance curves for the Baudouin serpent with all holes closed and with two holes open (Color figure online)

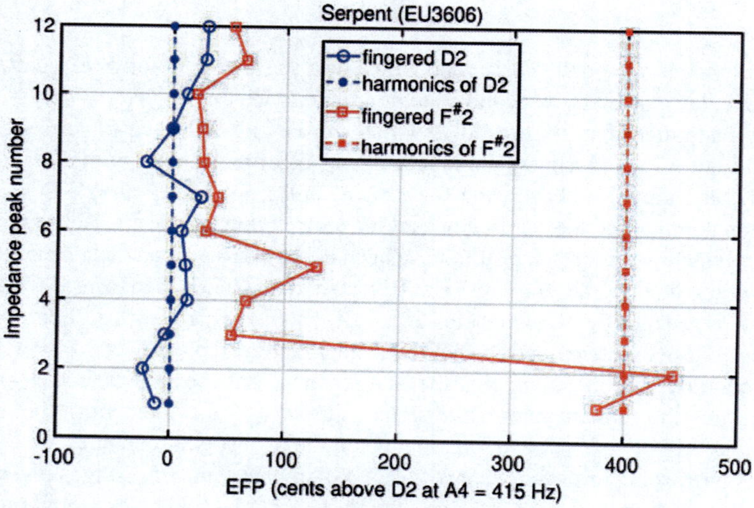

Fig. 7.63 EFP plots for the Baudouin serpent with all holes closed (open circles) and with holes 1 and 2 open (open squares). Also shown are harmonics of D2 (filled circles) and F♯2 (filled squares)

typically dominated by the first few harmonics, giving the instrument a characteristic mellow timbre. This is ideal for its original function of supporting male voices in cathedral choirs (Hostiou 2005), and its warm and blending sound has also been exploited by orchestral composers including Berlioz.

The steadily widening conical bore of the serpent results in a low value ($B = 0.29$) of the brassiness potential parameter. The relative unimportance of nonlinear propagation in serpent timbre means that the sound does not develop a brassy edge even when played at maximum volume.

7.8.3 *The Ophicleide*

The instrument illustrated in Fig. 7.57c is a nine key ophicleide (by Gautrot, Paris, mid-nineteenth century); its bore profile is shown by the magenta curve in Fig. 7.58. The first half of the tube has a conical profile similar to that of the serpent, as illustrated in Fig. 7.61, but the bore then flares out and ends in a 200 mm diameter bell. The nominal pitch of this ophicleide is C2, although closing all the keys lowers the pitch a further semitone to B1. The input impedance for the instrument with all holes closed is shown by the blue curve in Fig. 7.64; the red curve shows the input impedance with the seventh hole opened (the fingering for G2). The corresponding EFP plots are shown in Fig. 7.65.

The blue EFP plot in Fig. 7.65 reveals that when all the holes are closed the input impedance peaks from $n = 1$ to $n = 9$ are close to the target harmonic series based on B1. The $n = 1$ peak, however, is more than 100 cents too sharp to fit this series. Modelling studies have shown that this deviation in the first resonance is due to the flaring of the final section of the tube. The red EFP plot in Fig. 7.65 shows the situation when the G2 fingering pattern is employed. The increased size

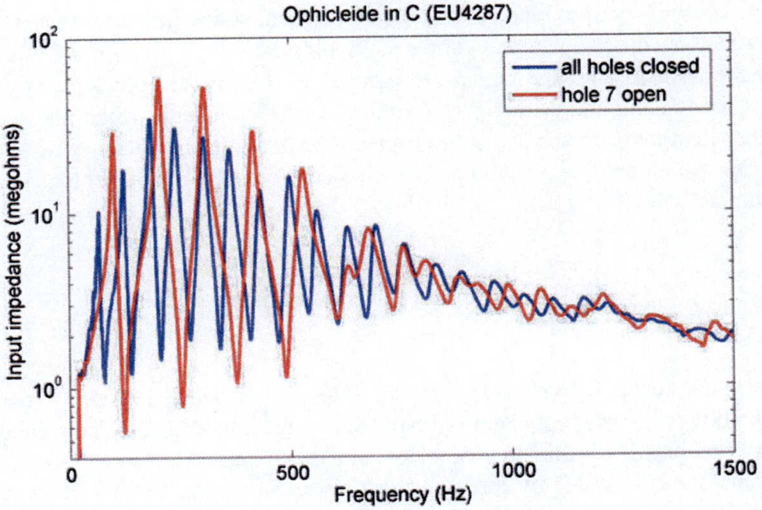

Fig. 7.64 Input impedance curves for the Gautrot ophicleide with all holes closed (blue) and with the seventh hole open (red)

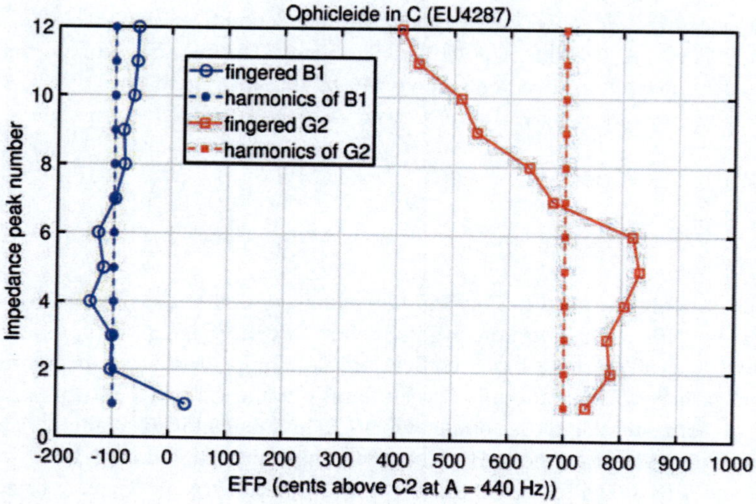

Fig. 7.65 EFP plots for the Gautrot ophicleide with all holes closed (open circles) and with the seventh hole open (open squares). Also shown are harmonics of B1 (filled circles) and G2 (filled squares)

and improved spacing of the keyed toneholes of the ophicleide result in a higher cutoff frequency than the serpent, but the lack of efficient venting is still noticeable in the flattening of the peaks from $n = 7$ upwards. The peaks from $n = 1$ to $n = 6$ are significantly sharper than would be ideal to play the target note G2; although accuracy of intonation is much easier on the ophicleide than on the serpent, the player is still likely to encounter this type of problem, which can be corrected by the careful embouchure adjustment known as 'lipping'.

The brassiness potential parameter value $B = 0.3$ for the ophicleide is similar to that for the serpent. Although the timbre of the ophicleide when played forte is richer than that of the serpent, both instruments are in the category of 'cuivres doux' for which nonlinear sound propagation does not play a major role in spectral enrichment.

7.8.4 The Bass Tuba

The original score of Berlioz's *Requiem*, published in 1838, included a part for 'bombardon or monster ophicleide with pistons' (Macdonald 2002). The bombardon had a bore profile similar to that of a contrabass ophicleide in F, but the key mechanism was replaced by three or more piston valves. In 1843 Berlioz visited several German cities and greatly admired the bass tubas with five valves designed by Wilhelm Wieprecht for use in Prussian military bands. A tuba of this type (by

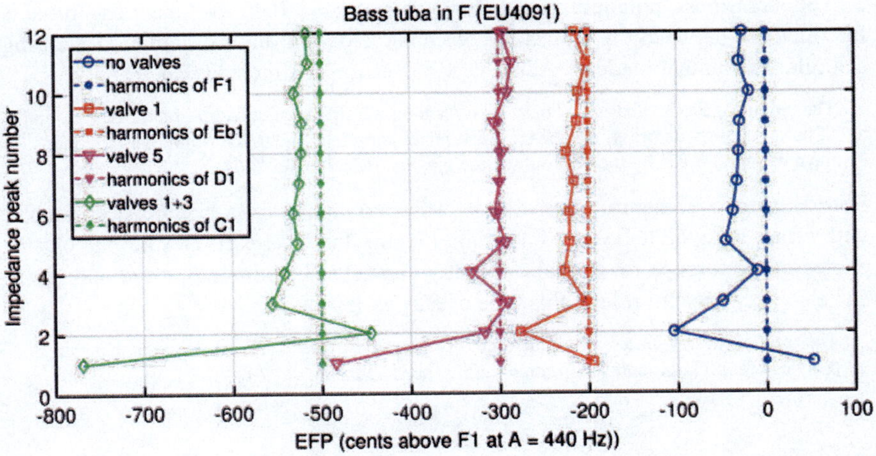

Fig. 7.66 EFP plots for the bass tuba shown in Fig. 7.57d. Valves operated for pitches F1 (open circles), E^b1 (open squares), D1 (open triangles), and C1 (open diamonds). Also shown are harmonics of F1 (filled circles), E^b1 (filled squares), D1 (filled triangles), and C1 (filled diamonds)

C.F. Zetsche Söhne, Berlin, probably 3rd quarter nineteenth century) is illustrated in Fig. 7.57d. The bore profile of this instrument with no valves activated is shown by the green curve in Fig. 7.58. This profile differs substantially from that of the ophicleide: over the first 1000 mm it remains almost cylindrical, and it then has a smooth and gradual flare towards the 270 mm diameter bell. The impedance peak frequencies for this instrument with no valves depressed are close to a series of harmonics with fundamental F1, as shown by the blue EFP plot in Fig. 7.66. Intonation problems associated with the use of valves in combination (Campbell and Greated 1987) were reduced by providing more than three valves. The additional tubing inserted by most valves can be adjusted by a separate tuning slide. The fifth valve, for example, can be tuned to reduce the pitch by a minor third, allowing the third valve to be tuned to give an accurate downward step of a perfect fourth in combination with the first valve. The EFP values in Fig. 7.66 were obtained from a reference measurement in which all the tuning slides were fully pushed in; for performance the slides would be pulled out to different extents to optimise the intonation.

Residual problems in pitching notes in the lowest register are revealed by the irregularity of the EFP plots for $n = 1$ and $n = 2$. The sharpening of the lowest resonance already noted on the ophicleide is evident in the plot with no valves depressed. As the proportion of cylindrical tubing in the bore profile is increased by activating valves, the flattening effect observed on the trombone comes to dominate. The EFP values for the fingering used for the note C1 are shown by the green plot in Fig. 7.66: the $n = 1$ peak is almost 3 semitones flatter than this nominal pitch.

The brassiness potential parameter value $B = 0.46$ for the bass tuba is intermediate between the values for the ophicleide and the trombone. Describing a similar instrument made by Adolphe Sax, Berlioz commented that

> The sound of Sax's tuba is not only steadier than that of the ophicleide, but stronger and of better quality. Its brassy resonance harmonises completely with the trombones and lacks the dull sound made by the best ophicleide players. (Macdonald 2002).

Here Berlioz is evidently referring to fortissimo chords by a trombone ensemble, with either an ophicleide or a bass tuba playing the bass part. Despite his clearly expressed preference for the tuba's brassy character, he admits that the ophicleide can also play a useful role in this type of ensemble:

> The tone of the low notes [on ophicleides] is rough, but they can work wonders beneath a body of brass instruments in certain circumstances. (Macdonald 2002).

7.8.5 Berlioz and Pedal Notes

Berlioz exploited the dramatic effect of a fortissimo trombone pedal note at several points in his symphonic works. In the *Requiem* he called for pedal notes at several different slide positions, commenting in his '*Treatise on Orchestration*' that he had to convince several of the trombonists in the first performance that these pedal notes were indeed playable (Macdonald 2002). Berlioz was a most imaginative and original orchestrator, with a particular fascination for the varied sonorities which could be found in combinations of brass instruments. The passage in the *Requiem* in which unison pedal notes on several tenor trombones are accompanied only by flutes playing around four octaves higher has frequently been criticised, but it is clear from the discussion of this passage in the '*Treatise*' that the admittedly bizarre result was carefully calculated and intentional.

Another example of the use of trombone pedal notes for dramatic effect is found in the *Marche au supplice* from the *Symphonie Fantastique* (Fig. 7.67). Berlioz explains his goal thus:

> I needed to devise low harmonies of exceptional savagery in an unusual sonority. I believe I achieved this . . . by this diminished seventh between an ophicleide and a tenor trombone's pedal A. (Macdonald 2002).

It is significant that, although a B$^\flat$ bass ophicleide could have played the lower line in Fig. 7.67, Berlioz assigned the line to unison tenor trombones. His demand for the magnificent snarling brassiness of a sforzando pedal note on the trombones was not understood by the editors of an early twentieth-century version of the score, in which the upper line was reassigned to a single trombone while the lower line was given to two tubas. The effect of this re-orchestration might have been impressive, but it was certainly far from the sound which Berlioz intended. It is good

Fig. 7.67 Excerpt from the original scoring of the 'Marche au supplice' from the *Symphonie Fantastique* by Hector Berlioz

to recognise that the growing interest in performing nineteenth-century music using period instruments is allowing the originality and resourcefulness of composers such as Berlioz to be once more heard and appreciated.

Chapter 8
How Brass Instruments Are Made

Musicians – not only players but also composers, arrangers and conductors – consciously or unconsciously assess brass instruments by certain criteria. These include compass, capability for intonation control, dynamic range, timbre, responsiveness, ergonomics and appearance. In this chapter, we discuss the processes of design and manufacture through which instrument makers can exercise control in order to meet the requirements, expectations and desires of musicians.

Brass instruments have for long been made in widely differing commercial enterprises, ranging from individual artisans (using mostly hand craft techniques), through medium-sized workshops (some operating a master-apprentice system), to factories able to adopt mass production and automated processes. Mechanised processes allow large numbers of instruments of the same model to be produced very economically, and the quality can be uniformly high given good design. Purely hand craft is employed in making historical instruments where it is believed that use of historical techniques gives the most faithful re-creations. A professional-quality instrument such as a french horn made in a medium-sized workshop can today take up to 50 h of work for a skilled operative, whereas a cheap mass-produced instrument can be turned out on an assembly line with less than 6 h of work.

8.1 Materials

Brass has for centuries been such an extensively common material for brasswind that the word has come to be used for the entire class, regardless of the specific material of construction. The properties of brass are such that it has been relatively economical for instrument makers to obtain, is readily pulled and beaten into shape and has produced instruments which are light enough to be held for use while being robust enough to be durable, and are of attractive appearance when polished.

M. Campbell et al., *The Science of Brass Instruments*, Modern Acoustics and Signal Processing, https://doi.org/10.1007/978-3-030-55686-0_8

The composition of the brass is a matter of individual preference (see Sects. 1.3.3 and 6.6); typical proportions for instrument bells are around 70% copper, 30% zinc (yellow brass); around 85% copper, 15% zinc ('gold' brass); and around 90% copper, 10% zinc ('red' brass). There are usually trace elements present in commercially available brass, importantly a small amount of lead which is necessary for turning in a lathe (von Steiger et al. 2013). The main parts of instruments are fabricated from sheet brass and seamless tubing bought in from brass foundries.

Although brass continues to be regarded as the most commonly used material, other metals have been used for brass instruments: bronze in the ancient world (see Sect. 1.1.2), silver and in the nineteenth century the alloy german silver (also known as white bronze, or Maillechort). The composition of german silver varies but is typically around 70% copper, 18% nickel and 12% zinc. Today german silver is more commonly used for garnishing such as ferrules, mouthpipe sheaths and bow guards on instruments primarily of brass. So-called 'beryllium' bells have a small percentage of beryllium added to copper or copper alloy.

Since the late twentieth century, the bells and larger sections of tubing of some sousaphones have been made of fibreglass, principally to reduce the weight of these instruments for marching band use. Carbon fibre, its strength superior to that of brass, is proving to be a satisfactory material for components such as bells and trombone outer slides for high-grade instruments. In recent years low-priced plastic instruments have been developed which have proved commercially successful: several firms offer trombones, trumpets and tubas. These instruments are mostly made from acrylonitrile butadiene styrene (ABS), but some parts such as slide stockings and valve components can be metal. There have been applications of 3D printing to the production of instruments, both for rapid prototyping and the production of replica instruments. Initial experiments have focussed on finger-hole brasswind such as cornetts and serpents which have been traditionally made of wood.

8.2 Design

In Chap. 7 we saw how the overall bore profile determines in broad terms how an instrument behaves. In Chaps. 4 and 6 we saw how details of the bore profile such as the bell flare can influence compass, capability for intonation control, dynamic range, timbre and responsiveness. With the understanding we now have of brasswind acoustics, it is possible to design an instrument with some confidence in predicting its behaviour in performance (Braden 2006). Nevertheless, the traditional procedure of trial and error continues to guide instrument design. One reason for this is that most new brass instruments brought to market are established kinds of brasswind: it is very rare for a completely new type of instrument to be devised in the way that Halari invented the ophicleide or Wieprecht and Moritz the bass tuba in the nineteenth century. Tools are available such as the Brass Instrument Analysis System (BIAS) (Artim 2020) with which makers can test instruments under development (see Sect. 4.2.1). The associated software can make suggestions for bore profile optimisation.

One feature of modern brass instruments is the great attention given to the first section of tubing following the mouthpiece, known as the mouthpipe or 'leadpipe' (Buick et al. 2002). In some models there is a step down in bore diameter positioned at the end of the mouthpiece: this reduces or eliminates the sudden expansion in bore where the mouthpiece stem abruptly ends. The bore constriction often continues for a couple of centimetres before an approximately conical expansion towards the tuning-slide or valves. In trombones there is often a cylindrical portion in the mouthpipe narrower than the main bore of the descending inner slide. This mouthpipe constriction is often termed the 'venturi', an inappropriate term since the Venturi effect concerns the steady flow of incompressible fluids through a constriction while air is compressible.

The term 'leadpipe' is often used as a synonym for 'mouthpipe' to denote the mouthpiece receiver together with the following section of instrument tubing. However, some high-quality instruments are made in which this initial section of tube is removable and interchangeable. Here the term 'leadpipe' is more usual. A detachable leadpipe is 150–250 mm long including the mouthpiece receiver and is inserted into an outer sheath which forms part of the body of the instrument (in the case of trombones, the leadpipe extends into the descending inner slide). Players of such instruments can select a leadpipe to optimise the instrument for their requirements or equip themselves with a number to choose from to suit a repertoire or performance situation. The bore profiles of alternative leadpipes supplied for a particular instrument have interior diameters that vary by a only few tenths of a millimetre, but even such small changes noticeably affect the response of the instrument.

8.3 Metal Forming

A cylindrical or conical tube can be formed from flat brass sheet. Before the nineteenth century, the tubing for instruments was made by hand from strips of brass rolled around a rod (known as a mandrel) and seamed by brazing; today factory-made tubing is used. Very few parts of brass instruments are purely conical: most sections if not cylindrical gradually increase in rate of expansion from one end to the other. This is most perceptible in bell flares, where the rate of expansion noticeably increases towards the bell end. The shape of actual instrument bells (and other sections) does not follow a simple mathematical formula. Gently expanding sections can be approximated by an exponential curve; flaring sections such as trumpet and trombone bells can be close to a so-called Bessel horn (Sect. 4.3.7). Although actual bells only rarely follow a Bessel horn over their entire length, a Bessel horn fit can be useful for representing bell flares over portions of their length (Braden 2006).

One factor which makers can control is wall thickness. We saw in Sect. 6.6 that the structures of brass instruments vibrate when played at loud dynamics, although

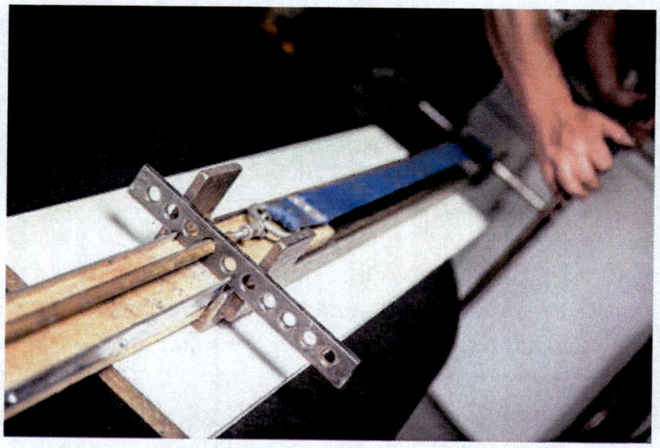

Fig. 8.1 Use of a draw-bench to fit tube to a mandrel. Photograph: Max Münkwitz, courtesy of the International Trumpet-making Workshop. Reproduced from Münkwitz et al. (2014)

authorities disagree on the significance of wall vibrations. At the very least, the walls of any wind instrument need to be rigid and massive enough to substantially maintain the shape of the air column at all dynamic levels. Beyond this, ergonomic and practical considerations dictate the weight (and thus the wall thickness) of an instrument. Traditionally instruments at risk from damage in handling such as marching band instruments are made of heavier gauge metal than instruments for professional use in the concert hall.

Tubing supplied to an instrument-making workshop is not necessarily the precise diameter and wall thickness required: generally it needs to be adjusted using a draw-bench (see Fig. 8.1). Cylindrical tubing can be brought to size by placing loosely over a mandrel (in this case a rod) and forcibly drawing it through a hole in a steel 'draw-plate'. This reduces the internal diameter of the tube to diameter of the mandrel and also lengthens the tube. Gently expanding sections can be produced by 'swaging': placing tubing (seamed or seamless) loosely over a mandrel and forcibly drawing it through a lead plate starting at the narrow end. This is also done on a draw-bench and forces the tubing to take up an interior shape corresponding to the exterior shape of the mandrel. It is also stretched and thinned in the process.

Any non-conical section of tubing formed from flat sheet requires working of the metal by stretching or contraction. Bell flares require considerable stretching of tubing made from flat metal: traditionally this was done by beating a heated workpiece on an anvil before bringing it to the correct shape by burnishing on a mandrel (see Fig. 8.2). Some maintain that hand beating permits finer control over wall thickness. Wide bell flares were sometimes made with a 'gusset', a triangular piece of sheet brass incorporated into an incomplete bell flare by seaming. Rather than being gusseted and seamed, some bells are now made by drawing the final flare from sheet brass and then cross-seam welding to the bell stem. Bell flares are now usually spun: a roughly-formed bell is placed on a mandrel and turned in a lathe (see Fig. 8.3). While being spun, it is worked to a uniform thickness and burnished and the rim turned over a wire reinforcement which is fixed by soft solder.

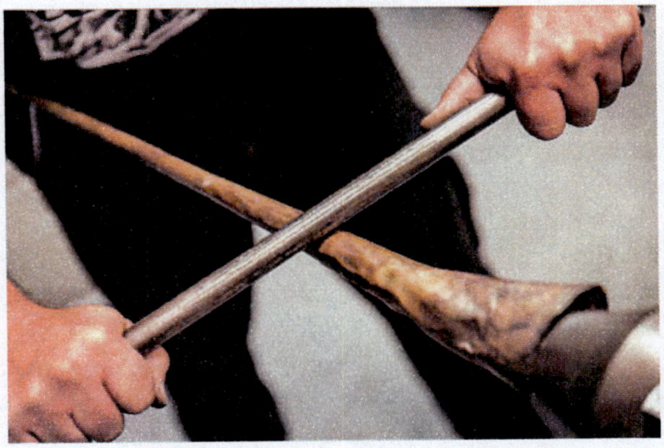

Fig. 8.2 Burnishing a bell by hand. Photograph: Max Münkwitz, courtesy of the International Trumpet-making Workshop. Reproduced from Münkwitz et al. (2014)

Fig. 8.3 Spinning a bell on a lathe. Photograph: Robert Pyle

In both manual processes and spinning, achieving the required internal shape depends on the mandrel. If not actually made by the instrument maker, mandrels are made to the maker's specification and form an essential (but often unwritten) part of the 'recipe' for making an instrument. Mandrels of iron or steel can be used for many years, allowing continuity in the production of individual models of brass instrument design over time.

Few instruments are completely straight: most require tubing to be bent. Tubing is sometimes prepared for bending by filling with a weak aqueous soap solution and then freezing. Low-melting point alloys and resin are other alternatives to the traditional filling with lead or pitch. A tube filled with an appropriately soft solid can be bent around a form without tearing or collapsing the walls; once bent the filler can be removed with gentle heat. Tube bent by such a process is slightly lengthened; it is also given a slightly elliptical cross-section which is not of major acoustical significance, although some manufacturers restore a circular cross-section by forcing steel balls through bent tubing.

In any process which involves changing the shape of metal, the workpiece becomes harder. In both hand craft and factory methods, the brass has to be annealed after each step to restore its softness and ductility. Annealing is the process of heating metal until it is red hot and then quenching in water. This changes the crystalline structure of metal which has been hardened by hammering, burnishing or bending and leaves it soft and malleable for further operations. One of the decisions a manufacturer has to make is whether to leave brass instrument bells finally work-hardened or annealed.

In some factories, small sections of tubing are formed from sheet metal by 'deep drawing' in presses: the brass sheet is placed between two steel dies and forced into the shape of a saucer; subsequent pairs of dies press the brass via a cup shape into a section of tube of the desired shape which is then cut to length at both ends. Sections of standard tube or deep-drawn pieces can be placed between external steel dies and expanded by hydraulic pressure to the required size and shape. These industrial techniques are expensive to tool up, but they allow large numbers of components to be produced very economically.

8.4 Valves

There is now little difference in performance between piston and rotary valves, and the use of one rather than the other is largely a matter of tradition. With a piston valve, the length of travel has to be at least one bore diameter. Because with rotary valves there is of necessity some system of levers to transform the linear motion of the touchpiece into a rotation of the rotor, the connectors can be designed so that the length of travel of the touchpiece is arbitrarily small. In practice, however, the travel of the player's finger and the restoring force of the spring are not greatly different for rotary and piston valves. The biggest distinction is that when a rotor is halfway between its two positions, the windway is almost blocked, whereas when a piston is half depressed, air can pass through both the direct passage and the valve loop, albeit with some constriction. With sufficient lip control, a player can 'half-valve' a piston instrument to produce a glissando between notes.

The most common purpose of the valve is to extend the tube length, lowering each of the series of notes available to the player by a semitone, two semitones or a larger interval, depending on the relationship between the air column lengths with and without the valve operated. By far the most widely used arrangement is for the valve operated by the first finger to give a pitch lowering of two semitones, the second finger to lower by one semitone and the third finger to give three semitones. The right hand is used for these three valves except with the french horn, where the right hand has maintained its position in the bell for hand-stopping and the left hand operates the valves. There have been, however, a significant number of instruments in Germany which have the role of the first and second fingers reversed and a significant number of French instruments where the third finger gives a pitch lowering of four semitones. It is equally easy to arrange the valve passages so that operating the valve cuts out the extra tubing of the valve loop rather than adding it. This 'ascending valve' was for many years used on french horns in France for the third valve: operating the ascending third valve raised the pitch by two semitones. These alternative configurations required different fingering techniques. Four-valve french horns generally have the fourth valve ascending from 12 ft F to 9 ft $B^{\flat}$ or from 9 ft $B^{\flat}$ to 6 ft F and arranged to be operated by the left thumb. Instruments such as tubas which commonly have four or more valves most usually have the fourth valve lowering by five semitones; the finger used for the fourth valve is the first finger of the left hand or the fourth of the right.

Some early valved instruments had only two valves. Two valves giving one and two semitones separately give three semitones in combination. With the longer tube length of the trumpet (typically 6 ft F or 7 ft D in the nineteenth century), two valves were entirely adequate to execute the repertoire. With the french horn, players used valves in combination with hand-stopping, and two valves were enough to give a complete chromatic compass; two-valved horns were made in France until the end of the nineteenth century. Even with the early cornet, a third valve was not always considered to be worth the extra expense. However, as the facility of valve technique led to greater demands from composers and arrangers, and the idioms of brass writing became more chromatic, three valves became normal for cornets, trumpets and band instruments; french horns for orchestral use increased to four valves ('double horns') in the first half of the twentieth century. Orchestral tubas, in order to have a chromatic compass down to the lowest notes players can effectively produce, have at least four, and in some models five or six valves.

An inherent problem with valve instruments is the intonation when valves are used in combination: if a valve adds the correct amount of tubing to lower the instrument by n semitones, it will not add enough tubing to lower the instrument by n semitones if another valve is in use at the same time. The calculations can be complex since valves increase the amount of cylindrical tubing in the windway and affect the bore profile as well as the air column length. Typically, a three-semitone third valve used in combination with a five-semitone fourth valve will when operated together give a lowering of nearer seven than eight semitones.

Various ways of overcoming this problem have been devised. With small instruments, it is often enough to tune the third valve to lower the pitch by slightly more than three semitones and then to avoid using it on its own – the player can then 'lip' any wayward notes up or down sufficiently for reasonably good intonation. It is common with trumpets and cornets to fit the main or third valve tuning-slide (sometimes the first) with a finger-ring or sprung lever so that it can be moved by the player, at least in slow-moving passages. Some tubas are designed so that a tuning slide can be manipulated in performance to adjust intonation. Some models of large instruments such as euphoniums and tubas have large numbers of valves (five or six) allowing the player some flexibility of fingering – there might, for instance, be two valves nominally giving a semitone but with one adding more tubing than the other so that the player can find a judicious valve combination and successfully lip any note in tune. The ‘compensating’ valve system automatically improves intonation by running the additional tubing of a master valve (that giving the greatest lowering of pitch) through separate passages in the other valves; these then add the required extra length of tubing when used in combination with the master. Compensating valves are used on some french horns and many euphoniums and tubas. The ‘double principle’ used in ‘full double’ french horns uses alternative rather than additional valve loops for the valves other than the master: this makes for a heavy instrument, since typically three valves each have two full-length valve loops; so as a result it has rarely been used on tubas, but it is the preferred valve pattern for french hornys.

The position of the valves on the tubing is a matter of design preference for instruments with a substantial amount of cylindrical tubing such as the french horn. Recent research by Gregor Widholm (2005) has shown that this can have a significant effect on the smoothness of slurred transitions from one note to another that involve a change in the valves operated.

Piston and rotary valve cases are turned on a lathe and machined to size, and then the ports are drilled. Small pieces of tubing which connect two valves, or each valve with the rest of the instrument (the ‘knuckles’), are inserted in the holes, and the whole assembly (‘valve cluster’) of three or four valve cases with connectors and knuckles is wired together and soldered at the same time. Valve pistons are made of hollow construction for lightness: the body is often produced by deep drawing, and the ports are then drilled to take the internal passages (‘coquilles’) which are inserted through the drilled holes and silver-soldered into place. Protruding tubing is then removed, and the piston is fitted by a process of reaming the inside of the case, grinding the piston and lapping. To prevent a piston from rotating, a lug (the ‘key’) projects from the piston body and runs in a vertical channel (the ‘keyway’) cut into the case interior. Valve rotors are made by a similar process, though on cheaper instruments they can be cut from solid brass or bronze.

8.5 Assembly

Before an instrument is assembled, the parts are cleaned. Small pressed parts are deburred manually or in a vibrating bath. The instrument is assembled starting from the bell or the next largest section and working towards the mouthpipe. Straight sections (‘branches’), U-bends (‘bows’), and other sections are joined by soldering; each joint is covered by a sleeve or ‘ferrule’. An instrument can be left with its brass polished (sometimes referred to as ‘raw’ brass) which over time gradually acquires a patina if not repeatedly polished, but most instruments are given some kind of surface protection after polishing to preserve their appearance longer. Silver-plating by electrolysis has been offered as a standard option by many makers since the mid-nineteenth century and gives a brilliant, easily-cleaned surface. This adds a protective layer of silver so thin that the acoustical properties of the instrument are not perceptibly affected. Other instruments are sprayed with clear or tinted lacquer, which can preserve the colour of brass but which adds a thin coating to the metal surface. This is thought by some to have an effect on the sound of the instrument; if it does it is very small.

Since the beginning of the nineteenth century, most brass instruments have been equipped with a tuning-slide: slide trombones were the last to receive this feature. In addition to the main tuning-slide, each valve loop has its own tuning-slide; in many cases, the outer slides are soldered directly to the knuckles.

Several manuals (Nödl 1970; Bahnert et al. 1986; Dullat 1989) provide detailed descriptions of brass instrument construction methods appropriate for medium-sized workshops. Barclay (1992) describes in detail the traditional hand craft materials, tools and techniques, while Münkwitz et al. (2014) outline practical procedures for employing traditional techniques today. The manufacturing techniques used in several present-day factories were described in a number of articles in the periodical *Brass Bulletin*.

Chapter 9
Looking Back and Looking Forward

The scientific study of brass instruments is a research field with a history stretching back several centuries, and many distinguished physicists, mathematicians and engineers have contributed to its development. The mathematical basis of our understanding of the operation of musical wind instruments was established in the seventeenth and eighteenth centuries by Marin Mersenne (1635) and Daniel Bernoulli (1762). In the last quarter of the nineteenth century, the great German physicist Hermann Helmholtz (1877) discussed the nature of the lip valve in brass instruments, and David James Blaikley (1878), Works Manager for the London brass instrument manufacturer Boosey & Co., carried out pioneering experiments on the internal resonances and playing properties of brass instruments. In France, important experimental and theoretical work on brass instrument acoustics was published by Henri Bouasse (1929). In the same year, the Acoustical Society of America was founded; the first volume of its Journal included a brief report of a scientific study on the sound output of various musical instruments including trumpet, trombone, french horn and tuba (Sivian et al. 1930, 1931). Since that time the pace of research on brass instrument acoustics has greatly accelerated; the present book has tried to give an outline of the major achievements of this work and an introduction to the current state of knowledge.

There have been two principal motivations for the worldwide research effort in brass instrument acoustics. One is the intellectual curiosity, shared by musicians and scientists alike, to understand better the strengths and limitations of the musical instruments which play such a vital role in our cultural life. The other is the desire of scientists to be able to offer useful advice and guidance to instrument makers, performers, teachers and composers based on solid experimental evidence and tested computational techniques for simulation and synthesis.

This final chapter offers a brief glimpse of two areas in which acoustical understanding and scientific techniques have recently found practical applications in the design, construction and playing of brass instruments. Section 9.1 describes some studies which have aided projects aimed at constructing playable reproductions

M. Campbell et al., *The Science of Brass Instruments*, Modern Acoustics and Signal Processing, https://doi.org/10.1007/978-3-030-55686-0_9

of ancient brass instruments such as the Celtic carnyx and the Etruscan cornu. Section 9.2 reviews developments using digital technology to extend the capabilities of existing brass instruments and to map the new world of virtual brass instrument sounds.

9.1 Brass Instruments in the Ancient World

The earliest known surviving brass instruments are two trumpets found in the burial site of Tutankhamun, King of Egypt in the fourteenth century BCE. Surviving instruments from Ancient Greek, Etruscan and Roman archaeological sites, together with evidence from tomb carvings, wall paintings and other iconographic sources, show that brass instruments also played important roles in the cultures of these societies (Ziolkowski 2002). Playing a 2000-year-old instrument is now considered to pose an unacceptable risk of damage, and in any cases the ravages of time have left most original specimens in an unplayable state. The natural desire to hear what the instruments might have sounded like has been satisfied by constructing accurate reproductions of the originals, on which modern brass players can investigate their playing properties. The following sections describe two such recent projects in which scientific modelling techniques have aided the development of working reproductions of ancient brass instruments.

9.1.1 Etruscan Cornu and Lituus

The wall painting reproduced in Fig. 9.1 shows two Etruscan musicians carrying brass instruments. The figure on the left is holding a cornu, a wide-bored approxi-

Fig. 9.1 Etruscan cornu and lituus (Hescana Tomb, Porano, Italy, fourth century BCE)

mately conical instrument curved into an almost complete circle with a diametral supporting strut. The musician in the centre is holding a lituus, whose shape resembles a shepherd's crook; a considerable section of the bore is approximately cylindrical, but the final section is a curved conical bell.

As part of the European Musical Archaeology Project (EMAP 2018), reproductions of the Etruscan cornu and lituus have been constructed. The fact that the two instruments are shown in close proximity in Fig. 9.1 and in other contemporary images suggests that they were probably played simultaneously. Part of the remit of the EMAP team was to organise public concerts in which the reproduced instruments could be demonstrated. Since there was some flexibility in the choice of absolute length scale of each instrument, it was requested by the musicians who were to play the instruments that they should be scaled to play 'in tune' (Campbell et al. 2017). But what does 'in tune' mean in this context?

The underlying problem is that the cornu has an almost conical bore throughout (like an alphorn), whereas a large fraction of the lituus is cylindrical (like a trombone). Figures 9.2 and 9.3 show the effect of increasing the conicity of an originally cylindrical tube 3 m long on the equivalent fundamental pitches of its acoustic modes (Sect. 4.3.4). The input radius is maintained constant at 10 mm; as the cone half-angle is increased from 0 to 1°, the EFP of the first tube resonance increases by 900 cents, while the EFP of the resonances from the sixth upwards are not significantly changed. It is therefore impossible to find a length-scaling factor which can bring a mostly conical instrument into tune with a mostly cylindrical instrument over its whole range of resonances. Shortening the cylinder would raise the lower resonance frequencies, bringing them closer to those of the cone, but the higher resonances of the cylinder would then be much higher than the corresponding resonances of the cone.

The EFP plots in Fig. 9.3 were derived from input impedance curves calculated numerically using the transfer matrix method (see Sect. 4.7). Applying the same

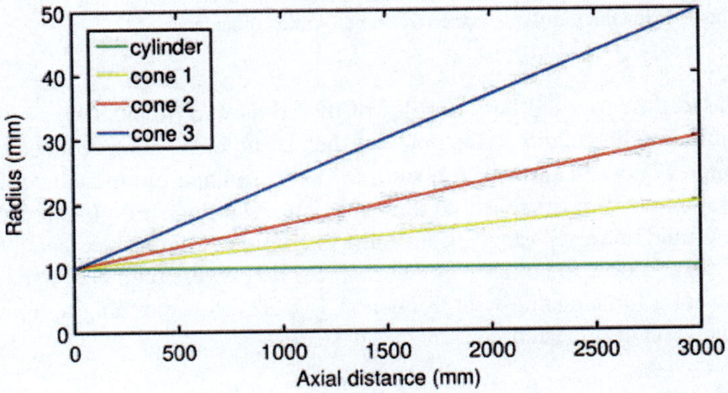

Fig. 9.2 A set of tubes, each with an input radius of 10 mm, a length of 3000 mm, and output radii ranging from 10 mm to 50 mm

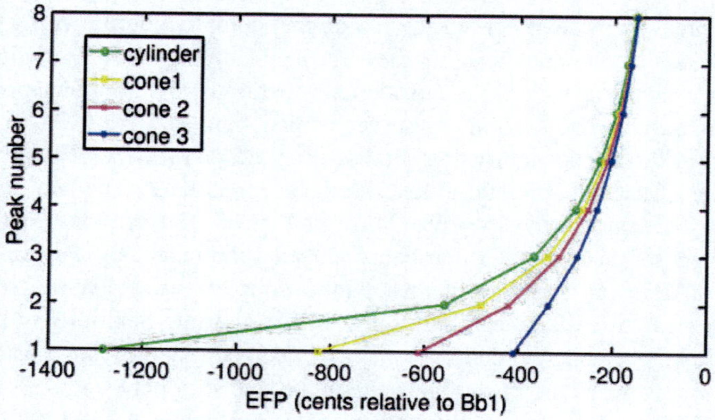

Fig. 9.3 Calculated EFP plots for the tubes shown in Fig. 9.2

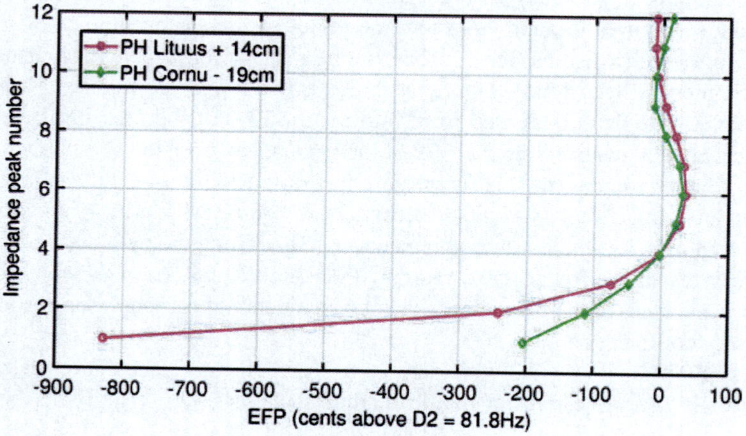

Fig. 9.4 Calculated EFP plots for the proposed reconstructions of cornu and lituus, with bore adjustments to bring the fourth resonance frequencies into coincidence

technique to the proposed bore profiles of the cornu and lituus, it was possible to identify minor adjustments to the bores of the instruments which reduced the pitch discrepancies between equivalent resonances on cornu and lituus to under 20 cents for natural notes from the fourth to the tenth. The resulting EFP plots are shown in Fig. 9.4. Within this pitch range, at least, the instruments could be expected to sound 'in tune'. In addition, the pitches of these natural notes were predicted to be close to harmonics of a fundamental with frequency 81.8 Hz, corresponding to a pitch with note name D on a suggested Ancient Greek scale.

9.1.2 The Celtic Carnyx

The Roman legions which conquered Gaul under Julius Caesar and invaded Britain under Claudius were faced by opposing armies which included players of an unusual and intimidating brass instrument called the carnyx. This was an instrument which had a straight bore around 2 m in length, terminated by an elaborate bell resembling the head of a wild boar. There are many depictions of carnyces on Roman coins and carved reliefs, but no original instrument has survived intact. In 1816 a bronze boar's head which was identified as the bell of a carnyx was discovered in a peat moss at Deskford in northern Scotland (Hunter 2019). This relic (Fig. 9.5a), now in the National Museum of Scotland, was the basis for the construction of a complete carnyx by the craftsman John Creed in 1992. The reconstructed boar's head is shown in Fig. 9.5b. The design of the rest of the instrument was guided by the illustration of three military carnyx players on the Gundestrup cauldron, a large silver bowl from the same period now in the National Museum of Denmark (Fig. 9.6a). The complete carnyx is shown in Fig. 9.6b; in this photograph it is being sounded by an artificial mouth in the laboratory of the Musical Acoustics Group at the University of Edinburgh, which provided acoustical guidance during the reconstruction. The virtuoso brass player John Kenny was involved in the reconstruction project and has subsequently made extensive use of the carnyx in musical performances.

In 2004 a remarkable archaeological discovery at the Gallo-Roman site at Tintignac, in the Corrèze region of central France, shed new light on the nature and methods of construction of the Gallic carnyx. Among more than 500 fragments of iron and bronze objects found in a pit were parts of seven carnyces, including one almost complete instrument (Fig. 9.7a). A research team led by the archaeologist Christophe Maniquet studied the Tintignac carnyx (Maniquet et al. 2011) and instituted a project to reconstruct a playable copy. The craftsman who carried out this work, Jean Boisserie, is shown holding the reconstructed instrument in Fig. 9.7b.

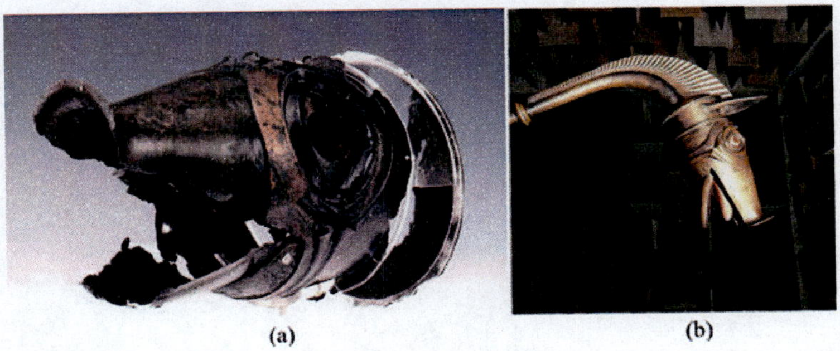

Fig. 9.5 (**a**) The bell of a carnyx, in the form of a boar's head, found in Deskford, Scotland, in 1816. Courtesy of Aberdeenshire Council Museums. Image: National Museums Scotland. (**b**) Reconstruction of the Deskford carnyx by John Creed, under test in the anechoic chamber at the University of Edinburgh

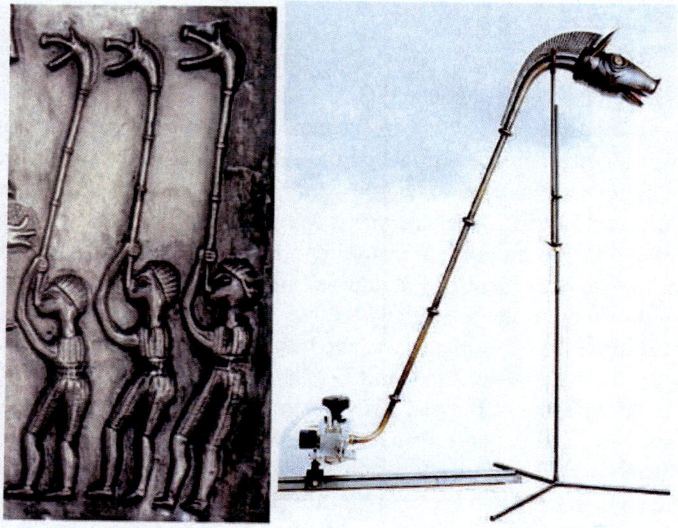

(a) (b)

Fig. 9.6 (**a**) Illustration of three carnyx players in a military procession (Gundestrup cauldron). Reproduced by permission of the National Museum of Denmark. (**b**) Reconstruction of Deskford carnyx played by artificial lips in the University of Edinburgh

Fig. 9.7 (**a**) Carnyx assembled from fragments found at Tintignac, France, in 2004. Photograph: Patrick Ernaux, Inrap; Tintignac, Naves, Corrèze, France. (**b**) Reconstruction of the Tintignac carnyx by Jean Boisserie

(a) (b)

Fig. 9.8 Measurement of input impedance in the anechoic chamber at Le Mans University (CNRS)

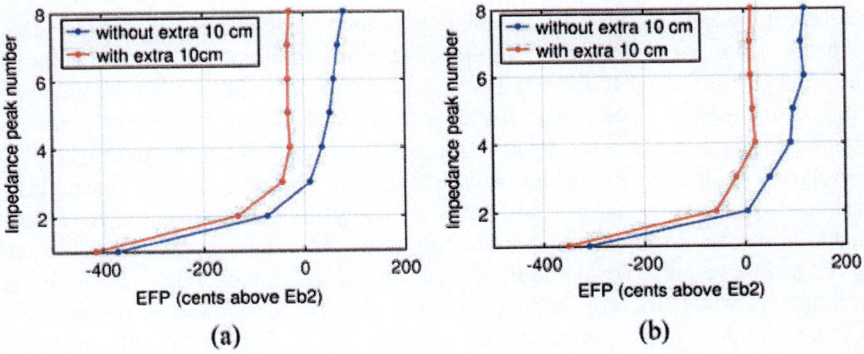

Fig. 9.9 (**a**) Calculated and (**b**) measured EFP plots for the reconstructed Tintignac carnyx

The reconstruction was tested in the acoustics laboratory of Le Mans University (Gilbert et al. 2012).

The input impedance of the Tintignac carnyx was calculated using the transfer matrix method, and from this the EFP plot shown in blue in Fig. 9.9a was derived. This is broadly in accord with the EFP plot shown in blue in Fig. 9.9b, obtained from an input impedance measurement carried out in the acoustics laboratory at Le Mans (Fig. 9.8). Since the bore profile of the main tube resembles a highly truncated cone, the resonances depart significantly from a harmonic series, especially at lower frequencies (Fig. 9.3).

Joël Gilbert and colleagues noted that the natural notes of the Tintignac carnyx would be significantly closer to a harmonic relationship if the length of the carnyx tube were slightly longer. The red curve in Fig. 9.9a shows the calculated EFP plot for an instrument identical to the Tintignac carnyx except that an additional section 10 cm long has been inserted near the mouthpiece end. The impedance peaks from the fourth to the eighth now lie very close to a vertical line representing perfect harmonicity on an EFP plot. There is of course no evidence that the repertoire of the original carnyx players included the well-tuned bugle calls possible on an instrument with harmonic resonances, but the centring and stability of notes on a brass instrument are improved when the upper resonances are close to harmonics. Consultation with Christophe Maniquet revealed that there was indeed another fragment in the Tintignac hoard, 11 cm long, which could well have been missed when the original instrument was reassembled. A second reconstruction was made, in which the additional length could be inserted; the measured EFP plot for the lengthened carnyx showed the expected improvement in harmonicity (Fig. 9.9b, red plot). The musician John Kenny reported that the extra section notably improved the performance of the carnyx.

9.2 Brass Instruments in the Digital World

In recent decades the speed of technological innovation has revolutionised many aspects of our lives, particularly in the areas of information retrieval and communication. Superficially it appears that the world of brass instrument manufacture and performance has been little affected by this whirlwind of change: the trumpets, trombones, french horns and tubas in modern bands and orchestras are remarkably similar to the instruments which would have been played in similar ensembles a century ago. During that century, however, many important advances have been made in our understanding of the scientific principles which govern the behaviour of brass instruments. Thanks to the endeavours of the international community of researchers whose theories and experiments have been reviewed in the previous chapters of this book, we have now reached a stage where manufacturers are increasingly using scientific measurement techniques and computer simulations in the design of instruments, composers and performers are exploring the possibilities of electronic enhancement and synthesis and teachers are turning to science for advice on novel pedagogical approaches.

As the great Danish physicist Niels Bohr warned, it is difficult to make predictions, especially about the future. We end our book with a brief and inevitably speculative glance at some areas in which science is already making a contribution to the brass community and exciting new developments seem to be in prospect.

9.2.1 Optimisation in Instrument Design

Modifications in the design of conventional brass instruments have traditionally employed a procedure which can be described as experience-guided trial and error. If a maker decides to change the shape of a trumpet bell, for example, knowledge gained in previous experiments or taught by past masters might suggest that complementary changes in other parts of the bore are necessary to preserve good intonation. A prototype of the resulting design is made and tested. If the intonation is not satisfactory, further small bore changes are made and another prototype constructed. Several cycles of this expensive and time-consuming process could be necessary before an acceptable design is achieved.

It is possible to reduce the necessity for manufacturing multiple prototype instruments by making use of the scientific understanding of brass acoustics reviewed in Chap. 4. The input impedance curve of the proposed instrument can be calculated from knowledge of the bore profile; the frequencies of the peaks in the impedance curve correspond closely to the natural notes of the instrument. If some of the natural notes of the virtual instrument are considered to be unacceptably sharp or flat, small changes can be made to the bore profile and the input impedance curve recalculated. Only when a satisfactory design is achieved is it necessary to manufacture a real instrument.

The trial and error process can also be automated using suitable computer software. Smith and Daniell (1976) proposed a method for correcting the intonation of wind instruments using perturbation theory. This mathematical technique effectively finds all possible modifications of the bore which give the desired changes in mode frequencies. Many of the possible bore shapes involve large and/or abrupt changes, and the calculation involves a minimisation process which selects the smoothest of the calculated bore profiles.

The dramatic growth of computational speed and power in the early years of the twenty-first century, together with refinements of the method of input impedance calculation, has made it possible to develop efficient algorithms for optimisation of wind instrument bore profiles using input impedance targets (Kausel 2001; Petiot and Tavard 2008; Braden et al. 2009; Macaluso and Dalmont 2011; Noreland et al. 2013). In these methods the continuous bore of the instrument is approximated by a finite number of short sections, which may be cylindrical, conical or flaring; discontinuities and toneholes can also be incorporated. The optimisation process starts with a 'best guess' bore profile and systematically explores the effect on the impedance curve of small changes in the parameters defining each section. A version of this type of optimisation software has been successfully used by brass manufacturers (Egger et al. 2005).

An optimisation method requires a clearly defined target. In algorithms based on the input impedance curve, the target is usually specified as a given set of input impedance peak frequencies (corresponding to the mode frequencies of the air column). In their pioneering 1976 paper, Smith and Daniell pointed out that defining an ideal set of mode frequencies for a brass instrument is not straightforward, since a specific mode can play a number of different roles. Each mode can support the fundamental frequency of a played note: it could then be expected that the ideal mode frequencies should be equal to the frequencies of the corresponding notes in an equally tempered scale. This requirement is however incompatible with the Bouasse-Benade prescription that a strong and well-centred played note requires a harmonically related set of modes: apart from the octave, none of the intervals in the harmonic series correspond exactly to intervals in the equally tempered scale. In addition, the assumption that brass instrumentalists always play in equal temperament is questionable (Farkas 1956; Kopiez 2003; Bronk 2010).

The ultimate judge of good intonation is of course the musician who has to play the instrument. From the performer's point of view, the optimum target should be specified in terms of the playing frequencies of the notes rather than the air column resonances. The frequency of a played note is the result of the nonlinear coupling of the lips and the multiple resonances of the air column, and depending on the player's choice of embouchure, it can differ significantly from the frequency of the lowest supporting air column mode (Eveno et al. 2014). Poirson et al. (2007) included playing tests by musicians in the definition of an impedance-based target for trumpet optimisation. A standard orchestral B^{b} trumpet was provided with an experimental leadpipe made up by coupling together four short conical sections, each of which could be chosen from a set with several slightly different values of the input and output radii. By selecting suitable combinations of the sections, twelve test trumpets with subtly different input impedance curves were created, and ten professional trumpet players were asked to evaluate the intonation of each test instrument. Using sophisticated statistical analysis, it was possible to find a model equation relating the intonation scores of the experts and the frequencies of the instrument air column; from this relationship, it was possible to define a target set of frequency ratios corresponding to the maximum intonation score. Interestingly, these ratios were close to but not identical with the integer ratios of the perfect harmonic series. An optimisation calculation was finally carried out to calculate the bore of the ideal leadpipe based on the target frequencies.

While the involvement of professional performers in the setting of the optimisation target helps to ensure that the results are musically meaningful, the use of human subjects in sensory evaluation is not straightforward and is impracticable for systematic large-scale testing of new instrument designs. An alternative method for defining the optimisation target in terms of playing frequency, using the physical modelling approach described in Chap. 5, has been described by Tournemenne et al. (2019). The system studied was a trumpet played by a 'virtual musician', whose embouchure was modelled as an outward-striking lip valve. In this case, the target was defined as a set of equally tempered playing frequencies. The procedure started with the initial bore profile of a trumpet, whose objective properties were defined by the impedance peak frequencies. Three of the parameters defining the vibrational behaviour of the lip valve (the mouth pressure p_{m}, the mass per unit area μ_l and the lip resonance frequency f_l) were treated as variables in a numerical simulation which searched for stable periodic solutions, using the harmonic balance method (Gilbert et al. 1989). Many embouchures satisfying this criterion were found; the playing frequencies of notes whose spectra satisfy an additional 'realistic sound' criterion were averaged to give an estimate of the playing frequency for that note. A derivative-free optimisation algorithm was then used to find the modification to the trumpet leadpipe bore which minimised the discrepancies between the equal temperament target pitches and the pitches of the simulated played notes.

The results of the optimisation based on physical modelling were encouraging, and the optimised instrument had significantly improved intonation. There is considerable scope for extensions of this approach to other aspects of brass instrument behaviour such as timbre and playability, since frequency spectra and threshold pressures can be extracted from the simulations. Further developments of the model, including more sophisticated treatment of the lip valve and the inclusion of windway resonances, will probably be necessary as the virtual trumpeter makes the long journey from beginner to expert.

9.2.2 Modification of Instruments Using Active Control

A performance on a brass instrument involves two partners: the player and the instrument. The simple model illustrated in Fig. 2.15 showed that these two partners are intimately linked in a feedback loop, each partner exerting an important influence on the other. In Sect. 9.2.1 an approach to optimising the design of a trumpet was described in which the player was replaced by a 'virtual musician' – a sound generator controlled by computer software which reproduced the behaviour of human lips in the interaction with the real instrument. In this section we describe some preliminary studies of the inverse process, in which the real musician is presented with an instrument whose acoustical behaviour is computer controlled.

The process of real-time modification of the resonances of an acoustical system through a feedback loop is described as active control (Nelson and Elliott 1991). The basic idea behind active control is simple, although its practical implementation is often far from straightforward. Consider, for example, an elementary wind instrument whose resonator is a cylinder closed at the far end. A resonant mode of the air column arises from the reflection of a forward-travelling wave at the closed end. If a loudspeaker at the end generates a signal which exactly cancels the backward-travelling reflected wave, the resonance will disappear; if, on the other hand, the loudspeaker signal reinforces the reflection, the strength of the resonant mode will be enhanced. Additional resonances can be introduced by suitably programming the software controlling the loudspeaker.

This approach was implemented in the creation of an active-controlled end-blown flute by Jean Guérard as part of his PhD studies at the Université Paris 6 (Guérard 1998; Guérard and Boutillon 1998). In Guérard's experimental setup, illustrated in Fig. 9.10, five miniature microphones mounted in the walls of the cylindrical tube were used to measure separately the forward- and backward-going waves (see Sect. 4.2.3). A suitable signal fed to the control loudspeaker cancelled the reflected wave; in this state the acoustical behaviour of the cylinder resembled that

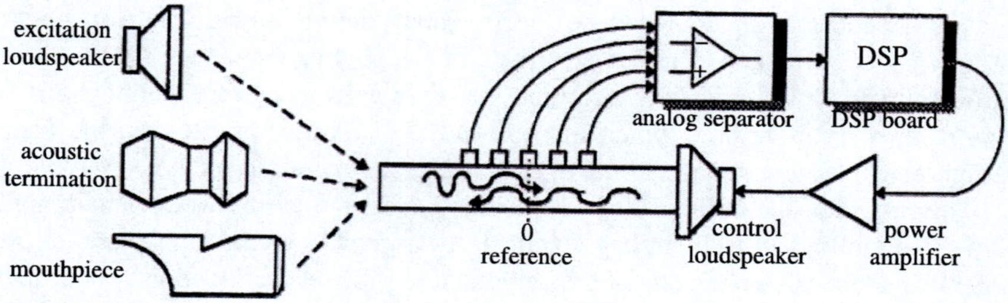

Fig. 9.10 Experimental setup for active controlled flute (Guérard and Boutillon 1998)

of an infinitely long pipe. Additional digital filters in the control software were used to reproduce the reflection functions corresponding to different patterns of open and closed holes on a real flute; in this way the flute could be 'played' over a pitch range of an octave.

Applications of active control methods in musical acoustics have been mainly confined to stringed and percussion instruments (Griffin et al. 2003; Boutin et al. 2015a; Benacchio et al. 2016), although the single-mode Helmholtz resonator designed by Chen and Weinreich (1996) to study the phase behaviour of the lips of trumpeters (see Sect. 5.2.1) could be considered as the first actively controlled brass instrument. The use of modal active control to change the playing properties of a simplified bass clarinet without holes has been described by Meurisse et al. (2015a). In this case the controlling feedback loop was provided by a single wall-mounted microphone and loudspeaker near the open end of the tube. By measuring and adjusting the transfer function between the speaker and microphone of the control setup, it was possible to modify the input impedance and the radiated sound of the instrument.

Improving the sound quality and playability of a brass instrument through the use of active control is an intriguing possibility, but its practical implementation is currently limited by technical problems such as latency in the control loop. One important application of active control to solve a common problem with brass instrument mutes has been described by Meurisse et al. (2015b). In Sect. 4.5.2 it was noted that when a straight mute is used with a brass instrument, the frequencies of its lowest impedance peaks are slightly modified, and an extra 'parasitic' peak appears. This peak affects the playability of the instrument, making some lower notes difficult or impossible to sound quietly (Velut et al. 2017b).

In the trombone mute with active control illustrated in Fig. 9.11, a small microphone inside the mute cavity senses the internal pressure. This signal is sent to the control software, which supplies a correcting signal with modified gain and phase to an internal loudspeaker. The experimental results in Fig. 9.12 show that with a phase inversion ($\Phi = \pi$) and a gain of 2 in the control loop, the parasitic peak can be almost completely suppressed, leading to a marked improvement in the playability of the affected note (in this case the pedal note $B^{\flat}1$).

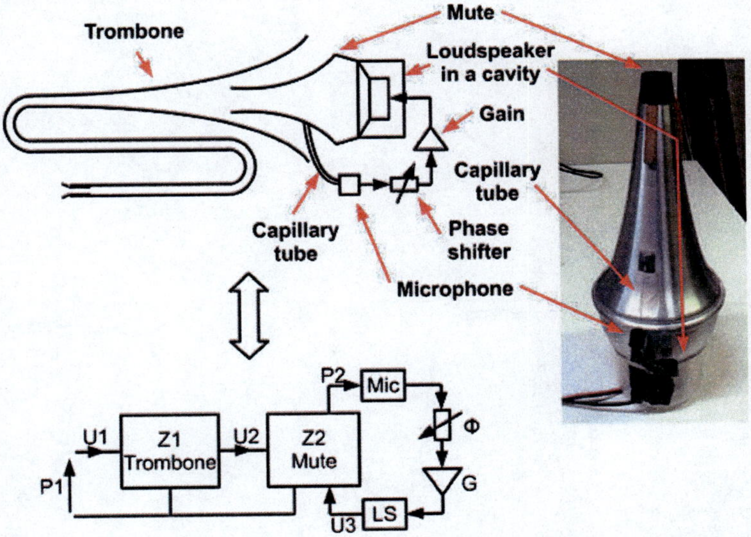

Fig. 9.11 (Top left) Schematic diagram of a straight mute with embedded microphone, speaker and control system, inserted into the bell of a trombone. (Right) Photograph of the active straight mute. (Bottom left) Equivalent electric circuit of the mute and control system coupled to the trombone, with U1 and P1, respectively, flow and pressure at the input of the trombone, Z1 impedance of the trombone, U2 flow at the input of the mute, P2 pressure measured inside the mute by the microphone (Mic), Z2 impedance of the mute, U3 flow induced by the loudspeaker (LS) inside the mute, Φ the phase shift and G the control gain. From Meurisse et al. (2015b) with the permission of the Acoustical Society of America

9.2.3 Live Electronics and Augmented Instruments

The term ‘live electronics’ refers to the real-time manipulation of sound by electronic means during a musical performance. From the 1960s onwards, composers such as Karlheinz Stockhausen made extensive use of techniques including spectral filtering, looping and spatialisation to modify the sounds of acoustic instruments in the course of performance. In the following decades, a number of pieces for solo brass instruments and live electronics appeared. A notable example is the composition *Post-Prae-Ludium per Donau* (1987) for six-valve F tuba and live electronics by Luigi Nono. In performance of this work, the signal from a microphone close to the bell of the tuba is subject to various modifications, including amplification, feedback, delay, reverberation and filtering, controlled in real time by two sound technicians. Four loudspeakers create a spatialised sound field which supplements the acoustic output of the instrument. Considerable freedom in the score encourages the performer to react and respond to the sound generated by the live electronics; the unpredictable nature of the interaction between tubist and sound technicians makes each performance a unique musical experience.

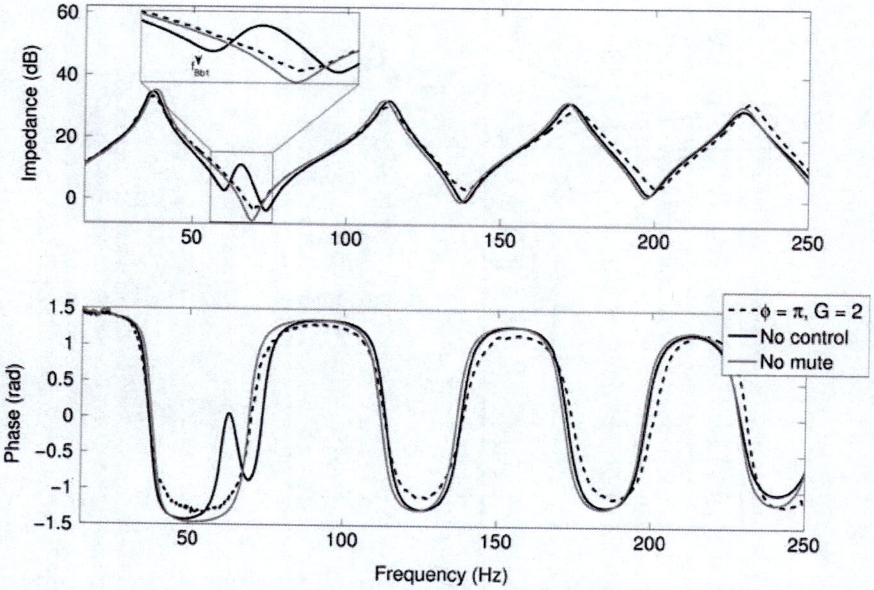

Fig. 9.12 (Top) Input impedance magnitude of a trombone in $B^{\flat}1$ position without a mute (solid grey line), with a normal mute (solid black line) and with an active mute with phaseshift $\Phi = \pi$ and gain $G = 2$ (dashed black line). Inset: zoom on the extra impedance peak. The vertical dotted arrow is at 58 Hz, the playing frequency of $B^{\flat}1$. (Bottom) Phases of the input impedance. From Meurisse et al. (2015b) with the permission of the Acoustical Society of America

In twenty-first-century performances of *Post-Prae-Ludium per Donau* and similar works, the live electronics is often implemented digitally on a laptop computer using Max/MSP software. The speed and flexibility of this package have led to the evolution of a new genre of compositions in which the player interacts with the computer, shaping its processing algorithms through modifications of the acoustically radiated sound. An example is *Music for Tuba and Computer* (2008) by Cort Lippe. In this piece, the computer tracks many parameters of the sound signal, including pitch, amplitude, spectrum and articulation; this information is used to modify in real time the synthesis and compositional algorithms in the Max/MSP software. The feedback loop coupling player and computer, in which an expressive gesture by one partner evokes a complementary response from the other, has some resemblance to the interaction between two human performers in a chamber music ensemble or jazz duet.

The examples of the use of live electronics described above do not involve any modification of the acoustic instrument, although the performer may employ techniques (such as breathing quietly through the instrument) which rely on electronic modification to be effective (Rodgers 2015). In 1989 Dexter Morrill and Perry Cook described a project in which a trumpet and a valve trombone were physically adapted by adding switches and linear potentiometers in the region around the valve casing and additional switches under the valves (Morrill and Cook 1989; Cook 2001). These switches and pots were operated by the fingers and thumbs of the player to generate MIDI control signals which could be routed through a computer to synthesisers, samplers and signal processors. Pressure sensors in the mouthpiece of the trumpet, and in a Harmon mute on the trombone, provided an acoustic signal for pitch tracking. Instruments modified in this way, with additional sensors which provide performers with the ability control extra sound or musical parameters, are described as hybrid or augmented instruments (Miranda and Wanderley 2006).

Following the pioneering work of Cook and Morrill, a number of augmented brass instruments have been constructed (Thibodeau and Wanderley 2013). In 1993 the trumpeter and composer Jonathan Impett and the electronics engineer Bert Bongers developed an instrument described as a meta-trumpet (Impett 1994). A conventional trumpet was substantially modified by fitting a range of physical sensors, the outputs of which were directed to MIDI-controlled instruments and processors. Magnetic sensors in extended valve caps tracked the movements of magnets mounted in the valve pistons, while two pressure sensors were accessible by otherwise unused fingers. Right-left inclination of the instrument was signalled by two mercury switches mounted below the centre of the bell, and its overall position in space was monitored by externally mounted ultrasonic receivers which tracked the signals from a cluster of transmitters on the bell. Six switches near the valves were available for modifying controller functions or triggering events. The main aim of the meta-trumpet project was to provide a close and dynamic relationship between the performer, the instrument and the improvised or composed musical material. The possibilities of this integrated compositional and performance system were first illustrated in Impett's *Mirror-Rite* (Impett 1996).

A striking example of a hybrid instrument combining innovations in both acoustical design and augmentation technology is the mutantrumpet, developed and played by the composer and performer Ben Neill (Fig. 9.13). In its original form, introduced by Neill in the early 1980s, the mutantrumpet was an expanded acoustic instrument providing the player with a combination of three trumpets and a trombone. Analog electronics was integrated into this remarkable device through a collaboration with Robert Moog, inventor of the Moog synthesiser. In 1992, while in residency at the STEIM research and development lab for new instruments in Amsterdam, Neill made the mutantrumpet fully computer interactive, and the instrument has been further developed in subsequent residencies at STEIM (Neill 2017).

Figure 9.14 illustrates Version 4.0 of the mutantrumpet, introduced in 2019. The instrument has six valves, three of which perform the same functions as those on a conventional trumpet. Two additional acoustic valves are provided to direct the sound output between the three separate bells, one of which is a piccolo trumpet bell incorporating a slide section on which glissandi can be executed. A third additional valve is used to create microtonal pitch shifts. As well as the customary switches and sliders, two joysticks allow for two-dimensional continuous parameter variation. Neill's performances involve a high degree of improvisation, in which the interaction between acoustic and electronic manipulation is used to build up

Fig. 9.13 The trumpeter Ben Neill playing the mutantrumpet. Courtesy Ben Neill

Fig. 9.14 Version 4.0 of the mutantrumpet. Courtesy of Ben Neill

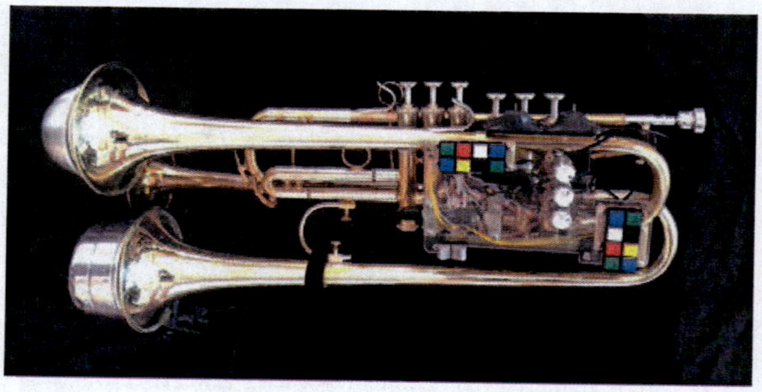

complex textures and sonorities. The mutantrumpet sensor outputs can also be mapped to the control parameters of a video projection to create a synthesised audiovisual performance.

The freedom of movement of a player is inevitably restricted when the instrument is connected by cabling to a computer interface, and augmented instruments have increasingly made use of wireless connectivity. One such instrument is the Electrumpet, which was invented in 2008 by Hans Leeuw and has subsequently undergone several stages of further development (Leeuw 2009, 2012). In the 2020 version illustrated in Fig. 9.15, the Electrumpet has a number of analog sensors in the form of potentiometers operated by pistons with the same mechanical response as normal trumpet valves. The aim of this design is to exploit the skill in valve manipulation which is part of an expert player's conventional technique. The signals from the sensors are digitised and sent by a 5 GHz WiFi link to the laptop running the sound processing software. Earlier versions included a second mouthpiece linked to a breath controller, but this was dispensed with in the 2020 version. A display of the current control parameters is provided by an iPhone mounted on the trumpet. The casing which holds the iPhone also includes a 3D touch sensor which allows hand

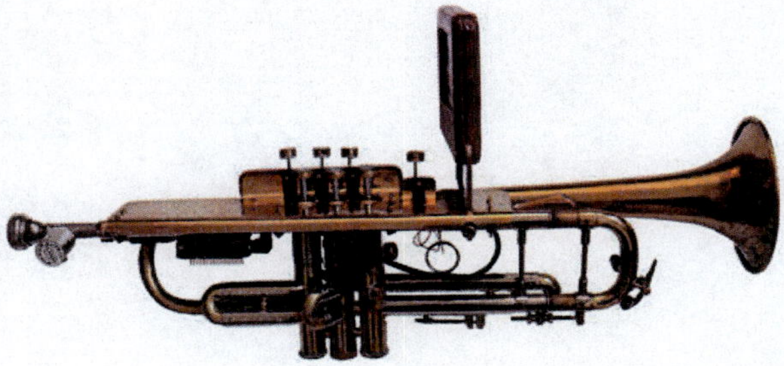

Fig. 9.15 The Electrumpet, invented by Hans Leeuw. Courtesy of Hans Leeuw

gestures to modulate the radiated sound in a way analogous to the use of a plunger mute.

Although individual performers have imaginatively exploited the creative possibilities offered by augmented brass instruments, the technology has as yet had little impact on mainstream brass playing. One reason for this is that many of the instruments have been developed to satisfy the specific techniques and aesthetic requirements of the creator, presenting other performers with a formidable learning curve. Another reason is that the augmentation process frequently requires expensive and irreversible modifications to the original acoustic instrument, together with a high degree of technical expertise in maintaining and operating the added hardware and software. To address these concerns, several simplified and readily demountable augmentation systems have been proposed. The Easily Removable Optical Sensing System (EROSS) introduced by Leonardo Jenkins et al. (2013) incorporates three infrared position sensors which sit under the valve bottom caps and track continuously the positions of the pistons. The sensors are mounted in a housing 3D printed to fit snugly around the valve casings without impeding normal playing and relay control information through a wireless transceiver.

The EROSS approach has been further developed by Sarah Belle Reid and colleagues in the Minimally Invasive Gesture Sensing Interface (MIGSI) for trumpet (Reid et al. 2016, 2019). The optical sensors tracking motion of the valve pistons are supplemented by an accelerometer sensing the pitch (raising or lowering the bell) and roll (tilting to left or right) of the instrument. Force sensors placed inside the hand guard covering the valve casings respond to pressure from the player's left hand thumb and fingers, providing additional control signals. The streams of data are sent wirelessly to the controlling computer. The MIGSI trumpet (Fig. 9.16) has been used successfully in many performances, and further hardware and software improvements are envisaged. A crucial aspect of the design philosophy has been to take advantage of the expertise of the professional performer by capturing already familiar gestures in ways which allow subtle and expressive modulations of the electronic contributions to the performance. The MIGSI augmentation equipment can be attached to a trumpet as easily as inserting a mute; devices like this may in time become as familiar items as mutes in the trumpet player's toolkit.

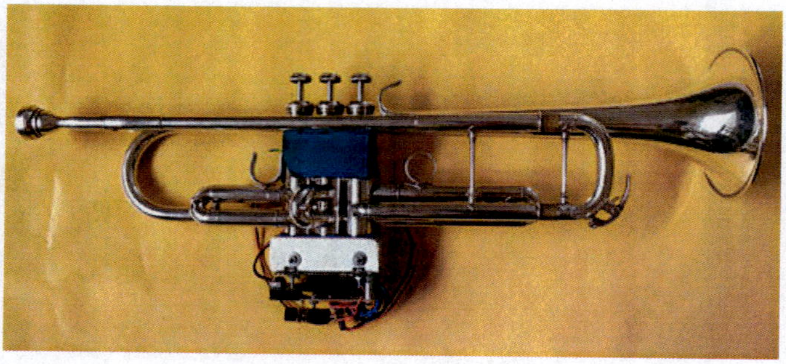

Fig. 9.16 The Minimally Invasive Gesture Sensing Interface (MIGSI) for trumpet. Courtesy of Sarah Belle Reid

9.2.4 Epilogue

Several millennia have elapsed since our ancestors first buzzed lips on the end of a hollow bone or a conch shell. Over the last few centuries, the pace of technological development has accelerated dramatically, and our scientific understanding of the interaction between brass player and instrument has advanced to the stage at which it is now able to offer makers reliable guidelines for the design of high-quality instruments. The digital revolution has led to the introduction of electronic techniques which can realistically synthesise in real time the sound of conventional brass instruments and which can offer the instrument designer the possibility of hearing a proposed new design of instrument which has not yet been constructed. Active control feedback systems have been used to correct problems in brass instrument mutes, and the use of instrumented mouthpieces in brass instrument teaching is being explored. The ready availability of 3D printing offers the musician the option of designing and manufacturing an individually tailored mouthpiece or even a complete instrument. Electronic augmentation of instruments will no doubt continue to evolve, with intriguing possibilities as the 'Internet of Things' spreads to the band and orchestra.

We end our book by saluting the performers from all times and places who have chosen to make music by sounding resonators with their lips, the craftsmen who have laboured to perfect the shapes and functioning of the resonators and the mathematicians, physicists, engineers and psychologists whose communal endeavours have helped us to understand better the science underlying the majesty and splendour of these wonderful instruments: the brasses.

References

Adachi, S. and Sato, M. (1995). Time-domain simulation of sound production in the brass instrument. J. Acoust. Soc. Am. **97**, 3850–3861, https://doi.org/10.1121/1.412398.

Adachi, S. and Sato, M. (1996). Trumpet sound simulation using a two-dimensional lip vibration model. J. Acoust. Soc. Am. **99**, 1200–1209, https://doi.org/10.1121/1.414601.

Aliverti, A. (1996). *Metodi et tecniche innovative per lo studio della meccanica respiratoria.* Doctoral dissertation, Politecnico di Milano, Milan, Italy.

Amir, N., Rosenhouse, G. and Shimony, U. (1995). A discrete model for tubular acoustic systems with varying cross-section – the direct and inverse problem. Parts I and II: theory and experiments. Acustica **81**, 450–474.

Amir, N., Pagneux, V. and Kergomard, J. (1997). A study of wave propagation in varying cross-section waveguides by modal decomposition. Part II. Results. J. Acoust. Soc. Am. **101**, 2504–2517, https://doi.org/10.1121/1.419306.

Artim GmbH (2020). A-2230 Gänserndorf, Ziehrergasse 4, Austria. http://artim.at. Accessed April 2020.

Auregan, Y. and Depollier, C. (1995). Snoring: linear stability analysis and in-vitro experiments. J. Sound Vib. **188**, 39–54.

Ayers, R. D. (1998). New perspectives on brass instruments. Proc. International Symposium on Musical Acoustics, Leavenworth, USA, 129–134.

Ayers, R. D. (2001). Basic tests for models of the lip reed. Proc. International Symposium on Musical Acoustics, Perugia, Italy, 83–86.

Backus, J. (1974). Input impedance curves for the reed woodwind instruments. J. Acoust. Soc. Am. **56**, 1266–1279, https://doi.org/10.1121/1.1903418.

Backus, J. (1976). Input impedance curves for the brass instruments. J. Acoust. Soc. Am. **60**, 470–480, https://doi.org/10.1121/1.381104.

Backus, J. and Hundley, T. C. (1966). Wall vibration in flue organ pipes and their effect on tone. J. Acoust. Soc. Am. **39**, 936–945, https://doi.org/0.1121/1.1909975.

Bahnert, H., Herzberg, T. and Schramm, H. (1986). *Metallblasinstrumente*, 2nd. Edn. Wilhelmshaven, Heinrichshofen.

Baines, A. (1976). *Brass Instruments: their History and Development*. London, Faber.

Balasubramanian, S., Chatziioannou, V. and Kausel, W. (2019). Analysis of axisymmetric structural vibrations in brass instruments. Acta Acust. united Ac. **105**, 506–515, https://doi.org/10.3813/AAA.919332.

Barbieri, P. (2013). *Physics of Wind Instruments and Organ Pipes 1100-2010*, Chap. C. Latina, Il Levante Libreria.

Barclay, R. (1992). *The art of the trumpet maker*. Oxford University Press.

M. Campbell et al., *The Science of Brass Instruments*, Modern Acoustics and Signal Processing, https://doi.org/10.1007/978-3-030-55686-0

Beauchamp, J. (1969). Nonlinear characteristics of brass tones. J. Acoust. Soc. Am. **46**, 98, https://doi.org/10.1121/1.1973734.

Beauchamp, J. (1980). Analysis of simultaneous mouthpiece and output waveforms. Audio Engineering Society preprint No. 1626.

Beauchamp, J. (2007). *Analysis, Synthesis, and Perception of Musical Sounds*. New York, Springer.

Benacchio, S., Mamou-Mani, A., Chomette, B. and Finel, V. (2016). Active control and sound synthesis – two different ways to investigate the influence of the modal parameters of a guitar on its sound. J. Acoust. Soc. Am. **139**, 1411–1419, https://doi.org/10.1121/1.4944572.

Benade, A. H. (1960). On the mathematical theory of woodwind finger holes. J. Acoust. Soc. Am. **32**, 1591–1608, https://doi.org/10.1121/1.1907968.

Benade, A. H. (1968). On the propagation of sound waves in a cylindrical conduit. J. Acoust. Soc. Am. **44**, 616–623, https://doi.org/10.1121/1.1911130.

Benade, A. H. (1973). The physics of brasses. Scientific American **229(1)**, 24–35.

Benade, A. H. (1976). *Fundamentals of musical acoustics*. New York, Oxford University Press. 2nd ed. Mineola, Dover, 1990.

Benade, A. H. (1983). Air column, reed, and player's windway interaction in musical instruments. In *Vocal Fold Physiology: Biomechanics, Acoustics, and Phonatory Control*, ed. J. R. Titze and R. C. Scherer, pp. 425–452, Denver Center for the Performing Arts, Denver.

Benade, A. H. and Gans, D. J. (1968). Sound production in wind instruments. Ann. NY. Acad. Sci. **155**, 247–263.

Benade, A. H. and Ibisi, M. I. (1987). Survey of impedance methods and a new piezo-disk-driven impedance head for air columns. J. Acoust. Soc. Am. **81**, 1152–1167, https://doi.org/10.1121/1.394636.

Benade, A. H. and Jansson, E. V. (1974). On plane and spherical waves in horns with non-uniform flare: I. Theory of radiation, resonance frequencies, and mode conversion. Acustica **31**, 79–98.

Berjamin, H., Lombard, B., Vergez, C. and Cottanceau, E. (2017). Time-domain numerical modeling of brass instruments including nonlinear wave propagation, viscothermal losses, and lips vibration. Acta Acust. united Ac. **103**, 117–131, https://doi.org/10.3813/AAA.919038.

Bergeot, B., Vergez, C. and Gazengel, B. (2013). Prediction of the dynamic oscillation threshold in a clarinet model with a linearly increasing blowing pressure. Nonlinear Dynam. **73**, 521–534.

Bernoulli, D. (1762). Physical, mechanical and analytical researches on sound and on the tones of differently constructed organ pipes (in French). Mem. Acad. R. Sci. **1764**, 431–485.

Bertsch, M. and Waldherr, K. (2005). Reliability of trumpet players' judgments during instrument blind tests. Proc. 2nd Congress of Alps-Adria Acoustics Association, EAA Symposium, Opatija, Croatia, 207–216.

Bertsch, M., Waldherr, K. and Kausel, W. (2005). Sensory evaluation testing of trumpets and correlation with acoustic Measurements. Proc. Forum Acusticum, Budapest, Hungary, 575–580.

Bevan, C. (2000) *The Tuba Family*, 2nd edn. Winchester, Piccolo Press.

Bilbao, S. (2009). *Numerical Sound Synthesis: Finite Difference Schemes and Simulation in Musical Acoustics*, New York, Wiley.

Bilbao, S. and Chick, J. (2013). Finite difference time domain simulation for the brass instrument bore. J. Acoust. Soc. Am. **134**, 3860–3871, https://doi.org/10.1121/1.4822479.

Bismarck, G. von (1974). Timbre of steady sounds: a factorial investigation of its verbal attributes. Acustica **30**, 146–159.

Blackstock, D. T. (1985). Generalized Burgers equation for plane waves. J. Acoust. Soc. Am. **77**, 2050–2053, https://doi.org/10.1121/1.391778.

Blaikley, D. J. (1878). On brass wind instruments as resonators. Philos. Mag. 5th Series, **VI**(July–Dec), 119–128.

Blevins, R. D. (1986) *Flow-induced vibrations*. Malabar, Krieger.

Bouasse, H. (1929). *Instruments à vent tomes I et II)*. Paris, Delagrave; repr. with additional material by Jean Kergomard, Paris, Blanchard (1986).

Bouhuys, A. (1964). Lung volumes and breathing patterns in wind-instrument players. J. Appl. Physiol. **19**, 967–975.

Bouhuys, A. (1965). Sound power production in wind instruments. J. Acoust. Soc. Am. **37**, 453–456, https://doi.org/10.1121/1.1909350.

Boutin, H., Fletcher, N., Smith, J. and Wolfe, J. (2013). Lip motion, the playing frequency of the trombone and the upstream and downstream impedances. Proc. Stockholm Musical Acoustics Conference (SMAC 2013), Stockholm, Sweden, 281–489.

Boutin, H., Besnainou, C. and Polack, J. D. (2015a). Modifying the resonances of a xylophone bar using active control. Acta Acust. united Ac. **101**, 408–420, https://doi.org/10.3813/AAA.918836.

Boutin, H., Fletcher, N., Smith, J. and Wolfe, J. (2015b). Relationships between pressure, flow, lip motion, and upstream and downstream impedances for the trombone. J. Acoust. Soc. Am. **137**, 1195–1209, https://doi.org/10.1121/1.4908236.

Bowsher, J. M. and Watkinson, P. S. (1982). Manufacturers' opinions about brass instruments. Brass Bulletin **38**, 25–30.

Braden, A. (2006). *Bore optimisation and impedance modelling of brass musical instruments* Ph.D. Thesis, University of Edinburgh.

Braden, A., Newton, M. and Campbell, D. M. (2009). Trombone bore optimization based on input impedance targets. J. Acoust. Soc. Am. **125**, 2404–2412, https://doi.org/10.1121/1.3087423.

Bromage, S. R. (2007). *Visualisation of the lip motion of brass instrument players, and investigations of an artificial mouth as a tool for comparative studies of instruments.* Ph.D. thesis, University of Edinburgh.

Bromage, S. R., Richards, O. F. and Campbell, D. M. (2003). Reproducibility and control of the embouchure of an artificial mouth for playing brass instruments: Proc. Stockholm Musical Acoustics Conference (SMAC03), Stockholm, Sweden, pp. 197–200.

Bromage, S., Campbell, M., Chick, J., Gilbert, J. and Stevenson, S. (2006). Motion of the brass player's lips during extremely loud playing Proc. 8e Congrès Francais d'Acoustique, Tours, France.

Bromage, S., Campbell, M. and Gilbert, J. (2010). Open areas of vibrating lips in trombone playing. Acta Acust. united Ac. **96**, 603–613, https://doi.org/10.3813/AAA.918315.

Bronk, B. (2010). It's Just Intonation. Bandmasters Review, Texas Bandmasters Association, Dec. 2010, 15–20.

Bruneau, M. (2006) *Fundamentals of Acoustics*. London, Wiley-ISTE.

Buick, J. M., Kemp, J., Sharp, D. B., van Walstijn, M., Campbell, D. M. and Smith, R. A. (2002). Distinguishing between similar tubular objects using pulse reflectometry: a study of trumpet and cornet lead pipes. Meas. Sci. Technol. **13**, 750–757. https://10.1088/0957-0233/13/5/313.

Cala, J., Kenyon, C. M., Ferrigno, G., Carnevali, P., Aliverti, A., Pedotti, A., Macklem, P. T. and Rochester, D. F. (1996). Chest wall and lung volume estimation by optical reflectance motion analysis. J. Appl. Physiol. **81**, 2680–2689.

Campbell, D. M. (1999). Non-linear dynamics of musical reed and brass wind instruments. Contemp. Phys. **40**, 415–431.

Campbell, M. (2004). Brass instruments as we know them today. Acta Acust. united Ac. **90**, 600–610.

Campbell, M. (2013). Objective evaluation of musical instrument quality: a grand challenge in musical acoustics. ICA2013 Montreal, Proc. Mtg. Acoust. **19**, 032003.

Campbell, M. (2014a). Musical acoustics for musicians. Proc. International Symposium on Musical Acoustics, Le Mans, France.

Campbell, M. (2014b). Evaluating musical instruments. Physics Today **67**(3), 35–40.

Campbell, M. and Greated, C. (1987). *The Musician's Guide to Acoustics*. Oxford University Press.

Campbell, D. M. and MacGillivray, T. (1999). Acoustics of the Carnyx. In Hickmann, E., Laufs, I. and Eichmann, R. (Eds) *Studies in Musical Archaeology II: Music Archaeology of Early Metal Ages*, pp. 357–363. Berlin, Deutsches Archaeologisches Institut.

Campbell, M., Greated, C. and Myers, A. (2004). *Musical Instruments: History, Technology and Performance of Instruments of Western Music*. Oxford University Press.

Campbell, D., Chick, J. and Myers, A. (2014a). Acoustical comparisons of sackbuts and trombones. Historic Brass Society Journal **26**, 63–78, https://doi.org/10.2153/0120140011005.

Campbell, D., Chick, J., Gilbert, J., Kemp, J., Myers, A. and Newton, M. (2014b). Spectral enrichment in brass instruments due to nonlinear sound propagation: a comparison of measurements and predictions. Proc. International Symposium on Musical Acoustics, Le Mans, France.

Campbell, D. M., Gilbert, J. and Holmes, P. (2017). Seeking the sound of ancient horns. J. Acoust. Soc. Am. **141**, 3935, https://doi.org/10.1121/1.4988906.

Campbell, D. M., Gilbert, J. and Myers, A. (2020). Brassiness and the understanding of brass instruments. Proc. e-Forum Acusticum, Lyon.

Carse, A. (1939). *Musical Wind Instruments*. London, Macmillan.

Carter, S. (2012). *The Trombone in the Renaissance*. Hillsdale, Pendragon.

Carter, S. (2019). Trombone. In *The Cambridge Encyclopedia of Brass Instruments*, eds. Herbert, T., Myers, A. and Wallace, J. Cambridge University Press.

Castellengo, M. (1981). Sons multiphoniques aux instruments à vent (multiphonic sounds on woodwind instruments). Paris, Rapports de l'IRCAM.

Caussé, R., Kergomard, J. and X. Lurton (1984). Input impedance of brass musical instruments—Comparison between experiment and numerical models. J. Acoust. Soc. Am. **75**, 241–254, https://doi.org/10.1121/1.390402.

Caussé, R., Eveno, P., Gilbert, J. and Petiot, J. F. (2013). What can we deduce from measured resonance frequencies of trumpets concerning their playing frequencies? Proc. Mtgs. Acoust. **19**, 035066.

Chang, Y. M. (1994). Reed stability. J. Fluid. Struct. **8**, 771–783.

Chaigne. A. and Kergomard, J. (2016). *Acoustics of Musical Instruments*. New York, Springer.

Chen, F. C. and Weinreich, G. (1996) Nature of the lip reed. J. Acoust. Soc. Am. **99**, 1227–1233, https://doi.org/10.1121/1.414604.

Chen, J. M., Smith, J. and Wolfe, J. (2008). Experienced saxophonists learn to tune their vocal tracts. Science **319**, 776.

Chen, J. M., Smith, J. and Wolfe, J. (2012). Do trumpet players tune resonances of the vocal tract? J. Acoust. Soc. Am. **131(1)**, 722–727, https://doi.org/10.1121/1.3651241.

Chick, J. P., Logie, S. M., Campbell, M. and Gilbert, J. (2012). Spectral enrichment and wall losses in trombones played at high dynamic levels. Proc. Acoustics 12 Nantes, France, 2783–2788.

Chung, J. Y. and Blaser, D. A. (1980). Transfer function method of measuring in-duct acoustic properties. I: Theory. II: Experiment. J. Acoust. Soc. Am. **68**, 907–921, https://doi.org/10.1121/1.385211.

Clinch, P. G., Troup, G. J. and Harris, L. (1982). The importance of vocal tract resonance in clarinet and saxophone performance—a preliminary account. Acustica **50**, 280–284.

Cochelin, B. and Vergez, C. (2009). A high-order purely frequency-based harmonic balance formulation for continuation of periodic solutions. J. Sound Vib. **324**, 243–262.

Collver, M. and Dickey, B. (1996). *A Catalog of Music for the Cornett*. Bloomington, Indiana University Press.

Cook, P. R. (2001). Principles for designing computer music controllers. Proc. International Conference New Interfaces for Musical Expression (NIME), 1–4.

Cook, P. R., Abel, J. S., Kolar, M. A., Huang, P., Huopaniemi, J., Rick, J. W., Chafe, C. and Chowning, J. M. (2010). Acoustic analysis of the Chavín pututus (*Strombus galeatus* marine shell trumpets). J. Acoust. Soc. Am. **128**, 2359, https://doi.org/10.1121/1.3508370.

Copley, D. C. and Strong, W. J. (1996). A stroboscopic study of lip vibrations in a trombone. J. Acoust. Soc. Am. **99**, 1219–1226, https://doi.org/10.1121/1.414603.

Cullen, J. S. (2000). *A study of brass instrument acoustics using an artificial lip reed mechanism, laser Doppler anemometry and other techniques*. Ph.D. thesis, University of Edinburgh.

Cullen, J., Gilbert, J. and Campbell, D. M. (2000). Brass instruments: linear stability analysis and experiments with an artificial mouth. Acta Acust united Ac **86**, 704–724.

Dalmont, J. P. (2001). Acoustic impedance measurement, Part I: a review. J. Sound. Vib. **243**, 427–439.

Dalmont, J. P. (2001) Acoustic impedance measurement, Part II: a new calibration method. J. Sound. Vib. **243**, 441–459.

Dalmont, J. P., Gazengel, B., Gilbert, J. and Kergomard, J. (1995). Some aspects of tuning and clean intonation in reed instruments. Appl. Acoust. **46(1)**, 19–60.

Dalmont, J. P., Gilbert, J. and Kergomard, J. (2000). Reed instruments, from small to large amplitude periodic oscillations and the Helmholtz motion analogy. Acta Acust. united Ac. **86**, 671–684.

Dalmont, J. P., Joly, N. and Nederveen, C. J. (2001). Radiation impedance of tubes with different flanges: numerical and experimental investigations. J. Sound Vib. **244**, 504–534.

Dalmont, J. P., Curtit, M. and Yahaya, A. F. (2012). On the accuracy of bore reconstruction from input impedance measurements: Application to bassoon crook measurements. J. Acoust. Soc. Am. **131**, 708–714, https://doi.org/10.1121/1.3651793.

Debut, V. and Kergomard, J. (2004). Analysis of the self-sustained oscillations of a clarinet as a Van der Pol oscillator. International Congress on Acoustics, Kyoto, Japan.

Del Mar, N. (1983). *Anatomy of the Orchestra*, 2nd Edn. London, Faber.

Deutsch, D. (2013). Grouping mechanisms in music. In D. Deutsch (Ed.), *The Psychology of Music*, 3rd Edn., 183–248, San Diego, Elsevier.

Dickens, P., Smith, J. R. and Wolfe, J. (2007). Improved precision in measurements of acoustic impedance spectra using resonance-free calibration loads and controlled error distribution. J. Acoust. Soc. Am. **121**, 1471–1481.

Doc, J. B., Vergez, C. and Missoum, S. (2014). A minimal model of a single-reed instrument producing quasi-periodic sounds. Acta Acust. united Ac. **100**, 543–554.

Doedel, E. J., Champneys, A. R., Fairgrieve, T. F., Kuznetsov, Y. A., Sandstede, B. and Wang, X. J. 1997. *AUTO97: Continuation and bifurcation software for ordinary differential equations (with HomCont), User's Guide.* Concordia University. http://indy.cs.concordia.ca/auto/. Accessed April 2020.

Dubos, V., Kergomard, J., Khettabi, A., Dalmont, J. P., Keefe, D. H. and Nederveen, C. (1999). Theory of sound propagation in a duct with a branched tube using modal decomposition. Acta Acust. united Ac. **85**, 153–169.

Dudgeon, R. T., Eastop, P., Herbert, T. and Wallace, J. (1997). Playing, learning and teaching brass. In Herbert, T. and Wallace, J. (Eds.), *The Cambridge Companion to Brass Instruments*, Cambridge University Press.

Dullat, G. (1989). *Metallblasinstrumentenbau: Entwicklungsstufen und Technologie*. Frankfurt, Bochinsky.

Earnshaw, S. (1860). On the mathematical theory of sound. Philos. T. R. Soc. Lond. **150**, 133–148.

Egger, R. and Kausel, W. (2005). The Brass-Wind-Instrument-Optimizer as a tool for instrument makers: A case study. In Kühnelt, H. and Widholm, G. (Eds.), Proc. EAA workshop on Wind Instrument Acoustics, ViennaTalk 2005—On the Future of Wind Instruments.

Ehara, F., Nagai, K. and Mizutani, K. (2001). Relationships between vibration of artificial lips and sound frequency of a trumpet. Proc. International Symposium on Musical Acoustics, Perugia, Italy, 513–516.

Elliott, S. J. and Bowsher, J. M. (1982). Regeneration in brass wind instruments. J. Sound Vib. **83**, 181–217.

EMAP (2018). European Music Archeology Project. http://www.emaproject.eu. accessed April 2020.

Evans, A., Ackermann, B. and Driscoll, T. (2010). Functional anatomy of the soft palate applied to wind playing. Med. Probl. Perform. Art **25(4)**, 183–189.

Eveno, P., Dalmont, J. P., Caussé, R. and Gilbert, J. (2012). Wave propagation and radiation in a horn: Comparisons between models and measurements. Acta Acust. united Ac. **98**, 158–165, https://doi.org/10.3813/AAA.918501.

Eveno, P. and Le Conte, S. (2013). The use of the input impedance for characterising historical serpents. Proc. Stockholm Musical Acoustics Conference (SMAC2013), Stockholm, Sweden, 495–501.

Eveno, P., Petiot, J. F., Gilbert, J., Kieffer, B. and Caussé, R. (2014). The relationship between bore resonance frequencies and playing frequencies in trumpets. Acta Acust. united Ac. **100**, 362–374.

Fabre, B., Gilbert, J. and Hirschberg, A. (2018). *Modeling of Wind Instruments*. In R. Bader (Ed.), *Springer Handbook of Systematic Musicology*, pp. 121–139. New York, Springer.

Farkas, P. (1956). *The Art of French Horn Playing*. Evanston, Summy-Birchard.

Farkas, P. (1962) *The Art of Brass Playing*, Rochester, Wind Music.

Farner, S., Vergez, C., Kergomard, J. and Aude Lizée 2006). Contribution to harmonic balance calculations of periodic oscillation for self-sustained musical instruments with focus on single-reed instruments. J. Acoust. Soc. Am. **119**, 1794–1804, https://doi.org/10.1121/1.2166602.

Fastl, H. and Zwicker, E. (2007). *Psychoacoustics*. New York, Springer.

Félix, S. and Pagneux, V. (2001). Sound propagation in rigid bends: A multimodal approach. J. Acoust. Soc. Am. **110**, 1329–1337, https://doi.org/10.1121/1.1391249.

Félix, S., Dalmont, J. P. and Nederveen, C. J. (2012). Effects of bending portions of the air column on the acoustical resonances of a wind instrument. J. Acoust. Soc. Am. **131**, 4164–4172, https://doi.org/10.1121/1.3699267.

Fletcher, N. H. (1979). Excitation mechanisms in woodwind and brass instruments. Acustica **43**, 63–72.

Fletcher, N. H. (1993). Autonomous vibration of simple pressure-controlled valves in gas flows. J. Acoust. Soc. Am. **93**, 2172–2180, https://doi.org/10.1121/1.406857.

Fletcher, N. H. and Rossing, T. D. (1998). *The Physics of Musical Instruments*, 2nd Ed. New York, Springer.

Fletcher, N. H. and Tarnopolsky, A. (1999). Blowing pressure, power, and spectrum in trumpet playing. J. Acoust. Soc. Am. **105**, 874–881, https://doi.org/10.1121/1.426276.

Fowler, J. E. (1874). Improvements in mouth-pieces for musical instruments. U.S. Patent No. 147759, Feb. 24 1874.

Fréour, V. (2013). *Acoustic and Respiratory Pressure Control in Brass Instrument Performance.* Ph.D. thesis, McGill University, Montreal, Canada.

Fréour, V. and Scavone, G. P. (2013). Acoustical interaction between vibrating lips, downstream air column, and upstream airways in trombone performance. J. Acoust. Soc. Am. **134**, 3887–3898, https://doi.org/10.1121/1.4823847.

Fréour, V., Lopes, N., Hélie, T., Caussé, R. and Scavone, G. P. (2015). In-vitro and numerical investigations of the influence of a vocal-tract resonance on lip auto-oscillations in trombone performance, Acta Acust. united Ac **101**, 256–269, https://doi.org/10.3813/AAA.918824.

Fréour, V., Guillot, L., Masuda, H., Usa, S., Tominaga, E., Tohgi, Y., Vergez, C. and Cochelin, B. (2020). Numerical continuation of a physical model of brass instruments: Application to trumpet comparisons, J. Acoust. Soc. Am. **148(5)**, 748–758, https://doi.org/10.1121/10.0001603.

Fritz, C. and Wolfe, J. (2005). How do clarinet players adjust the resonances of their vocal tracts for different playing effects? J. Acoust. Soc. Am. **118(5)**, 3306–3315, https://doi.org/10.1121/1.2041287.

Fritz, C., Curtin, J., Poitevineau, J., Morrel-Samuels, P. and Tao, F. C. (2012). Player preferences among new and old violins. Proc. Natl. Acad. Sci. USA **109(3)**, 760–763.

Fritz, C., Curtin, J., Poitevineau, J., Borsarello, H., Wollman, I. and Tao, F. C. (2014). Soloist evaluations of six Old Italian and six new violins. Proc. Natl. Acad. Sci. USA **111(20)**, 7224–7229.

Gautier, F. and Tahani, N. (1998). Vibroacoustic behavior of a simplified musical wind instrument. J. Sound Vib. **213**, 107–125.

Gautier, F., Gilbert, J., Dalmont, J. P. and Pico Vila, R. (2007). Wave propagation in a fluid filled rubber tube: theoretical and experimental results for the Korteweg's wave. Acta Acust. united Ac. **93**, 333–344.

Gautier, F., Nief, G., Gilbert, J. and Dalmont, J. P. (2012). Vibro-acoustics of organ pipes – Revisiting the Miller experiment. J. Acoust. Soc. Am. **131**, 737–738.

Gautier, F., Sécail-Géraud, M. and Gilbert, J. (2013). Analysis of vibroacoustics of trombone bells thanks to an adaptation of the Miller experiment. Proc. Mtgs. Acoust. **19**, 035072, https://doi.org/10.1121/1.4799444.

Gazengel, B., Gilbert, J. and Amir, N. (1995). From the measured input impedance to the synthesis signal: where are the traps? Acta Acustica **3**, 445–472.

Gibiat, V. (1988). Phase space representations of musical acoustical signals. J. Sound Vib. **123**, 529–536.

Gibiat, V. and Castellengo, M. (2000). Period doubling occurences in wind instruments musical performance. Acta Acust. united Ac. **86**, 746–754.

Gibiat, V. and Laloë, F. (1990). Acoustical impedance measurements by the two-microphone-three-calibration (TMTC) method. J. Acoust. Soc. Am. **88**, 2534–2545, https://doi.org/10.1121/1.399975.

Gibiat, V. and Padilla, A. (2005). The 'trumpet marine', an instrument between a trumpet and a violin. Proc. 4th Forum Acusticum, Budapest.

Gilbert, J. and Petiot, J. F. (1997). Brass instruments: some theoretical and experimental results. Proc. International Symposium on Musical Acoustics, Edinburgh, UK: in Proc. Ins. Ac. **19(5)**, 391–400.

Gilbert, J., Kergomard, J. and Ngoya, E. (1989). Calculation of the steady-state oscillations of a clarinet using the harmonic balance technique. J. Acoust. Soc. Am. **86**, 35–41, https://doi.org/10.1121/1.398352.

Gilbert, J., Ponthus, S. and Petiot, J. F. (1998). Artificial buzzing lips and brass instruments: experimental results. J. Acoust. Soc. Am. **104**, 1627–1632. https://doi.org/10.1121/1.424375.

Gilbert, J., Simon, L. and Terroir, J. (2005). Vibrato of saxophones. J. Acoust. Soc. Am. **118**, 2649–2655, https://doi.org/10.1121/1.2036027.

Gilbert, J., Leboso Ruiz, L. M. and Gougeon, S. (2006). Influence de la température sur la justesse d'un instrument à vent. Proc. 8ème Congrès Français d'Acoustique (CFA2006), Tours, France.

Gilbert, J., Campbell, D., Myers. A. and Pyle, R. (2007). Difference between brass instruments arising from variations in brassiness due to non-linear propagation. Proc. International Symposium on Musical Acoustics (ISMA), Barcelona.

Gilbert, J., Menguy, L. and Campbell, D. (2008). A simulation tool for brassiness studies. J. Acoust. Soc. Am. **123**, 1854–1857, https://doi.org/10.1121/1.2872342.

Gilbert, J., Libouban, L. and Dalmont, J. P. (2010). La clarinette cuivre-t-elle? Pertinence de la prise en compte des phénomènes de propagation non-linéaires pour la modélisation de la clarinette. Proc. 10ème Congrès Français d'Acoustique (CFA2010), Lyon, France.

Gilbert, J., Brasseur, E., Dalmont, J. P. and Maniquet, C. (2012). Acoustical evaluation of the Carnyx of Tintignac. Proc. Acoustics2012, 3955–3959, Nantes, France.

Gilbert, J., Dalmont, J. P., Potier, R. and Reby, D. (2014). Is nonlinear propagation responsible for the brassiness of the elephant trumpet calls? Acta Acust. united Ac **100**, 734–738, https://doi.org/10.3813/AAA.918752.

Gilbert, J., Leblanc, L. and Vergez, C. (2018). L'analyse de stabilité linéaire pour évaluer la facilité d'émission des cuivres. Etude comparative de trombones ténor et basse. Proc. 14e Congrès Français d'Acoustique, Le Havre, France.

Gilbert, J., Maugeais, S. and Vergez, C. (2020). Minimal blowing pressure allowing periodic oscillations in a simplified reed musical instrument model: Bouasse-Benade prescription assessed through numerical continuation. Acta Acustica **4**, 27, https://doi.org/10.1051/aacus/2020026.

Giordano, N. (2017). Lip dynamics in a physical model of the trumpet. Proc. International Symposium on Musical Acoustics, Montreal, Canada, 101–104.

Gray, C. D. (2005) *Acoustic pulse reflectometry for the measurement of the vocal tract, with application to voice synthesis*. Ph.D. thesis University of Edinburgh.

Grey, J. M. and Moorer, J. A. (1977). Perceptual evaluation of synthesised musical instrument tones. J. Acoust. Soc. Am. **61**, 1270–1277, https://doi.org/10.1121/1.381508.

Griffin, S., Lane, S. A. and Clark, R. L. (2003). The application of smart structures toward feedback suppression in amplified acoustic guitars. J. Acoust. Soc. Am. **113(6)**, 3188–3196, https://doi.org/10.1121/1.1562649.

Guckenheimer, J. and Holmes, P. (1983). *Nonlinear oscillations, dynamical systems, and bifurcations systems, and bifurcations of vector fields*. New York, Springer.

Guérard, J. (1998). *Modelisation numerique et simulation experimentale de systemes acoustique. Application aux instruments de musique*. Ph.D. thesis, Université Paris 6.

Guérard, J. and Boutillon, X. (1998). Real-time wave separation in a cylindrical pipe with applications to reflectometry, echo cancellation, and a hybrid musical instrument. Proc. 16th International Congress on Acoustics, Seattle, 2261–2262.

Guillemain, P., Kergomard, J. and Voinier, T. (2005). Real-time synthesis of clarinet-like instruments using digital impedance models. J. Acoust. Soc. Am. **118**, 483–494, https://doi.org/10.1121/1.1937507.

Guillemain, P., Vergez, C. and Farcy, A. (2010), An instrumented saxophone mouthpiece and its use to understand how an experienced musician plays. Acta Acust. united Ac. **96**, 622–634, https://doi.org/10.3813/AAA.918317.

Hall, W. M. (1932). Comments on the theory of horns. J. Acoust. Soc. Am. **3**, 552–561, https://doi.org/10.1121/1.1915578.

Hamilton, M. F. and Blackstock, D. T. (1988). *Nonlinear Acoustics*. New York, Academic Press.

Harrison, R. (2018). *Physical modelling of brass instruments using finite-difference time-domain methods*. Ph.D. thesis, University of Edinburgh.

Harrison, R., Bilbao, S., and Perry, J. (2015). An algorithm for a valved brass instrument synthesis environment using finite-difference time-domain methods with performance optimisation. Proc. 18th International Conference on Digital Audio Effects, Trondheim, Norway.

Harrison, R., Bilbao, S., Perry, J. and Wishart, T. (2016). An environment for physical modeling of articulated brass instruments. Comput. Music J. **39**, 80–95.

Hartmann, W. M. (2004). *Signals, Sound, and Sensation*, 5th corr. printing. New York, Springer.

Hartmann, W. M. (2013). *Principles of Musical Acoustics*. New York, Springer.

Haynes, B. (2002). *A History of Performing Pitch: the Story of 'A'*. Lanham, Scarecrow Press.

Helmholtz, H. L. F. (1877). *Die Lehre von den Tonempfingungen*, 4th Ed. Braunschweig, Friedrich Vieweg. English translation with additional material: Ellis, A. J. *On the Sensations of Tone*, 2nd Ed. London, Longman, Green and Co. (1885); repr. New York, Dover (1954)

Hélie, T. and Rodet, X. (2003). Radiation of a pulsating portion of a sphere: application to horn radiation. Acta Acust. united Ac. **89**, 565–577.

Henrich, N., Smith, J. and Wolfe, J. (2011), Vocal tract resonances in singing: Strategies used by sopranos, altos, tenors, and baritones. J. Acoust. Soc. Am. **129(2)**, 1024–1035, https://doi.org/10.1121/1.3518766.

Hendrie, D. (2007). *Development of bore reconstruction technique applied to the study of brass wind instruments*. Ph.D. thesis, University of Edinburgh.

Herbert, T. (1997). *Brass bands and other vernacular brass traditions*. In T. Herbert and J. Wallace (Eds.): *The Cambridge Companion to Brass Instruments*. Cambridge University Press.

Herbert, T., Myers. A., and Wallace, J. (Eds.). (2019). *The Cambridge Encyclopedia of Brass Instruments*. Cambridge University Press.

Hézard, T., Fréour, V., Caussé, R., Hélie, T, and Scavone, G. (2014). Synchronous multimodal measurements on lips and glottis: comparison between two human-valve oscillating systems. Acta Acust. united Ac. **100**, 1172–1185, https://doi.org/10.3813/AAA.918796.

Hirschberg, A., Van de Laar, R. W. A., Marrou-Maurières, J. P., Wijnands, A. P. J., Dane, J. H., Kruijswijk, S. G. and Houtsma, A. J. M. (1990). A quasi-stationary model of air flow in the reed channel of single-reed woodwind instruments. Acustica **70**, 146–154.

Hirschberg, A., Kergomard, J., and Weinreich, G. (Eds.). (1995). *Mechanics of musical instruments*. New York, Springer.

Hirschberg, A., Pelorson, X. and Gilbert, J. (1996a). Aeroacoustics of Musical Instruments. Meccanica **31**, 131–141.

Hirschberg, A., Gilbert, J., Msallam, R. and Wijnands, A. P. J. (1996b). Shock waves in trombones. J. Acoust. Soc. Am. **99**, 1754–1758, https://doi.org/10.1121/1.414698.

Hoekje, P. (1986). *Intercomponent energy exchange and upstream/downstream symmetry in nonlinear self-sustained oscillations of reed instruments*. Ph.D. thesis, Case Western Reserve University.

Hoekje, P. (2003). Dayton C. Miller. Echoes **13(1)**, 1–7.

Holmes, P. J. (1977). Bifurcations to divergence in flow induced oscillations: finite dimensional analysis. J. Sound Vib. **53**, 471–503.

Hostiou, V. (2005). Contribution of acoustic in playing early instruments in agreement with historical usage: the serpent as example. Proc. Forum Acusticum, Budapest, 399–403.

Hunter, F. (2019). *The carnyx in Iron Age Europe: the Deskford carnyx in its European context.* Monographien Band **146**, Vols. 1 and 2. Mainz, Römisch-Germanisches Centralmuseum.

Iltis, P. W., Schoonderwaldt, E., Zhang, S., Frahm, J. and Altenmüller, E. (2015). Real-time MRI comparisons of brass players: A methodological pilot study. Hum. Movement Sci. **42**, 132–135.

Impett, J. (1994). A meta-trumpet(er). Proc. International Computer Music Conference, Aarhus, Denmark, 147–150.

Impett, J. (1996). Projection and interactivity of musical structures in *Mirror-Rite*. Organ. Sound **1(3)**, 203–211.

Ishizaka, K. and Flanagan, J. (1972). Synthesis of voiced sounds from a two-mass model of the vocal cords. Bell Syst. Tech. J. **51**, 1233–1268.

Jansson, E. V. and Benade, A. H. (1974). On plane and spherical waves in horns with non-uniform flare: II. Prediction and measurements of resonance frequencies and radiation losses. Acustica **31**, 185–202.

Jenkins, L., Page, W., Trail, S., Tzanetakis, G. and Driessen, P. (2013). An easily removable, wireless optical sensing system (EROSS) for the trumpet. Proc. International Conference New Interfaces for Musical Expression (NIME), 352–357.

Kaburagi, T., Yamada, N., Fukui, T. and Minamiya. E. (2011). A methodological and preliminary study on the acoustic effect of a trumpet player's vocal tract. J. Acoust. Soc. Am. **130**, 536–545, https://doi.org/10.1121/1.3596471.

Karkar, S., Cochelin, B. and Vergez, C. (2013). A high-order, purely frequency based harmonic balance formulation for continuation of periodic solutions: The case of non-polynomial nonlinearities. J. Sound Vib. **332**, 968–977.

Kausel, W. (2001). Optimization of brasswind instruments and its application in bore reconstruction. J. New Music Res. **30**, 69–82.

Kausel, W., Mayer, A. and Nachtmann, G. (2008). More experimental evidence favouring the hypothesis of significant wall vibration influence on radiated horn sound. J. Acoust. Soc. Am. **123**, 3120, https://doi.org/10.1121/1.2933036.

Kausel, W., Zietlow, D. W. and Moore, T. R. (2010). Influence of wall vibrations on the sound of brass wind instruments. J. Acoust. Soc. Am. **128**, 3161–3174, https://doi.org/10.1121/1.3493457.

Kausel, W., Chatziioannou, V., Moore, T. R., Gorman, B. R. and Rokni, M. (2015). Axial vibrations of brass wind instrument bells and their acoustical influence: Theory and simulations. J. Acoust. Soc. Am. **137**, 3149–3162, https://doi.org/10.1121/1.4921270.

Keefe, D. H. (1982). Experiments on the single woodwind tone hole. J. Acoust. Soc. Am. **72**, 688–699, https://doi.org/10.1121/1.388249.

Keefe, D. H. (1984). Acoustical wave propagation in cylindrical ducts: Transmission line parameter approximations for isothermal and nonisothermal boundary conditions. J. Acoust. Soc. Am. **75**, 58–62, https://doi.org/10.1121/1.390300.

Keefe, D. H. (1990). Woodwind air column models. J. Acoust. Soc. Am. **88**, 35–51, https://doi.org/10.1121/1.399911.

Kelly, K. (1998). The dynamics of breathing. The Instrumentalist, December issue.

Kemp, J. A. (2002). *Theoretical and experimental study of wave propagation in brass musical instruments*. Ph.D. thesis, University of Edinburgh.

Kemp, J. A., van Walstijn, M., Campbell, D. M., Chick, J. P. and Smith, R. A. (2010). Time domain wave separation using multiple microphones. J. Acoust. Soc. Am. **128**, 195–205, 10.1121/1.3392441.

Kemp, J., López-Carromero, A. and Campbell, M. (2017). Pressure fields in the vicinity of brass musical instrument bells measured using a two dimensional grid array and comparison with multimodal models. Proc. 24th International Congress on Sound and Vibration, London.

Kent, E. L. (1956). *The Inside Story of Brass Instruments*. Elkhart, C. G. Conn Ltd.

Kergomard, J. and Caussé, R. (1986). Measurement of acoustic impedance using a capillary: An attempt to achieve optimization. J. Acoust. Soc. Am. **79**, 1129–1140, https://doi.org/10.1121/1.393385.

Kinsler, L. E., Frey, A. R., Coppens, A. B. and Sanders, J. V. (1999). *Fundamentals of Acoustics*, 4th Ed. New York, Wiley.

Klaus, S. K. and Pyle, R. W. (2015). Trumpet mute pitch: an analysis of three historic trumpet mutes. Proc. Third Vienna Talk on Musical Acoustics, University of Music and Performing Arts, Vienna, 88–91.

Kolar, M. A. (2014) Pututus, resonance and beats: Acoustic wave interference effects at Ancient Chavín de Huántar, Peru. J. Acoust. Soc. Am. **136**, 2270, https://doi.org/10.1121/1.4900202. Lay language version: https://ccrma.stanford.edu/groups/chavin/ASA2014.html.

Kolar, M. A., Rick, J. W., Cook, P. R. and Abel, J. S. (2012). Ancient Pututus contextualized: integrative archaeoacoustics at Chavín de Huántar, Perú. In M. Stöckli and A. Adje Both (Eds.), *Flower World—Music Archaeology of the Americas*, Vol.1, pp. 23–54. Berlin, Ekho.

Kopiez, R. (2003). Intonation of harmonic intervals: Adaptability of expert musicians to equal temperament and just intonation. Music Percept. **20(4)**, 383–410.

Krehl, P. and Engemann, S. (1995). August Toepler – the first who visualized shock waves. Shock Waves **5**, 1–18.

Kühnelt, H. (2007). Vortex sound in recorder- and flute-like instruments: Numerical simulation and analysis. Proc. International Symposium on Musical Acoustics, Barcelona, 1-S1-6.

Kuhl, P. K. (2020). Demonstration of the McGurk Effect. https://auditoryneuroscience.com/vocalizations-speech/mcgurk-effect. Accessed April 2020.

Leeuw, H. (2009). The Electrumpet, a hybrid electro-acoustic instrument. Proc. International Conference New Interfaces for Musical Expression (NIME), 193–198.

Leeuw, H. (2012). The electrumpet, additions and revisions. Proc. International Conference New Interfaces for Musical Expression (NIME), 258–259.

Lefebvre, A. (2010). *Computational acoustic methods for the design of woodwind instruments.* Ph.D. thesis, McGill University, Montreal.

Leissa, A. (1973). Vibrations of shells. Acoustical Society of America.

Levine, H. and Schwinger, J. (1948). On the radiation of sound from an unflanged pipe. Phys. Rev. **73**, 383–406.

Li, A., Sharp, D. B. and Forbes, B. J. (2005). Increasing the axial resolution of bore profile measurements using acoustic pulse reflectometry. Meas. Sci. Technol. **16**, 2001–2019, doi:10.1088/0957-0233/16/10/017.

Libin, L. (Ed.) (2014) *The Grove Dictionary of Musical Instruments*, 2nd Edn. Oxford University Press.

Logie, S. (2012). *An Acoustical Study of the Playing Characteristics of Brass Wind Instruments.* Ph.D. thesis, University of Edinburgh.

Lokki, T. (2014). Tasting music like wine – sensory evaluation of concert halls. Physics Today **67(1)**, 27.

López, I., Hirschberg, A., van Hirtum, A., Ruty, N. and Pelorson, X. (2006). Physical modeling of buzzing artificial lips: The effect of acoustical feedback, Acta Acust. united Ac. **92**, 1047–1059.

López-Carromero, A. (2018). *Experimental investigation of acoustic characteristics of radiation and playing gestures for lip-excited musical instruments.* Ph.D. thesis, University of Edinburgh.

López-Carromero, A., Campbell, D. M., Kemp, J. and Rendon, P. L. (2016). Validation of brass wind instrument radiation models in relation to their physical accuracy using an optical Schlieren imaging setup. Proc. Mtgs. Acoust. 28, 035003.

Lous, N. J. C., Hofmans, G. C. J., Veldhuis, R. N. J. and Hirschberg, A. (1999). A symmetrical two-mass vocal fold model coupled to vocal tract and trachea, with application to prosthesis design, Acta Acust. united Ac, **84**, 1135–1150.

Luce, D. and Clark, M. (1967). Physical correlates of brass instrument tones, J. Acoust. Soc. Am. **42**, 1232–1243, https://doi.org/10.1121/1.1910710.

Lurton, X. (1981). Etude analytique de l'impédance d'entrée des instruments à embouchure. Acustica **49**, 142–151.

Macaluso, C. A. and Dalmont, J. P. (2011) Trumpet with near-perfect Harmonicity: Design and acoustic results, J. Acoust. Soc. Am. **129**, 404–414. https://doi.org/10.1121/1.3518769.

Macdonald, H. (2002). *Berlioz's Orchestration Treatise: a Translation and Commentary*. Cambridge University Press.

Maganza, C., Caussé, R. and Laloë, F. (1986). Bifurcations, period doubling and chaos in clarinet like systems. EPL **1**, 295–302.

Maniquet, C,. Lejars, T., Armbruster, B., Pernot, M., Drieux-Daguerre, M., Mora, P. and Espinasse, L. (2011). Le carnyx et le casque-oiseau celtiques de Tintignac (Naves-Corrèze). Description et étude technologique. Aquitania **27**, 63–150.

Manneville, P. (2010). *Instabilities, Chaos and Turbulence*, 2nd. Ed., London, Imperial College Press.

Martin, D. W. (1942). Lip vibrations in a cornet mouthpiece. J. Acoust. Soc. Am. **13**, 305–308, https://doi.org/10.1121/1.1902242.

Maugeais, S. and Gilbert, J. (2017). Nonlinear acoustic propagation applied to brassiness studies, a new simulation tool in the time domain. Acta Acust. United Ac. **103**, 67–79, https://doi.org/10.3813/AAA.919034.

McGurk, H. and MacDonald, J. (1976). Hearing lips and seeing voices. Nature **264**, 746–748.

McIntyre, M. E., Schumacher, R. T. and Woodhouse, J. (1983). On the oscillations of musical instruments. J. Acoust. Soc. Am. **74**, 1325–1345, https://doi.org/10.1121/1.390157.

Menguy, L. and Gilbert, J. (2000). Weakly non-linear gas oscillations in air-filled tubes; solutions and experiments. Acta Acust. United Ac. **86**, 798–810.

Mersenne, M. (1635). *Harmonicorum libri XII*, Paris.

Meurisse, T., Mamou-Mani, A., Benacchio, S., Chomette, B., Finel, V., Sharp, D. and Caussé, R. (2015). Experimental demonstration of the modification of the resonances of a simplified self-sustained wind instrument through modal active control. Acta Acust. United Ac. **101**, 581–593, https://doi.org/10.3813/AAA.918854.

Meurisse, T., Mamou-Mani, A., Caussé, R., Sluchin, B. and Sharp, D. (2015). An active mute for the trombone. J. Acoust. Soc. Am. **138**, 3539–3548, https://doi.org/10.1121/1.4936901.

Meyer, J. (2009). *Acoustics and the Performance of Music*, 5th Edn., English transl. Hansen, U. New York, Springer.

Meyer, J. and Wogram, K. (1969). Die Richtcharakteristiken des Hornes. Das Musikinstrument **18(6)**, 1.

Meyer, J. and Wogram, K. (1970). Die Richtcharakteristiken von Trompete, Posaune und Tuba. Das Musikinstrument **19**, 171.

Miklós, A., Angster, J., Pitsch, S. and Rossing, T. D. (2003). Reed vibration in lingual organ pipes without the resonators. J. Acoust. Soc. Am. **113**, 1081–1091, https://doi.org/10.1121/1.1534101.

Miller, D. C. (1909). The influence of the material of wind instruments on the tone quality. Science **29**, 151–171.

Miranda, E. R. and Wanderley, M. M. (2006). *New Digital Instruments: Control and Interaction Beyond the Keyboard*. Middleton, Wisconsin, A-R Editions.

Moore, T. R., Kaplon, J., McDowall, G. and Martin, K. (2002). Vibrational modes of trumpet bells. J. Sound Vib. **254**, 777–786.

Moore, T. R., Shirley, E. T., Codrey, I. E. W. and Daniels, A. E. (2005). The effects of bell vibrations on the sound of the modern trumpet. Acta Acust. United Ac. **91**, 578–589.

Moore, T. R., Gorman, B. R., Rokni, M., Kausel, W. and Chatziioannou, V. (2015). Axial vibrations of brass wind instrument bells and their acoustical influence: Experiments. J. Acoust. Soc. Am. **138**, 1233–1240, https://doi.org/10.1121/1.4928138.

Morrill, D. and Cook, P. R. (1989). Hardware, software, and composition tools for a real-time improvised solo trumpet work. Proc. 1989 International Computer Music Conference, 211–214.

Msallam, R., Dequidt, S., Caussé, R. and Tassart, S. (2002). Physical model of the trombone including nonlinear effects. Application to sound synthesis of loud tones. Acta Acust. United Ac. **86**, 725–736.

Münkwitz, M., Seraphinoff, R. and Barclay, R. (2014). *Making a Natural Trumpet – Herstellung einer Naturtrompete*. Ottawa, Loose Cannon Press.

Myers, A. (1994). Nominal Pitch, *FoMRHI Quarterly* **75**, April, Comm.1247, 24–25.

Myers, A. (1997a) The horn function and brass instrument character. In Carter, S. (Ed.) *Perspectives in Brass Scholarship*: Proc. International Symposium on Historic Brass Instruments, Amherst, 1995. New York, Pendragon, 239–262.

Myers, A. (1997b). Design, technology and manufacture since 1800. In Herbert, T. and Wallace, J. (Eds.), *Cambridge Companion to Brass Instruments*, Cambridge University Press, 115–130.

Myers, A. (2016). The 'Concert' or 'Vocal' Horn. Galpin Soc. J. **LXIX**, 168–180.

Myers, A. and Campbell, D. M. (2006). Trumpet design and acoustical characteristics. In Carter, S. (Ed.), *Brass Scholarship in Review*: Proceedings of Historic Brass Society Conference, Cité de la Musique, Paris 1999. New York, Pendragon, ISBN 978-1-57647-105-0, pp. 251–274.

Myers, A., Pyle, R. W., Gilbert, J., Campbell, D. M., Chick, J. and Logie, S. (2012). Effects of nonlinear sound propagation on the characteristic timbres of brass instruments. J. Acoust. Soc. Am. **131**, 678–688, https://doi.org/10.1121/1.3651093.

Myers, A. (2019). The Typology and Timbre of the Tuba. In Philipsen, C. (Ed,) *Vom Serpent zur Tuba: Entwicklung und Einsatz der tiefen Polsterzungen-Instrumente mit Grifflöchern und Ventilen*. (Michaelsteiner Konferenzberichte, Band 83). Augsburg, Wissner; Blankenburg, Harz: Stiftung Kloster Michaelstein, pp. 159–174.

Nayfeh, A. H. (1995). *Nonlinear Oscillations*. New York, Wiley.

Neal, M. A., Richards, O., Campbell, D. M. and Gilbert, J. (2001). Study of the reed mechanism of brass instruments using an artificial mouth. Proc. International Symposium on Musical Acoustics, Perugia, Italy, 99–102.

Neal, M. A. (2002). *A study of the brass instrument lip reed mechanism using artificial lips and Lattice Boltzmann flow simulations*. Ph.D. thesis, University of Edinburgh.

Nederveen, C. J. (1969). *Acoustical Aspects of Woodwind Instruments*. Amsterdam, Fritz Knuf.

Nederveen, C. J. (1998a). *Acoustical Aspects of Woodwind Instruments*, 2nd. Ed. with additional material, Northern Illinois University, 1998.

Nederveen, C. J. (1998b). Influence of a toroidal bend on wind instrument tuning. J. Acoust. Soc. Am. **104**, 1616–1626, https://doi.org/10.1121/1.424374.

Nederveen, C. J. and Dalmont, J. P. (2004). Pitch and level changes in organ pipes due to wall resonances. J. Sound Vib. **271**, 227–239.

Neill, B. (2017). The Mutantrumpet. Proc. International Conference New Interfaces for Musical Expression (NIME), 489–490.

Nelson, P. A. and Elliott, S. J. (1991). *Active Control of Sound*. Academic Press.

Newton, M., Campbell, D. M. and Gilbert, J. (2008). Mechanical response measurements of real and artificial brass player's lips. J. Acoust. Soc. Am. **123**, EL14–EL20, https://doi.org/10.1121/1.2805042.

Nief, G., Dalmont, J.-P., Gautier, F., Gilbert, J. and Picó Vila. R. (2006). Influence des vibrations de parois d'un tuyau sur son impédance d'entrée: application aux instruments à vent. Proc. 8e Congrès Francais d'Acoustique, Tours, France, 379–382.

Nief, G., Gautier, F., Dalmont, J. P. and Gilbert, J. (2008). Influence of wall vibrations on the behaviour of a simplified wind instrument. J. Acoust. Soc. Am. **124**, 1320–1331, https://doi.org/10.1121/1.2945157.

Nödl, K. (1970). *Metallblasinstrumentenbau*, Frankfurt, Das Musikinstrument.

Noreland, D. (2002). A numerical method for acoustic waves in horns. Acta Acust. United Ac. **88**, 576–586.

Noreland, D., Kergomard, J., Laloë, F., Vergez, C., Guillemain, P. and Guilloteau, A. (2013). The logical clarinet: Numerical optimization of the geometry of woodwind instruments. Acta Acust. United Ac. 99, 615–628, https://doi.org/10.3813/AAA.918641, https://doi.org/10.3813/AAA.918316.

Norman, L., Chick, J. P., Campbell, D. M., Myers, A. and Gilbert, J. (2010). Player control of 'brassiness' at intermediate dynamic levels in brass instruments. Acta Acust. United Ac. **96**, 614–621, https://doi.org/10.3813/AAA.918316.

Ossman, T., Pichler, H. and Widholm, G. (1989). BIAS: A computer-aided test system for brass wind instruments. Audio Engineering Society Preprint No. 2834.

Otondo, F. and Rindel, J. H. (2004). The influence of the directivity of musical instruments in a room. Acta Acust. United Ac. **90**, 1178–1184.

Otondo, F. and Rindel, J. H. (2005). A new method for the radiation representation of musical instruments in auralizations. Acta Acust. United Ac. **91**, 902–906.

Pagneux, V., Amir, N. and Kergomard, J. (1996). A study of wave propagation in varying cross-section waveguides by modal decomposition. Part I. Theory and validation. J. Acoust. Soc. Am. **100**, 2034–2048, https://doi.org/10.1121/1.417913.

Pandya, B. H., Settles, G. S. and Miller, J. D. (2003). Schlieren imaging of shock waves from a trumpet. J. Acoust. Soc. Am. **114**, 3363–3367, https://doi.org/10.1121/1.1628682.

Pätynen, J. 2011. *A Virtual Symphony Orchestra for Studies of Concert Hall Acoustics*. Ph.D. thesis, Aalto University.

Pätynen, J. and Lokki, T. (2010). Directivities of symphony orchestra instruments. Acta Acust. United Ac. **96**, 138–187, https://doi.org/10.3813/AAA.918265.

Pelorson, X., Hirschberg, A., van Hassel, R. R., Wijnands, A. P. J. and Auregan, Y. (1994). Theoretical and experimental study of quasisteady-flow separation within the glottis during phonation. Application to a modified two-mass model. J. Acoust. Soc. Am. **96**, 3416–3431, https://doi.org/10.1121/1.411449.

Pelzer, S., Pollow, M. and Vorländer, M. 2012. Auralization of a virtual orchestra using directivities of measured symphonic instruments. Proc. Acoustics12, Nantes, France, 2379–2384.

Petiot, J. F. (2003). Measurement of the force applied to the mouthpiece during brass instrument playing. Proc. Stockholm Musical Acoustics Conference SMAC03, Stockholm, Sweden, 225–228.

Petiot, J. F. and Gilbert, J. (2013). Comparison of trumpet sounds played by a musician and simulated by physical modelling. Acta Acust. United Ac. **99**, 629–641, https://doi.org/10.3813/AAA.918642.

Petiot, J. F., Teissier, F., Gilbert, J. and Campbell, M. (2003). Comparative analysis of brass wind instruments using an artificial mouth: first results. Acta Acust. United Ac. **89**, 974–979.

Petiot, J. F. and Tavard, F. (2008). Design and manufacturing of an artificial marine conch by bore optimization. Proc. Acoustics 2008, Paris, France.

Petiot, J. F., Kersaudy, P., Scavone, G., MacAdams, S. and Gazengel, B. (2017). Investigations of the relationships between perceived qualities and sound parameters of saxophone reeds. Acta Acust. United Ac. **103**, 812–829, https://doi.org/10.3813/AAA.919110.

Pico Villa, R. and Gautier, F. (2007). Vibroacoustics of slightly distorted cylindrical shells; model of the acoustic input impedance. J. Sound Vib., **302**, 18–38.

Pierce, A. D. (1989). *Acoustics*, 2nd Ed. Acoustical Society of America, Woodbury, NY.

Poirson, E., Petiot, J. F. and Gilbert, J. (2005). Study of the brightness of trumpet tones. J. Acoust. Soc. Am. **118**, 2656–2666, https://doi.org/10.1121/1.2006007.

Poirson, E., Petiot, J. F. and Gilbert, J. (2007). Integration of user perceptions in the design process: Application to musical instrument optimization. J. Mech. Design **129**, 1206–1214.

Praetorius, M. (1619). *Syntagma Musicum* Vol. 2, Wolfenbüttel; repr. Bärenreiter, 1958.

Pratt, R. L., Elliott, S. J. and Bowsher, J. M. (1977). The measurement of the acoustic impedance of brass instruments. Acta Acust. United Ac. **38**, 236–246.

Pyle, R. W. (1975). Effective length of horns. J. Acoust. Soc. Am. **57**, 1309–1317, https://doi.org/10.1121/1.380607.

Pyle, R. W. (1991). A computational model of the Baroque trumpet and mute. Historic Brass Society Journal **3**, 79–97.

Pyle, R. W. and Myers, A. (2006). Scaling of brasswind instruments. J. Acoust. Soc. Am. **119(5)**, 3259.

Rayleigh, Lord (1894). *The Theory of Sound*. London, Macmillan; repr. Dover, 1945.

Reid, S., Gaston, R., Honigman, C. and Kapur, A. (2016). Minimally Invasive Gesture Sensing Interface (MIGSI) for Trumpet. Proc. International Conference New Interfaces for Musical Expression (NIME), Brisbane, Australia, 352–357.

Reid, S., Gaston, R. and Kapur, A. (2019). Perspectives on time: performance practice, mapping strategies, and composition with MIGSI. In Proc. International Conference on New Interfaces for Musical Expression (NIME) Porto Alegre, Brazil, 234–239.

Richards, O. F. (2003). *Investigation of the lip reed using computational modelling and experimental studies with an artificial mouth*. Ph.D. thesis, University of Edinburgh.

Richards, O. F., Campbell, D. M. and Gilbert, J. (2003). Modelling the lip reed: computational and experimental investigations of two-mode inward/outward striking behaviour. Proc. Stockholm Musical Acoustics Conference (SMAC03), Stockholm, Sweden, 233–235.

Ricot, D., Caussé, R. and Misdariis, N. (2005). Aerodynamic excitation and sound production of blown-closed free reeds without acoustic coupling; the example of the accordion reed. J. Acoust. Soc. Am. **117**, 2279–2290, https://doi.org/10.1121/1.1852546.

Rodgers, G. (2018). Partial filter: tuba and electronics. https://polarpatterns.wordpress.com/2015/03/21/partial-filter-tuba-and-electronics/, retrieved July 2020.

Roederer, J. (2009). *The Physics and Psychophysics of Music*. New York, Springer.

Saneyoshi, J., Teramura, H. and Yoshikawa, H. (1987). Feedback oscillations in reed woodwind and brasswind instruments. Acustica **62**, 194–210.

Sax, A. (1850). Note pour Messieurs les Conseilleurs [of the court of Rouen]. Translated in Heyde, H. (2001), Methods of organology and proportions in brass wind instrument making. Historic Brass Society Journal **13**, 1–51.

Scavone, G. P., Lefebvre, A. and da Silva, A. R. (2008). Measurement of vocal-tract influence during saxophone performance. J. Acoust. Soc. Am. **123(4)**, 2391–2400, https://doi.org/10.1121/1.2839900.

Schumacher, R. T. (1978). Self sustained oscillations of the clarinet: an integral equation approach. Acustica **40**, 298–309.

Schumacher, R. T. (1981). Ab initio calculations of the oscillation of a clarinet. Acustica **48**, 71–85.

Schwab, B. and Schultze-Florey, A. (2004).Velopharyngeal insufficiency in woodwind and brass players. Med. Probl. Perform. Arts **19**, 21–25.

Scott, A., Chick, J. and Myers, A. (2019). The *Cor Solo*: History and Characteristics. Historic Brass Society Journal **31**, 1–34. https://doi.org/10.2153/0120190011001.

Sécail-Géraud, M., Leblanc, L., Gilbert, J., Gautier, F. and Hoekje, P. (2018). Analyse de couplages vibroacoustiques sur un double pavillon de trombone en situation de jeu. Proc. 14e Congrès Français d'Acoustique, Le Havre, France.

Settles, G. S. (2001). *Schlieren and shadowgraph techniques: visualizing phenomena in transparent media*. New York, Springer.

Sharp, D. (1996). *Acoustic pulse reflectometry for the measurement of musical wind instruments*. Ph.D. thesis, University of Edinburgh.

Sharp, D. B., Mamou-Mani, A. and van Walstijn, M. (2011). A single microphone capillary based system for measuring the complex input impedance of musical wind instruments. Acta Acust. United Ac. **97**, 819–829, https://doi.org/10.3813/AAA.918462.

Silva, F., Vergez, C. and Kergomard, J. (2007). Oscilllation thresholds for striking outward reeds coupled to a resonator. Proc. International Symposium on Musical Acoustics, Barcelona.

Silva, F., Vergez, C., Guillemain, P., Kergomard, J. and Debut, V. (2014). MoReeSC: a framework for the simulation and analysis of sound production in reed and brass instruments. Acta Acust. United Ac. **100**, 126–138, https://doi.org/10.3813/AAA.918693.

Singh, R. and Schary, M. (1978). Acoustic impedance measurement using sine sweep excitation and known volume velocity technique. J. Acoust. Soc. Am. **64**, 995–1003, https://doi.org/10.1121/1.382061.

Sivian, L. J., Dunn, H. K. and White, S. D. (1930). Absolute amplitudes and spectra of certain musical instruments. J. Acoust. Soc. Am. **1**, 172, https://doi.org/10.1121/1.1901894.

Sivian, L. J., Dunn, H. K. and White, S. D. (1931). Absolute amplitudes and spectra of certain musical instruments and orchestras. J. Acoust. Soc. Am. **2**, 330–371, https://doi.org/10.1121/1.1915260.

Sluchin, B. (1995). Jeu et chant simultanés sur les cuivres (Playing and singing simultaneously on brass instruments). Paris, Editions Musicales Européennes.

Sluchin, B. and Caussé, R. (1991). *Sourdines des Cuivres*. Paris, Editions de la Maison des Sciences de l'Homme.

Smith, J. O. (2010). *Physical Audio Signal Processing*. http://ccrma.stanford.edu/~jos/pasp/, online book, 2010 edition, accessed April, 2020.

Smith, R. A. (1986). The effect of material in brass instruments: a review. Proc. Ins. Ac. **18**, 91–96.

Smith, R. A. and Daniell, G. J. (1976). Systematic approach to the correction of intonation in wind instruments. Nature **262**, 761–765.

Sommerfeldt, S. D. and Strong, W. J. (1988). Simulation of a player-clarinet system. J. Acoust. Soc. Am. **83**, 1908–1918, https://doi.org/10.1121/1.396475.

Stevenson, R. L. (1881). *Virginibus Puerisque and Other Papers*. London, Kegan Paul.

Stevenson, S. (2009). *Experimental Investigations of Lip Motion in Brass Instrument Playing*. Ph.D. thesis, University of Edinburgh.

Story, B. H., Titze, I. R. and Hoffman, E. A. (1996). Vocal tract area functions from magnetic resonance imaging. J. Acoust. Soc. Am. **100**, 537–554, https://doi.org/10.1121/1.415960.

Stradner, G. (2015). Transposing mutes for trumpets. Proc. Third Vienna Talk on Musical Acoustics, 16–19 Sept., University of Music and the Performing Arts, Vienna, 92–25.

Strong, W. J. and Dudley, J. D. (1993). Simulation of a player-trumpet system, Proc. Stockholm Musical Acoustics Conference (SMAC93), Stockholm, Sweden, 520–524.

Tarnopolsky, A., Fletcher, N., Hollenberg, L., Lange, L., Smith, J. and Wolfe, J. (2005). The vocal tract and the sound of a didgeridoo. Nature **436**, 39.

Tarnopolsky, A. Z., Fletcher, N. H., Hollenberg, L. C. L., Lange, B. D., Smith, J. and Wolfe, J. (2006). Vocal tract resonances and the sound of the Australian didjeridu (yidaki) I. Experiment. J. Acoust. Soc. Am. **119**, 1194–1204, https://doi.org/10.1121/1.2146089.

Thibodeau, J. and Wanderley, M. M. (2013). Trumpet Augmentation and Technological Symbiosis. Comput. Music J. **37(3)**, 12–25.

Thompson, M. W. and Strong, W. J. (2001). Inclusion of wave steepening in a frequency-domain model of trombone sound production. J. Acoust. Soc. Am. **110**, 556–562, https://doi.org/10.1121/1.1371759.

Thomsen, J. J. (1997). *Vibrations and stability*. London, McGraw-Hill.

Titze, I. R. (1988). The physics of small-amplitude oscillation of the vocal folds. J. Acoust. Soc. Am. **83**, 1536–1552, https://doi.org/10.1121/1.395910.

Tournemenne, R., Petiot, J. F., Talgorn, B., Kokkolaras, M. and Gilbert, J. (2019). Sound simulation based design optimization of brass wind instruments. J. Acoust. Soc. Am. **145**, 3795–3804, https://doi.org/10.1121/1.5111346.

van Walstijn, M. O., Cullen, J. S. and Campbell, D. M. (1997). Modelling viscothermal wave propagation in wind instrument air columns. Proc. International Symposium on Musical Acoustics, Edinburgh: in Proc. Institute of Acoustics 19(5), 413–418, 1997.

van Walstijn, M., Campbell, M., Kemp, J., Sharp, D. (2005). Wideband measurement of the acoustic impedance of tubular objects. Acta Acust. United Ac. **91**, 590–604.

Vasari, G. (1568). *Le vite de' più eccellenti pittori, scultori, e architettore*, 2nd Ed. Florence, Giunti.

Velut, L., Vergez, C. and Gilbert, J. (2016). Measurements and time-domain simulations of multiphonics in the trombone. J. Acoust. Soc. Am. **140**, 2876–2887, https://doi.org/10.1121/1.4964634.

Velut, L., Vergez, C., Gilbert, J. and Djahanbani, M. (2017a). How well can Linear Stability Analysis predict the behaviour of an outward valve brass instrument model? Acta Acoust. United Ac. **103**, 132–148, https://doi.org/10.3813/AAA.919039.

Velut, L., Vergez, C. and Gilbert, J. (2017b). Physical modelling of trombone mutes, the pedal note issue. Acta Acust. United Ac., **103**, 668–675, https://doi.org/10.3813/AAA.919095.

Vereecke, H. and Carter, S. (2014). Mouthpieces for brasswinds in the writings of Victor-Charles Mahillon: a historical and analytical review. Historic Brass Society Journal **26**, 43–62, 10.2153/0120140011004.

Vergez, C. and Rodet, X. (1997). Model of the trumpet functioning: real time simulation and experiments with an artificial mouth model. Proc. International Symposium on Musical Instruments, Edinburgh.

Vergez, C. and Rodet, X. (2001a). Trumpet and trumpeter: physical modelling for sound synthesis. Proc. International Congress on Acoustics, Rome.

Vergez, C. and Rodet, X. (2001b). Trumpet and trumpet player: a highly nonlinear interaction studied in the framework of nonlinear dynamics. Int. J. Bifurcat. Chaos **11**, 1801–1814.

Vilain, C. E., Pelorson, X., Hirschberg, A., Le Marrec, L., Op't Root, W. and Willems, J. (2003). Contributions to the physical modelling of the lips. Influence of the mechanical boundary conditions. Acta Acust. United Ac. **89**, 882–887.

von Steiger, A., Senn, M., Tuchschmid, M., Leber, H. J., Lehmann, E. and Mannes, D. (2013). Can we look over the shoulders of historical brasswind instrument makers? Aspects of the materiality of nineteenth-century brass instruments in France. Historic Brass Society Journal **25**, 21–38, 10.2153/0120130011002.

Walther, J. G. (1732). *Musicalisches Lexicon*. Leipzig, Wolfgang Deer.

Watkinson, P. S. and Bowsher, J. M. (1982). Vibration characteristics of brass instrument bells, J. Sound Vib. **85**, 1–17.

Watkinson, P. S., Shepherd, R. and Bowsher, J. M. (1982). Acoustic energy losses in brass instruments. Acustica **51**, 213–221.

Watson, A. (2019). Breathing. In *The Cambridge Encyclopedia of Brass Instruments*, eds. Herbert, T., Myers, A. and Wallace, J. Cambridge University Press.

Webster, J. C. (1947). An electrical method of measuring the intonation of cup-mouthpiece instruments. J. Acoust. Soc. Am. **19**, 902–906.

Whitehouse, J. W., Sharp, D. B. and Hill, T. J. W. (2003). Wall vibrations in wind instruments. IOA Acoustics Bulletin **28**, 20–23.

Widholm, G. (1995). Brass wind instrument quality measured and evaluated by a new computer system. In Proc. 15th International Congress on Acoustics, Trondheim, Norway, Vol. III, 517–520.

Widholm, G. (2005). The Vienna horn – a historical relict successfully used by top orchestras of the 21st century. Proc. Forum Acusticum, Budapest, 441–445.

Wilson, T. A. and Beavers, G. S. (1974). Operating modes of the clarinet. J. Acoust. Soc. Am. **56**, 653–658, https://doi.org/10.1121/1.1903304.

Wogram, K. (1972). *Ein Beitrag zur Ermittlung der Stimmung von Blechblasinstrument* Dr.-Ing. diss., Technischen Universitat Carolo-Wilhelmina zu Braunschweig.

Wolfe, J. and Smith, J. (2008). Acoustical coupling between lip valves and vocal folds. Acoust. Aust. **36**, 23–27.

Wolfe, J., Fletcher, N. H. and Smith, J. (2015). The interactions between wind instruments and their players. Acta Acust. United Ac. **101**, 211–233, https://doi.org/10.3813/AAA.918820.

Woodhouse, J. (1993). On the playability of violins. Part I: Reflection functions. Acustica **78**, 125–136.

Worman, W.E. (1971). *Self-sustained non-linear oscillations of medium amplitude in clarinet-like systems*. Ph.D. thesis, Case Western Reserve University, Cleveland.

Yoshikawa, S. (1995) Acoustical behaviour of brass player's lips. J. Acoust. Soc. Am. **97**, 1929–1939, https://doi.org/10.1121/1.412066.

Yoshikawa, S. and Muti, Y. (2003). Lip wave generation in horn players and the estimation of lip-tissue elasticity. Acta Acust. United Ac. **89**, 145–162.

Yoshikawa, S. and Nobara, N. (2017). Acoustical modeling of mutes for brass instruments. In Schneider, A. (Ed.), *Studies in Musical Acoustics and Psychoacoustics*, Current Research in Systematic Musicology 4. New York: Springer, 143–186, https://doi.org/10.1007/978-3-319-47292-8.

Ziolkowski, J. (2002). The Roman Bucina: a distinct musical instrument? Historic Brass Society Journal **14**, 31–58.

Zorumski, W. E. (1973). Generalized radiation impedances and reflection coefficients of circular and annular ducts. J. Acoust. Soc. Am. **54**, 1667–1673, https://doi.org/10.1121/1.1914466.

Zwicker, G. and Kosten, C. (1949) *Sound Absorbing Materials*. Amsterdam, Elsevier.

Index

M. Campbell et al., *The Science of Brass Instruments*, Modern Acoustics and Signal Processing, https://doi.org/10.1007/978-3-030-55686-0

S

T